U-Statistics in Banach Spaces

U-STATISTICS IN BANACH SPACES

Yu. V. Borovskikh

///VSP///
Utrecht, The Netherlands, 1996

VSP BV
P.O. Box 346
3700 AH Zeist
The Netherlands

First published in 1996

ISBN 90-6764-200-2

CIP-DATA KONINKLIJKE BIBLIOTHEEK, DEN HAAG

Borovskikh, Yuri V.

U-statistics in Banach spaces / Yuri V. Borovskikh. -
Utrecht : VSP
With index, ref.
ISBN 90-6764-200-2 bound
NUGI 815
Subject headings: U-statistics / Banach spaces

Printed in The Netherlands by Koninklijke Wöhrmann BV, Zutphen.

To

My grandson Anton

God, Thou great symmetry,
Who put a biting lust in me
From whence my sorrows spring,
For all the frittered days
That I have spent in shapeless ways
Give me one perfect thing.

Anna Wickham (*"Envoi"*, 1921)

Contents

Preface

U-statistics are universal objects of modern probabilistic summation theory. They appear in various statistical problems and have very important applications. The mathematical nature of this class of random variables has a functional character, and therefore, leads to the investigation of probabilistic distributions in infinite-dimensional spaces. The situation when the kernel of a U-statistic takes values in a Banach space, turns out to be the most natural and interesting.

In this case the construction of the probabilistic theory of such U-statistics is based on methods of probability theory in Banach spaces. The mathematical beauty and universal attractiveness of U-statistics is clearly explained by the universal idea of symmetry lying in its definition. The idea of symmetry is dominant in modern probability theory in Banach spaces. That is why U-statistical objects are natural in all aspects of probability theory in Banach spaces. In concentrated form this idea is revealed in the representation of every U-statistic as an integral functional with respect to a permanent random measure. Asymptotic properties of random permanents tell us that a U-statistic is a statistical image of an element of chaos, and a family of U-statistics serves as a statistical model of chaos processes.

In this book I present in a systematic form the probabilistic theory of U-statistics with values in Banach spaces (UB-statistics), which has been developed up to the present time. The exposition of the material in this book is based around the following topics:

- algebraic and martingale properties of U-statistics;
- inequalities;
- law of large numbers;
- the central limit theorem;
- weak convergence to a Gaussian chaos and multiple stochastic integrals;
- invariance principle and functional limit theorems;
- estimates of the rate of weak convergence;
- asymptotic expansion of distributions;
- large deviations;
- law of iterated logarithm;
- dependent variables;

– relation between Banach-valued U-statistics and functionals from permanent random measures.

All asymptotic properties of UB-statistics depend on Banach space B, rank r, kernel Φ and moment limitations for Φ. For $r = 1$ (non-degenerate kernel) the asymptotic behaviour of UB-statistics can essentially be reduced to the asymptotic behaviour of sums of B-valued random elements.

If $r \geq 2$ (degenerate kernel), then the family of possible limit distributions for UB-statistics is extended by the distributions of functionals of infinite-dimensional Gaussian vectors defined by kernel Φ, distributions of multiple stochastic integrals and functionals of them, distributions of Gaussian chaos and Poisson chaos and so on.

I hope that this book will serve to introduce many people to the beautiful theory of U-statistics in Banach spaces. Many people have helped me in the writing of this book, either through discussions or by reading one of the versions of the manuscript. I am extremely grateful to V.S. Korolyuk for his continuous support in my efforts and his diligence in exploring stochastic laws of U-statistics during the last twenty years.

My deep gratitude is expressed to O.A. Cherkashina (Adamenko) (Institute of Mathematics, Kiev) and E.V. Cherkashin (International Mathematical Center, Kiev) for their activity in publishing mathematical manuscripts, and especially for their editorial work on several versions of this book. I would also like to thank Joseph Thomas (British) for his English language editing of the text.

Komarovo,
near St Petersburg,
Russia
8th May, 1995

Yuri V. Borovskikh

Introduction

In this book we consider U-statistics

$$U_n = n^{-[m]} \sum_{1 \le i_1 \ne \cdots \ne i_m \le n} \Phi(X_{i_1}, \ldots, X_{i_m})$$

with Banach-valued kernels $\Phi : X^m \to B$. The exploration of the stochastic laws of random elements of this type is accompanied by all the problems which are inherent in the investigations of stochastic properties of sums of independent Banach-valued random vectors (this corresponds to the case $m = 1$).

In Chapter 1 we present preliminary information which includes the definition of canonical functions g_c; rank r of kernel Φ, $1 \le r \le m$; canonical Hoeffding representation; multiple stochastic integrals and Banach-valued polynomial chaos. Naturally, our treatment of these definitions is relevant to UB-statistics.

Chapter 2 is devoted to inequalities. Here we have moment inequalities related to the geometry of Banach spaces and maximal inequalities following from martingale properties of UB-statistics. Decoupling inequalities, consisting in the comparison between $E\Psi(\|U_n\|)$ and $E\Psi(\|U_n^{\mathrm{dec}}\|)$ play a very important role in what follows. Here

$$U_n^{\mathrm{dec}} = n^{-[m]} \sum_{1 \le i_1 \ne \cdots \ne i_m \le n} \Phi(X_{i_1}^1, \ldots, X_{i_m}^m)$$

is a decoupling version of U_n, and random variables $\{X_1^j, \ldots, X_n^j\}_{j=1}^m$ are independent copies of $\{X_1, \ldots, X_n\}$.

Laws of large numbers for UB-statistics and for Banach-valued von Mises' statistics are considered in Chapter 3. The estimates of the rate of convergence in these theorems are obtained by martingale inequalities.

Chapter 4 is devoted to the study of the convergence in distribution of UB-statistics and the different functionals of them. Conditions of the theorems are formulated in terms of canonical functions g_c, $c = r, \ldots, m$, and are dependent on rank r, $1 \le r \le m$. In the non-degenerate case ($r = 1$) in limit we have asymptotic normality. If kernel Φ is degenerate ($r \ge 2$), then Gaussian chaos arises if the considered Banach space is of type 2. Under additional moment conditions for canonical functions g_c these statements hold true for multi-sample UB-statistics and symmetric statistics. We also consider weak convergence under the conditions of Poisson approximation, stable approximation and with increasing degrees. The results for U-statistics with varying kernels and for weighted U-statistics are adduced as well.

Weak convergence of probability measures generated by U-statistics in functional spaces is considered in Chapter 5. Random U-processes of a stochastic sequence $\{U_n(\Phi)\}$ can be defined in different ways. It is possible to consider a partial sum process similar to $(U_{[nt]}(\Phi),\ 0 \le t \le 1)$ as a sequence of processes indexed by parameter t when Φ is fixed. Another interpretation of $(U_n(\Phi), \Phi \in \mathcal{F})$ is a stochastic process indexed by family $\mathcal{F}$ consisting of measurable symmetric kernels. In this chapter, for such U-processes we study the weak convergence in non-degenerate and in degenerate cases, and in Poisson approximation schemes. Martingale functional limit theorems are applied to U-statistics with varying kernels.

In Chapter 6, the rate of weak convergence is investigated, and various approximation estimates depending on a Banach space B, a kernel Φ and a distribution P are obtained. In different situations we apply different methods of analysis. For the estimation of the rate of Gaussian convergence we apply the characteristic function method and the method of compositions. In Poisson approximation cases we apply the Chen–Stein method. Rank r of kernel Φ has an effect upon the form of the limit distribution. That is why it is natural that the rate of convergence depends on this rank r. We consider nonuniform estimates and estimates with increasing degree both in Gaussian approximation and in Poisson approximation schemes. The most interesting estimates are obtained for UH-statistics.

Chapter 7 is devoted to the problem of the refinement of the limit theorems of the weak convergence of probability distributions generated by UB-statistics. This problem is connected with asymptotic expansion and clarification of the conditions for it to be true. These conditions can depend on the method of investigations. The method of characteristic functions demands the well known Cramer condition and the conditions of the moments of individual terms. In this problem, rank r of kernel Φ has an essential effect. Depending on different values of r this problem has different solutions. But an asymptotic expansion can be written for every r and the coefficients of this expansion can be represented in the form of multiple stochastic integrals if kernel Φ satisfies some additional conditions.

In Chapter 8 we consider the behaviour of probability

$$P(n^{r/2}\|U_n - \theta\| > x)$$

in the region $x = O(n^{\beta})$ with some $0 \le \beta \le 1$. First we obtain some exponential inequalities for all $x > 0$, and then we study the large deviations of this probability when $x = x(n) \to \infty$ as $n \to \infty$. The results depend on rank r of kernel Φ and the properties of a Banach space B. For UH-statistics the estimations of the rate of convergence in the regions of large deviations are obtained.

Chapter 9 is devoted to the law of iterated logarithm for UB-statistics. For U-statistics with values in Banach spaces of type 2, the validity of bounded LIL, compact LIL and functional LIL is proved. Corresponding theorems include conditions for canonical functions g_c.

Chapter 10 serves as an introduction to the theory of UB-statistics when the random variables $X_1, \ldots, X_n$ are dependent. Theoretically, for this class of random B-valued elements, we can consider the whole spectrum of asymptotic and nonasymptotic problems which was considered for independent random variables. The author confines himself to the variables for which he has a personal preference: symmetrically dependent random variables, weakly dependent random variables, bootstrap variables and order statistics.

The bibliographical supplements and comments contain remarks on the scientific works which were used during the writing of this book, and those which influenced the author.

Additionally in the bibliography, the principal papers where the exploration of the stochastic laws of U-statistics were actively conducted, can be found at the beginning with the appearance of the fundamental work of Hoeffding (1948a).

Chapter 1

Basic definitions

1.1 One-sample *UB*-statistics

Definition of *UB*-statistics. Let B be a real separable Banach space with a norm $\|\cdot\|$ and let B^* be the dual to space B. Denote by $x^*(x)$ the value of functional $x^* \in B^*$ at $x \in B$. Let $X_1, \ldots, X_n$ be independent random variables taking values in the measurable space $(X, \mathfrak{X})$ and all with identical distribution P. Consider a Bochner integrable symmetric function $\Phi : X^m \to B$ of m variables given on X^m and taking values in B. We define a *U-statistic* as follows

$$U_n = \binom{n}{m}^{-1} \sum_{1 \le i_1 < \ldots < i_m \le n} \Phi(X_{i_1}, \ldots, X_{i_m}). \tag{1.1.1}$$

It is clear that $U_n \in B$. Using terminology of Borovskikh (1986) U-statistic (1.1.1) with a B-valued *kernel* Φ is called a *UB-statistic*. In particular, if $B = R$ it is called a *UR-statistic* and if $B = H$, where H is a real separable Hilbert space, it is called a *UH-statistic*.

Let $\mathcal{P} = \{P\}$ be a class of probability distributions on $(X, \mathfrak{X})$. By $\theta : P \to \theta(P)$, we denote a functional given on $\mathcal{P}$ and taking values in B where

$$\theta(P) = \int \ldots \int \Phi(x_1, \ldots, x_m) P(dx_1) \ldots P(dx_m).$$

If $E\|\Phi\| < \infty$, then U_n is an unbiased estimate of the B-valued element $\theta(P) = E\Phi(X_1, \ldots, X_m)$.

In particular, if in (1.1.1), $m = 1$, then $U_n = n^{-1}(\Phi(X_1) + \cdots + \Phi(X_n))$ is the sum of independent identically distributed (i.i.d.) B-valued random variables. Therefore, U-statistic (1.1.1) can be treated as a functional-algebraic generalization of usual mean.

Without loss of generality, we suppose that Φ is a symmetric function of its arguments, since otherwise we can use a symmetric kernel $S_m\Phi$ according to the formula

$$S_m\Phi(x_1,\dots,x_m) = \frac{1}{m!}\sum \Phi(x_{i_1},\dots,x_{i_m})$$

where the summation is carried out over all $m!$ permutations $(i_1,\dots,i_m)$ of the numbers $(1,\dots,m)$. Then U-statistic (1.1.1) is defined as follows

$$U_n = n^{-[m]} \sum_{1\le i_1\ne\cdots\ne i_m\le n} \Phi(X_{i_1},\dots,X_{i_m}), \tag{1.1.2}$$

where $n^{[m]} = n(n-1)\cdots(n-m+1)$.

Note, that for all $m \le n$, we have

$$\sum_{\sigma} \Phi(X_{i_1},\dots,X_{i_m}) = (n-m)! \sum_{1\le i_1\ne\cdots\ne i_m\le n} \Phi(X_{i_1},\dots,X_{i_m}),$$

where the symbol $\sum_\sigma$ denotes the summation over all $n!$ permutations $\{i_1, i_2,\dots,i_n\}$ of the numbers $\{1,2,\dots,n\}$. Therefore the following formula holds

$$U_n = \frac{1}{n!}\sum_{\sigma} \Phi(X_{i_1},\dots,X_{i_m}). \tag{1.1.3}$$

Hence, the U-statistic defined according to (1.1.1)–(1.1.3) is a symmetrized version of the unbiased estimate $\Phi(X_1,\dots,X_m)$ of parameter θ.

Description of structure. In the definitions of (1.1.1)–(1.1.3) the integer m is called a *degree* of kernel Φ (or of a U-statistic) and $m \ge n$. Consequently, for different values of parameter m we have

1) *if* $m = 1$, *then* $U_n = n^{-1}\big(\Phi(X_1) + \cdots + \Phi(X_n)\big)$,

2) *if* $m = 2$, *then* $U_n = \binom{n}{2}^{-1} \sum_{1\le i<j\le n} \Phi(X_i, X_j)$,

. . . .

n-1) *if* $m = n-1$, *then* $U_n = n^{-1}\sum_{j=1}^{n} \Phi(X_1,\dots,X_{j-1},X_{j+1},\dots,X_n)$,

n) *if* $m = n$, *then* $U_n = \Phi(X_1, X_2,\dots,X_n)$.

Relations 1)–n) show that U-statistic (1.1.1) contains a wide range of random variables as m takes different values from 1 to n: we can begin from sums of independent random variables ($m = 1$), different forms of dependence ($2 \le m \le n-1$), and continue to the symmetric function ($m = n$) with a sufficiently general structure, because there are not any restrictions to function Φ in definition (1.1.1).

Canonical functions. Given $E\|\Phi\| < \infty$, we define the functions $\Phi_c : X^c \to B$, for $c = 0, 1, \dots, m$, by

$$\Phi_c(x_1, \dots, x_c) = E\left(\Phi(X_1, \dots, X_m)\Big|X_1 = x_1, \dots, X_c = x_c\right),$$

setting $\Phi_0 = \theta$ and $\Phi_m = \Phi$. One can easily see that

$$\Phi_c(x_1, \dots, x_c) = E\Phi_{c+1}(x_1, \dots, x_c, X_{c+1})$$

for $1 \le c \le m-1$. Denote

$$\tilde{\Phi} = \Phi - \theta, \quad \tilde{\Phi}_c = \Phi_c - \theta, \quad 1 \le c \le m.$$

The kernel $\Phi(x_1, \dots, x_m)$ in (1.1.1) can be represented in the form

$$\Phi(x_1, \dots, x_m) = \int \dots \int \Phi(y_1, \dots, y_m) \prod_{s=1}^{m} \delta_{x_s}(dy_s), \tag{1.1.4}$$

where

$$\delta_x(A) = \begin{cases} 1, & \text{if } x \in A \\ 0, & \text{if } x \notin A \end{cases}$$

for any $x \in X$ and $A \in \mathfrak{X}$.

This relation enables us to use the multiplicative properties of the integrand in the following way. It is well-known that for arbitrary numbers $(a_1, \dots, a_m)$ and $(b_1, \dots, b_m)$

$$\prod_{s=1}^{m}(a_s + b_s) = \prod_{s=1}^{m} a_s + \sum_{c=1}^{m} \sum_{1 \le j_1 < \dots < j_c \le m} \prod_{s=1}^{c} b_{j_s} \prod_{s=1}^{m-c} a_{i_s} \tag{1.1.5}$$

where $\{i_1, \dots, i_{m-c}\} = \{1, \dots, m\} \setminus \{j_1, \dots, j_c\}$.

Note that

$$\delta_{x_s}(dy_s) = P(dy_s) + (\delta_{x_s}(dy_s) - P(dy_s)), \quad s = 1, \dots, m.$$

In (1.1.5), we set

$$a_s = P(dy_s), \quad b_s = \delta_{x_s}(dy_s) - P(dy_s), \quad s = 1, \dots, m,$$

and then, we substitute this relation into (1.1.4).

As a result, taking into account the symmetry properties of function Φ_c, we obtain the formula

$$\Phi(x_1, \dots, x_m) - \theta =$$

$$= \sum_{c=1}^{m} \sum_{1 \le j_1 < \dots < j_c \le m} \int \dots \int \Phi_c(y_1, \dots, y_c) \prod_{s=1}^{c} (\delta_{x_{j_s}}(dy_s) - P(dy_s)).$$

For $c = 1, \dots, m$, we denote

$$g_c(x_1, \dots, x_c) = \int \dots \int \Phi_c(y_1, \dots, y_c) \prod_{s=1}^{c} (\delta_{x_s}(dy_s) - P(dy_s)). \qquad (1.1.6)$$

Obviously, functions g_c are symmetric with respect to their arguments $x_1, \dots, x_c$. Since $E\delta_{X_s}(dy_s) = P(dy_s)$, $s = 1, \dots, m$, it follows from (1.1.6) that functions g_c possess the property of *complete degeneracy*

$$\begin{aligned} &Eg_1(X_1) = 0\,, \\ &Eg_2(x_1, X_2) = 0\,, \\ &\dots\dots\dots\dots \\ &Eg_m(x_1, \dots, x_{m-1}, X_m) = 0\,. \end{aligned} \qquad (1.1.7)$$

Functions g_c, which are defined by (1.1.6) and having property (1.1.7), are called *canonical* or *P-canonical functions.* It follows that, for all $c = 1, \dots, m$, we have

$$\tilde{\Phi}_c(x_1, \dots, x_c) = \sum_{d=1}^{c} \sum_{1 \le i_1 < \dots < i_d \le c} g_d(x_{i_1}, \dots, x_{i_d})\,.$$

If (1.1.5) is satisfied for $a_s = -P(dy_s)$, $b_s = \delta_{x_s}(dy_s)$, $s = 1, \dots, m$, we can write (1.1.6) in the form

$$g_c(x_1, \dots, x_c) = \sum_{d=1}^{c} (-1)^{c-d} \sum_{1 \le j_1 < \dots < j_d \le c} \tilde{\Phi}_d(x_{j_1}, \dots, x_{j_d})\,.$$

Hoeffding representation. Let us write U-statistic U_n with a Bochner integrable kernel Φ in the form

$$U_n - \theta = \binom{n}{m}^{-1} S_n$$

where

$$S_n = \sum_{1 \le i_1 < \dots < i_m \le n} \tilde{\Phi}(X_{i_1}, \dots, X_{i_m})\,.$$

Denote

$$S_{nc} = \sum_{1 \le i_1 < \dots < i_c \le n} g_c(X_{i_1}, \dots, X_{i_m}) \qquad (1.1.8)$$

for $c = 1, \dots, m$.

Then we obtain

$$S_n = S_{nm} + \sum_{c=1}^{m-1} \sum_{1 \le i_1 < \dots < i_m \le n} \sum_{1 \le j_1 < \dots < j_c \le m} g_c(X_{i_{j_1}}, \dots, X_{i_{j_c}})\,.$$

For $c = 1$, the term on the right-hand side can be written as follows

$$\sum_{1 \le i_1 < \dots < i_m \le n} \sum_{j=1}^{m} g_1(X_{i_j})\,.$$

In this sum each $g_1(X_i)$, $1 \le i \le n$, appears the same number of times. Since the sum contains $\binom{n}{m} m$ terms, each term is encountered $n^{-1}\binom{n}{m} m$ times.

Therefore, the sum $S_{n1} = g_1(X_1) + \cdots + g_1(X_n)$ appears $\binom{n}{1}^{-1}\binom{n}{m}\binom{m}{1}$ times. By repeating this argument for $c = 2, \ldots, m-1$, we obtain

$$S_n = \sum_{c=1}^{m} \binom{n}{c}^{-1} \binom{n}{m} \binom{m}{c} S_{nc} .$$

This implies the following formula

$$U_n - \theta = \sum_{c=1}^{m} \binom{m}{c} \binom{n}{c}^{-1} S_{nc} , \tag{1.1.9}$$

which is called the *Hoeffding representation* (or a *canonical decomposition*) of a U-statistic.

Rank of *UB*-statistics. Let $r \ge 1$ be the first integer for which the relations

$$g_1 = \cdots = g_{r-1} = 0, \quad g_r \ne 0 \tag{1.1.10}$$

hold true. Clearly, r takes the values $1, 2, \ldots, m$. An integer r satisfying (1.1.10) is called a *rank* of the *UB*-statistic (or of its kernel Φ). Note that the symbol 0, in (1.1.10), denotes the zero element of the Banach space B. If $r = 1$, then a U-statistic (or its kernel Φ) is called *non-degenerate*. For $r \ge 2$, a U-statistic (or its kernel Φ) is called *degenerate,* and r is the *order of degeneracy*. If $r = m$, then we say that the kernel Φ possesses the property of *complete degeneracy,* or that Φ is a *P-canonical function* (*kernel*). For example, by virtue of (1.1.7), the function $g_c(x_1, \ldots, x_c)$ in (1.1.6) possesses this property for all $c = 1, \ldots, m$. We denote

$$U_{nc} = \binom{n}{c}^{-1} \sum_{1 \le i_1 < \ldots < i_c \le n} g_c(X_{i_1}, \ldots, X_{i_c}) \tag{1.1.11}$$

for $c = 1, \ldots, m$. Then the Hoeffding representation (1.1.9) can be written as

$$U_n - \theta = \sum_{c=r}^{m} \binom{m}{c} U_{nc} , \tag{1.1.12}$$

i.e. any *UB*-statistic (1.1.1) is a linear combination of *UB*-statistics (1.1.11) whose kernels possess the property of complete degeneracy (1.1.7), and the integer r denotes the rank in (1.1.10).

Martingale structure of *UB*-statistics. Assume that a family of σ-algebras $\mathfrak{F}_n$, $n \ge 0$, is given on the probability space $(\Omega, \mathfrak{F}, P)$ and that this family is such that

$$\{\varnothing, \Omega\} = \mathfrak{F}_0 \subseteq \mathfrak{F}_1 \subseteq \ldots \subseteq \mathfrak{F}_n \subseteq \mathfrak{F} .$$

Suppose that the sequence of B-valued random variables $\xi_0, \xi_1, \ldots$ is defined on $(\Omega, \mathfrak{F}, P)$ and that for every $n \geq 2$ the variable ξ_n is $\mathfrak{F}_n$-measurable.

Definition 1.1.1. A stochastic sequence $(\xi_n, \mathfrak{F}_n)$, $n \geq 0$, is called a B-valued *martingale* if $E\|\xi_n\| < \infty$ and for $n \geq 0$

$$E(\xi_{n+1}|\mathfrak{F}_n) = \xi_n \quad \text{a.s.}$$

Assume that on the probability space $(\Omega, \mathfrak{F}, P)$ another B-valued sequence $\eta_0, \eta_1, \ldots$ is given such that for every $n \geq 0$ the random variable η_n is $\mathfrak{F}_n$-measurable.

Definition 1.1.2. A stochastic sequence $(\eta_n, \mathfrak{F}_n)$, $n \geq 0$, is called a B-valued *martingale difference* if $E\|\eta_n\| < \infty$ and for all $n \geq 0$

$$E(\eta_{n+1}|\mathfrak{F}_n) = 0 \quad \text{a.s.}$$

The connection between martingales and martingale differences is obvious. Indeed, if $(\xi_n, \mathfrak{F}_n)$ is a martingale, then $(\eta_n, \mathfrak{F}_n)$ with $\eta_0 = \xi_0$ and $\eta_n = \xi_n - \xi_{n-1}$, $n \geq 1$, is a martingale difference. And vice versa, if $(\eta_n, \mathfrak{F}_n)$ is a martingale difference, then $(\xi_n, \mathfrak{F}_n)$ with $\xi_n = \eta_0 + \cdots + \eta_n$ is a martingale.

Example 1.1.1. Let random variable ζ_n be measurable with respect to the σ-algebra $\mathfrak{F}_n$, $n = 1, 2, \ldots$, such that $E\|\zeta_n\| < \infty$.

Denote

$$\xi_{nk} = E(\zeta_n|\mathfrak{F}_k) - E(\zeta_n|\mathfrak{F}_{k-1}), \quad k = 1, \ldots, n.$$

Then the following representation holds

$$\zeta_n - E\zeta_n = \xi_{n1} + \cdots + \xi_{nn},$$

where sequence $(\xi_{nk}, \mathfrak{F}_k)$, $k \geq 1$, forms a martingale difference for every n (the scheme of the series).

Suppose that on the probability space $(\Omega, \mathfrak{F}, P)$, a nonincreasing family of σ-algebras $\mathfrak{B}_1 \supseteq \mathfrak{B}_2 \supseteq \ldots$ is given such that the random variables ξ_n are measurable with respect to $\mathfrak{B}_n$ for all n.

Definition 1.1.3. A stochastic sequence $(\xi_n, \mathfrak{B}_n)$, $n \geq 1$, is called a B-valued *reversed martingale* if $E\|\xi_n\| < \infty$ and for all $n \geq 1$

$$E(\xi_n|\mathfrak{B}_{n+1}) = \xi_{n+1} \quad \text{a.s.}$$

It is well-known that in this case ξ_n converges as $n \to \infty$ to the random variable $\xi_\infty = E(\xi_1|\mathfrak{B}_\infty)$ with probability 1, i.e.

$$\xi_n \to \xi_\infty \quad \text{a.s.}$$

where $\mathfrak{B}_\infty = \cap_{n=1}^{\infty} \mathfrak{B}_n$.

Denote $\eta_n = \xi_n - \xi_{n+1}$, $n \geq 1$. Thus, η_n is measurable with respect to $\mathfrak{B}_n$. Since $(\xi_n, \mathfrak{B}_n)$ is a reversed martingale, we have

$$E(\eta_n | \mathfrak{B}_{n+1}) = 0 \quad \text{a.s.}$$

Consequently, the sequence $(\eta_n, \mathfrak{B}_n)$, $n \geq 1$, forms a B-valued reversed martingale difference, and

$$\xi_n - \xi_\infty = \sum_{j=n}^{\infty} \eta_j \,.$$

Definition 1.1.4. A B-valued reversed martingale $(\xi_n, \mathfrak{B}_n)$, $n \geq 1$, is called a *regular reversed martingale* if a Bochner integrable random variable ξ exists such that

$$\xi_n = E(\xi | \mathfrak{B}_n) \,.$$

The following assertion is due to Hoeffding. It establishes the martingale property of sequence (S_{nc}), $n \geq c$, in (1.1.8), for every $c = 1, \ldots, m$.

Lemma 1.1.1. *Let $E\|\Phi\| < \infty$ and let $\mathfrak{F}_k = \sigma\{\omega : X_1, \ldots, X_k\}$ be a σ-algebra generated by $X_1, \ldots, X_k$, $\mathfrak{F}_0 = \{\varnothing, \Omega\}$. Then*

$$E(S_{nc} | \mathfrak{F}_k) = S_{kc}, \quad c \leq k \leq n,$$

for every $c = 1, \ldots, m$, i.e. the stochastic sequence $(S_{nc}, \mathfrak{F}_n)$, $n \geq c$, forms a martingale.

Proof. By property (1.1.7), we have

$$E(g_c(X_{i_1}, \ldots, X_{i_c}) | X_1, \ldots, X_k) = 0$$

if at least one of the subscripts $i_1, \ldots, i_c$ is not contained in $\{1, \ldots, k\}$. For example, if $i_1 \notin \{1, \ldots, k\}$, then

$$\begin{aligned} &E(g_c(X_{i_1}, \ldots, X_{i_c}) | X_1, \ldots, X_k) = \\ &= E\{E[g_c(X_{i_1}, \ldots, X_{i_c}) | X_1, \ldots, X_k, X_{i_2}, \ldots, X_{i_c}] | X_1, \ldots, X_k\} = \\ &= E\{E[g_c(X_{i_1}, \ldots, X_{i_c}) | X_{i_2}, \ldots, X_{i_c}] | X_1, \ldots, X_k\} = \\ &= E\{0 | X_1, \ldots, X_k\} = 0 \,. \end{aligned}$$

Thus,

$$\begin{aligned} E(S_{nc} | X_1, \ldots, X_k) &= \sum_{1 \leq i_1 < \ldots < i_c \leq k} g_c(X_{i_1}, \ldots, X_{i_c}) \\ &= S_{kc} \,. \end{aligned}$$

Now we represent UB-statistics as a sum of B-valued martingale differences. According to Example 1.1.1, we obtain

$$U_n - \theta = \xi_{n1} + \cdots + \xi_{nn}, \tag{1.1.13}$$

where

$$\xi_{nk} = E(U_n|X_1, \ldots, X_k) - E(U_n|X_1, \ldots, X_{k-1}).$$

From (1.1.9) and Lemma 1.1.1, one can find that

$$E(U_n|X_1, \ldots, X_k) = \theta + \sum_{c=1}^{m} \binom{m}{c} \binom{n}{c}^{-1} S_{kc}.$$

Thus, in this case we have

$$\xi_{nk} = \sum_{c=1}^{m} \binom{m}{c} \binom{n}{c}^{-1} (S_{kc} - S_{k-1,c}).$$

For $k \geq c$, $k = 1, \ldots, n$, $c = 1, \ldots, m$, denote

$$\zeta_{kc} = \sum_{1 \leq i_1 < \ldots < i_{c-1} \leq k-1} g_c(X_{i_1}, \ldots, X_{i_{c-1}}, X_k),$$

where $\zeta_{k1} = g_1(X_k)$, $k = 1, \ldots, n$, by definition. Since $S_{kc} - S_{k-1,c} = \zeta_{kc}$, then

$$\xi_{nk} = \sum_{c=1}^{m} \binom{m}{c} \binom{n}{c}^{-1} \zeta_{kc}.$$

In particular, for $m = 2$, we have

$$\xi_{nk} = 2n^{-1} g_1(X_k) + 2n^{-1}(n-1)^{-1} \sum_{j=1}^{k-1} g_2(X_j, X_k).$$

It is obvious that the sequence $(\xi_{nk}, \mathfrak{F}_k)$, $k \geq 1$, forms a Banach-valued martingale difference, where $\mathfrak{F}_k = \sigma\{\omega : X_1, \ldots, X_k\}$.

Let us formulate these reasonings in the following form.

Lemma 1.1.2. *If $E\|\Phi\| < \infty$, then UB-statistics can be represented as sum* (1.1.13) *of Banach-valued martingale differences $(\xi_{nk}, \mathfrak{F}_k)$, $k \geq 1$.*

Show that UB-statistics are Banach-valued reversed regular martingales for a suitably chosen nonincreasing flow of σ-algebras. This assertion is due to Berk. Denote by

$$\mathfrak{B}_n = \sigma\{\omega : U_n, U_{n+1}, \ldots\} \tag{1.1.14}$$

a σ-algebra generated by the sequence of U-statistics $U_n, U_{n+1}, \ldots$ where $n \geq m$. It is clear that $\mathfrak{B}_n \supseteq \mathfrak{B}_{n+1}$ for all $n = m, m+1, \ldots$ and U_n is a $\mathfrak{B}_n$-measurable B-valued random element.

Lemma 1.1.3. *Stochastic sequence $(U_n, \mathfrak{B}_n)$, $n \geq m$, forms a B-valued reversed regular martingale and*

$$U_n = E(\Phi(X_1, \dots, X_m)|\mathfrak{B}_n) \tag{1.1.15}$$

for all $n = m, m+1, \dots$ where $\mathfrak{B}_n$ are taken from (1.1.14).

Proof. The sequence $U_n, U_{n+1}, \dots$ remains unchanged under any permutation of random variables $X_1, \dots, X_n$ because U-statistics are symmetric functions. Taking into account that X_i, $i = 1, \dots, n$, are independent and identically distributed, we obtain

$$E(\Phi(X_{i_1}, \dots, X_{i_m})|\mathfrak{B}_n) = E(\Phi(X_1, \dots, X_m)|\mathfrak{B}_n)$$

for any $i_1, \dots, i_m$ such that $1 \leq i_1 < \dots < i_m \leq n$. Carrying out the summation over all subscripts $1 \leq i_1 < \dots < i_m \leq n$ on both the left-hand side and right-hand side of this relation, we obtain

$$E(U_n|\mathfrak{B}_n) = E(\Phi(X_1, \dots, X_m)|\mathfrak{B}_n).$$

Since $E(U_n|\mathfrak{B}_n) = U_n$, this yields (1.1.15). In (1.1.15), we take the conditional expectation with respect to the σ-algebra $\mathfrak{B}_{n+1}$. Then

$$\begin{aligned} E(U_n|\mathfrak{B}_{n+1}) &= E(E(\Phi(X_1, \dots, X_m)|\mathfrak{B}_n)|\mathfrak{B}_{n+1}) \\ &= E(\Phi(X_1, \dots, X_m)|\mathfrak{B}_{n+1}), \end{aligned}$$

where the second equality is true due to the property of conditional expectation and the inclusion $\mathfrak{B}_n \supseteq \mathfrak{B}_{n+1}$. Together with (1.1.15), this gives $E(U_n|B_{n+1}) = U_{n+1}$, i.e. the property of a reversed martingale for the sequence $(U_n, \mathfrak{B}_n)$, $n \geq m$.

Formula (1.1.15) is valid not only for the σ-algebra $\mathfrak{B}_n$ from (1.1.14) but also for a wider class of sets which are called *symmetric* or *permutable*. In order to use (1.1.15) in different situations, we must define the notion of a symmetric event. Let R^∞ be a space of ordered numerical sequences $(x_1, x_2, \dots)$ and let $\mathfrak{B}(R^\infty)$ be a σ-algebra of Borel sets in R^∞.

Definition 1.1.5. Suppose that $\xi_1, \xi_2, \dots$ is a sequence of random variables and that $\mathfrak{A} = \sigma\{\omega : \xi_1, \xi_2, \dots\}$ is a σ-algebra generated by this sequence. The event $A \in \mathfrak{A}$ is called *symmetric* (or *permutable*) if there exists a Borel set $B \in \mathfrak{B}(R^\infty)$ such that the equality

$$\begin{aligned} A &= \{\omega : \ (\xi_1, \xi_2, \dots, \xi_n, \xi_{n+1}, \dots) \in B\} \\ &= \{\omega : \ (\xi_{i_1}, \xi_{i_2}, \dots, \xi_{i_n}, \xi_{n+1}, \dots) \in B\} \end{aligned}$$

holds for any $n \geq 1$ and any permutation $(i_1, \dots, i_n)$ of the numbers $(1, \dots, n)$.

Example 1.1.2. Let $X_1, X_2, \dots$ be a sequence of random variables with values in an arbitrary measurable space $(X, \mathfrak{X})$, and let $\mathfrak{B}_n$ be a σ-algebra generated by an unordered set of random variables $(X_1, \dots, X_n)$ and by a sequence $X_{n+1}, X_{n+2}, \dots$, i.e.

$$\mathfrak{B}_n = \sigma\{\omega : (X_1, \dots, X_n), X_{n+1}, X_{n+2}, \dots\}.$$

It is clear that $\mathfrak{B}_n \supseteq \mathfrak{B}_{n+1}$ for all $n = 1, 2, \dots$. For the stochastic sequence $(U_n, \mathfrak{B}_n)$, Lemma 1.1.3 holds.

Example 1.1.3. Let $X = R$ and $X_{n1} \le X_{n2} \le \dots \le X_{nn}$ be ordered statistics corresponding to random variables $X_1, \dots X_n$. Denote by

$$\mathfrak{B}_n = \sigma\{\omega : X_{n1}, \dots, X_{nn}, X_{n+1}, X_{n+2}, \dots\}$$

the σ-algebra generated by $X_{n1}, \dots, X_{nn}, X_{n+1}, X_{n+2}, \dots$.

It is clear that $\mathfrak{B}_n \supseteq \mathfrak{B}_{n+1}$ for every $n \ge 1$. Therefore

$$\begin{aligned} U_n &= E(\Phi(X_{i_1}, \dots, X_{i_m})|\mathfrak{B}_n) \\ &= E(\Phi(X_{i_1}, \dots, X_{i_m})|X_{n1}, \dots, X_{nn}) \end{aligned}$$

for all $1 \le i_1 < \dots < i_m \le k$ and $m \le k \le n$.

Hoeffding formula. Let $k = [n/m]$ be the integral part of the number n/m. We set

$$\begin{aligned} \varphi_n(x_1, \dots, x_n) &= k^{-1}(\Phi(x_1, \dots, x_m) + \\ &\quad +\Phi(x_{m+1}, \dots, x_{2m}) + \dots + \Phi(x_{km-m+1}, \dots, x_{km})). \end{aligned}$$

Let $\sum_{\sigma}$ be the sum over all $n!$ permutations $(i_1, \dots, i_n)$ of the numbers $(1, \dots, n)$ and let $\sum_{\mathbf{c}}$ denote the sum over all $\binom{n}{m}$ combinations $(i_1, \dots, i_m)$ constructed from the numbers $(1, \dots, n)$. Then

$$\sum_{\sigma} \varphi_n(X_{i_1}, \dots, X_{i_n}) = m!(n-m)! \sum_{\mathbf{c}} \Phi(X_{i_1}, \dots, X_{i_m})$$

or

$$\sum_{\sigma} \varphi_n(X_{i_1}, \dots, X_{i_n}) = m!(n-m)! \binom{n}{m} U_n .$$

Therefore, for *UB*-statistics, the *Hoeffding formula*

$$U_n = \frac{1}{n!} \sum_{\sigma} \varphi_n(X_{i_1}, \dots, X_{i_n}) \tag{1.1.16}$$

is true. This means that *UB*-statistics can be expressed as the mean value of $n!$ terms, each being the mean value of k independent identically distributed Banach-valued random variables.

Note that, according to Lemma 1.1.3 and (1.1.16)

$$U_n = E(\varphi_n | \mathfrak{B}_n), \tag{1.1.17}$$

where σ-algebra $\mathfrak{B}_n$ is taken from Example 1.1.2.

1.2 Multi-sample *UB*-statistics

Assume that c independent series $\{X_{j1}, \dots, X_{jn_j}\}$, $j = 1, \dots, c$, of independent random variables are given. All variables in the jth series take values in a measurable space $(X_j, \mathfrak{X}_j)$ and have the same distribution P_j on it.

Let $\Phi(x_{ji},\, i = 1, \dots, m_j,\, j = 1, \dots, c)$ be a measurable function symmetric with respect to m_j arguments of the jth set, $m_j \geq 1$, $j = 1, \dots, c$, and taking its values in Banach space B. A *multi-sample UB-statistic* is defined as follows

$$U_{\bar{n}} = \prod_{j=1}^{c} \binom{n_j}{m_j}^{-1} \sum \Phi(X_{j\alpha},\, \alpha = i_{j1}, \dots, i_{jm_j},\, j = 1, \dots, c), \tag{1.2.1}$$

where $\bar{n} = (n_1, \dots, n_c)$ and the summation is carried out over all $1 \leq i_{j1} < \dots < i_{jm_j} \leq n_j$, $m_j \leq n_j$, $j = 1, \dots, c$. As in a one-sample case, $U_{\bar{n}}$ is an unbiased estimate of B-valued element

$$\theta = \theta(P) = E\Phi(X_{ji},\, i = 1, \dots, m_j,\, j = 1, \dots, c),$$

where $P = (P_1, \dots, P_c)$. A numerical vector $\bar{m} = (m_1, \dots, m_c)$ is called the *degree* of *U*-statistic (or of the kernel Φ), and the integer c is called the *order* of *U*-statistic (1.2.1).

Canonical functions. For every vector $\bar{d} = (d_1, \dots, d_c)$, $0 \leq d_j \leq m_j$, $1 \leq j \leq c$, under condition $E\|\Phi\| < \infty$, we define the functions

$$\Phi_{\bar{d}}(x_{ji},\, i = 1, \dots, d_j,\, 1 \leq j \leq c) =$$

$$= E\Phi(x_{j1}, \dots, x_{jd_j}, X_{jd_{j+1}}, \dots, X_{jm_j},\, 1 \leq j \leq c)$$

such that $\Phi_{\bar{0}} = \theta$, $\Phi_{\bar{m}} = \Phi$, where $\bar{0} = (0, \dots, 0)$, $\bar{m} = (m_1, \dots, m_c)$.

Taking into account the expression for $\Phi_{\bar{d}}$, we now define the canonical functions

$$g_{\bar{d}}(x_{ji},\, i = 1, \dots, d_j,\, 1 \leq j \leq c) =$$

$$= \sum_{\bar{r}=\bar{0}}^{\bar{d}} (-1)^{d_1+\dots+d_c-r_1-\dots-r_c} \sum_{1\le i_{j1}<\dots<i_{jr_j}\le d_j} \Phi_{\bar{d}}(x_{ji_{js}},\, 1\le s\le r_j,\, 1\le j\le c)\,.$$

$g_{\bar{d}}$ can be represented in the integral form

$$g_{\bar{d}}(x_{ji},\, i=1,\dots,d_j,\, 1\le j\le c) =$$

$$= \int \dots \int \Phi_{\bar{d}}(y_{ji},\, i=1,\dots,d_j,\, 1\le j\le c) \prod_{j=1}^{c}\prod_{s=1}^{d_j} (\delta_{x_{js}}(dy_{js}) - P_j(dy_{js}))\,. \quad (1.2.2)$$

Because $E\delta_{X_{js}}(dy_{js}) = P_j(dy_{js})$, $s=1,\dots,m_j$, $j=1,\dots,c$, then from (1.2.2) it can be clearly seen that function $g_{\bar{d}}$ possesses the property of complete degeneracy

$$\begin{aligned} &Eg_{\bar{d}}(\dots;X_{j1};\dots) = 0\,,\\ &Eg_{\bar{d}}(\dots;x_{j1},X_{j2};\dots) = 0\,,\\ &\dots\dots\dots\dots\dots\dots\\ &Eg_{\bar{d}}(\dots;x_{j1},x_{j2},\dots,x_{j\,m_j-1},X_{jm_j};\dots) = 0 \end{aligned} \quad (1.2.3)$$

with respect to the variables of every jth set.

Functions $g_{\bar{d}}$ in (1.2.2) with property (1.2.3) are called *canonical functions*.

The reverse to formula (1.2.2) has the following form

$$\Phi_{\bar{d}}(x_{ji},\, i=1,\dots,d_j,\, 1\le j\le c) =$$

$$= \sum_{\bar{r}=\bar{0}}^{\bar{d}} \sum_{1\le i_{j1}<\dots<i_{jr_j}\le d_j} g_{\bar{r}}(x_{ji_{js}},\, 1\le s\le r_j,\, 1\le j\le c)\,.$$

Putting $\bar{d}=\bar{m}$, we obtain the formula for kernel Φ from (1.2.1)

$$\Phi(x_{ji},\, i=1,\dots,m_j,\, j=1,\dots,c) =$$

$$= \sum_{\bar{r}=\bar{0}}^{\bar{m}} \sum_{1\le i_{j1}<\dots<i_{jr_j}\le m_j} g_{\bar{r}}(x_{ji_{js}},\, 1\le s\le r_j,\, 1\le j\le c)\,. \quad (1.2.4)$$

Canonical Hoeffding representation. Substituting (1.2.4) into (1.2.1) and then processing the sum with respect to every jth block of variables in the same way as with the deduction of (1.1.9), we obtain

$$U_{\bar{n}} = \sum_{\bar{d}=\bar{0}}^{\bar{m}} \prod_{j=1}^{c} \binom{m_j}{d_j} U_{\bar{n}\bar{d}}\,, \quad (1.2.5)$$

where by $U_{\bar{n}\bar{d}}$ the multi-sample *UB*-statistic with kernel $g_{\bar{d}}$ from (1.2.2) is denoted, and $U_{\bar{n}\bar{0}} = \theta$.

Particularly, when $c = 2$ and $m_1 = m_2 = 2$ from (1.2.5) it follows that

$$U_{\bar{n}} - \theta =$$

$$= \frac{2}{n_1} \sum_{i=1}^{n_1} g_{10}(X_{1i}) + \frac{2}{n_2} \sum_{j=1}^{n_2} g_{01}(X_{2j}) +$$

$$+ \frac{4}{n_1 n_2} \sum_{i=1}^{n_1} \sum_{j=1}^{n_2} g_{11}(X_{1i}, X_{2j}) +$$

$$+ \binom{n_1}{2}^{-1} \sum_{1 \le i_1 < i_2 \le n_1} g_{20}(X_{1i_1}, X_{1i_2}) +$$

$$+ \binom{n_2}{2}^{-1} \sum_{1 \le j_1 < j_2 \le n_2} g_{02}(X_{2j_1}, X_{2j_2}) +$$

$$+ \frac{2}{n_2} \binom{n_1}{2}^{-1} \sum_{1 \le i_1 < i_2 \le n_1} \sum_{j=1}^{n_2} g_{21}(X_{1i_1}, X_{1i_2}, X_{2j}) +$$

$$+ \frac{2}{n_1} \binom{n_2}{2}^{-1} \sum_{i=1}^{n_1} \sum_{1 \le j_1 < j_2 \le n_2} g_{12}(X_{1i}, X_{2j_1}, X_{2j_2}) +$$

$$+ \binom{n_1}{2}^{-1} \binom{n_2}{2}^{-1} \sum_{1 \le i_1 < i_2 \le n_1} \sum_{1 \le j_1 < j_2 \le n_2} g_{22}(X_{1i_1}, X_{1i_2}, X_{2j_1}, X_{2j_2}) . \qquad (1.2.6)$$

Rank vector. Define function d for every vector $\bar{d} = (d_1, d_2, \dots, d_c)$ such that $0 \le d_j \le m_j$, $j = 1, \dots, c$, as a sum of its coordinates

$$d = d_1 + d_2 + \cdots + d_c .$$

It is obvious that d takes values $0, 1, \dots, m$, where $m = m_1 + m_2 + \cdots + m_c$.

For the set of functions $\{g_{\bar{d}}, \bar{0} \le \bar{d} \le \bar{m}\}$ define vector $\bar{r} = (r_1, \dots, r_c)$ which satisfy the following relations:

$$g_{\bar{d}} = 0, \quad g_{\bar{r}} \ne 0 \qquad (1.2.7)$$

for every vector $\bar{d} = (d_1, \dots, d_c)$ such that $r = d + 1$. Such a vector $\bar{r} = (r_1, \dots, r_c)$ is called a *rank vector* and the sum $r = r_1 + r_2 + \cdots + r_c$ is called a *rank sum*. If $r = 1$, then *UB*-statistic (1.2.1) (or a kernel Φ) is called non-degenerate. When $r \ge 2$ such a *UB*-statistic (or kernel Φ) is called degenerate. If $r = m$, that is $\bar{r} = \bar{m}$, then kernel Φ is said to possess the property of complete degeneracy.

If $g_{\bar{d}}$ has property (1.2.3), then the *canonical Hoeffding representation* (1.2.5) can be written as

$$U_{\bar{n}} - \theta = \sum_{d=r}^{m} \sum_{d_1 + \cdots + d_c = d} \prod_{j=1}^{c} \binom{m_j}{d_j} U_{\bar{n}\bar{d}} \qquad (1.2.8)$$

when $0 \le d_j \le m_j$, $j = 1, \dots, c$.

Martingale structure. Let N denote the set of positive integers and let N^c be the set of c-dimensional vectors of the form $\bar{n} = (n_1, \dots, n_c)$, $n_i \in N$, $\bar{n} \in N^c$.

Definition 1.2.1. Let $(\mathfrak{F}_{\bar{m}})$, $\bar{m} \in N^c$ be a nondecreasing flow of σ-algebras, i.e. $\mathfrak{F}_{\bar{m}} \subseteq \mathfrak{F}_{\bar{n}}$, as $\bar{m} \le \bar{n}$, and let $\xi_{\bar{m}}$ be a $\mathfrak{F}_{\bar{m}}$-measurable B-valued random variable. A stochastic sequence $(\xi_{\bar{m}}, \mathfrak{F}_m)$, $\bar{m} \in N^c$, is called a *$*$-martingale,* if $E\,\|\xi_{\bar{m}}\| < \infty$ and for all $\bar{m}$, $\bar{n} \in N^c$ the equality

$$E(\xi_{\bar{m}} | \mathfrak{F}_{\bar{n}}) = \xi_{\bar{m} \wedge \bar{n}}, \qquad a.s.$$

holds, where $\bar{m} \wedge \bar{n} = \min(\bar{m}, \bar{n})$ is a coordinate-wise minimum.

Definition 1.2.2. Let $(\mathfrak{B}_{\bar{m}})$, $\bar{m} \in N^c$ be a nonincreasing flow of σ-algebras, i.e. $\mathfrak{B}_{\bar{m}} \supseteq \mathfrak{B}_{\bar{n}}$, as $\bar{m} \le \bar{n}$, and let $\xi_{\bar{m}}$ be a $\mathfrak{B}_{\bar{m}}$-measurable B-valued random variable. A stochastic sequence $(\xi_{\bar{m}}, \mathfrak{B}_{\bar{m}})$ is called a *reversed $*$-martingale,* if $E\,\|\xi_{\bar{m}}\| < \infty$ and for all $\bar{m}$, $\bar{n} \in N^c$ the equality

$$E(\xi_{\bar{m}} | \mathfrak{B}_{\bar{n}}) = \xi_{\bar{m} \vee \bar{n}} \qquad a.s.$$

holds, where $\bar{m} \vee \bar{n} = \max(\bar{m}, \bar{n})$ is a coordinate-wise maximum.

Multi-sample Hoeffding lemma. Given $\bar{k} = (k_1, \dots, k_c) \in N^c$, define σ-algebras

$$\mathfrak{F}_{jk_j} = \sigma\big\{\omega\colon X_{j1}, \dots, X_{jk_j}\big\}, \qquad j = 1, \dots, c,$$

and denote by $\mathfrak{F}_{\bar{k}}$ the product of these σ-algebras, i.e.

$$\mathfrak{F}_{\bar{k}} = \mathfrak{F}_{1k_1} \times \mathfrak{F}_{2k_2} \times \cdots \times \mathfrak{F}_{ck_c} = \sigma\left\{\omega\colon X_{j1}, \dots, X_{jk_j};\ j = 1, \dots, c\right\}.$$

For $\bar{d} \le \bar{m}$ denote

$$S_{\bar{n}\bar{d}} = \sum_{\substack{1 \le i_{j1} < \cdots < i_{jd_j} \le n_j \\ j=1,\dots,c}} g_{\bar{d}}(X_{j\alpha},\ \alpha = i_{j1}, \dots, i_{jd_j},\ j = 1, \dots, c), \tag{1.2.9}$$

i.e. $S_{\bar{n}\bar{d}} = \prod_{j=1}^{c} \binom{n_j}{d_j} U_{\bar{n}\bar{d}}$.

Lemma 1.2.1. *Let $E\,\|g_{\bar{d}}\| < \infty$. Then for all $\bar{k} \ge \bar{d}$ and $\bar{n} \ge \bar{d}$ the relation*

$$E(S_{\bar{n}\bar{d}} | \mathfrak{F}_{\bar{k}}) = S_{\bar{n} \wedge \bar{k}, \bar{d}}$$

holds, i.e. the stochastic sequence $(S_{\bar{n}\bar{d}}, \mathfrak{F}_{\bar{n}})$, $\bar{n} \ge \bar{d}$ is a $*$-martingale.

Proof. By degeneracy property (1.2.3), we have

$$E\Big(g_{\bar{d}}(X_{j\alpha},\ \alpha = i_{j1},\dots,i_{jd_j},\ j=1,\dots,c) \mid X_{j1},\dots,X_{jk_j},\ j=1,\dots,c\Big) = 0$$

if at least one of the indices $i_{j1}, \dots, i_{jd_j}$ is not contained in $\{1,\dots,k_j\}$.

Thus,

$$E\Big(S_{\bar{n}\bar{d}} \mid X_{j1},\dots,X_{jk_j},\ j=1,\dots,c\Big) =$$

$$= \sum_{\substack{1\le i_{j1}<\cdots<i_{jd_j}\le n_j\wedge k_j \\ j=1,\dots,c}} g_{\bar{d}}(X_{j\alpha},\ \alpha = i_{j1},\dots,i_{jd_j},\ j=1,\dots,c) =$$

$$= S_{\bar{n}\wedge\bar{k},\bar{d}}$$

and the lemma is proved.

Multi-sample Berk lemma. For $\bar{k} = (k_1,\dots,k_c) \in N^c$, let $\mathfrak{B}_{jk_j}$ denote a σ-algebra generated by an unordered collection of random variables $(X_{j1},\dots,X_{jk_j})$ and by the sequence $X_{j,k_j+1}, X_{j,k_j+2}, \dots$, i.e.

$$\mathfrak{B}_{jk_j} = \sigma\Big\{\omega\colon (X_{j1},\dots,X_{jk_j}), X_{j,k_j+1}, X_{j,k_j+2},\dots,\Big\}, \qquad j=1,\dots,c.$$

It is obvious that $\mathfrak{B}_{jn} \supseteq \mathfrak{B}_{j,n+1}$ for all $n = 1, 2, \dots$. Denote by $\mathfrak{B}_{\bar{k}}$ the product of these σ-algebras, i.e.

$$\mathfrak{B}_{\bar{k}} = \mathfrak{B}_{1k_1} \times \mathfrak{B}_{2k_2} \times \cdots \times \mathfrak{B}_{ck_c} =$$
$$= \sigma\Big\{\omega\colon (X_{j1},\dots,X_{jk_j}), X_{j,k_j+1}, X_{j,k_j+2},\dots,; j=1,\dots,c\Big\}.$$

Lemma 1.2.2. *Let $E\|\Phi\| < \infty$. Then the stochastic sequence $(U_{\bar{n}}, \mathfrak{B}_{\bar{n}})$, $\bar{n} \ge \bar{m}$ forms a reversed $*$-martingale and the relation*

$$U_{\bar{n}} = E\Big(\Phi(X_{ij},\ i=1,\dots,m_j,\ j=1,\dots,c) \mid \mathfrak{B}_{\bar{n}}\Big) \qquad (1.2.10)$$

holds for all $\bar{n} \ge \bar{m}$.

Proof. To begin with, we establish (1.2.10). Note that any permutation of the variables $X_{j1}, \dots, X_{jk_j}$ does not affect the σ-algebra $\mathfrak{B}_{jk_j}$. This fact together with the independence and the identical distribution of $X_{j1}, \dots, X_{jn_j}$ enables one to derive the relation

$$E\Big(\Phi(X_{j\alpha},\ \alpha = i_{j1},\dots,i_{jd_j},\ j=1,\dots,c) \mid \mathfrak{B}_{\bar{n}}\Big) =$$
$$= E\Big(\Phi(X_{ij},\ i=1,\dots,m_j,\ j=1,\dots,c) \mid \mathfrak{B}_{\bar{n}}\Big)$$

which holds for all $i_{j1}, \dots, i_{jm_j}$ such that $1 \le i_{j1} < \cdots < i_{jm_j} \le n_j$. Here, the fact is also taken into account that by definition $\mathfrak{B}_{\bar n}$ is a product of independent σ-algebras.

Further, by summing on both sides of the latter equality by all indices $1 \le i_{j1}, \cdots < i_{jm_j} \le n_j$, $j = 1, \dots, c$, we obtain

$$E(U_{\bar n}|\mathfrak{B}_{\bar n}) = E\Big(\Phi(X_{ji},\ i = 1,\dots,m_j,\ j = 1,\dots,c) \mid \mathfrak{B}_{n}\Big).$$

Since $E(U_{\bar n}|\mathfrak{B}_{\bar n}) = U_{\bar n}$, the obtained relation implies (1.2.10). In (1.2.10), we take the conditional expectation with respect to the σ-algebra $\mathfrak{B}_{\bar k}$.

Then we obtain

$$\begin{aligned} E(U_{\bar n}|\mathfrak{B}_{\bar k}) &= E\Big(E(\Phi(X_{ji},\ i = 1,\dots,m_j,\ j = 1,\dots,c) \mid \mathfrak{B}_{\bar n}) \Big|\mathfrak{B}_{\bar k}\Big) \\ &= E\Big(\Phi(X_{ji},\ i = 1,\dots,m_j,\ j = 1,\dots,c) \,|\mathfrak{B}_{\bar n\vee\bar k}\Big) \\ &= U_{\bar n\vee \bar k}\,, \end{aligned}$$

where the second equality holds since $\mathfrak{B}_{\bar n} \cap \mathfrak{B}_{\bar k} = \mathfrak{B}_{\bar n\vee\bar k}$.

Thus, the stochastic sequence $(U_{\bar n}, \mathfrak{B}_{\bar n})$, $\bar n \ge \bar m$ is a reversed $*$-martingale.

Hoeffding formula. Denote by

$$\begin{aligned} \varphi_{\bar n}(x_{j1},\dots,x_{jn_j},\ j = 1,\dots,c) = k^{-1}\Big(&\Phi(x_{jm_j},\ j = 1,\dots,c) \\ &+ \Phi(x_{j,m_j+i},\ i = 1,\dots,m_j,\ j = 1,\dots,c) + \cdots \\ &\cdots\cdots\cdots\cdots\cdots\cdots \\ &+ \Phi(x_{j,km_j-m_j+i},\ i = 1,\dots,m_j,\ j = 1,\dots,c)\Big)\,, \end{aligned}$$

where $k = \min\big([n_1/m_1], [n_2/m_2], \dots, [n_c/m_c]\big)$.

Then UB-statistic (1.2.1) can be represented as

$$U_{\bar n} = \prod_{j=1}^{c} \frac{1}{n_j!} \sum \varphi_{\bar n}(X_{ji_{j1}},\dots,X_{ji_{jn_j}},\ j = 1,\dots,c)\,,$$

where the summation is carried out over all permutations $(i_{j1},\dots,i_{jn})$, $j = 1, \dots, c$, of the numbers $(1,\dots,n_j)$, $j = 1, \dots, c$, respectively. The proof of this formula is similar to the proof of (1.1.16).

Also we have the following formula which is similar to formula (1.1.17)

$$U_{\bar n} = E(\varphi_{\bar n}|\mathfrak{B}_{\bar n})\,, \tag{1.2.11}$$

where the σ-algebra $\mathfrak{B}_{\bar n}$ comes from Lemma 1.2.2.

1.3 Von Mises' statistics

One-sample Banach-valued von *Mises' statistics* are defined as follows

$$V_n = n^{-m} \sum_{i_1=1}^{n} \cdots \sum_{i_m=1}^{n} \Phi(X_{i_1}, \ldots , X_{i_m}) \tag{1.3.1}$$

with the symmetric kernel $\Phi \colon X^m \to B$, where $X_1, \ldots , X_n$ are independent random variables with values on a measurable space $(X, \mathfrak{X})$ and with the same distribution P.

With the help of *Dirac's δ-measure,* we define the *empirical measure* P_n by the formula

$$P_n(A) = n^{-1}\left(\delta_{X_1}(A) + \cdots + \delta_{X_n}(A)\right), \qquad A \in \mathfrak{X}.$$

By using P_n, we can write (1.3.1) in the integral form

$$V_n = \int \cdots \int \Phi(x_1, \ldots , x_m) \prod_{j=1}^{c} P_n(dx_j) .$$

Hence, $V_n = \theta(P_n)$, where Banach-valued element $\theta(P)$ is defined in Section 1.1. Though P_n is an unbiased estimate of P, the statistic V_n is not an unbiased estimate of $\theta(P)$.

Representation via canonical functions. For $c = 1, \ldots , m$, we set

$$V_{nc} = \int \cdots \int \Phi_c(x_1, \ldots , x_c) \prod_{j=1}^{c} (P_n(dx_j) - P(dx_j)) ,$$

or, in the notation of (1.1.6),

$$V_{nc} = n^{-c} \sum_{i_1=1}^{n} \cdots \sum_{i_c=1}^{n} g_c(X_{i_1}, \ldots , X_{i_c}) .$$

The element V_{nc} is a von Mises' statistic with kernel g_c, possessing the property of complete degeneracy (1.1.7). According to the argument presented in Section 1.1, we can write the representation of V_n via canonical functions g_c

$$\theta(P_n) - \theta(P) = \sum_{c=r}^{m} \binom{m}{c} V_{nc} , \tag{1.3.2}$$

in which r is a rank defined in (1.1.10). Here, we also take into account that $V_n = \theta(P_n)$.

Representation via *U*-statistics. For $c = 1, \ldots , m$ we introduce *UB*-statistics

$$U_{nc} = \binom{n}{c}^{-1} \sum_{1 \le i_1 < \cdots < i_c \le n} \Phi_{mc}(X_{i_1}, \ldots , X_{i_c})$$

with symmetric kernels

$$\Phi_{mc}(x_1,\dots,x_c)=\sum_{\substack{\nu_1+\dots+\nu_c=m\\ \nu_j\ge 1}}\frac{m!}{\nu_1!\dots\nu_c!}\,\Phi(x_1^{\nu_1},\dots,x_c^{\nu_c}),$$

where $x_j^{\nu_j}=(x_j,\dots,x_j)$ is a vector, all ν_j coordinates of which are equal to x_j.

Let us write this kernel Φ_{mc} for some m. We have, for $m=2$,

$$\begin{aligned}\Phi_{21}(x_1)&=\Phi(x_1,x_1),\\ \Phi_{22}(x_1,x_2)&=2\,\Phi(x_1,x_2);\end{aligned}$$

for $m=3$,

$$\begin{aligned}\Phi_{31}(x_1)&=\Phi(x_1,x_1,x_1),\\ \Phi_{32}(x_1,x_2)&=3\left(\Phi(x_1,x_1,x_2)+\Phi(x_1,x_2,x_2)\right),\\ \Phi_{33}(x_1,x_2,x_3)&=6\,\Phi(x_1,x_2,x_3);\end{aligned}$$

and for $m=4$,

$$\begin{aligned}\Phi_{41}(x_1)&=\Phi(x_1,x_1,x_1,x_1),\\ \Phi_{42}(x_1,x_2)&=4\,\Phi(x_1,x_1,x_1,x_2)+4\,\Phi(x_2,x_2,x_2,x_1)\\ &\quad+6\,\Phi(x_1,x_1,x_2,x_2),\\ \Phi_{43}(x_1,x_2,x_3)&=12\left(\Phi(x_1,x_1,x_2,x_3)\right.\\ &\quad\left.+\Phi(x_2,x_2,x_1,x_3)+\Phi(x_3,x_3,x_1,x_2)\right),\\ \Phi_{44}(x_1,x_2,x_3,x_4)&=24\,\Phi(x_1,x_2,x_3,x_4).\end{aligned}$$

For $n\ge m$, we have the following representation

$$V_n=\sum_{c=1}^{m}\binom{n}{c}\,n^{-m}\,U_{nc}\,, \tag{1.3.3}$$

which indicates that V_n is a linear combination of the U-statistics.

The validity of (1.3.3) follows from the simple combinatorial formula

$$\sum_{i_1=1}^{n}\cdots\sum_{i_m=1}^{n}\Phi(x_{i_1},\dots,x_{i_m})=$$

$$=\sum_{c=1}^{m\wedge n}\sum\nolimits_1\sum\nolimits_2\frac{m!}{\nu_1!\dots\nu_c!}\,\Phi(\overbrace{x_{i_1},\dots,x_{i_1}}^{\nu_1\text{ times}},\dots,\overbrace{x_{i_c},\dots,x_{i_c}}^{\nu_c\text{ times}}) \tag{1.3.4}$$

where the summation in $\sum_1$ is carried out over all indices $1\le i_1<\dots<i_c\le n$, $1\le c\le m\wedge n$, and the summation in $\sum_2$ is carried out over all vectors $(\nu_1,\dots,\nu_c)$ such that $\nu_1+\dots+\nu_c=m$, $\nu_j\ge 1$, $j=1,\dots,c$, and $m\ge 1$, $n\ge 1$.

In particular, if in (1.3.3), $m = 1$, then

$$V_n = \frac{1}{n}\sum_{j=1}^{n}\Phi(X_j)$$

is the sum of independent identically distributed B-valued random variables and if in (1.3.3), $m = 2$, then

$$V_n = \frac{1}{n^2}\sum_{j=1}^{n}\Phi(X_j, X_j) + \frac{2}{n^2}\sum_{1\le i<j\le n}\Phi(X_i, X_j). \tag{1.3.5}$$

Multi-sample von Mises' statistics. In the case of c samples, the generalized version of (1.3.1) has the form

$$V_{\bar{n}} = \prod_{j=1}^{c} n_j^{-m_j}\sum\Phi(X_{j\alpha},\ \alpha = i_{j1},\dots,i_{jm_j},\ j = 1,\dots,c), \tag{1.3.6}$$

where the summation is carried out over all indices $1 \le i_{jk} \le n_j$, $1 \le k \le m_j$, $j = 1$, $\dots$, c.

With the help of the empirical measure, we can write (1.3.6) in the integral form

$$V_{\bar{n}} = \int\cdots\int\Phi(x_{ij},\ i = 1,\dots,m_j,\ j = 1,\dots,c)\prod_{j=1}^{c}\prod_{i=1}^{m_j}P_{nj}(dx_{ji}),$$

where P_{nj} is the empirical measure corresponding to the j-s sample.

1.4 Banach-valued symmetric statistics

Let X_1, $\dots$, X_n be independent random variables taking values in a measurable space $(X, \mathfrak{X})$ and having on it the same distribution P. Let us consider *B-valued symmetric statistics* $S = S(X_1,\dots,X_n)$ (*SB-statistics*), i.e. the values of S are invariant with respect to an arbitrary permutation of the arguments X_1, $\dots$, X_n. Under the condition $E\,\|S\| < \infty$, we now derive the canonical Hoeffding representation for S.

According to (1.1.4) we can write

$$S(x_1,\dots,x_m) = \int\cdots\int S(y_1,\dots,y_n)\prod_{j=1}^{n}\delta_{x_j}(dy_j). \tag{1.4.1}$$

Denote

$$S_{nc}(x_1,\dots,x_c) = E\left(S(X_1,\dots,X_n) \mid X_1 = x_1,\dots,\ X_c = x_c\right),$$

where $c = 1, \dots, n$.

Setting $m = n$ and

$$a_s = P(dy_s), \quad b_s = \delta_{x_s}(dy_s) - P(dy_s), \qquad s = 1, \dots, n$$

in (1.1.5) and inserting this formula into (1.4.1), we obtain

$$S(X_1, \dots, X_n) - E\,S(X_1, \dots, X_n) =$$
$$= \sum_{c=1}^{n} \sum_{1 \le j_1 < \cdots < j_c \le n} \int \cdots \int S_{nc}(y_1, \dots, y_c) \prod_{s=1}^{c} \big(\delta_{X_{j_s}}(dy_s) - P(dy_s)\big).$$

If we denote

$$g_{nc}(x_1, \dots, x_c) = \int \cdots \int S_{nc}(y_1, \dots, y_c) \prod_{s=1}^{c} \big(\delta_{x_s}(dy_s) - P(dy_s)\big)$$

then this formula can be rewritten in the form

$$S(X_1, \dots, X_n) - E\,S(X_1, \dots, X_n) = \sum_{c=1}^{n} \sum_{1 \le j_1 < \cdots < j_c \le n} g_{nc}(X_{j_1}, \dots, X_{j_c}). \qquad (1.4.2)$$

Obviously, the functions g_{nc} are symmetric with respect to their arguments x_1, $\dots$, x_c and possess the property of complete degeneracy. Relation (1.4.2) shows that if $E\,\|S\| < \infty$, then the symmetric statistic $S = S(X_1, \dots, X_n)$ is a linear combination of U-statistics whose kernels satisfy condition (1.1.7). An integer $r \ge 1$ is called a rank of a statistic S if it satisfies the conditions

$$g_{n1} = \cdots = g_{n,r-1} = 0, \quad g_{nr} \ne 0.$$

Clearly, $r \in \{1, 2, \dots, n\}$. If, for the symmetric statistic S with rank r, there exists an integer $m \le n$ such that

$$S(X_1, \dots, X_n) - E\,S(X_1, \dots, X_n) = \sum_{c=r}^{m} \sum_{1 \le j_1 < \cdots < j_c \le n} g_{nc}(X_{j_1}, \dots, X_{j_c}),$$

then we say that S has the m-th *degree*. In particular, the U-statistic has degree m. If $S = U_n$, then in (1.4.2) we have

$$g_{nc}(x_1, \dots, x_c) = \binom{m}{c} \binom{n}{c}^{-1} g_c(x_1, \dots, x_c).$$

1.5 Permanent symmetric statistics

Definition of permanent. Consider a matrix

$$A_n^m = [a_{ij};\ 1 \le i \le m,\ 1 \le j \le n]$$

whose dimension is $m \times n$ $(m \leq n)$ and its elements a_{ij} are real-valued (or complex-valued) numbers.

The *permanent* of a matrix A_n^m is defined as a sum of products

$$\operatorname{Per} A_n^m = \sum_{1 \leq j_1 \neq \cdots \neq j_m \leq n} \prod_{s=1}^{m} a_{sj_s} \,. \tag{1.5.1}$$

Each term in (1.5.1) is a product of the matrix elements taken one by one from each row and each column. The number of terms in (1.5.1) is equal to the permanent of the matrix with dimension $m \times n$ and with all its elements equal to 1.

Random measures. Let $\zeta_j(B)$, $1 \leq j \leq n$, be a family of orthogonal random measures on a measurable space $(X, \mathfrak{X})$ which are mutually independent and identically distributed with the structure function $P(B \cap B') = E\,\zeta_j(B)\,\zeta_j(B')$, B, $B' \in \mathfrak{X}$. Here $P(B)$ is a structure measure on $(X, \mathfrak{X})$.

The *permanent random measure* on a measurable space $(X, \mathfrak{X})$ is determined by the following formula

$$\pi_n^\zeta(B_1, \ldots, B_m) := n^{-[m]} \operatorname{Per} A_n^m(\zeta(B)) \,, \tag{1.5.2}$$

where

$$A_n^m(\zeta(B)) = [a_{ij} = \zeta_j(B_i);\ 1 \leq i \leq m,\ 1 \leq j \leq n] \,,$$

or in another form

$$\pi_n^\zeta(B_1, \ldots, B_m) = n^{-[m]} \sum_{1 \leq j_1 \neq \cdots \neq j_m \leq n} \prod_{s=1}^{m} \zeta_{j_s}(B_s) \,. \tag{1.5.3}$$

Here $n^{[m]} = n(n-1)\ldots(n-m+1)$.

Example 1.5.1. The *Dirac permanent measure* is defined by a sequence of independent identically distributed random variables ξ_i, $i = 1, \ldots, n$ with distribution function $P(B) = P(\xi_j \in B)$ according to the following formula

$$\pi_n^\delta(B_1, \ldots, B_m) = n^{-[m]} \operatorname{Per} A_n^m(\delta_\xi(B)) \,,$$

where

$$A_n^m(\delta_\xi(B)) = \left[a_{ij} = \delta_{\xi_j}(B_i);\ 1 \leq i \leq m,\ 1 \leq j \leq n\right] .$$

In the specific case of $m = 1$, the Dirac permanent measure

$$\pi_n^\delta(B) = n^{-1} \sum_{j=1}^{n} \delta_{\xi_j}(B)$$

is an empirical distribution of the sample ξ_j, $j = 1, \ldots, n$.

Note that by the definition of the Dirac permanent measure,

$$E\,\pi_n^{\delta}(B_1,\dots,B_m) = \prod_{s=1}^{m} P(B_s)\,.$$

Example 1.5.2. The *Gaussian permanent measure* is defined as follows

$$\pi_n^{W}(B_1,\dots,B_m) = n^{-[m]}\,\mathrm{Per}\,A_n^m(W(B)),$$

where

$$A_n^m(W(B)) = \left[a_{ij} = W_j(B_i);\ 1 \le i \le m,\ 1 \le j \le n\right]$$

and $W_j(B)$, $1 \le j \le n$, are Gaussian random measures which are mutually independent and identically distributed with zero mean and covariance function $P(B \cap B') = E\,W_j(B)\,W_j(B')$. Note that $E\,\pi_n^{W}(B_1,\dots,B_m) = 0$.

Example 1.5.3. The *Poisson permanent measure* is defined by the formula

$$\pi_n^{\nu}(B_1,\dots,B_m) = n^{-[m]}\,\mathrm{Per}\,A_n^m(\nu(B)),$$

where

$$A_n^m(\nu(B)) = \left[a_{ij} = \nu_j(B_i);\ 1 \le i \le m,\ 1 \le j \le n\right]$$

with the Poisson random measures $\nu_j(B)$, $j = 1, \dots, n$, $B \in \mathfrak{X}$, which are mutually independent and identically distributed according to the Poisson law with the intensity measure $P(B) = E\,\nu_j(B)$, $1 \le j \le n$, $B \in \mathfrak{X}$. Note that

$$E\,\pi_n^{\nu}(B_1,\dots,B_m) = \prod_{s=1}^{m} P(B_s).$$

Banach-valued permanent symmetric statistics. Let $\Phi\colon X^m \to B$ be a Bochner integrable symmetric function of m variables taking X^m into Banach space B. *Permanent symmetric statistics* of degree m are determined by the integral over the permanent random measure

$$U_n^{\zeta}(\Phi) = \int_X \cdots \int_X \Phi(x_1,\dots,x_m)\,\pi_n^{\zeta}(dx_1,\dots,dx_m) \tag{1.5.4}$$

or, using representation (1.5.3) for a permanent random measure, permanent symmetric statistic (1.5.4) can be represented in the following form

$$U_n^{\zeta}(\Phi) = n^{-[m]} \sum_{1 \le j_1 \ne \dots \ne j_m \le n} \Phi(\zeta_{j_1},\dots,\zeta_{j_m})\,, \tag{1.5.5}$$

where

$$\Phi(\zeta_{j_1},\dots,\zeta_{j_m}) = \int_X \cdots \int_X \Phi(x_1,\dots,x_m) \prod_{s=1}^{m} \zeta_{j_s}(dx_s)\,.$$

Note that the arguments ζ_{j_s}, $1 \le s \le m$, in (1.5.5) are unreal and they are real only in the case of the Dirac measure

$$U_n^\delta(\Phi) = n^{-[m]} \sum_{1 \le j_1 \ne \cdots \ne j_m \le n} \Phi(X_{j_1}, \dots, X_{j_m}) \tag{1.5.6}$$

defined by the given sample X_j, $1 \le j \le n$. Symmetric statistic (1.5.6) is a usual U-statistic.

Permanent U-process. Let $\mathcal{F}$ be a class of real-valued symmetric functions on X^m. We define a *permanent U-process*

$$\{U_n^\zeta(\Phi) \colon \Phi \in \mathcal{F}\} \tag{1.5.7}$$

as a collection of permanent symmetric statistics indexed by a family of symmetric kernels. Nolan and Pollard first (1987, 1988) treated $\{U_n^\delta(\Phi), \Phi \in \mathcal{F}\}$ as a stochastic process indexed by $\mathcal{F}$. They called it a *U-process.*

1.6 Multiple stochastic integrals

Wiener–Itô integral. Let X_1, X_2, ... be independent random variables with values in a measurable space $(X, \mathfrak{X})$ and with the same distribution P. *Single Wiener integrals* are defined as a Gaussian family of random variables $\{I_1(\varphi),\ \varphi \in L^2(X, \mathfrak{X}, P)\}$ with moments

$$E\, I_1(\varphi) = 0, \quad E\, I_1(\varphi)\, I_1(\psi) = \int_X \varphi(x)\, \psi(x)\, P(dx)\,.$$

A subfamily $\{W(A) = I_1(I(A)),\ A \in \mathfrak{X}\}$ is called a Gaussian random measure on $(X, \mathfrak{X}, P)$, where $I(A)$ denotes the indicator function of the set A. The random variable $I_1(\varphi)$ is called a single Wiener integral of the function φ with respect to the Gaussian random measure $W(\cdot)$ and is denoted by

$$I_1(\varphi) = \int_X \varphi(x)\, W(dx)\,.$$

Therefore, a single Wiener integral $I_1(\varphi)$ coincides in distribution with a normal random variable having mean value zero and the variance $\sigma_\varphi^2 = E\, \varphi^2(X_1)$.

Let $L^2(X^m)$ be a space of symmetric functions $f \colon X^m \to R$ such that $E\, f^2(X_1, \dots, X_m) < \infty$. The *multiple Wiener–Itô integral* of order m is a linear mapping $I_m(\cdot)$ from the space $L^2(X^m)$ into the space of random variables which are functionals of the Gaussian family $I_1(\varphi)$. This mapping is defined uniquely by the following two conditions:

(i) for functions of the form $f_\varphi = \varphi(x_1)\cdots\varphi(x_m)$,

$$I_m(f_\varphi) = \sigma_\varphi^m\, H_m(\sigma_\varphi^{-1}\, I_1(\varphi)),$$

where H_m is a *Hermite polynomial* of degree m with leading coefficient 1.

(ii) formula

$$EI_m(f)\, I_m(g) = m! \int_X \cdots \int_X f(x_1,\dots,x_m)\, g(x_1,\dots,x_m) P(dx_1)\dots P(dx_m)$$

holds for arbitrary functions $f, g \in L^2(X^m)$.

The multiple stochastic Wiener–Itô integral of the mth order is denoted symbolically by

$$I_m(f) = \int_X \cdots \int_X f(x_1,\dots,x_m) W(dx_1)\dots W(dx_m).$$

If it is necessary to emphasize the dependence of $I_m(f)$ on the random measure $W(\cdot)$ we will write $I_m^W(f)$ instead of $I_m(f)$.

Properties of the Wiener–Itô integral. Let $\varphi_1,\dots,\varphi_m,\dots$ be an orthonormal set of functions in L^2, and let $j_1 \geq 0,\dots,\ j_m \geq 0$, be the integers such that $j_1 + \cdots + j_m = k$. For all $i = 1,\dots,k$, we define the functions $\psi_i = \varphi_s$ for indices $j_1 + \cdots + j_{s-1} < i \leq j_1 + \cdots + j_s$. Then the following *Itô formula* is true

$$H_{j_1}(I_1(\varphi_1))\cdots H_{j_m}(I_1(\varphi_m)) =$$
$$= \int_X \cdots \int_X \psi_1(x_1)\cdots\psi_k(x_k)\, W(dx_1)\dots W(dx_k).$$

Denote

$$\tau_j = \int_X \varphi_j(x)\, W(dx)$$

for $j = 1,2,\dots$. Evidently, $\tau_1, \tau_2, \dots$ are independent standard normal variables. Further, define the *Wick polynomials* $: \tau_{j_1}\cdots\tau_{j_m} :$ by the formula of Itô (1951)

$$: \tau_{j_1}\cdots\tau_{j_m} : = \int_X \cdots \int_X \varphi_{j_1}(x_1)\cdots\varphi_{j_m}(x_m)\, W(dx_1)\dots W(dx_m).$$

If $f \in L^2(X^m)$, then

$$f(x_1,\dots,x_m) = \sum c_{j_1,\dots,j_m}\, \varphi_{j_1}(x_1)\cdots\varphi_{j_m}(x_m),$$

$$\int_X \cdots \int_X f(x_1,\dots,x_m) W(dx_1)\dots W(dx_m) = \sum c_{j_1,\dots,j_m} : \tau_{j_1}\cdots\tau_{j_m} : \qquad (1.6.1)$$

Let η be a random functional which is measurable with respect to the σ-algebra generated by the single Wiener integrals $\{I_1(\varphi),\ \varphi \in L^2(X, \mathfrak{X}, P)\}$. Then, if $E\eta^2 < \infty$, this functional η has a unique representation of the form

$$\eta = \sum_{k=0}^{\infty} I_k(f_k)/k! \qquad (1.6.2)$$

where

$$\sum_{k=0}^{\infty} Ef_k^2(X_1, \dots, X_k)/k! < \infty .$$

For example, if

$$\eta = \exp\{I_1(\varphi) - \frac{1}{2}\sigma_\varphi^2\}$$

then we have the expression for the moment generating function of the Hermite polynomials

$$\exp\{I_1(\varphi) - \frac{1}{2}\sigma_\varphi^2\} = \sum_{k=0}^{\infty} I_k(f_\varphi)/k!,$$

where $f_\varphi = \varphi(x_1) \cdots \varphi(x_k)$.

For any function $f \in L^2(X^m)$, there exist constants $c > 0$ and $x_0 > 0$ such that

$$P(|I_m(f)| > x) \leq \exp(-cx^{2/m}) \tag{1.6.3}$$

for all $x > x_0$.

There are several different approaches for constructing multiple stochastic integrals with respect to the appropriate random measure (see, e.g. Engel (1982), Janicki and Weron (1994), Major (1981)).

1.7 *B*-valued polynomial chaos

Let $X_1, \dots, X_n$ be independent non-degenerate real-valued random variables and let Q be a polynomial of degree m of n real variables with values in a Banach space B. Then, random variable $Q(X_1, \dots, X_n)$ is called a *B-valued polynomial chaos* of degree m based on the sequence $X_1, \dots, X_n$ (see, Kwapień and Woyczynski (1992)).

Such a polynomial chaos permits a representation

$$Q(X_1, \dots, X_n) = \sum_{c=0}^{m} \sum_{1 \leq i_1 \leq \dots \leq i_c \leq n} q_{i_1, \dots, i_c} X_{i_1} \cdots X_{i_c} ,$$

where $g_{i_1, \dots, i_c} \in B$ are the coefficients of Q.

If Q is a homogeneous polynomial of degree m then $Q(X_1, \dots, X_n)$ is said to be a *multilinear form* of degree m of random variables $X_1, \dots, X_n$ (or a *homogeneous chaos* of degree m).

Such a homogeneous chaos permits a representation

$$Q(X_1, \dots, X_n) = \sum_{1 \leq i_1 \leq \dots \leq i_m \leq n} q_{i_1, \dots, i_m} X_{i_1} \cdots X_{i_m}$$

where $q_{i_1, \dots, i_m} \in B$, and if

$$Q(X_1, \dots, X_n) = \sum_{c=0}^{m} \sum_{1 \leq i_1 < \dots < i_c \leq n} q_{i_1, \dots, i_c} X_{i_1} \cdots X_{i_c} , \tag{1.7.1}$$

then Q is called a *tetrahedral form* (Kwapień and Woyczynski (1992)).

Let $Q(X_1, X_2, \dots)$ be a *tetrahedral polynomial chaos* of degree m, with coefficients in a Banach space B and based on the infinite sequence of random variables $X_1, X_2, \dots$ and defined as

$$Q(X_1, X_2, \dots) = \lim_{n \to \infty} Q(X_1, \dots, X_n) . \tag{1.7.2}$$

Further we will see that a *Gaussian polynomial chaos* $Q(\tau_1, \tau_2, \dots)$ of the infinite sequence of Gaussian random variables $\tau_1, \tau_2, \dots$ appears to be a weak limit of *UB*-statistics. Under different conditions on a kernel Φ it appears as a weak limit of either Poisson polynomial chaos or stable polynomial chaos.

The polynomial chaos $Q(X_1, \dots, X_n)$ of the form (1.7.1) is a special case of the *generalized UB-statistics* of the type

$$U_n = \sum_{1 \leq i_1 \neq \dots \neq i_m \leq n} \Phi_{i_1, \dots, i_m}(X_{i_1}, \dots, X_{i_m}) . \tag{1.7.3}$$

Chapter 2

Inequalities

2.1 Inequalities based on the Hoeffding formula

In this section we obtain some estimates for moments $E\|U_n - \theta\|^q$ for any $q \geq 1$, using the Hoeffding formula (1.1.17). In the following we give some well-known inequalities, which will be needed for the proofs of the theorems.

Jensen's inequality. Let X be a Banach-valued random variable with $E\|X\| < \infty$ and assume $\Psi: B \to R$ is a convex function such that $|E\Psi(X)| < \infty$. Then

$$\Psi(EX) \leq E\Psi(X).$$

For example, $\Psi(x) = \|x\|^q$ if $q \geq 1$.

The conditional Jensen's inequality. Let X be a Banach valued random variable with $E\|X\| < \infty$ and a measurable according to σ-algebra $\mathfrak{F}$. Let $\Psi: B \to R$ be a convex function such that $|E\Psi(X)| < \infty$. Then

$$\Psi(E(X|\mathfrak{F})) \leq E(\Psi(X)|\mathfrak{F}).$$

For example, $\Psi(x) = \exp(\|x\|^q)$ if $q \geq 1$.

Inequalities for norms. For a sequence of independent, zero-mean random variables $X_1, \dots, X_n$ with values in a Banach space, we denote

$$S_n = \sum_{i=1}^{n} X_i, \quad S_n^* = \max_{1 \leq i \leq n} \|S_i\|, \quad X_n^* = \max_{1 \leq i \leq n} \|X_i\|,$$

$$\|S_n\|_q = (E\|S_n\|^q)^{1/q}, \quad q > 0.$$

The integrability of S_n has been studied by a number of authors. The most general result in that direction is the following inequality obtained by Talagrand (1989):

$$\|S_n\|_q \le \frac{cq}{1+\ln q}(\|S_n\|_1 + \|X_n^*\|_q) \tag{2.1.1}$$

for some universal constant $c > 0$, and all $q \ge 1$.

Inequality (2.1.1) is very close to Theorem 3.5.1 of Kwapień and Woyczynski (1992): if $q > 4$ and $c_q = 1 + q/\ln(q/2)$ then

$$\|S_n\|_q \le c_q(\|S_n\|_2 + \|X_n^*\|_q)\,. \tag{2.1.2}$$

The order $q/\ln q$ for the constant c_q is optimal even in the real case. This can be seen in the following example.

Example 2.1.1. Let $X_1, \dots, X_n$ be real independent identically distributed random variables such that

$$\begin{aligned} P(X_1 = +1) &= P(X_1 = -1) \\ &= \ln q/q, \\ P(X_1 = 0) &= 1 - 2\ln q/q\,. \end{aligned}$$

Then

$$\begin{aligned} (E|S_n|^q)^{1/q} &\ge n(P(S_n = n))^{1/q} \\ &= n(\ln q/q)^{n/q}\,, \\ ES_n^* &= 2n\ln q/q \quad \text{and} \quad X_n^* \le 1\,. \end{aligned}$$

Therefore, in (2.1.2) we necessarily have that, for each $n = 1, 2, \dots$

$$n(\ln q/q)^{n/q} \le c_q((2n\ln q/q)^{1/2} + 1)\,.$$

Choosing n such that $n - 1 \le q/\ln q < n$, we get $c_q \ge cq/\ln q$, where c is a universal constant.

Now we will apply Hoeffding formulae (1.1.17) and (1.2.11) and also (2.1.1)–(2.1.2) in order to obtain some inequalities for *UB*-statistics.

Theorem 2.1.1. *Suppose that kernel Φ possesses the property of complete degeneracy: $E\Phi(x_1, \dots, x_{m-1}, X_m) = 0$ a.s.*

If $E\|\Phi\|^q < \infty$ for some $q \ge 1$, then

$$\|U_n\|_q \le \frac{cq}{1+\ln q}k^{-1}(\|S_k(\Phi)\|_1 + \|\Phi^*\|_q)\,, \tag{2.1.3}$$

where

$$S_k(\Phi) = \sum_{i=1}^{k} \Phi(X_{im-m+1}, \ldots, X_{im}),$$

$$\Phi_k^* = \max_{1 \le i \le k} \|\Phi(X_{im-m+1}, \ldots, X_{im})\|,$$

$k = [n/m]$ is the integral part of the number n/m (greatest integer, less than or equal to n/m) and c is a universal constant from (2.1.1).

Proof. With help of the Hoeffding formula (1.1.17) and the conditional Jensen's inequality we have

$$E\|U_n\|^q \le k^{-q} E\|S_k(\Phi)\|^q.$$

From this inequality and (2.1.1), we obtain estimate (2.1.3).

For multi-sample *UB*-statistics the following inequality is valid.

Theorem 2.1.2. *Let $E\|\Phi\|^q < \infty$ for some $q \ge 1$ and assume kernel Φ possesses the property of complete degeneracy* (1.2.3). *Then*

$$\|U_{\bar{n}}\|_q \le \frac{Cq}{1+\ln q} k^{-1}(\|S_{kc}(\Phi)\|_1 + \|\Phi_{kc}^*\|_q),$$

where

$$S_{kc}(\Phi) = \sum_{i=1}^{k} \Phi(X_{j,im_j-m_j+s},\ s = 1, \ldots, m_j,\ j = 1, \ldots, c),$$

$$\Phi_{kc}^* = \max_{1 \le i \le k} \|\Phi(X_{j,im_j-m_j+s},\ s = 1, \ldots, m_j,\ j = 1, \ldots, c)\|,$$

$k = \min([n_1/m_1], \ldots, [n_c/m_c])$ *and the constant C coincides with the universal constant c from* (2.1.1).

Majorization inequalities. Here we will use the notion of majorization in the following definition.

Definition 2.1.1. Let X and Y be two random variables with values in a Banach space B. We say that X is *majorized* (*dominated*) by Y (in short $X \prec Y$) if for each convex continuous function $\Psi: B \to R^+$ the following is valid

$$E\Psi(X) \le E\Psi(Y).$$

Our goal with the help of majorization is to reduce the analysis of sums of Banach-valued random variables to the analysis of real-valued random variables.

Theorem 2.1.3. *Let $X_1, \dots, X_n$ be independent random variables with values in B and let $\varepsilon_1, \dots, \varepsilon_n$ be an independent sequence of Bernoulli random variables, $P(\varepsilon_i = 1) = P(\varepsilon_i = -1) = 1/2$ which is independent of $X_1, \dots, X_n$. Then*

$$\Big\|\sum_{i=1}^n X_i\Big\| - E\Big\|\sum_{i=1}^n X_i\Big\| \prec 2\sum_{i=1}^n \varepsilon_i \|X_i\|,$$

i.e. for each convex continuous function $\Psi: R \to R^+$,

$$E\Psi\Big(\Big\|\sum_{i=1}^n X_i\Big\| - E\Big\|\sum_{i=1}^n X_i\Big\|\Big) \le E\Psi\Big(2\sum_{i=1}^n \varepsilon_i\|X_i\|\Big). \tag{2.1.4}$$

Proof. Denote $\mathfrak{F}_0 = \{\varnothing, \Omega\}$ and $\mathfrak{F}_i = \sigma(\omega : X_1, \dots, X_i, \varepsilon_1, \dots, \varepsilon_i)$. Define, for $k = 1, \dots, n$,

$$\xi_k := E\Big(\Big\|\sum_{i=1}^n X_i\Big\| \Big| \mathfrak{F}_k\Big) - E\Big(\Big\|\sum_{i=1}^n X_i\Big\| \Big| \mathfrak{F}_{k-1}\Big)$$

and

$$\eta_k := 2\varepsilon_k \|X_k\|.$$

It is sufficient to show that for each convex continuous function $\Psi: R \to R^+$,

$$E(\Psi(\xi_k)|\mathfrak{F}_{k-1}) \le E(\Psi(\eta_k)|\mathfrak{F}_{k-1}) \tag{2.1.5}$$

for $k = 1, 2, \dots, n$. In fact, (2.1.5) has as a consequence for each $i = 1, \dots, n$

$$E(\Psi(\xi_1 + \dots + \xi_{i-1} + \xi_i + \eta_{i+1} + \dots + \eta_n)|\mathfrak{F}_{i-1}) \le$$
$$\le E(\Psi(\xi_1 + \dots + \xi_{i-1} + \eta_i + \eta_{i+1} + \dots + \eta_n)|\mathfrak{F}_{i-1})$$

since $\eta_{i+1}, \dots, \eta_n$ is independent of $\mathfrak{F}_i$, and then we have

$$E\Psi(\xi_1 + \dots + \xi_{i-1} + \xi_i + \eta_{i+1} + \dots + \eta_n) \le$$
$$\le E\Psi(\xi_1 + \dots + \xi_{i-1} + \eta_i + \eta_{i+1} + \dots + \eta_n)$$

for each $i = 1, \dots, n$. Summing over all $i = 1, \dots, n$ on the left and on the right-hand side of this inequality we get (2.1.4).

Now we shall show that (2.1.5) takes place. For this purpose let us fix k, and let X_k^1 be a copy of X_k which is independent of $X_1, \dots, X_n$, and let $\mathfrak{F}_k^1 = \sigma(\mathfrak{F}_k, X_k^1)$. Define, for $k = 1, \dots, n$,

$$\xi_k^1 := E\Big(\Big\|\sum_{i=1}^n X_i\Big\| \Big| \mathfrak{F}_k\Big) - E\Big(\Big\|\sum_{i=1, i\ne k}^n X_i + X_k^1\Big\| \Big| \mathfrak{F}_k^1\Big).$$

Since, a.s.

$$E\Big(\Big\|\sum_{i=1, i\ne k}^n X_i + X_k^1\Big\| \Big| \mathfrak{F}_k\Big) = E\Big(\Big\|\sum_{i=1}^n X_i\Big\| \Big| \mathfrak{F}_{k-1}\Big),$$

and since $\mathfrak{F}_k^1 \supset \mathfrak{F}_k$, the Jensen inequality implies that

$$E(\Psi(\xi_k)|\mathfrak{F}_{k-1}) \le E(\Psi(\xi_k^1)|\mathfrak{F}_{k-1}).$$

So, to finish the proof of (2.1.5), it is sufficient to show that

$$E(\Psi(\xi_k^1)|\mathfrak{F}_{k-1}) \le E(\Psi(\eta_k)|\mathfrak{F}_{k-1}). \tag{2.1.6}$$

Since η_k and ξ_k^1 are $\mathfrak{F}_{k-1}$-conditionally symmetric, it suffices to check (2.1.6) for even convex continuous functions $\Psi\colon R \to R^+$. But in this case

$$\begin{aligned}\Psi(\xi_k^1) &\le \frac{1}{2}\Psi\Big(2\Big|E\Big(\Big\|\sum_{i=1}^n X_i\Big\|\Big|\mathfrak{F}_k\Big) - E\Big(\Big\|\sum_{i=1,i\ne k}^n X_i\Big\|\Big|\mathfrak{F}_k\Big)\Big|\Big) + \\ &+ \frac{1}{2}\Psi\Big(2\Big|E\Big(\Big\|\sum_{i=1,i\ne k}^n X_i + X_k^1\Big\|\Big|\mathfrak{F}_k^1\Big) - E\Big(\Big\|\sum_{i=1,i\ne k}^n X_i\Big\|\Big|\mathfrak{F}_k\Big)\Big|\Big) \le \\ &\le \frac{1}{2}\Psi(2\|X_k\|) + \frac{1}{2}\Psi(2\|X_k^1\|),\end{aligned}$$

because Ψ is nondecreasing on R^+. So,

$$E(\Psi(\xi_k^1)|\mathfrak{F}_{k-1}) \le E\Big(\frac{1}{2}\Psi(2\|X_k\|) + \frac{1}{2}\Psi(2\|X_k^1\|)|\mathfrak{F}_{k-1}\Big) = E(\Psi(\eta_k)|\mathfrak{F}_{k-1}).$$

This is (2.1.6) and hence we obtain (2.1.4).

If $B = H$ is a Hilbert space and the random variables involved are symmetric then the conclusion of Theorem 2.1.3 can be considerably strengthened (see Kwapień and Woyczynski (1992)).

Theorem 2.1.4. *Let $X_1, \ldots, X_n$ be independent, symmetric random variables with values in Hilbert space H, and let $\varepsilon_1, \ldots, \varepsilon_n$ be independent Bernoulli random variables, independent of $X_1, \ldots, X_n$. Then for each convex, continuous and symmetric function $\Psi\colon R \to R^+$, with non-negative fourth derivative,*

$$E\Psi\Big(\Big\|\sum_{i=1}^n X_i\Big\|\Big) \le E\Psi\Big(\sum_{i=1}^n \varepsilon_i\|X_i\|\Big). \tag{2.1.7}$$

Theorem 2.1.5. *Let $X_1, \ldots, X_n$ be independent random variables with values in Banach space B and assume $\varepsilon_1, \ldots, \varepsilon_n$ is a Bernoulli sequence independent of $X_1, \ldots, X_n$. Then for each convex, continuous $\Psi\colon R \to R^+$ the following inequality*

$$E\Psi(\|X_1 + \cdots + X_n\|) \le \frac{1}{2}\Psi(2E\|X_1 + \cdots + X_n\|) + \frac{1}{2}E\Psi\Big(4\sum_{i=1}^n \varepsilon_i\|X_i\|\Big) \tag{2.1.8}$$

takes place.

Proof. From the convexity of Ψ, it follows that

$$E\Psi(\|X_1+\cdots+X_n\|)\le\frac{1}{2}\Psi(2E\|X_1+\cdots+X_n\|)+$$
$$+\frac{1}{2}E\Psi(2(\|X_1+\cdots+X_n\|-E\|X_1+\cdots+X_n\|))$$

[by using Theorem 2.1.3 on the second term]

$$\le\frac{1}{2}\Psi(2E\|X_1+\cdots+X_n\|)+\frac{1}{2}E\Psi\Big(4\sum_{i=1}^{n}\varepsilon_i\|X_i\|\Big).$$

This proves (2.1.8).

Corollary 2.1.1. *Let $\Psi(x)=|x|^p$, $p\ge 1$, and assume $X_1,\dots,X_n$ are independent random variables with values in B. Then*

$$E\|X_1+\cdots+X_n\|^p\le$$
$$\le 2^{p-1}(E\|X_1+\cdots+X_n\|)^p+2^{2p-1}\max(1,(p-1)^{p/2})\,E\Big(\sum_{i=1}^{n}\|X_i\|^2\Big)^{p/2}.$$

Proof. It is sufficient to apply Khintchine's inequality

$$E_\varepsilon\Big|\sum_{i=1}^{n}\varepsilon_i\|X_i\|\Big|^p\le\max(1,(p-1)^{p/2})\Big(\sum_{i=1}^{n}\|X_i\|^2\Big)^{p/2},\qquad p\ge 1$$

on the second term on the right-hand side of (2.1.8).

Corollary 2.1.2. *Let $\Psi(x)=|x|^p$, $p\ge 1$, and assume $X_1,\dots,X_n$ are independent random variables with values in B. Then*

$$E\|X_1+\cdots+X_n\|^p\le$$
$$\le 2^{p-1}(E\|X_1+\cdots+X_n\|)^p+2^{2p-1}\max(1,(p-1)^{p/2})\,n^{\frac{p}{2}\vee 1-1}\sum_{i=1}^{n}E\|X_i\|^p.$$

Proof. We apply Corollary 2.1.1 and the inequality

$$\Big(\sum_{i=1}^{n}\|X_i\|^2\Big)^{p/2}\le n^{\frac{p}{2}\vee 1-1}\sum_{i=1}^{n}\|X_i\|^p,\qquad p\ge 1. \tag{2.1.9}$$

Theorem 2.1.6. *Suppose that $E\|\Phi\|^p<\infty$ for some $p\ge 1$. Then*

$$E\|U_n-\theta\|^p\le 2^{p-1}k^{-p}(E\|S_k(\Phi)\|)^p+$$
$$+2^{2p-1}\max(1,(p-1)^{p/2})k^{-((p-1)\wedge(p/2))}E\|\Phi-\theta\|^p, \tag{2.1.10}$$

where

$$S_k(\Phi) = \sum_{i=1}^{k}(\Phi(X_{(i-1)m+1}, \dots, X_{im}) - \theta)$$

and $k = [n/m]$ *is the integral part of the number* n/m.

Proof. By the Hoeffding formula and by using the conditional Jensen's inequality, we obtain

$$E\|U_n - \theta\|^p \le k^{-p} E\|S_k(\Phi)\|^p.$$

With help of this inequality and Corollary 2.1.2 we have (2.1.10).

The inequality analogous to (2.1.10) for multi-sample *UB*-statistics follows from the Hoeffding formula (1.2.11) and Corollary 2.1.2.

2.2 Martingale moment inequalities

Geometry and martingales in Banach spaces. Probability theory in Banach spaces is closely connected with geometric theory of Banach spaces. The study of the properties of the sums of independent random elements leads to the concept of B-convexity and to the concepts of type and cotype for Banach spaces. The Radon–Nikodym property and the concept of p-uniform smoothness (q-uniform convexity) may be characterized by the properties of convexity, and by the inequalities for martingales with values in Banach spaces.

Let B be a separable real Banach space with norm $\|\cdot\|$. For any $t \ge 0$, we define a modulus of smoothness of B as follows

$$\rho(t) = \frac{1}{2}\sup\{\|x + ty\| + \|x - ty\| - 2\colon x, y \in B, \|x\| = \|y\| = 1\}.$$

Banach space B is called p-uniformly smooth, $1 < p \le 2$, if $\rho(t) = O(t^p)$ as $t \to 0$. For example, L^r and l_r, $1 < r < \infty$, are $\min(r, 2)$-uniformly smooth. It is well-known that, for every Banach space B and any $t \ge 0$, the smoothness module $\rho(t)$ satisfies the inequality $\rho(t) \ge (1 + t^2)^{1/2} - 1$, where the right-hand side expression is the smoothness module of the Hilbert space H, i.e. this space is 2-uniformly smooth. The concept of smoothness of Banach space B plays an important role in the theory of B-valued martingales. For the B-valued martingale $(M_n, \mathfrak{F}_n)$, $n = 0, 1, \dots$, we denote $m_n = M_n - M_{n-1}$, $n = 1, 2, \dots$, and set $M_0 = 0$ a.s. The sequence $(m_n, \mathfrak{F}_n)$ forms a B-valued martingale difference, since $E(m_n|\mathfrak{F}_{n-1}) = 0$ a.s. As is known (see Woyczynski (1975)), Banach space B is p-uniformly smooth if and only if, for every $q \ge 1$, there exists a constant $\alpha_q = \alpha(B, q)$ such that, for any B-valued martingale $(M_n, \mathfrak{F}_n)$ the inequality

$$E\|M_n\|^q \le \alpha_q E\Big(\sum_{i=1}^{n}\|m_i\|^p\Big)^{q/p} \tag{2.2.1}$$

holds. Therefore, the property of p-uniform smoothness is characterized by Burkholder inequality (2.2.1).

If kernel Φ belongs to a p-uniformly smooth Banach space then with the help of (2.2.1) we can be obtain the inequalities for the moments of UB-statistics with sharper estimates than (2.1.3).

Inequalities for one-sample *UB*-statistics. Consider the element S_{nc} in (1.1.8). We can write it in the form

$$S_{nc} = \sum_{i_c=c}^{n} \sum_{i_{c-1}=c-1}^{i_c-1} \cdots \sum_{i_1=1}^{i_2-1} g_c(X_{i_1}, \ldots, X_{i_c}).$$

If $c = 1$ and rank $r = 1$, then S_{n1} is a sum of independent identically distributed random elements.

Theorem 2.2.1. *Assume that Banach space B is p-uniformly smooth and $E\|g_c\|^q < \infty$ for $q \geq 1$.*

Then the inequality

$$E\|S_{nc}\|^q \leq \alpha_q^c \binom{n}{c} E\|g_c\|^q \tag{2.2.2}$$

holds for $q \leq p$ and all $n \geq c$, $c = 1, \ldots, m$, where α_q is a constant in (2.2.1).

Proof. Let us apply martingale inequality (2.2.1). This is possible in view of Lemma 1.1.1. First, by using (2.2.1), we can write

$$E\|S_{nc}\|^q \leq \alpha_q E\Big(\sum_{i_c=c}^{n} \Big\| \sum_{i_{c-1}=c-1}^{i_c-1} \cdots \sum_{i_1=1}^{i_2-1} g_c(X_{i_1}, \ldots, X_{i_c})\Big\|^p\Big)^{q/p}.$$

Since

$$\|a_1 + \cdots + a_n\|^\gamma \leq \max(1, n^{\gamma-1})(\|a_1\|^\gamma + \cdots + \|a_n\|^\gamma), \tag{2.2.3}$$

$a_i \in B$, $\gamma \geq 0$ then the expression in the right-hand side does not exceed the expression

$$\sum_{i_c=c}^{n} E\Big\| \sum_{i_{c-1}=c-1}^{i_c-1} \cdots \sum_{i_1=1}^{i_2-1} g_c(X_{i_1}, \ldots, X_{i_c})\Big\|^q,$$

when $\gamma = q/p \leq 1$.

To estimate the expectation, we apply (2.2.1) once again and take (1.1.7) into account. As a result, we get

$$E\Big\| \sum_{i_{c-1}=c-1}^{i_c-1} \cdots \sum_{i_1=1}^{i_2-1} g_c(X_{i_1}, \ldots, X_{i_c})\Big\|^q \leq$$

$$\leq \alpha_q E\Big(\sum_{i_{c-1}=c-1}^{i_c-1}\Big\|\sum_{i_{c-2}=c-2}^{i_{c-1}-1}\cdots\sum_{i_1=1}^{i_2-1} g_c(X_{i_1},\ldots,X_{i_c})\Big\|^p\Big)^{q/p}.$$

The expectation on the right-hand side does not exceed the expression

$$\sum_{i_{c-1}=c-1}^{i_c-1} E\Big\|\sum_{i_{c-2}=c-2}^{i_{c-1}-1}\cdots\sum_{i_1=1}^{i_2-1} g_c(X_{i_1},\ldots,X_{i_c})\Big\|^q.$$

If we continue this recurrent process, by using (2.2.1) at each step, and combining all the estimates, we obtain (2.2.2).

Corollary 2.2.1. *Under the conditions of Theorem* 2.2.1, *relations* (1.1.12) *and* (2.2.2) *imply the inequality*

$$E\|U_n-\theta\|^q \leq (m-r+1)^{q-1}\sum_{c=r}^{m}\alpha_q^c\binom{m}{c}^q\binom{n}{c}^{-q+1}E\|g_c\|^q,$$

where r denotes the rank in (1.1.10). *For example, if $B=H$ or $B=R$, then one can set*

$$\alpha_q=\sup_x\Big(|x|^{-q}(|1+x|^q-1-qx)\Big)$$

for $1\leq q\leq 2$; here, $\alpha_q\leq 2^{2-q}$.

Theorem 2.2.2. *Let Banach space B be p-uniformly smooth and $E\|g_c\|^q<\infty$ for $q\geq 1$. Then the inequality*

$$E\|S_{nc}\|^q \leq \alpha_q^c\binom{n}{c}n^{(q/p-1)c}E\|g_c\|^q \tag{2.2.4}$$

holds for $p\leq q$ and all $n\geq c$, $=1,\ldots,m$, where α_q is the constant in (2.2.1).

Proof. We apply (2.2.1) again. Now we apply (2.2.3) at $\gamma=q/p\geq 1$. Therefore $\max(1,n^{\gamma-1})=n^{q/p-1}$. As a result of reducing the recurrent estimates, we obtain (2.2.4).

Corollary 2.2.2. *Under the conditions of Theorem* 2.2.2, *relations* (1.1.12) *and* (2.2.4) *imply the inequality*

$$E\|U_n-\theta\|^q \leq (m-r+1)^{q-1}\sum_{c=r}^{m}\alpha_q^c\binom{m}{c}^q\binom{n}{c}^{-q+1}n^{(q/p-1)c}E\|g_c\|^q,$$

where r denotes the rank in (1.1.10). *In particular, if $B=H$, then $p=2$ and this inequality can be reduced to*

$$E\|U_n-\theta\|^q \leq (m-r+1)^{q-1}\sum_{c=r}^{m}\alpha_q^c\binom{m}{c}^q\binom{n}{c}^{-q+1}n^{(q-2)c/2}E\|g_c\|^q$$

for all $q \geq 2$, where one can choose

$$\alpha_q = \left(18\sqrt{2}\right)^q q^q.$$

Note that if $B = R$, then the module of smoothness $\rho(t)$ is equal to $\max(0, t-1)$. Therefore, $\rho(t) = O(t^2)$, $t \to \infty$, and we can assume that the space R is 2-uniformly smooth. Thus, for the case $B = R$ Corollary 2.2.2 also holds.

Corollary 2.2.3. *Assume that Banach space B is 2-uniformly smooth, $r = m$, and $E\|\Phi\|^q < \infty$ for some $q \geq 1$. Then*

$$E\|U_n\|^q \leq \alpha_q^m \binom{n}{m}^{-q+1} E\|\Phi\|^q,$$

for $1 \leq q \leq 2$, and

$$E\|U_n\|^q \leq \alpha_q^m \binom{n}{m}^{-q+1} n^{(q-2)m/2} E\|\Phi\|^q$$

for $q \geq 2$.

Inequalities for multi-sample *UB*-statistics. Consider the random element $S_{\bar{n}\bar{d}}$ in (1.2.9).

Theorem 2.2.3. *Let Banach space B be p-uniformly smooth and $E\|g_{\bar{d}}\|^q < \infty$ for $q \geq 1$. Then the inequality*

$$E\|S_{\bar{n}\bar{d}}\|^q \leq \alpha_q^d \prod_{j=1}^{c} \binom{n_j}{d_j} E\|g_{\bar{d}}\|^q \tag{2.2.5}$$

is valid for $q \leq p$ and for all $\bar{n} \geq \bar{d}$, $d \geq r$ where $d = d_1 + \cdots + d_c$, $r = r_1 + \cdots + r_c$ and r is the rank vector in (1.2.7); α_q is the constant in (2.2.1).

Proof. We can write $S_{\bar{n}\bar{d}}$ in (1.2.9) as follows

$$S_{\bar{n}\bar{d}} = \sum_{1 \leq i_{11} < \cdots < i_{1d_1} \leq n_1} \cdots \sum_{1 \leq i_{c1} < \cdots < i_{cd_c} \leq n_c} g_{\bar{d}}(X_{j\alpha}, \alpha = i_{j1}, \ldots, i_{jd_j}, j = 1, \ldots, c).$$

The expectation of the first sum with respect to the summation index $1 \leq i_{11} < \cdots < i_{1d_1} \leq n_1$ is estimated with the help of (2.2.2). Then, for the conditional expectation we have

$$E(\|S_{\bar{n}\bar{d}}\|^q | X_{ji} = x_{ji}, i = 1, \ldots, n_j, j = 2, \ldots, c) \leq$$

$$\leq \alpha_q^{d_1} \binom{n_1}{d_1} E\Big\| \sum_{1 \leq i_{j1} < \cdots < i_{jd_1} \leq n_j} g_{\bar{d}}(X_{11}, \ldots, X_{1d_1}; x_{j\alpha}, \alpha = i_{j1}, \ldots, i_{jd_j}, j = 2, \ldots, c)\Big\|^q.$$

Here, to estimate the expectation with respect to the summation index $1 \le i_{21} < \cdots < i_{2d_2} \le n_2$ we employ once again (2.2.2) and after this we obtain

$$E(\|S_{\bar{n}\bar{d}}\|^q | X_{ji} = x_{ji},\ i = 1, \ldots, n_j, j = 3, \ldots, c) \le$$
$$\le \alpha_q^{d_1+d_2} \binom{n_1}{d_1} \binom{n_2}{d_2} E\Big\| \sum_{1 \le i_{j1} < \cdots < i_{jd_1} \le n_j} g_{\bar{d}}(X_{11}, \ldots, X_{1d_1}; X_{21}, \ldots, X_{2d_2}; x_{j\alpha},$$
$$\alpha = i_{j1}, \ldots, i_{jd_j},\ j = 3, \ldots, c)\Big\|^q.$$

Continuing this induction process of estimation up to index $j = c$, we finally arrive at (2.2.5).

Corollary 2.2.4. *Let $E\|\Phi\|^q < \infty$, $q \ge 1$. If Banach space B is p-uniformly smooth, then the inequality*

$$E\|U_{\bar{n}} - \theta\|^q \le (m - r + 1)^{q-1} \sum_{d=r}^{m} \alpha_q^d \beta_d^{q-1} \sum_{d_1+\cdots+d_c=d} \prod_{j=1}^{c} \binom{m_j}{d_j}^q \binom{n_j}{d_j}^{-q+1} E\|g_{\bar{d}}\|^q,$$

is valid for $q \le p$, where $r = r_1 + \cdots + r_c$ is the rank sum and

$$\beta_d = \sum_{\substack{d_1+\cdots+d_c=d,\\ 0 \le d_j \le m_j, j=1,\ldots,c}} 1$$

is the number of solutions of the Diophantine equation $d_1 + \cdots + d_c = d$, which satisfies the conditions $0 \le d_j \le m_j$, $j = 1, \ldots, c$.

Proof. At first, with the help of (1.2.8), we write

$$E\|U_n - \theta\|^q \le (m - r + 1)^{q-1} \sum_{d=r}^{m} E\Big\| \sum_{d_1+\cdots+d_c=d} \prod_{j=1}^{c} \binom{m_j}{d_j} U_{\bar{n}\bar{d}}\Big\|^q.$$

Here, we again apply inequality (2.2.3). We employ (2.2.3) once again. As a result, we obtain

$$E\Big\| \sum_{d_1+\cdots+d_c=d} \prod_{j=1}^{c} \binom{m_j}{d_j} U_{\bar{n}\bar{d}}\Big\|^q \le \beta_d^{q-1} \sum_{d_1+\cdots+d_c=d} \prod_{j=1}^{c} \binom{m_j}{d_j}^q \binom{n_j}{d_j}^{-q} E\|S_{\bar{n}\bar{d}}\|^q.$$

Together with (2.2.5), this completes the proof of the corollary.

Theorem 2.2.4. *Let B be a p-uniformly smooth Banach space and $E\|g_{\bar{d}}\|^q < \infty$ for $q \ge 1$. Then the inequality*

$$E\|S_{\bar{n}\bar{d}}\|^q \le \alpha_q^d \prod_{j=1}^{c} \binom{n_j}{d_j} n_j^{(q/p-1)d_j} E\|g_{\bar{d}}\|^q \tag{2.2.6}$$

holds for $p \le q$ and all $\bar{n} \ge \bar{d}$, $d \ge r$, where $r = r_1 + \cdots + r_c$ is the rank sum and α_q is the constant in (2.2.1).

Proof. It is sufficient to repeat the arguments of reducing in the proof of Theorem 2.2.3, but instead of Theorem 2.2.1 it is necessary to use Theorem 2.2.2.

Corollary 2.2.5. *Let $E\|\Phi\|^q < \infty$ for $q \geq 1$. If Banach space B is p-uniformly smooth, then the inequality*

$$E\|U_n - \theta\|^q \leq$$

$$\leq (m - r + 1)^{q-1} \sum_{d=r}^{m} \alpha_q^d \beta_d^{q-1} \sum_{d_1+\cdots+d_c=d} \prod_{j=1}^{c} \binom{m_j}{d_j}^q \binom{n_j}{d_j}^{-q+1} n_j^{(q/p-1)d_j} E\|g_{\bar{d}}\|^q$$

holds for $p \leq q$.

Proof. It is sufficient to apply (1.2.8) and (2.2.6).

Corollary 2.2.6. *Let Banach space B be p-uniformly smooth and $\bar{r} = \bar{m}$, $E\|\Phi\|^q < \infty$ for $q \geq 1$.*

Then, if $q \leq p$,

$$E\|U_{\bar{n}}\|^q \leq \alpha_q^m \prod_{j=1}^{c} \binom{n_j}{m_j}^{-q+1} E\|\Phi\|^q$$

and, if $p \leq q$,

$$E\|U_{\bar{n}}\|^q \leq \alpha_q^m \prod_{j=1}^{c} \binom{n_j}{m_j}^{-q+1} n_j^{(q/p-1)m_j} E\|\Phi\|^q .$$

2.3 Maximal inequalities

One-sample *UB*-statistics. Assume that the B-valued kernel Φ has rank r. For $c \geq r$, we first consider the martingale $(S_{kc}, \mathfrak{F}_k)$, $k = c, c+1, \ldots$.

Theorem 2.3.1. *Let $E\|g_c\|^q < \infty$, for $q \geq 1$, and let $\{a_k\}$, $k \geq 1$, be a nonincreasing sequence of nonnegative numbers. Then*

$$P(\max_{c \leq k \leq n} a_k\|S_{kc}\| \geq t) \leq t^{-q}\Big\{a_n^q E\|S_{nc}\|^q + \sum_{k=c}^{n-1}(a_k^q - a_{k+1}^q)\, E\|S_{kc}\|^q\Big\} \quad (2.3.1)$$

for all $n \geq c$ and $t \geq 0$.

Proof. By Lemma 1.1.1, we have

$$S_{kc} = E(S_{nc}|\mathfrak{F}_k)\,,$$

for $c \leq k \leq n$. By virtue of the Hölder inequality, this yields

$$\|S_{kc}\|^q \leq E(\|S_{nc}\|^q|\mathfrak{F}_k)\,,$$

i.e. the stochastic sequence $(\|S_{kc}\|^q, \mathfrak{F}_k)$, $k = c, c+1, \ldots$, is an R-valued submartingale. Hence, (2.3.1) follows from the general maximal martingale inequality (see, e.g. Borovskikh and Korolyuk (1994a), Theorem 2.2.2, p. 77).

Corollary 2.3.1. *For $a_k = 1$, $c \le k \le n$, relation* (2.3.1) *implies the inequality*

$$P(\max_{c \le k \le n} \|S_{kc}\| \ge t) \le t^{-q} E\|S_{nc}\|^q$$

for all $c = r, \ldots, m$.

Consider the reverse martingale $(U_{kc}, \mathfrak{B}_k)$, $k = c, c+1, \ldots$.

Theorem 2.3.2. *Assume that $E\|g_c\|^q < \infty$, for $q \ge 1$, and that $\{b_k\}$, $k \ge 1$, is a nondecreasing sequence of positive numbers. Then*

$$P(\max_{n \le k \le N} b_k \|U_{kc}\| \ge t) \le t^{-q} \Big\{ b_n^q E\|U_{nc}\|^q + \sum_{k=n+1}^{N} (b_k^q - b_{k-1}^q) E\|U_{kc}\|^q \Big\} \quad (2.3.2)$$

for all $N \ge n \ge c$ and $t > 0$.

Proof. By Lemma 1.1.3, for $k \ge c$, we have

$$U_{k+1,c} = E(U_{kc} | \mathfrak{B}_{k+1}).$$

This implies that the stochastic sequence $(\|U_{kc}\|^q, \mathfrak{B}_k)$ is an R-valued reverse submartingale, i.e.

$$\|U_{k+1,c}\|^q \le E(\|U_{kc}\|^q | \mathfrak{B}_{k+1}).$$

Hence, (2.3.2) is contained in the general maximal martingale inequality (see, e.g. Borovskikh and Korolyuk (1994a), Theorem 2.2.3, p. 78) as a special case.

Corollary 2.3.2. *For $b_k = 1$, relation* (2.3.2) *implies the inequality*

$$P(\sup_{k \ge n} \|U_{kc}\| \ge t) \le t^{-q} E\|U_{nc}\|^q.$$

For the reverse martingale $(U_n, \mathfrak{B}_n)$, $n = r, r+1, \ldots$, the following assertions are valid, which are similar to Theorem 2.3.2 and Corollary 2.3.2.

Theorem 2.3.3. *Assume that $E\|\Phi\|^q < \infty$ for $q \ge 1$, and that $\{b_k\}$, $k \ge 1$, is a nondecreasing sequence of positive numbers. Then*

$$P(\max_{n \le k \le N} b_k \|U_k - \theta\| \ge t) \le$$

$$\le t^{-q} \Big\{ b_n^q \, E\|U_n - \theta\|^q + \sum_{k=n+1}^{N} (b_k^q - b_{k-1}^q) \, E\|U_k - \theta\|^q \Big\}$$

for all $N \geq n$ *and* $t > 0$; *in particular,*

$$P(\sup_{k \geq n} \|U_k - \theta\| \geq t) \leq t^{-q} E\|U_n - \theta\|^q .$$

Multi-sample *UB*-statistics. Consider the stochastic sequence $(U_{\bar{n}}, \mathfrak{B}_{\bar{n}})$, $\bar{n} \geq \bar{m}$.

Theorem 2.3.4. *The following inequalities are valid:*

$$P(\sup_{\bar{k} \geq \bar{n}} \|U_{\bar{k}} - \theta\| \geq t) \leq t^{-1}\{\alpha_{c-1} + \beta_{c-1} E(\|U_{\bar{n}} - \theta\| (\ln^+ \|U_n - \theta\|)^{c-1})\}$$

for any $t \geq 0$ *and*

$$E(\sup_{\bar{k} \geq \bar{n}} \|U_{\bar{k}} - \theta\|^q) \leq \gamma_{cq} E\|U_{\bar{n}} - \theta\|^q, \qquad q > 1$$

where $\ln^+ z = \max(1, \ln z)$, $z > 0$;

$$\alpha_c = c\Big(\frac{e}{e-1}\Big)^c, \quad \beta_c = c!\Big(\frac{e}{e-1}\Big)^c, \quad \gamma_{cq} = \Big(\frac{q}{q-1}\Big)^{cq}.$$

Proof. By Lemma 1.2.2 for $\bar{n} \geq \bar{k}$

$$E(U_{\bar{n}} - \theta | \mathfrak{B}_{\bar{k}}) = U_{\bar{k}} - \theta.$$

Hence, we obtain

$$\|U_{\bar{k}} - \theta\|^q \leq E(\|U_{\bar{n}} - \theta\|^q \,|\mathfrak{B}_{\bar{k}})$$

for any $q \geq 1$. Thus, the stochastic sequence $(\|U_{\bar{k}} - \theta\|^q, \mathfrak{B}_{\bar{k}})$ is an R-valued reverse submartingale with multivariate index $\bar{k}$. Therefore, the conclusion of the present theorem comes from the known inequalities for such submartingales (see, e.g. Fazekas (1983, 1985)).

2.4 Contraction and symmetrization inequalities

Contraction and symmetrization inequalities provide a comparison between moments (and tails) of sums $\sum_{i=1}^n X_i$, and moments (and tails) of $\sum_{i=1}^n \varepsilon_i X_i$, in conditions when the multiplier sequence $\varepsilon_1, \ldots, \varepsilon_n$ is subject to boundedness constraints. It is possible to apply this technique in *UB*-statistics. We start with the well-known Lévy inequalities, which are needed in several proofs below.

Lemma 2.4.1. *If* $X_1, \ldots, X_n$ *are independent and symmetric random variables with values in* B *and* $S_n = X_1 + \cdots + X_n$, *then for every* $t > 0$,

$$P(\max_{1 \leq k \leq n} \|S_k\| > t) \leq 2P(\|S_n\| > t).$$

The proof of Lévy's inequalities are in many papers (see, for example, Araujo and Giné (1980), Theorem 2.6).

Lemma 2.4.2. *Let $\Psi: B \to R$ be a convex function on B. If X is independent of Y and $EY = 0$ in the sense of Bochner's integration, then*

$$E\Psi(X) \leq E\Psi(X+Y).$$

In particular,

$$E\|X\|^p \leq E\|X+Y\|^p, \qquad p \geq 1.$$

Proof. For every $x \in B$ we have $x = E(x+Y)$ and

$$\begin{aligned} \Psi(x) &= \Psi(E(x+Y)) \qquad \text{[by Jensen's inequality]} \\ &\leq E\Psi(x+Y). \end{aligned}$$

Then apply Fubini's theorem.

The following inequality is an obvious and simple consequence of Lemma 2.4.1 and Lemma 2.4.2.

Lemma 2.4.3. *Let $X_1, \ldots, X_n$ be independent zero-mean random Banach-valued elements. Let $\Psi(x)$, $x \geq 0$, be a convex increasing function. Then*

$$E\Psi\left(\max_{1\leq j\leq n}\left\|\sum_{i=1}^{j} X_i\right\|\right) \leq 2E\Psi\left(2\left\|\sum_{i=1}^{n} X_i\right\|\right).$$

Proof. Let $\{X_i^1\}$ be an independent copy of $\{X_i\}$. Then $\{X_i - X_i^1\}$ is a sequence of independent symmetric random elements. Further,

$$\begin{aligned} E\Psi\left(\max_{1\leq j\leq n}\left\|\sum_{i=1}^{j} X_i\right\|\right) &\leq \qquad \text{[by Lemma 2.4.2]} \\ &\leq E\Psi\left(\max_{1\leq j\leq n}\left\|\sum_{i=1}^{j}(X_i - X_i^1)\right\|\right) \qquad \text{[by Lemma 2.4.1]} \\ &\leq 2E\Psi\left(\left\|\sum_{i=1}^{n}(X_i - X_i^1)\right\|\right) \qquad \text{[because of convexity]} \\ &\leq 2\left(\frac{1}{2}E\Psi\left(2\left\|\sum_{i=1}^{n} X_i\right\|\right) + \frac{1}{2}E\Psi\left(2\left\|\sum_{i=1}^{n} X_i^1\right\|\right)\right) \\ &\leq 2E\Psi\left(2\left\|\sum_{i=1}^{n} X_i\right\|\right). \end{aligned}$$

Lemma 2.4.4. *Let $X_1, \ldots, X_n$ be independent and symmetric random variables with values in B. Then, for each sequence $\alpha_1, \ldots, \alpha_n \in R$, and any $t > 0$,*

$$P\Big(\Big\|\sum_{i=1}^{n} \alpha_i X_i\Big\| > t\Big) \le 2P\Big(\max_{1\le i\le n} |\alpha_i| \Big\|\sum_{i=1}^{n} X_i\Big\| > t\Big).$$

Proof. Without any loss of generality we can assume that $0 \le \alpha_n \le \alpha_{n-1} \le \cdots \le \alpha_1 = 1$. Then we have

$$\begin{aligned}\sum_{i=1}^{n} \alpha_i X_i &= \sum_{i=1}^{n} \alpha_i (S_i - S_{i-1}) \\ &= \sum_{i=1}^{n} (\alpha_i - \alpha_{i+1}) S_i,\end{aligned}$$

with the understanding that $\alpha_{n+1} = 0$. Since $\sum_{i=1}^{n} (\alpha_i - \alpha_{i+1}) = 1$,

$$\Big\{\omega\colon \Big\|\sum_{i=1}^{n} (\alpha_i - \alpha_{i+1}) S_i\Big\| > t\Big\} \subset \Big\{\omega\colon \max_{1\le j\le n} \Big\|\sum_{i=1}^{j} X_i\Big\| > t\Big\}.$$

From here, by Lemma 2.4.1, we get the desired inequalities.

Symmetrization lemma for Banach-valued sums.

Lemma 2.4.5. *Let $X_1, \ldots, X_n$ be a sequence of independent random elements with values in Banach space B with $EX_i = 0$ for all i. Let $\{\varepsilon_i\}$ be a sequence of independent Bernoulli random variables independent of $\{X_i\}$ with law*

$$P(\varepsilon_i = 1) = P(\varepsilon_i = -1) = 1/2.$$

Then, for any convex increasing function Ψ,

$$\begin{aligned}E\Psi\Big(\frac{1}{2}\Big\|\sum_{i=1}^{n} \varepsilon_i X_i\Big\|\Big) &\le E\Psi\Big(\Big\|\sum_{i=1}^{n} X_i\Big\|\Big) \\ &\le E\Psi\Big(2\Big\|\sum_{i=1}^{n} \varepsilon_i X_i\Big\|\Big).\end{aligned}$$

Proof. These relations follow easily from known symmetrization results (see, for example, Araujo and Giné (1980), Lemma 2.13).

Symmetrization inequality for *UB*-statistics. Let $X_1, \ldots, X_n$ be a sequence of independent random variables taking values in a measurable space $(X, \mathfrak{X})$ and having an identical distribution P on a set X. Let $\{X_i^1\}$ be an independent copy of $\{X_i\}$. Denote by $\mathcal{F}$ a class of Banach-valued functions on X^m, $m \geq 1$. For each $\Phi \in \mathcal{F}$, define

$$\Phi(X_{\mathbf{i}}, X_{\mathbf{i}}^1) = \Phi(X_{i_1}, \dots, X_{i_m}) - \Phi(X_{i_1}^1, X_{i_2}, \dots, X_{i_m}) - \\ - \Phi(X_{i_1}, X_{i_2}^1, \dots, X_{i_m}) + \cdots + (-1)^m \Phi(X_{i_1}^1, \dots, X_{i_m}^1),$$

where $\mathbf{i} = (i_1, \dots, i_m)$ ranges over the $n^{[m]} = n(n-1)\cdots(n-m+1)$ ordered m-tuples of distinct integers I_n^m from the set $\{1, \dots, n\}$. Then for each $\Phi \in \mathcal{F}$, we define the symmetrized U-process as the set $\{U_n(\varepsilon\Phi) \colon \Phi \in \mathcal{F}\}$ where

$$U_n(\varepsilon\Phi) = n^{-[m]} \sum_{\mathbf{i}} \varepsilon_{i_1} \cdots \varepsilon_{i_m} \Phi(X_{\mathbf{i}}, X_{\mathbf{i}}^1) \tag{2.4.1}$$

and $\{\varepsilon_i\}$ is a sequence of Bernoulli.

Lemma 2.4.6. *Let $\mathcal{F}$ be a class of P-canonical functions $\Phi \colon X^m \to B$, $m \geq 1$. If Ψ is a convex function of $[0, \infty)$, then*

$$E\Psi(\sup_{\Phi\in\mathcal{F}} \|U_n(\Phi)\|) \leq E\Psi(\sup_{\Phi\in\mathcal{F}} \|U_n(\varepsilon\Phi)\|). \tag{2.4.2}$$

Proof. For $i = 1, \dots, n$ let

$$Y_i = I(\varepsilon_i = 1)X_i + I(\varepsilon_i = -1)X_i^1,$$
$$Y_i^1 = I(\varepsilon_i = -1)X_i + I(\varepsilon_i = 1)X_i^1, \tag{2.4.3}$$

where $I(A)$ is the indicator function of set A. The distribution of a random variable Y_i^1 coincides with the distribution of a random variable X_i for each i. Similarly the distribution of a random variable Y_i is the same as X_i^1. In addition, the random variables Y_i and Y_i^1 are independent.

Let E^Y denote expectation over the variables Y_i^1, conditional on the variables Y_i. The property of degeneracy implies that $\Phi(Y_{i_1}, \dots, Y_{i_m})$ can be replaced by a sum of 2^m terms,

$$\Phi(Y_{i_1}, \dots, Y_{i_m}) - E^Y\Phi(Y_{i_1}^1, Y_{i_2}, \dots, Y_{i_m}) - E^Y\Phi(Y_{i_1}, Y_{i_2}^1, \dots, Y_{i_m}) + \cdots \\ \cdots (-1)^m E^Y\Phi(Y_{i_1}^1, \dots, Y_{i_m}^1)$$

without changing the left-hand side of (2.4.2). Note that when E^Y is taken out of the sum, and the sup applied, by virtue of Jensen's inequality, Ψ increases the value of the left-hand side of (2.4.2) to

$$E\Psi\Big(\sup_{\mathcal{F}}\Big\|\sum_{\mathbf{i}}[\Phi(Y_{i_1}, \dots, Y_{i_m}) - \cdots (-1)^m\Phi(Y_{i_1}^1, \dots, Y_{i_m}^1)]\Big\|\Big). \tag{2.4.4}$$

Represent (2.4.4) in terms of X_i's, X_i^1's and ε_i's.

Fix $\mathbf{i}$. Consider the contribution of i_1. The sum of 2^m terms in (2.4.4) can be grouped as a sum of 2^{m-1} pairs

$$\pm[\Phi(Y_{i_1},\dots) - \Phi(Y^1_{i_1},\dots)].$$

Here the terms differ only in the first position. If $\varepsilon_{i_1} = 1$, replace Y_{i_1} by X_{i_1} and $Y^1_{i_1}$ by $X^1_{i_1}$; if $\varepsilon_{i_1} = -1$, replace Y_{i_1} by $X^1_{i_1}$ and $Y^1_{i_1}$ by X_{i_1}. The term equals

$$\pm\varepsilon_{i_1}[\Phi(X_{i_1},\dots) - \Phi(X^1_{i_1},\dots)],$$

that is, (2.4.4) is unchanged if we replace Y_{i_1} by X_{i_1} and $Y^1_{i_1}$ by $X^1_{i_1}$; then compensate with factor ε_{i_1}. In a similar way, replace all the other Y_i by X_i and Y^1_i by X^1_i, transforming (2.4.4) into

$$E\Psi\Big(\sup_{\mathcal{F}}\Big\|\sum_{\mathbf{i}}\varepsilon_{i_1}\cdots\varepsilon_{i_m}[\Phi(X_{i_1},\dots,X_{i_m}) - \cdots (-1)^m\Phi(X^1_{i_1},\dots,X^1_{i_m})]\Big\|\Big).$$

This expectation is the same as the quantity on the right-hand side of (2.4.4).

2.5 Decoupling inequalities

One-sample *UB*-statistics of degree 2. Let $X_1,\dots,X_n$ be a sequence of independent random variables in a measurable space $(X,\mathfrak{X})$. Let $\mathcal{F}_{ij}$ be families of Bochner integrable functions of two variables taking X^2 into Banach space B. Let $\Phi = (\Phi_{ij},\ 1\le i\ne j\le n)$ and $\mathcal{F} = (\mathcal{F}_{ij},\ 1\le i\ne j\le n)$, where $\Phi_{ij}\in\mathcal{F}_{ij}$. Assume that Ψ is an arbitrary convex increasing function for $x\ge 0$. Set

$$U_n = \sum_{1\le i\ne j\le n}\Phi_{ij}(X_i,X_j). \tag{2.5.1}$$

It is evident that the usual *UB*-statistics can be obtained from (2.5.1) by setting $\Phi_{ij} = \Phi/\binom{n}{2}$.

A *decoupled* version of U_n is

$$U_n^{\mathrm{dec}} = \sum_{1\le i\ne j\le n}\Phi_{ij}(X_i,X^1_j), \tag{2.5.2}$$

where $\{X^1_i\}$ is an independent copy of $\{X_i\}$.

Decoupling inequalities consist of comparing $E\Psi(\|U_n\|)$ to $E\Psi(\|U_n^{\mathrm{dec}}\|)$. The basic merit of decoupling inequalities is contained in the property that expectation $E\Psi(\|U_n^{\mathrm{dec}}\|)$ should be easier to compute than expectation $E\Psi(\|U_n\|)$.

Theorem 2.5.1. *Let $X_1,\dots,X_n$ be a sequence of independent random variables in a measurable space $(X,\mathfrak{X})$ and assume $\{X^1_i\}_{i=1}^n$ is an independent copy of $\{X_i\}_{i=1}^n$.*

Let $\mathcal{F}_{ij}$ be families of Bochner integrable functions $\Phi_{ij}\colon X^2 \to B$. For $x \geq 0$, let $\Psi(x)$ be any convex increasing function such that

$$\max_{1\leq i\neq j\leq n} E\Psi(\sup_{\Phi_{ij}\in\mathcal{F}_{ij}} \|\Phi_{ij}(X_i, X_j)\|) < \infty.$$

Then, for $\Phi = (\Phi_{ij},\, 1 \leq i \neq j \leq n)$, $\mathcal{F} = (\mathcal{F}_{ij},\, 1 \leq i \neq j \leq n)$,

$$E\Psi\Big(\sup_{\Phi\in\mathcal{F}}\Big\| \sum_{1\leq i\neq j\leq n} \Phi_{ij}(X_i, X_j)\Big\|\Big) \leq E\Psi\Big(7\sup_{\Phi\in\mathcal{F}}\Big\| \sum_{1\leq i\neq j\leq n} \Phi_{ij}(X_i, X_j^1)\Big\|\Big). \tag{2.5.3}$$

In the case of the kernels satisfying the degeneracy property

$$E(\Phi_{ij}(X_i, X_j)|X_i) = E(\Phi_{ij}(X_i, X_j)|X_j) = 0,$$

then

$$E\Psi\Big(\sup_{\Phi\in\mathcal{F}}\Big\| \sum_{1\leq i\neq j\leq n} \Phi_{ij}(X_i, X_j)\Big\|\Big) \leq E\Psi\Big(4\sup_{\Phi\in\mathcal{F}}\Big\| \sum_{1\leq i\neq j\leq n} \Phi_{ij}(X_i, X_j^1)\Big\|\Big). \tag{2.5.4}$$

If $\Phi_{ij} \in \mathcal{F}_{ij}$ satisfy the symmetry conditions

$$\Phi_{ij} = \Phi_{ji} \quad \textit{and} \quad \Phi_{ij}(X_i, X_j) = \Phi_{ij}(X_j, X_i),$$

then the reverse bound

$$E\Psi\Big(\frac{1}{3}\sup_{\Phi\in\mathcal{F}}\Big\| \sum_{1\leq j\neq j\leq n} \Phi_{ij}(X_i, X_j^1)\Big\|\Big) \leq E\Psi\Big(\sup_{\Phi\in\mathcal{F}}\Big\| \sum_{1\leq i\neq j\leq n} \Phi_{ij}(X_i, X_j)\Big\|\Big) \tag{2.5.5}$$

holds.

Proof. Firstly, the following lemma is required.

Lemma 2.5.1. *Let $Z_1, \ldots, Z_n$ be a sequence of independent random vectors with $P(Z_i = (X_i, X_i^1)) = P(Z_i = (X_i^1, X_i)) = 1/2$. More formally (see (2.4.3)),*

$$Z_i = I(\varepsilon_i = 1)(X_i, X_i^1) + I(\varepsilon_i = -1)(X_i^1, X_i)$$

where ε_i are symmetric Bernoulli random variables independent of each other and of $\{X_i\}$, $\{X_i^1\}$. Then

$$\begin{aligned} E^{\mathfrak{Z}}\Phi_{ij}(X_i, X_j) &= E^{\mathfrak{Z}}\Phi_{ij}(X_i, X_j^1) = \\ &= \frac{1}{4}(\Phi_{ij}(X_i, X_j) + \Phi_{ij}(X_i, X_j^1) + \Phi_{ij}(X_i^1, X_j) + \Phi_{ij}(X_i^1, X_j^1)), \end{aligned} \tag{2.5.6}$$

where $\mathfrak{Z} = \sigma(Z_i,\, i = 1, \ldots, n)$ is the σ-algebra, generated by $Z_1, \ldots, Z_n$ and $E^{\mathfrak{Z}}(\xi) = E(\xi|\mathfrak{Z})$ designates a conditional expectation of random variable ξ with respect to $\mathfrak{Z}$.

Proof. The first equality in (2.5.6) is correct because the conditional distributions of $\Phi_{ij}(X_i, X_j)$ and $\Phi_{ij}(X_i, X_j^1)$ with given $\mathfrak{Z}$ are the same. The second equality follows by noticing that the sum of the four terms is measurable with respect to $\mathfrak{Z}$.

Let $\mathfrak{X}^1 = \sigma(X_1, \ldots, X_n)$ be the σ-algebra, generated by $X_1, \ldots, X_n$ and in what follows, let sup denote $\sup_{\Phi \in \mathcal{F}}$. We have the following formula

$$\sum_{1 \le i \ne j \le n} \Phi_{ij}(X_i, X_j) =$$
$$= \sum_{1 \le i \ne j \le n} \Big(E^{\mathfrak{X}^1} \Phi_{ij}(X_i, X_j) + E^{\mathfrak{X}^1} \Phi_{ij}(X_i, X_j^1) +$$
$$+ E^{\mathfrak{X}^1} \Phi_{ij}(X_i^1, X_j) + E^{\mathfrak{X}^1} \Phi_{ij}(X_i^1, X_j^1) \Big) -$$
$$- \sum_{1 \le i \ne j \le n} \Big(E^{\mathfrak{X}^1} \Phi_{ij}(X_i, X_j^1) + E^{\mathfrak{X}^1} \Phi_{ij}(X_i^1, X_j) + E^{\mathfrak{X}^1} \Phi_{ij}(X_i^1, X_j^1) \Big).$$

From this formula, it follows that

$$E\Psi\Big(\sup\Big\| \sum_{1 \le i \ne j \le n} \Phi_{ij}(X_i, X_j) \Big\|\Big) =$$
$$= E\Psi\Big(\sup\Big\| \sum_{1 \le i \ne j \le n} E^{\mathfrak{X}^1} \Big(\Phi_{ij}(X_i, X_j) + \Phi_{ij}(X_i, X_j^1) + \Phi_{ij}(X_i^1, X_j) +$$
$$+ \Phi_{ij}(X_i^1, X_j^1) \Big) - \sum_{1 \le i \ne j \le n} E^{\mathfrak{X}^1} \Big(\Phi_{ij}(X_i, X_j^1) + \Phi_{ij}(X_i^1, X_j) + \Phi_{ij}(X_i^1, X_j^1) \Big) \Big\| \Big) =$$

[by formula (2.5.6) on the first term]

$$= E\Psi\Big(\sup\Big\| 4 \sum_{1 \le i \ne j \le n} E^{\mathfrak{X}^1} E^{\mathfrak{Z}} \Phi_{ij}(X_i, X_j^1) - \sum_{1 \le i \ne j \le n} E^{\mathfrak{X}^1} \Phi_{ij}(X_i, X_j^1) -$$
$$- \sum_{1 \le i \ne j \le n} E^{\mathfrak{X}^1} \Phi_{ij}(X_i^1, X_j) - \sum_{1 \le i \ne j \le n} E^{\mathfrak{X}^1} \Phi_{ij}(X_i^1, X_j^1) \Big\| \Big) \le$$

[since Ψ is increasing]

$$\le E\Psi\Big(4 \sup\Big\| \sum_{1 \le i \ne j \le n} E^{\mathfrak{X}^1} E^{\mathfrak{Z}} \Phi_{ij}(X_i, X_j^1) \Big\| + \sup\Big\| \sum_{1 \le i \ne j \le n} E^{\mathfrak{X}^1} \Phi_{ij}(X_i, X_j^1) \Big\| +$$
$$+ \sup\Big\| \sum_{1 \le i \ne j \le n} E^{\mathfrak{X}^1} \Phi_{ij}(X_i^1, X_j) \Big\| + \sup\Big\| \sum_{1 \le i \ne j \le n} E^{\mathfrak{X}^1} \Phi_{ij}(X_i^1, X_j^1) \Big\| \Big) \le$$

[by the convexity of Ψ]

$$\le \frac{4}{7} E\Psi\Big(7 \sup\Big\| \sum_{1 \le i \ne j \le n} E^{\mathfrak{X}^1} E^{\mathfrak{Z}} \Phi_{ij}(X_i, X_j^1) \Big\| \Big) +$$
$$+ \frac{1}{7} E\Psi\Big(7 \sup\Big\| \sum_{1 \le i \ne j \le n} E^{\mathfrak{X}^1} \Phi_{ij}(X_i, X_j^1) \Big\| \Big) +$$
$$+ \frac{1}{7} E\Psi\Big(7 \sup\Big\| \sum_{1 \le i \ne j \le n} E^{\mathfrak{X}^1} \Phi_{ij}(X_i^1, X_j) \Big\| \Big) +$$

$$+\frac{1}{7}E\Psi\Big(7\sup\Big\|\sum_{1\le i\ne j\le n}E^{\mathfrak{X}^1}\Phi_{ij}(X_i^1,X_j^1)\Big\|\Big)\le$$

[by the conditional Jensen's inequality and identity

$E\Phi_{ij}(X_i^1,X_j^1)=E\Phi_{ij}(X_i,X_j^1)$ in the last term]

$$\le\frac{4}{7}E\Psi\Big(7\sup\Big\|\sum_{1\le i\ne j\le n}\Phi_{ij}(X_i,X_j^1)\Big\|\Big)+\frac{1}{7}E\Psi\Big(7\sup\Big\|\sum_{1\le i\ne j\le n}\Phi_{ij}(X_i,X_j^1)\Big\|\Big)+$$

$$+\frac{1}{7}E\Psi\Big(7\sup\Big\|\sum_{1\le i\ne j\le n}\Phi_{ij}(X_i,X_j^1)\Big\|\Big)+\frac{1}{7}E\Psi\Big(7\sup\Big\|\sum_{1\le i\ne j\le n}\Phi_{ij}(X_i,X_j^1)\Big\|\Big)=$$

$$=E\Psi\Big(7\sup\Big\|\sum_{1\le i\ne j\le n}\Phi_{ij}(X_i,X_j^1)\Big\|\Big).$$

This completes the proof of inequality (2.5.3).

For the proof of inequality (2.5.4) we note that in the conditions of degeneracy by Lemma 2.5.1 the following identity

$$\Phi_{ij}(X_i,X_j)=4E^{\mathfrak{X}^1}E^3\Phi_{ij}(X_i,X_j^1)$$

holds. Then,

$$E\Psi\Big(\sup\Big\|\sum_{1\le i\ne j\le n}\Phi_{ij}(X_i,X_j)\Big\|\Big)=E\Psi\Big(4\sup\Big\|\sum_{1\le i\ne j\le n}E^{\mathfrak{X}^1}E^3\Phi_{ij}(X_i,X_j^1)\Big\|\Big)\le$$

[by the conditional Jensen's inequality]

$$\le E\Psi\Big(4\sup\Big\|\sum_{1\le i\ne j\le n}\Phi_{ij}(X_i,X_j^1)\Big\|\Big).$$

We now show that inequality (2.5.5), containing the lower bound, is satisfied. Under the symmetry assumptions on $\{\Phi_{ij}\}$,

$$\sum_{1\le i\ne j\le n}\Phi_{ij}(X_i,X_j^1)=\frac{1}{2}\sum_{1\le i\ne j\le n}(\Phi_{ij}(X_i,X_j^1)+\Phi_{ij}(X_i^1,X_j)).$$

By (2.5.6)

$$\Phi_{ij}(X_i,X_j^1)+\Phi_{ij}(X_i^1,X_j)=4E^3\Phi_{ij}(X_i,X_j)-\Phi_{ij}(X_i,X_j)-\Phi_{ij}(X_i^1,X_j^1).$$

From the preceding, it follows that

$$E\Psi\Big(\frac{1}{3}\sup\Big\|\sum_{1\le i\ne j\le n}\Phi_{ij}(X_i,X_j^1)\Big\|\Big)=$$

$$=E\Psi\Big(\frac{1}{6}\sup\Big\|\sum_{1\le i\ne j\le n}\big(4E^3\Phi_{ij}(X_i,X_j)-\Phi_{ij}(X_i,X_j)-\Phi_{ij}(X_i^1,X_j^1)\big)\Big\|\Big)\le$$

[by the convex increasing of Ψ]

$$\le\frac{4}{6}E\Psi\Big(\sup\Big\|\sum_{1\le i\ne j\le n}E^3\Phi_{ij}(X_i,X_j)\Big\|\Big)+\frac{1}{6}E\Psi\Big(\sup\Big\|\sum_{1\le i\ne j\le n}\Phi_{ij}(X_i,X_j)\Big\|\Big)+$$

$$+\frac{1}{6}E\Psi\Big(\sup\Big\|\sum_{1\le i\ne j\le n}\Phi_{ij}(X_i^1,X_j^1)\Big\|\Big)\le$$

[by the conditional Jensen's inequality on the first term and regrouping]

$$\le\frac{4}{6}E\Psi\Big(\sup\Big\|\sum_{1\le i\ne j\le n}\Phi_{ij}(X_i,X_j)\Big\|\Big)+\frac{2}{6}E\Psi\Big(\sup\Big\|\sum_{1\le i\ne j\le n}\Phi_{ij}(X_i,X_j)\Big\|\Big)=$$

$$=E\Psi\Big(\sup\Big\|\sum_{1\le i\ne j\le n}\Phi_{ij}(X_i,X_j)\Big\|\Big).$$

This completes the proof of inequality (2.5.5).

The following theorem contains decoupling inequalities for centered UB-statistics of a generalized form (2.5.1).

Theorem 2.5.2. *Under the conditions of Theorem* 2.5.1 *the following inequalities hold:*

$$E\Psi\Big(\frac{1}{3}\sup\Big\|\sum_{1\le i\ne j\le n}\big(\Phi_{ij}(X_i,X_j^1)-E\Phi_{ij}(X_i,X_j^1)\big)\Big\|\Big)\le$$

[*under symmetry conditions*]

$$\le E\Psi\Big(\sup\Big\|\sum_{1\le i\ne j\le n}\big(\Phi_{ij}(X_i,X_j)-E\Phi_{ij}(X_i,X_j)\big)\Big\|\Big)\le$$

[*without the assumption of the symmetry conditions*]

$$\le E\Psi\Big(6\sup\Big\|\sum_{1\le i\ne j\le n}\big(\Phi_{ij}(X_i,X_j^1)-E\Phi_{ij}(X_i,X_j^1)\big)\Big\|\Big).\qquad(2.5.7)$$

Proof. It is sufficient to apply the method of the proof of Theorem 2.5.1 with functions $\Phi_{ij}(X_i,X_j)-E\Phi_{ij}(X_i,X_j)$.

Remark 2.5.1. The fact that lower bounds do not hold for general Φ_{ij} follows evidently by using

$$\Phi_{ij}(X_i,X_j)=\Phi(X_i,X_j)=\varphi(X_j)-\varphi(X_i),$$

because $\sum_{1\le i\ne j\le n}\Phi_{ij}(X_i,X_j)=0$. Here $\varphi(x)$ is some Banach-valued function.

Now we present Theorem 2.5.1 in a more general form in which it is still valid.

Theorem 2.5.3. *Let $X_1,\dots,X_n$ be a sequence of independent random variables in a measurable space $(X,\mathfrak{X})$ and suppose that $\{X_i^1\}_{i=1}^n$ is an independent copy of $\{X_i\}_{i=1}^n$. Let $\mathcal{F}_{ij}^k$ be families of Bochner integrable functions Φ_{ij}^k such that $\Phi_{ij}^k\in\mathcal{F}_{ij}^k$ maps X^2 into a Banach space B. Let N_n be an arbitrary subset of $\{1,2,\dots,n\}$. For $x\ge 0$, let $\Psi(x)$ be any convex increasing function such that*

$$\max_{1\le i\ne j\le n}E\Psi(\max_{k\in N_n}\sup_{\Phi_{ij}^k\in\mathcal{F}_{ij}^k}\|\Phi_{ij}^k(X_i,X_j)\|)<\infty.$$

Then, for $\Phi^k = (\Phi^k_{ij}, 1 \le i \ne j \le k)$, $\mathcal{F}^k = (\mathcal{F}^k_{ij}, 1 \le i \ne j \le k)$,

$$E\Psi\Big(\max_{k\in N_n} \sup_{\Phi^k\in\mathcal{F}^k} \Big\| \sum_{1\le i\ne j\le k} \Phi^k_{ij}(X_i, X_j)\Big\|\Big) \le$$
$$\le E\Psi\Big(7 \max_{k\in N_n} \sup_{\Phi^k\in\mathcal{F}^k} \Big\| \sum_{1\le i\ne j\le k} \Phi^k_{ij}(X_i, X^1_j)\Big\|\Big).$$

In the case of the kernels satisfying the degeneracy property,

$$E\Psi\Big(\max_{k\in N_n} \sup_{\Phi^k\in\mathcal{F}^k} \Big\| \sum_{1\le i\ne j\le k} \Phi^k_{ij}(X_i, X_j)\Big\|\Big) \le$$
$$\le E\Psi\Big(4 \max_{k\in N_n} \sup_{\Phi^k\in\mathcal{F}^k} \Big\| \sum_{1\le i\ne j\le k} \Phi^k_{ij}(X_i, X^1_j)\Big\|\Big).$$

If $\Phi^k_{ij} = \mathcal{F}^k_{ij}$ *satisfy the symmetry conditions*

$$\Phi^k_{ij} = \Phi^k_{ji} \quad \text{and} \quad \Phi^k_{ij}(X_i, X_j) = \Phi^k_{ij}(X_j, X_i),$$

then the reverse bound holds:

$$E\Psi\Big(\frac{1}{3} \max_{k\in N_n} \sup_{\Phi^k\in\mathcal{F}^k} \Big\| \sum_{1\le i\ne j\le k} \Phi^k_{ij}(X_i, X^1_j)\Big\|\Big) \le$$
$$\le E\Psi\Big(\max_{k\in N_n} \sup_{\Phi^k\in\mathcal{F}^k} \Big\| \sum_{1\le i\ne j\le k} \Phi^k_{ij}(X_i, X_j)\Big\|\Big).$$

Proof. It is possible to repeat almost verbatim the proof of Theorem 2.5.1 with some nonessential complications.

The following corollary is an extension and decoupling version of Lemma 2.4.3.

Corollary 2.5.1. *Suppose that kernels* Φ_{ij} *are degenerate, that is*

$$E(\Phi_{ij}(X_i, X_j)|X_i) = E(\Phi_{ij}(X_i, X_j)|X_j) = 0\,.$$

Then,

$$E\Psi\Big(\max_{1\le k\le n} \Big\| \sum_{1\le i\ne j\le k} \Phi_{ij}(X_i, X_j)\Big\|\Big) \le 2E\Psi\Big(8\Big\| \sum_{1\le i\ne j\le n} \Phi_{ij}(X_i, X^1_j)\Big\|\Big).$$

If, in addition, kernels Φ_{ij} *are symmetric, i.e.*

$$\Phi_{ij} = \Phi_{ji} \quad \text{and} \quad \Phi_{ij}(X_i, X_j) = \Phi_{ij}(X_j, X_i)$$

then

$$E\Psi\Big(\max_{1\le k\le n} \Big\| \sum_{1\le i\ne j\le k} \Phi_{ij}(X_i, X_j)\Big\|\Big) \le 2E\Psi\Big(24\Big\| \sum_{1\le i\ne j\le n} \Phi_{ij}(X_i, X_j)\Big\|\Big).$$

Proof. From Theorem 2.5.3 it follows that

$$E\Psi\Big(\max_{1\le k\le n}\Big\|\sum_{1\le i\ne j\le k}\Phi_{ij}(X_i,X_j)\Big\|\Big)\le E\Psi\Big(4\max_{1\le k\le n}\Big\|\sum_{1\le i\ne j\le k}\Phi_{ij}(X_i,X_j^1)\Big\|\Big)\le$$

[by conditioning on $X_1,\ldots,X_n$ and using Lemma 2.4.3]

$$\le 2E\Psi\Big(8\Big\|\sum_{1\le i\ne j\le n}\Phi_{ij}(X_i,X_j^1)\Big\|\Big)\le$$

[by using inequality (2.5.5)]

$$\le 2E\Psi\Big(24\Big\|\sum_{1\le i\ne j\le n}\Phi_{ij}(X_i,X_j)\Big\|\Big).$$

One-sample *UB*-statistics of degree *m*. Now, we will give a generalization of Theorem 2.5.3 for one-sample *UB*-statistics of arbitrary degree $m\ge 2$.

Theorem 2.5.4. *Let $X_1,\ldots,X_n$ be a sequence of independent random variables in a measurable space $(X,\mathfrak{X})$, and suppose that $\{X_i^j,\ldots,X_n^j\}_{j=1}^m$ are independent copies of $\{X_i\}_{i=1}^n$. Let $\mathcal{F}^k_{i_1,\ldots,i_m}$ be families of Bochner integrable functions $\Phi^k_{i_1,\ldots,i_m}$ such that $\Phi^k_{i_1,\ldots,i_m}\in\mathcal{F}^k_{i_1,\ldots,i_m}$ maps X^m into a Banach space B. Let N_n be an arbitrary subset of $\{1,2,\ldots,n\}$. For $x\ge 0$, let $\Psi(x)$ be any convex increasing function such that*

$$\max_{1\le i_1\ne\cdots\ne i_m\le n}E\Psi\Big(\max_{k\in N_n}\sup_{\Phi^k_{i_1,\ldots,i_m}\in\mathcal{F}^k_{i_1,\ldots,i_m}}\big\|\Phi^k_{i_1,\ldots,i_m}(X_{i_1},\ldots,X_{i_m})\big\|\Big)<\infty.$$

Then for $\Phi^k=(\Phi^k_{i_1,\ldots,i_m},1\le i_1\ne\cdots\ne i_m\le k)$, $\mathcal{F}^k=(\mathcal{F}^k_{i_1,\ldots,i_m},1\le i_1\ne\cdots\ne i_m\le k)$, $k\in N_n$,

$$E\Psi\Big(\max_{k\in N_n}\sup_{\Phi^k\in\mathcal{F}^k}\Big\|\sum_{1\le i_1\ne\cdots\ne i_m\le k}\Phi^k_{i_1,\ldots,i_m}(X_{i_1},\ldots,X_{i_m})\Big\|\Big)\le$$
$$\le E\Psi\Big(A_m\max_{k\in N_n}\sup_{\Phi^k\in\mathcal{F}^k}\Big\|\sum_{1\le i_1\ne\cdots\ne i_m\le k}\Phi_{i_1,\ldots,i_m}(X^1_{i_1},\ldots,X^m_{i_m})\Big\|\Big)$$

with

$$A_m=\prod_{p=2}^m(2p^p-1).$$

In the case of the kernels satisfying the degeneracy property

$$E^{\mathfrak{X}^i}\Phi^k_{i_1,\ldots,i_m}(X^1_{i_1},\ldots,X^m_{i_m})=0,$$

where $\mathfrak{X}^i=\sigma(X^i_1,\ldots,X^i_n)$, then the constant A_m can be taken to be m^m.

If $\Phi^k_{i_1,\ldots,i_m}\in\mathcal{F}^k_{i_1,\ldots,i_m}$ satisfy the symmetry conditions

$$\Phi^k_{i_1,\ldots,i_m}=\Phi^k_{i_{s_1},\ldots,i_{s_m}}\quad\text{and}\quad\Phi_{i_1,\ldots,i_m}(X_{i_1},\ldots,X_{i_m})=\Phi_{i_1,\ldots,i_m}(X_{i_{s_1}},\ldots,X_{i_{s_m}})$$

for all permutations $(i_{s_1},\dots,i_{s_m})$ *of* $(i_1,\dots,i_m)$, *then the reverse bound for* $m\geq 2$ *holds:*

$$E\Psi\Big(B_m \max_{k\in N_n}\sup_{\Phi^k\in\mathcal{F}^k}\Big\|\sum_{1\leq i_1\neq\cdots\neq i_m\leq k}\Phi^k_{i_1,\dots,i_m}(X^1_{i_1},\dots,X^m_{i_m})\Big\|\Big)\leq$$

$$\leq E\Psi\Big(\max_{k\in N_n}\sup_{\Phi^k\in\mathcal{F}^k}\Big\|\sum_{1\leq i_1\neq\cdots\neq i_m\leq k}\Phi^k_{i_1,\dots,i_m}(X_{i_1},\dots,X_{i_m})\Big\|\Big),$$

where

$$B_m=\Big(\sum_{p=1}^m\frac{p^m}{(m-p)!\,p!}\Big)^{-1}.$$

Proof. In the following we will use $\sum$ instead of $\sum_{1\leq i_1\neq\cdots\neq i_m\leq k}$. Instead of $\Phi_{i_1,\dots,i_m}$ we will use just Φ. Instead of $\max_{k\in N_n}\sup_{\Phi^k\in\mathcal{F}^k}$ we will use just max sup. The proof requires the following lemma. It is a generalization of Lemma 2.5.1.

Lemma 2.5.2. *Let* $\mathfrak{Z}_l=\sigma(Z_i,\ i=1,\dots,n)$ *be* σ*-algebra, generated by* $Z_1,\dots,Z_n$ *where* $\{Z_i\}$ *are independent random vectors with* $Z_i=(X_i^{j_1},X_i^{j_2},\dots,X_i^{j_l})$ *with probability* $1/l!$ *for each permutation* $(j_1,j_2,\dots,j_l)$ *of* $(1,2,\dots,l)$, *where* $l=2,3,\dots,m$. *It means that* $\mathfrak{Z}_l$ *is the analogue of* σ*-algebra in Lemma* 2.5.1.

Then, for $l=2,3,\dots,m$,

$$E^{\mathfrak{Z}_l}\Phi(X^1_{i_1},X^1_{i_2},\dots,X^1_{i_m})=E^{\mathfrak{Z}_l}\Phi(X^1_{i_1},X^2_{i_2},\dots,X^m_{i_m})=$$

$$=l^{-m}\sum_{j_1=1}^{l}\cdots\sum_{j_m=1}^{l}\Phi(X^{j_1}_{i_1},X^{j_2}_{i_2},\dots,X^{j_m}_{i_m}).\qquad(2.5.8)$$

Proof. By formula (1.1.4) and Fubini's theorem we have

$$E^{\mathfrak{Z}_l}\Phi(X^1_{i_1},X^2_{i_2},\dots,X^m_{i_m})=$$
$$=E^{\mathfrak{Z}_l}\int_X\cdots\int_X\Phi(y_1,\dots,y_m)\,\delta_{X^1_{i_1}}(dy_1)\,\delta_{X^2_{i_2}}(dy_2)\cdots\delta_{X^m_{i_m}}(dy_m)=$$
$$=\int_X\cdots\int_X\Phi(y_1,\dots,y_m)\,E^{\mathfrak{Z}_l}\,\delta_{X^1_{i_1}}(dy_1)\,\delta_{X^2_{i_2}}(dy_2)\cdots\delta_{X^m_{i_m}}(dy_m).$$

Here

$$E^{\mathfrak{Z}_l}\delta_{X^1_{i_1}}(dy_1)\,\delta_{X^2_{i_2}}(dy_2)\dots\delta_{X^m_{i_m}}(dy_m)=$$
$$=\frac{1}{l}(\delta_{X^1_{i_1}}(dy_1)+\cdots+\delta_{X^l_{i_1}}(dy_1))\frac{1}{l}(\delta_{X^1_{i_2}}(dy_2)+\cdots+\delta_{X^l_{i_2}}(dy_2))\times\cdots$$
$$\cdots\times\frac{1}{l}(\delta_{X^1_{i_m}}(dy_m)+\cdots+\delta_{X^l_{i_m}}(dy_m))=$$
$$=\prod_{s=1}^m\Big(\frac{1}{l}\sum_{j=1}^l\delta_{X^j_{i_s}}(dy_s)\Big)=l^{-m}\sum_{j_1=1}^l\cdots\sum_{j_m=1}^l\delta_{X^{j_1}_{i_1}}(dy_1)\cdots\delta_{X^{j_m}_{i_m}}(dy_m).$$

From formula (1.1.4) and the preceding we obtain (2.5.8).

In the following, we will prove the upper and lower bounds of this theorem separately.

Proof of the upper bound. We will use certain recurrence relations over m. The technical difficulties of the proof of the upper bound are united in the following lemma. Note, that the claim of this lemma has an independent interest.

Lemma 2.5.3. *For all $n \geq m \geq 2$ and fixed $1 \leq i_1 \neq \cdots \neq i_m \leq n$, $i_s \neq i_r$ is $s \neq r$ we have the following relation*

$$\sum_{c=0}^{m}(m-1)^{m-c} \sum_{1\leq j_1<\cdots<j_c\leq m} \Phi_c(X^1_{i_{j_1}},\ldots,X^1_{i_{j_c}}) =$$
$$= m^m E^{\mathfrak{X}^1} E^{\mathfrak{Z}_m} \Phi(X^1_{i_1},X^2_{i_2},\ldots,X^m_{i_m}), \quad (2.5.9)$$

where

$$\Phi_c(X^1_{i_{j_1}},\ldots,X^1_{i_{j_c}}) = \int_X\cdots\int_X \Phi(X^1_{i_{j_1}},\ldots,X^1_{i_{j_c}},y_{c+1},\ldots,y_m)\prod_{s=c+1}^{m} P_{i_{j_s}}(dy_s)$$

with

$$\Phi_m(X^1_{i_{j_1}},\ldots,X^1_{i_{j_m}}) = \Phi(X^1_{i_1},\ldots,X^1_{i_m}), \qquad \Phi_0 = E\Phi(X^1_{i_1},X^2_{i_2},\ldots,X^m_{i_m})$$

and $\{j_{c+1},\ldots,j_m\} = \{1,\ldots,m\}\setminus\{j_1,\ldots,j_c\}$.

Proof. By Lemma 2.5.2, it follows that

$$m^m E^{\mathfrak{Z}_m}\Phi(X^1_{i_1},X^2_{i_2},\ldots,X^m_{i_m}) = \sum_{j_1=1}^{m}\cdots\sum_{j_m=1}^{m}\Phi(X^{j_1}_{i_1},X^{j_2}_{i_2},\ldots,X^{j_m}_{i_m}) =$$

[by formula (1.1.4)]

$$= \sum_{j_1=1}^{m}\cdots\sum_{j_m=1}^{m}\int_X\cdots\int_X \Phi(y_1,\ldots,y_m)\prod_{s=1}^{m}\delta_{X^{j_s}_{i_s}}(dy_s) =$$

[after the permutation of the order of summation and integration]

$$= \int_X\cdots\int_X \Phi(y_1,\ldots,y_m)\prod_{s=1}^{m}\Big(\sum_{j_s=1}^{m}\delta_{X^{j_s}_{i_s}}(dy_s)\Big) =$$
$$= \int_X\cdots\int_X \Phi(y_1,\ldots,y_m)\prod_{s=1}^{m}\Big(\sum_{j=1}^{m}\delta_{X^{j}_{i_s}}(dy_s)\Big).$$

From the preceding it follows that

$$m^m E^{\mathfrak{X}^1} E^{\mathfrak{Z}_m}\Phi(X^1_{i_1},X^2_{i_2},\ldots,X^m_{i_m}) = E^{\mathfrak{X}^1}\int_X\cdots\int_X \Phi(y_1,\ldots,y_m)\prod_{s=1}^{m}\Big(\sum_{j=1}^{m}\delta_{X^{j}_{i_s}}(dy_s)\Big) =$$

[by Fubini's theorem and the independence of $\{X^j_i\}$]

$$= \int_X \cdots \int_X \Phi(y_1, \ldots, y_m) \prod_{s=1}^{m} (\delta_{X^1_{i_s}}(dy_s) + (m-1)P_{i_s}(dy_s)) =$$

[by formula (1.1.5) at $a_s = \delta_{X^1_{i_s}}(dy_s)$ and $b_s = (m-1)P_{i_s}(dy_s)$]

$$= \sum_{c=0}^{m} (m-1)^{m-c} \sum_{1 \le j_1 < \cdots < j_c \le m} \int_X \cdots \int_X \Phi(X^1_{i_{j_1}}, \ldots, X^1_{i_{j_c}}, y_{c+1}, \ldots, y_m) \times$$

$$\times \prod_{s=c+1}^{m} P_{i_{j_s}}(dy_s) = \text{[with the preceding notation]}$$

$$= \sum_{c=0}^{m} (m-1)^{m-c} \sum_{1 \le j_1 < \cdots < j_c \le m} \Phi_c(X^1_{i_{j_1}}, \ldots, X^1_{i_{j_c}}).$$

Now formula (2.5.9) is proved.

Relation (2.5.9) immediately gives the following formula

$$\Phi(X^1_{i_1}, X^1_{i_2}, \ldots, X^1_{i_m}) = m^m E^{\mathfrak{X}^1} E^{\mathfrak{Z}_m} \Phi(X^1_{i_1}, X^2_{i_2}, \ldots, X^m_{i_m}) -$$

$$-(m-1)^m E\Phi(X^1_{i_1}, X^2_{i_2}, \ldots, X^m_{i_m}) - \sum_{c=1}^{m-1} (m-1)^{m-c} \sum_{1 \le j_1 < \cdots < j_c \le m} \Phi_c(X^1_{i_{j_1}}, \ldots, X^1_{i_{j_c}}),$$

i.e. we separate on the left-hand side of (2.5.9) the terms corresponding to $c = 0$ and $c = m$. On the right-hand side of this formula the number of terms is equal to $2m^m - 1$, including the zero terms.

From the preceding and the convexity of Ψ, it follows that

$$E\Psi\Big(\sup\Big\|\sum \Phi(X^1_{i_1}, X^1_{i_2}, \ldots, X^1_{i_m})\Big\|\Big) \le$$

$$\le (2m^m - 1)^{-1} m^m E\Psi\Big((2m^m - 1)\sup\Big\|\sum E^{\mathfrak{X}^1} E^{\mathfrak{Z}_m} \Phi(X^1_{i_1}, X^2_{i_2}, \ldots, X^m_{i_m})\Big\|\Big) +$$

$$+ (2m^m - 1)^{-1} (m-1)^m E\Psi\Big((2m^m - 1)\sup\Big\|\sum E\Phi(X^1_{i_1}, X^2_{i_2}, \ldots, X^m_{i_m})\Big\|\Big) +$$

$$+ (2m^m - 1)^{-1} \sum_{c=1}^{m-1} (m-1)^{m-c} \times$$

$$\times \sum_{1 \le j_1 < \cdots < j_c \le m} E\Psi\Big((2m^m - 1)\sup\Big\|\sum \Phi_c(X^1_{i_{j_1}}, \ldots, X^1_{i_{j_c}})\Big\|\Big) \le$$

[because of Jensen's inequalities]

$$\le (2m^m - 1)^{-1}(m^m + (m-1)^m) \times$$

$$\times E\Psi\Big((2m^m - 1)\sup\Big\|\sum \Phi(X^1_{i_1}, X^2_{i_2}, \ldots, X^m_{i_m})\Big\|\Big) +$$

$$+ (2m^m - 1)^{-1} \sum_{c=1}^{m-1} (m-1)^{m-c} \times$$

$$\times \sum_{1 \le j_1 < \cdots < j_c \le m} E\Psi\Big((2m^m - 1)\sup\Big\|\sum \Phi_c(X^1_{i_{j_1}}, \ldots, X^1_{i_{j_c}})\Big\|\Big). \tag{2.5.10}$$

Further, we consider the term in (2.5.10) corresponding to $c = m-1$, that is

$$(2m^m - 1)^{-1}(m-1) \sum_{1\le j_1<\cdots<j_{m-1}\le m} E\Psi\Big((2m^m-1)\sup\Big\|\sum \Phi_{m-1}(X^1_{i_{j_1}},\dots,X^1_{i_{j_{m-1}}})\Big\|\Big).$$

Here $\Phi_{m-1}(X^1_{i_{j_1}},\dots,X^1_{i_{j_{m-1}}})$ is a function of $m-1$ variables. Now we apply formula (2.5.9) at $m-1$ instead of m. And from this step we get, along with (2.5.10) the following inequality

$$\begin{aligned} & E\Psi\Big((2m^m-1)\sup\Big\|\sum \Phi_{m-1}(X^1_{i_{j_1}},\dots,X^1_{i_{j_{m-1}}})\Big\|\Big) \le \\ & \quad \le (2(m-1)^{m-1}-1)^{-1}\Big((m-1)^{m-1}+(m-2)^{m-1}\Big)\times \\ & \quad\quad \times E\Psi\Big((2m^m-1)(2(m-1)^{m-1}-1)\sup\|\Phi(X^1_{i_1},X^2_{i_2},\dots,X^m_{i_m})\|\Big) + \\ & \quad\quad + (2(m-1)^{m-1}-1)^{-1}\sum_{c=1}^{m-2}(m-2)^{m-1-c}\times \\ & \quad\quad \times \sum_{1\le k_1<\cdots<k_c\le m-1} E\Psi\Big((2m^m-1)(2(m-1)^{m-1}-1)\sup\Big\|\sum \Phi_c(X^1_{i_{k_1}},\dots,X^1_{i_{k_c}})\Big\|\Big). \end{aligned} \tag{2.5.11}$$

Further, in (2.5.9) and (2.5.10) we separate the terms, corresponding to the index $c = m-2$. And then we apply again formula (2.5.9). As a result we derive an inequality of type (2.5.10) and (2.5.11), where the expectations with the functions $\Phi_{m-3}, \Phi_{m-4}, \dots, \Phi_2$ remain. If we continue this process of applying this Lemma 2.5.3, we derive at the final step the upper bound

$$\sum c_i E\Psi\Big(A_m \sup\Big\|\sum \Phi(X^1_{i_1},X^2_{i_2},\dots,X^m_{i_m})\Big\|\Big),$$

where $\sum c_i$ is equal to 1 since it came from the convexity arguments of function Ψ. This completes the proof of the upper bound mentioned in the theorem.

Proof of the lower bound. In this case we assume that the symmetry conditions on $\{\Phi_{i_1,\dots,i_m}\}$ hold. Under these conditions

$$\sum \Phi(X^1_{i_1},X^2_{i_2},\dots,X^m_{i_m}) = \frac{1}{m!}\sum_{I^m_n}\sum_{\sigma}\Phi(X^{j_1}_{i_1},X^{j_2}_{i_2},\dots,X^{j_m}_{i_m}), \tag{2.5.12}$$

where the right-hand side summation is over $(j_1,\dots,j_m)\in\sigma =$ all permutations of $(1,\dots,m)\}$ and $I^m_n = \{(i_1,\dots,i_m)\colon 1\le i_1\ne\cdots\ne i_m\le n,\ i_s\ne i_r,\ \text{if } s\ne r\}$.

In the following lemma we represent the expression on the right-hand side of (2.5.12) via the permanent random measure (see formula (1.5.2)).

Lemma 2.5.4. *Let*

$$A^m_m(B_1,\dots,B_m) = \Big[a_{sj} = \delta_{X^j_{i_s}}(B_s);\ \ 1\le s\le m,\ 1\le j\le m\Big]$$

be a matrix whose dimension is $m \times m$ and elements a_{sj} are Dirac measures, and $B_s \in \mathfrak{X}$, $B_s \cup B_r = \varnothing$, if $s \neq r$.

Then for fixed $1 \le i_1 \neq \cdots \neq i_m \le n$ the following representation is valid:

$$\sum_\sigma \Phi(X_{i_1}^{j_1}, \ldots, X_{i_m}^{j_m}) = \int_X \cdots \int_X \Phi(y_1, \ldots, y_m) \operatorname{Per} A_m^m(dy_1, \ldots, dy_m) . \tag{2.5.13}$$

Proof. We have

$$\sum_\sigma \Phi(X_{i_1}^{j_1}, \ldots, X_{i_m}^{j_m}) =$$

[by formula (1.1.4)]

$$= \sum_\sigma \int_X \cdots \int_X \Phi(y_1, \ldots, y_m) \prod_{s=1}^m \delta_{X_{i_s}^{j_s}}(dy_s) =$$

[after the permutation of the order of summation and integration]

$$= \int_X \cdots \int_X \Phi(y_1, \ldots, y_m) \sum_\sigma \prod_{s=1}^m \delta_{X_{i_s}^{j_s}}(dy_s) =$$

[by virtue of the definition of permanent random measures]

$$= \int_X \cdots \int_X \Phi(y_1, \ldots, y_m) \operatorname{Per} A_m^m(dy_1, \ldots, dy_m) .$$

This gives the required result.

Further, we will apply Lemma 2.5.4 in this way. It is known (see, for example, Jacobs (1983), Satz 3.13, page 31) that

$$\operatorname{Per} A_m^m = \sum_{l=1}^m (-1)^{m-l} \sum_{|J|=l} \prod_{s=1}^m \Big(\sum_{j \in J} a_{sj} \Big), \tag{2.5.14}$$

where the second sum is taken over all subsets J of $\{1, \ldots, m\}$, such that $|J| = l$.

Let in (2.5.14), $a_{sj} = \delta_{X_{i_s}^j}(B_s)$ just as in Lemma 2.5.4. Then from (2.5.13) and (2.5.14) we obtain

$$\sum_\sigma \Phi(X_{i_1}^{j_1}, \ldots, X_{i_m}^{j_m}) = \sum_{l=1}^m (-1)^{m-l} \sum_{|J|=l} \sum_{j_1 \in J} \cdots \sum_{j_m \in J} \Phi(X_{i_1}^{j_1}, \ldots, X_{i_m}^{j_m}) =$$

[by Lemma 2.5.2]

$$= \sum_{l=1}^m (-1)^{m-l} l^m \sum_{|J|=l} E^{\mathfrak{Z}_1} \Phi(X_{i_1}^1, \ldots, X_{i_m}^1) .$$

From here taking account of (2.5.12) we get the formula

$$\sum \Phi(X_{i_1}^1, X_{i_2}^2, \ldots, X_{i_m}^m) = \frac{1}{m!} \sum_{l=1}^m (-1)^{m-l} l^m \sum_{|J|=l} E^{\mathfrak{Z}_1} \Phi(X_{i_1}^1, X_{i_2}^1, \ldots, X_{i_m}^1). \tag{2.5.15}$$

Note that

$$\frac{1}{m!}\sum_{l=1}^{m} l^m \sum_{|J|=l} 1 = B_m^{-1}. \tag{2.5.16}$$

From formula (2.5.15) and the convexity of Ψ, it follows that

$$E\Psi\Big(B_m \sup\Big\|\sum \Phi(X_{i_1}^1,\dots,X_{i_m}^m)\Big\|\Big) \le$$

$$\le \frac{B_m}{m!}\sum_{l=1}^{m} l^m \sum_{|J|=l} E\Psi\Big(\sup\Big\|\sum E^{\mathfrak{Z}_l}\Phi(X_{i_1}^1,\dots,X_{i_m}^1)\Big\|\Big) \le$$

[by (2.5.16) and the conditional Jensen's inequality]

$$\le E\Psi\Big(\sup\Big\|\sum \Phi(X_{i_1}^1,\dots,X_{i_m}^1)\Big\|\Big).$$

This completes the proof of the lower bound.

Remark 2.5.2. In the upper bound of Theorem 2.5.4 we have the constant

$$A_m = \prod_{p=1}^{m}(2p^p - 1).$$

V. de la Peña (1992) got the same upper bound but with another constant

$$C_m = 2^m \prod_{p=2}^{m}(p^p - 1).$$

It is easy to see that

$$A_2 = 7, \quad A_3 = 371, \quad A_4 = 193581,$$
$$C_2 = 12, \quad C_3 = 624, \quad C_4 = 318240.$$

In the general case the following is valid

$$A_m < C_m$$

for all $m \ge 2$. We shall prove this inequality. It is obvious that it is sufficient to prove that

$$\prod_{p=2}^{m}\Big((2p^p - 1)/2(p^p - 1)\Big) < 2.$$

Further, with a view to proving this last inequality, we write

$$\prod_{p=2}^{m}\Big((2p^p - 1)/2(p^p - 1)\Big) = \prod_{p=2}^{m}\Big(1 + \frac{1}{2(p^p - 1)}\Big) \le$$

[since $2(p^p - 1) > p^p \ge 2^p$]

$$\le \prod_{p=2}^{m}(1+p^{-p}) \le \prod_{p=2}^{m}(1+2^{-p}) \le$$

[since $1+x \le \exp(x)$, $x \ge 0$]

$$\le \exp\Big(\sum_{p=2}^{m} 2^{-p}\Big) \le \exp(1/2) < 2.$$

This gives the desired inequality.

Remark 2.5.3. In the lower bound of Theorem 2.5.4 we have the constant

$$B_m = 1\Big/\Big(\sum_{p=1}^{m} \frac{p^m}{(m-p)!\,p!}\Big).$$

V. de la Peña (1992) proved the same lower bound but with another constant

$$D_m = 1\Big/\Big(2^{2m-2}(m-1)!\Big).$$

Obviously

$$B_2 = \frac{1}{3}, \quad B_3 = \frac{1}{9}, \quad B_4 = \frac{3}{85},$$
$$D_2 = \frac{1}{4}, \quad D_3 = \frac{1}{32}, \quad D_4 = \frac{1}{384}.$$

In the general case the inequality

$$D_m < B_m$$

takes place as $m \ge 2$. We shall prove this inequality. It is sufficient to prove that

$$\sum_{p=1}^{m} \frac{p^m}{(m-p)!\,p!} < 2^{2m-2}(m-1)!\,.$$

We use induction on m. For $m = 2$ this inequality holds. Suppose, that the inequality is valid for $2, 3, \ldots, m$. Then we write

$$\sum_{p=1}^{m+1} \frac{p^{m+1}}{(m+1-p)!\,p!} = \sum_{p=1}^{m} \frac{p^{m+1}}{(m+1-p)!\,p!} + \frac{(m+1)^{m+1}}{(m+1)!} \le$$

[since $p \le m$ and $(m+1-p)! \ge (m-p)!$ for $1 \le p \le m$]

$$\le m \sum_{p=1}^{m} \frac{p^m}{(m-p)!\,p!} + \frac{(m+1)^m}{m!} \le$$

[using the induction hypothesis]

$$\le m\, 2^{2m-2}(m-1)! + \frac{(m+1)^m}{m!} \le$$

[by the Stirling inequality $m! > \sqrt{2\pi m}\, m^m e^{-m}$]

$$\leq 2^{2m-2}m! + \frac{e^{m+1}}{\sqrt{2\pi m}} \leq 2^{2m}\, m!\,.$$

This completes the proof of the required assertion.

Thus, the constants we get are better, than de la Peña's (1992) constants (see also the decoupling inequality for U-statistics in de la Peña *et al.* (1994)).

Corollary 2.5.2. *In the notation of Theorem* 2.5.4 *assume that the* Φ*'s satisfy the symmetry conditions and suppose* $\{\varepsilon_i^1\}_1^n$, $\{\varepsilon_i^2\}_1^n$, ... ,$\{\varepsilon_i^m\}_1^n$ *are independent copies of a sequence of independent Bernoulli random variables* $\{\varepsilon_i\}_1^n$ *with* $P(\varepsilon_i = 1) = P(\varepsilon_i = -1) = 1/2$, *where the* ε*'s are taken to be independent of the* X*'s. Then*

$$E\Psi\Big(B_m\Big\| \sum_{1\leq i_1\neq,\dots,\neq i_m\leq n} \Phi_{i_1\dots i_m}(X_{i_1}^1,\dots,X_{i_m}^m)\,\varepsilon_{i_1}^1\cdots\varepsilon_{i_m}^m\Big\|\Big) \leq$$

$$\leq E\Psi\Big(\Big\| \sum_{1\leq i_1\neq,\dots,\neq i_m\leq n} \Phi_{i_1\dots i_m}(X_{i_1}^1,\dots,X_{i_m}^1)\,\varepsilon_{i_1}^1\cdots\varepsilon_{i_m}^1\Big\|\Big) \leq$$

$$\leq E\Psi\Big(m^m\Big\| \sum_{1\leq i_1\neq,\dots,\neq i_m\leq n} \Phi_{i_1\dots i_m}(X_{i_1}^1,\dots,X_{i_m}^m)\,\varepsilon_{i_1}^1\cdots\varepsilon_{i_m}^m\Big\|\Big).$$

Proof. Taking $\Phi^m_{i_1,\dots,i_m}(X_{i_1}^1,\dots,X_{i_m}^1) \times \varepsilon_{i_1}^1\cdots\varepsilon_{i_m}^1$ instead of $\Phi^m_{i_1,\dots,i_m}(X_{i_1}^1,\dots,X_{i_m}^1)$ we get this result directly from Theorem 2.5.4.

Corollary 2.5.3. *Let in Theorem* 2.5.4 $N_n = \{n\}$, $\mathcal{F}_{i_1,\dots,i_m} = \mathcal{F}$, $\Phi^n_{i_1,\dots,i_m} = \Phi/\binom{n}{m}$, $\mathcal{F} = \left\{\Phi/\binom{n}{m}\right\}$, *where kernel* Φ *is a symmetric function with rank* m. *Then for UB-statistics* (1.1.1) *the following inequalities are valid:*

$$E\Psi(B_m\|U_n^{\text{dec}}\|) \leq E\Psi(\|U_n\|) \leq E\Psi(m^m\|U_n^{\text{dec}}\|),$$

where $\Psi(x)$ *is any convex increasing function.*

Corollary 2.5.4. *If kernel* Φ *has rank* r, $1 \leq r \leq m$, *then in the notation of Corollary* 2.5.3 *the following inequality is valid:*

$$E\Psi(\|U_n - \theta\|) \leq (m-r+1)^{-1}\sum_{c=r}^{m} E\Psi\Big(\binom{m}{c}c^c\|U_{nc}^{\text{dec}}\|\Big),$$

where U_{nc}^{dec} *is a decoupled version of* U_{nc} *defined in* (1.1.11).

Proof. By the Hoeffding representation (1.1.12)

$$E\Psi(\|U_n - \theta\|) \leq E\Psi\Big(\sum_{c=r}^{m}\binom{m}{c}\|U_{nc}\|\Big) \leq$$

[by the convexity of Ψ]

$$\leq (m-r+1)^{-1}\sum_{c=r}^{m} E\Psi\Big(\binom{m}{c}\|U_{nc}\|\Big) \leq$$

[because of the upper bound of Corollary 2.5.3]

$$\leq (m-r+1)^{-1}\sum_{c=r}^{m} E\Psi\Big(\binom{m}{c}c^c\|U_{nc}^{\mathrm{dec}}\|\Big).$$

Theorem 2.5.5. *For $m \geq 2$ let $\Phi_{i_1,\ldots,i_m}$ be an array of functions from R^m into R. Then for every finite t the following inequality takes place:*

$$E\exp\Big(t\sum\nolimits_{1\leq i_1<\cdots<i_m\leq n}\Phi_{i_1,\ldots,i_m}(X_{i_1},\ldots,X_{i_m})\Big) \leq$$
$$\leq \Big(E\exp\Big(2^{m-1}t\sum\nolimits_{1\leq i_1<\cdots<i_m\leq n}\Phi_{i_1,\ldots,i_m}(X_{i_1}^1,\ldots,X_{i_m}^m)\Big)\Big)^{1/2^{m-1}}$$

Proof. For the proof of this theorem see V. de la Peña (1994).

Multi-sample *UB*-statistics of arbitrary degree. Now we consider decoupling inequalities for *UB*-statistics of type (1.2.1). With the preceding notation of Section 1.2, we introduce the independent copies $\{X_{j1}^r,\ldots,X_{jn_j}^r\}_{r=1}^{m_j}$ of the sequences of independent random variables $X_{j1},\ldots,X_{jn_j}$, $j=1,\ldots,c$, respectively. Then, a decoupled version of $U_{\bar{n}}$ in (1.2.1) is

$$U_{\bar{n}}^{\mathrm{dec}} = \prod_{j=1}^{c}\binom{n_j}{m_j}^{-1}\sum\Phi(X_{ji_{jr}}^r,\ r=1,\ldots,m_j,\ j=1,\ldots,c), \qquad (2.5.17)$$

where the summation is carried out over all $1 \leq i_{j1} < \cdots < i_{jm_j} \leq n_j$, $m_j \leq n_j$, $j=1,\ldots,c$.

·In the following theorem we restrict only the upper bound in the decoupling inequalities for these *UB*-statistics.

Theorem 2.5.6. *Let for any convex increasing function Ψ :*

$$E\Psi(\|\Phi\|) < \infty.$$

Then the following upper bound is valid:

$$E\Psi(\|U_{\bar{n}}-\theta\|) \leq b_c^{-1}\sum_{d=r}^{m}\ \sum_{d_1+\cdots+d_c=d} E\Psi\Big(\prod_{j=1}^{c}\binom{m_j}{d_j}d_j^{d_j}\|U_{\bar{n}\bar{d}}^{\mathrm{dec}}\|\Big), \qquad (2.5.18)$$

where $U_{\bar{n}\bar{d}}^{\mathrm{dec}}$ is a decoupled version of $U_{\bar{n}\bar{d}}$, defined in (1.2.5) *and*

$$b_c = \sum_{d=r}^{m}\beta_d.$$

In addition β_d was determined in Corollary 2.2.4.

In particular, if rank vector $\bar{r} = \bar{m}$, *then inequality* (2.5.18) *is reduced to*

$$E\Psi(\|U_{\bar{n}}\|) \leq E\Psi\Big(\prod_{j=1}^{c} m_j^{m_j} \|U_{\bar{n}}^{\mathrm{dec}}\|\Big). \tag{2.5.19}$$

Proof. From the Hoeffding representation (1.2.8) and the convexity of Ψ, it follows that

$$E\Psi(\|U_{\bar{n}} - \theta\|) \leq b_c^{-1} \sum_{d=r}^{m} \sum_{d_1+\cdots+d_c=d} E\Psi\Big(\prod_{j=1}^{c} \binom{m_j}{d_j} \|U_{\bar{n}\bar{d}}\|\Big) \leq$$

[by the upper bound of Corollary 2.5.3, applied c times for conditional expectations (see the proof of Theorem 2.2.3)]

$$\leq b_c^{-1} \sum_{d=r}^{m} \sum_{d_1+\cdots+d_c=d} E\Psi\Big(\prod_{j=1}^{c} \binom{m_j}{d_j} d_j^{d_j} \|U_{\bar{n}\bar{d}}^{\mathrm{dec}}\|\Big).$$

This proves (2.5.18). Inequality (2.5.19) is an obvious consequence of (2.5.18).

Note, that it is possible to formulate decoupling inequalities for multi-sample *UB*-statistics with the same generality as in Theorem 2.5.4 for one-sample *UB*-statistics with two-sided bounds.

Symmetric statistics. Now we give a decoupling inequality for symmetric statistics that were considered in Section 1.4.

Theorem 2.5.7. *Let*

$$E\Psi(\|S\|) < \infty$$

for any convex increasing function Ψ. *Then the following upper bound holds:*

$$E\Psi(\|S - ES\|) \leq n^{-1} \sum_{c=1}^{n} E\Psi\Big(c^c \Big\| \sum_{1 \leq i_1 < \cdots < i_c \leq n} h_{nc}(X_{i_1}^1, \ldots, X_{i_c}^c)\Big\|\Big).$$

Proof. This inequality follows from (1.4.2) and the preceding arguments.

2.6 Hypercontractive method in moment inequalities

The hypercontractive method is useful in comparing the moments of sums of random variables, and if it is applicable, the constants in the resulting inequalities are much better than those that are obtainable with the help of other approaches, applied for this comparing.

Hypercontractive domination.

Definition 2.6.1. Let $0 < q < p < \infty$. For a couple of random variables X and Y with values in a Banach space B we shall say that X is $[p,q]$-*hypercontractively dominated* by Y ($X \prec_{[p,q]} Y$, in short) if, for every $x \in B$

$$(E\|x+X\|^p)^{1/p} \le (E\|x+Y\|^q)^{1/q}.$$

The usefulness of the hypercontractive domination relation is contained in the fact that it is inherited by sums from their independent summands. To prove that this domination is hereditary in this sense we will need the convexity-type inequality

$$\Big(\int_S\Big(\int_T |f(s,t)|^q\,\mu(dt)\Big)^{p/q}\nu(ds)\Big)^{1/p} \le \Big(\int_T\Big(\int_S |f(s,t)|^p\,\nu(ds)\Big)^{q/p}\mu(dt)\Big)^{1/q}, \tag{2.6.1}$$

which holds for every positive measure space $(S, B(S), \mu)$, $(T, B(T), \nu)$, and any measurable function $f(s,t)$ whenever $0 < q < p < \infty$. Inequality (2.6.1) can easily be seen since, letting $r = p/q > 1$, we have

$$\Big(\int_S\Big(\int_T |f(s,t)|^q\,\mu(dt)\Big)^{r}\Big)^{1/r} = \sup_g \int_S\int_T |f(s,t)|^q\,\mu(dt)\,g(s)\,\nu(ds) \le$$

[here the supremum is taken over g such that
$\int_S |g(s)|^{r'}\,\nu(ds) \le 1$ and $1/r + 1/r' = 1$]

$$\le \int_T\Big(\sup_g \int_S |f(s,t)|^q g(s)\,\nu(ds)\Big)\mu(dt) = \int_T\Big(\int_S |f(s,t)|^{qr}\,\nu(ds)\Big)^{1/r}\mu(dt).$$

If $q < p$ and (X_1, Y_1) and (X_2, Y_2) are independent random vectors such that, for each $i = 1, 2$

$$X_i \prec_{[p,q]} Y_i\,,$$

then the hypercontractive property of pairs X_1, Y_1 and X_2, Y_2 holds:

$$X_1 + X_2 \prec_{[p,q]} Y_1 + Y_2\,.$$

To prove this, we put $E^i = E(\cdot\,|X_i, Y_i)$, and applying inequality (2.6.1) combined with Fubini's theorem, we write

$$\begin{aligned}(E\|x+X_1+X_2\|^p)^{1/p} &= (E^1E^2\|x+X_1+X_2\|^p)^{1/p} \le \\ \le (E^1(E^2\|x+Y_1+X_2\|^q)^{p/q})^{1/p} &\le (E^2(E^1\|x+Y_1+X_2\|^p)^{q/p})^{1/q} \le \\ \le (E^2E^1\|x+Y_1+Y_2\|^q)^{1/q} &= (E\|x+Y_1+Y_2\|^q)^{1/q}.\end{aligned}$$

Thus, for a sequence of independent couples of random variables such that $X_i \prec_{[p,q]} Y_i$, $i = 1, \dots, n$, we have the hypercontractive property

$$\sum_{i=1}^n X_i \prec_{[p,q]} \sum_{i=1}^n Y_i\,.$$

In particular, we obtain the following comparison inequality for moments. If there exists a constant $\sigma > 0$ such that $\sigma X_i \prec_{[p,q]} X_i$, $i = 1, \ldots, n$, then

$$\Big(E\Big\|\sum_{i=1}^{n} X_i\Big\|^p\Big)^{1/p} \leq \sigma^{-1}\Big(E\Big\|\sum_{i=1}^{n} X_i\Big\|^q\Big)^{1/q}.$$

This observation justifies the following definition.

Definition 2.6.2. A real random variable θ is said to be in the *hypercontractive class* $HC(p,q,\sigma,B)$, if $\sigma y\theta \prec_{[p,q]} y\theta$ for all $y \in B$, or equivalently,

$$(E\|x + \sigma y\theta\|^p)^{1/p} \leq (E\|x + y\theta\|^q)^{1/q} \tag{2.6.2}$$

for all $x, y \in B$.

Remark 2.6.1. It follows easily from the preceding arguments that whenever (2.6.2) holds, and if $\{\theta_i\}$ is a sequence of independent copies of θ, then

$$\Big(E\Big\|\sum_{i=1}^{n} y_i\theta_i\Big\|^p\Big)^{1/p} \leq \sigma^{-1}\Big(E\Big\|\sum_{i=1}^{n} y_i\theta_i\Big\|^q\Big)^{1/q} \tag{2.6.3}$$

for all $y_i \in B$. Also, if $\xi_1, \xi_2, \ldots$ are random variables such that $\xi_k \in HC(p,q,\sigma,B)$ for $k = 1, 2, \ldots$ and $\xi_1, \xi_2, \ldots$ converge in distribution to ξ then $\xi \in HC(p,q,\sigma,B)$.

Hypercontractivity of Bernoulli random variable. The Bernoulli random variables present the simplest but far not trivial, example of hypercontractive random variables. It is very important in those methods, which use symmetrization techniques.

Lemma 2.6.1. *If ε is a Bernoulli random variable with distribution $P(\varepsilon = \pm 1) = 1/2$, and $1 < q < p < \infty$ then for any Banach space B*

$$\varepsilon \in HC\Big(p, q, \sqrt{(q-1)/(p-1)}, B\Big).$$

From Lemma 2.6.1 and Remark 2.6.1 it follows that the hypercontractivity of Gaussian random variable takes place with the constant, identical to the hypercontractivity constant of the Bernoulli random variables.

Corollary 2.6.1. *Let $1 < q < p < \infty$. If τ is an $N(0,1)$ Gaussian random variable then, for any Banach space B,*

$$\tau \in HC\Big(p, q, \sqrt{(q-1)/(p-1)}, B\Big).$$

Directly from (2.6.3) and Lemma 2.6.1 follows the following version of Khintchine's inequality: for x_i in any Banach space

$$\left(E\Big\|\sum_{i=1}^{n} x_i\varepsilon_i\Big\|^p\right)^{1/p} \le \left(\frac{p-1}{q-1}\right)^{1/2}\left(E\Big\|\sum_{i=1}^{n} x_i\varepsilon_i\Big\|^q\right)^{1/q} \tag{2.6.4}$$

for all $1 < q \le p < \infty$.

The same arguments mentioned in the proof of (2.6.3) yield Borell inequality

$$\left(E\Big\|\sum_{1\le i_1\ne\cdots\ne i_m\le n} x_{i_1,\ldots,i_m}\,\varepsilon_{i_1}\cdots\varepsilon_{i_m}\Big\|^p\right)^{1/p} \le$$
$$\le \left(\frac{p-1}{q-1}\right)^{m/2}\left(E\Big\|\sum_{1\le i_1\ne\cdots\ne i_m\le n} x_{i_1,\ldots,i_m}\,\varepsilon_{i_1}\cdots\varepsilon_{i_m}\Big\|^q\right)^{1/q} \tag{2.6.5}$$

for all elements $x_{i_1,\ldots,i_m}$ in a Banach space and for all $1 < q \le p < \infty$.

Hypercontractivity in action: sums of independent random variables. We will apply hypercontraction ideas and domination relations in comparing the moments of different orders for sums of independent random variables. In the first place we are interested in the following result.

Theorem 2.6.1. *Let $p \ge 1$. If $X_1, \ldots, X_n$ is a sequence of independent, zero-mean random variables with values in Banach space B, and $\varepsilon_1, \ldots, \varepsilon_n$ are independent Bernoulli random variables, independent of $X_1, \ldots, X_n$, then*

$$6^{-p}E\left(E_\varepsilon\Big\|\sum_{i=1}^{n}\varepsilon_i X_i\Big\|^2\right)^{p/2} \le E\Big\|\sum_{i=1}^{n} X_i\Big\|^p \le$$
$$\le 2^p\max(1,(p-1)^{p/2})\,E\left(E_\varepsilon\Big\|\sum_{i=1}^{n}\varepsilon_i X_i\Big\|^2\right)^{p/2}. \tag{2.6.6}$$

Proof. Let $p \ge 2$. By Lemma 2.4.5

$$E\Big\|\sum_{i=1}^{n} X_i\Big\|^p \le 2^p\Big\|\sum_{i=1}^{n}\varepsilon_i X_i\Big\|^p \le$$

[by using (2.6.4) with $q = 2$, $2 \le p < \infty$]

$$\le 2^p(p-1)^{p/2}E\left(E_\varepsilon\Big\|\sum_{i=1}^{n}\varepsilon_i X_i\Big\|^2\right)^{p/2}$$

which is an upper bound in (2.6.6). For proof of the lower bound we first write at first

$$E_\varepsilon\Big\|\sum_{i=1}^{n}\varepsilon_i X_i\Big\|^2 = E_\varepsilon\Big\|\sum_{i=1}^{n}\varepsilon_i X_i\Big\|^{\frac{2}{3}+\frac{4}{3}} \le$$

[by Hölder's inequality]

$$\leq \Big(E_\varepsilon\Big\|\sum_{i=1}^n \varepsilon_i X_i\Big\|\Big)^{2/3}\Big(E_\varepsilon\Big\|\sum_{i=1}^n \varepsilon_i X_i\Big\|^4\Big)^{1/3} \leq$$

[by using (2.6.4) with $q = 2$, $p = 4$]

$$\leq \Big(E_\varepsilon\Big\|\sum_{i=1}^n \varepsilon_i X_i\Big\|\Big)^{2/3}\Big(3E_\varepsilon\Big\|\sum_{i=1}^n \varepsilon_i X_i\Big\|^2\Big)^{2/3}.$$

Hence,

$$E_\varepsilon\Big\|\sum_{i=1}^n \varepsilon_i X_i\Big\|^2 \leq 3^2\Big(E_\varepsilon\Big\|\sum_{i=1}^n \varepsilon_i X_i\Big\|\Big)^2. \tag{2.6.7}$$

Further taking account of (2.6.7) for any $p \geq 1$, we write

$$\begin{aligned} E\Big(E_\varepsilon\Big\|\sum_{i=1}^n \varepsilon_i X_i\Big\|^2\Big)^{p/2} &\leq 3^p E\Big(E_\varepsilon\Big\|\sum_{i=1}^n \varepsilon_i X_i\Big\|\Big)^p \quad \text{[by Jensen's inequality]} \\ &\leq 3^p E\Big\|\sum_{i=1}^n \varepsilon_i X_i\Big\|^p \quad \text{[by Lemma 2.4.5]} \\ &\leq 6^p E\Big\|\sum_{i=1}^n X_i\Big\|^p. \end{aligned}$$

This completes the proof of the lower bound of inequality (2.6.6) for all $p \geq 1$.

Now let $1 \leq p \leq 2$. According to the preceding arguments we obtain the upper bound:

$$E\Big\|\sum_{i=1}^n X_i\Big\|^p \leq 2^p\, E\Big\|\sum_{i=1}^n \varepsilon_i X_i\Big\|^p \leq 2^p\, E\Big(E_\varepsilon\Big\|\sum_{i=1}^n \varepsilon_i X_i\Big\|^2\Big)^{p/2}.$$

This proves the theorem.

Corollary 2.6.2. *Let $B = H$ be a Hilbert space. Then from* (2.6.6) *it follows that*

$$6^{-p}E\Big(\sum_{i=1}^n \|X_i\|^2\Big)^{p/2} \leq E\Big\|\sum_{i=1}^n X_i\Big\|^p \leq 2^p \max(1, (p-1)^{p/2})E\Big(\sum_{i=1}^n \|X_i\|^2\Big)^{p/2}$$

for all $p \geq 1$.

Proof. Note that in (2.6.6)

$$E_\varepsilon\Big\|\sum_{i=1}^n \varepsilon_i X_i\Big\|^2 = \sum_{i=1}^n \|X_i\|^2.$$

Corollary 2.6.3. *Let $p \geq 1$ and $B = H$ be a Hilbert space. Then*

$$6^{-1}\max\Big\{\Big(\sum_{i=1}^n E\|X_i\|^2\Big)^{p/2},\ \sum_{i=1}^n E\|X_i\|^p\Big\} \leq E\Big\|\sum_{i=1}^n X_i\Big\|^p,$$

$$E\Big\|\sum_{i=1}^{n} X_i\Big\|^p \le 2^p \max(1,(p-1)^{p/2})\, n^{\frac{p}{2}\vee 1-1} \sum_{i=1}^{n} E\|X_i\|^p. \tag{2.6.8}$$

Proof. By the results of Jonson *et al.* (1985) for all independent nonnegative random variables $\xi_1,\dots,\xi_n$ with finite p-th moment, $1 \le p < \infty$, the following is valid

$$\max\Big\{\Big(\sum_{i=1}^{n} E\xi_i\Big)^p, \sum_{i=1}^{n} E\xi_i^p\Big\} \le E\Big(\sum_{i=1}^{n} \xi_i\Big)^p \le \Big(\frac{2p}{\log p}\Big)^p \max\Big\{\Big(\sum_{i=1}^{n} E\xi_i\Big)^p, \sum_{i=1}^{n} E\xi_i^p\Big\}.$$

Let $\xi_i = \|X_i\|^2$. Then the first inequality (2.6.8) follows from the lower bound of this inequality and Corollary 2.6.2. The second inequality in (2.6.8) follows from the upper bound of Corollary 2.6.2 and inequality (2.1.9).

Hypercontractivity in action: *UB*-statistics. Now we will apply the hypercontractive majorization in comparing moments of *UB*-statistics. We have the following result.

Theorem 2.6.2. *Let $X_1,\dots,X_n$ be a sequence of independent random variables in a measurable space $(X,\mathfrak{X})$ having the same P and suppose $\{X_1^j,\dots,X_n^j\}_{j=1}^m$ are independent copies of $\{X_i\}_1^n$. Let $\{\varepsilon_1^j,\dots,\varepsilon_n^j\}_{j=1}^m$ be independent copies of a sequence of independent Bernoulli random variables $\varepsilon_1,\dots,\varepsilon_n$ with $P(\varepsilon_i=1)=P(\varepsilon_i=-1)=1/2$, where the ε's are taken to be independent of the X's. Suppose that kernel Φ is a symmetric function, having rank m and $E\|\Phi\|^p<\infty$, $p\ge 1$. Then*

$$\Big(\frac{B_m}{6m}\Big) E\Big(E_\varepsilon\Big\| \sum_{1\le i_1\ne\cdots\ne i_m\le n} \Phi(X_{i_1}^1,\dots,X_{i_m}^m)\,\varepsilon_{i_1}\cdots\varepsilon_{i_m}\Big\|^2\Big)^{p/2} \le$$

$$\le E\Big\| \sum_{1\le i_1\ne\cdots\ne i_m\le n} \Phi(X_{i_1},\dots,X_{i_m})\Big\|^p \le$$

$$\le (2m^2)^{mp}(1\vee(p-1)^{mp/2})\, E\Big(E_\varepsilon\Big\| \sum_{1\le i_1\ne\cdots\ne i_m\le n} \Phi(X_{i_1}^1,\dots,X_{i_m}^m)\,\varepsilon_{i_1}\cdots\varepsilon_{i_m}\Big\|^2\Big)^{p/2}. \tag{2.6.9}$$

Proof. Let in Corollary 2.5.3, $\Psi(x)=x^p$, $x\ge 0$, $p\ge 1$, so that

$$E\Big\| \sum_{1\le i_1\ne\cdots\ne i_m\le n} \Phi(X_{i_1},\dots,X_{i_m})\Big\|^p \le m^{mp} E\Big\| \sum_{1\le i_1\ne\cdots\ne i_m\le n} \Phi(X_{i_1}^1,\dots,X_{i_m}^m)\Big\|^p \le$$

[Lemma 2.4.5 applied m times]

$$\le (2m)^{mp} E\Big\| \sum_{1\le i_1\ne\cdots\ne i_m\le n} \Phi(X_{i_1}^1,\dots,X_{i_m}^m)\,\varepsilon_{i_1}^1\cdots\varepsilon_{i_m}^m\Big\|^p \le$$

[by Theorem 2.5.4, conditioning on $\{X_1^j,\dots,X_n^j\}$, $j=1,\dots,m$]

$$\le (2m^2)^{mp} E\Big\| \sum_{1\le i_1\ne\cdots\ne i_m\le n} \Phi(X_{i_1}^1,\dots,X_{i_m}^m)\,\varepsilon_{i_1}\cdots\varepsilon_{i_m}\Big\|^p \le$$

[by using Hölder's inequality for $1 \le p \le 2$ and

Borell's inequality (2.6.5) with $q = 2$, $p \ge 2$]

$$\le (2m^2)^{mp}(1 \vee (p-1)^{mp/2})\, E\Big(E_\varepsilon\Big\| \sum_{1\le i_1\ne\cdots\ne i_m\le n} \Phi(X^1_{i_1},\ldots,X^m_{i_m})\,\varepsilon_{i_1}\cdots\varepsilon_{i_m}\Big\|^2\Big)^{p/2}.$$

This proves the upper bound in (2.6.9).

Next we consider the lower bound in (2.6.9). In the first place we note that by analogy with (2.6.7) the following takes place

$$E_\varepsilon\Big\| \sum_{1\le i_1\ne\cdots\ne i_m\le n} \Phi(X^1_{i_1},\ldots,X^m_{i_m})\,\varepsilon_{i_1}\cdots\varepsilon_{i_m}\Big\|^2 \le$$

$$\le 3^{2m}\Big(E_\varepsilon\Big\| \sum_{1\le i_1\ne\cdots\ne i_m\le n} \Phi(X^1_{i_1},\ldots,X^m_{i_m})\Big\|\Big)^2.$$

Taking account of this inequality we write for any $p \ge 1$

$$E\Big(E_\varepsilon\Big\| \sum_{1\le i_1\ne\cdots\ne i_m\le n} \Phi(X^1_{i_1},\ldots,X^m_{i_m})\,\varepsilon_{i_1}\cdots\varepsilon_{i_m}\Big\|^2\Big)^{p/2} \le$$

$$\le 3^{mp}\, E\Big(E_\varepsilon\Big\| \sum_{1\le i_1\ne\cdots\ne i_m\le n} \Phi(X^1_{i_1},\ldots,X^m_{i_m})\,\varepsilon_{i_1}\cdots\varepsilon_{i_m}\Big\|\Big)^p \le$$

[by Jensen's inequality]

$$\le 3^{mp}\, E\Big\| \sum_{1\le i_1\ne\cdots\ne i_m\le n} \Phi(X^1_{i_1},\ldots,X^m_{i_m})\,\varepsilon_{i_1}\cdots\varepsilon_{i_m}\Big\|^p \le$$

[by Theorem 2.5.4, conditioning on $\{X^j_1,\ldots,X^j_n\}$, $j = 1,\ldots,m$]

$$\le (3m)^{mp}\, E\Big\| \sum_{1\le i_1\ne\cdots\ne i_m\le n} \Phi(X^1_{i_1},\ldots,X^m_{i_m})\,\varepsilon^1_{i_1}\cdots\varepsilon^m_{i_m}\Big\|^p \le$$

[Lemma 2.4.5 applied m times]

$$\le (6m)^{mp}\, E\Big\| \sum_{1\le i_1\ne\cdots\ne i_m\le n} \Phi(X^1_{i_1},\ldots,X^m_{i_m})\Big\|^p \le$$

[by the lower bound of Corollary 2.5.3 with $\Psi(x) = x^p$, $x \ge 0$, $p \ge 1$]

$$\le (6mB_m^{-1})^{mp}\, E\Big\| \sum_{1\le i_1\ne\cdots\ne i_m\le n} \Phi(X_{i_1},\ldots,X_{i_m})\Big\|^p.$$

This completes the proof.

Corollary 2.6.4. *Let $B = R$ or $B = H$ be a Hilbert space. Then*

$$\Big(\frac{B_m}{6m}\Big)^{mp} E\Big(\sum_{1\le i_1\ne\cdots\ne i_m\le n} \|\Phi(X^1_{i_1},\ldots,X^m_{i_m})\|^2\Big)^{p/2} \le$$

$$\le E\Big\| \sum_{1\le i_1\ne\cdots\ne i_m\le n} \Phi(X_{i_1},\ldots,X_{i_m})\Big\|^p \le$$

$$\le (2m^2)^{mp}(1 \vee (p-1))^{mp/2}\, E\Big(\sum_{1\le i_1\ne\cdots\ne i_m\le n} \|\Phi(X^1_{i_1},\ldots,X^m_{i_m})\|^2\Big)^{p/2}. \quad (2.6.10)$$

Corollary 2.6.5. *Let $B = R$ or $B = H$ be a Hilbert space. Then under the conditions of Theorem 2.6.2 the following inequality for any $p \geq 1$ is valid*

$$E\Big\| \sum_{1\leq i_1\neq\cdots\neq i_m\leq n} \Phi(X_{i_1},\ldots,X_{i_m})\Big\|^p \leq$$
$$\leq (2m^2)^{mp}(1 \vee (p-1)^{mp/2})(n^{[m]})^{1\vee(p/2)}\, E\|\Phi(X_1,\ldots,X_m)\|^p \quad (2.6.11)$$

where $n^{[m]} = n(n-1)\cdots(n-m+1)$.

Proof. On the right-hand side of (2.6.10)

$$\Big(\sum_{1\leq i_1\neq\cdots\neq i_m\leq n} \|\Phi(X^1_{i_1},\ldots,X^m_{i_m}\|^2\Big)^{p/2} \leq (n^{[m]})^{\frac{p}{2}\vee 1-1} \sum_{1\leq i_1\neq\cdots\neq i_m\leq n} \|\Phi(X^1_{i_1},\ldots,X^m_{i_m}\|^p.$$

From here (2.6.11) follows since $X_1,\ldots,X_n$ have identical distribution.

2.7 Moment inequalities in Banach spaces of type p

Definition 2.7.1. Let $1 \leq p \leq 2$ and $2 \leq q \leq \infty$. A real separable Banach space is said to be of *type p* if a constant $A = A(B,p)$ exists so that

$$E\Big\|\sum_{i=1}^n \varepsilon_i x_i\Big\|^p \leq A\sum_{i=1}^n \|x_i\|^p \quad (2.7.1)$$

for any $n \geq 1$ and any $x_1,\ldots,x_n \in B$.

It is said to be of *cotype q*, if a constant $K = K(B,q)$ exists so that

$$E\Big\|\sum_{i=1}^n \varepsilon_i x_i\Big\|^q \geq K\sum_{i=1}^n \|x_i\|^q$$

for any $n \geq 1$ and any $x_1,\ldots,x_n \in B$.

Every Banach space is of type 1 and of cotype $+\infty$. Banach space L^p is of type $\min(2,p)$ and of cotype $\max(2,p)$. It is well known that finite-dimensional spaces and Hilbert spaces are type 2 spaces.

Let $1 \leq p \leq 2$. Then the statement: *B is of type p* is equivalent to the statement: a constant $C = C(p,B)$ exists so that

$$E\Big\|\sum_{i=1}^n X_i\Big\|^p \leq C\sum_{i=1}^n E\|X_i\|^p \quad (2.7.2)$$

for all independent random variables $X_1,\ldots,X_n$ with zero-mean and finite p-th moment. Moreover, from Definition 2.7.1 and the upper bound of Lemma 2.4.5 for $\Psi(x) = x^p$ it follows that in (2.7.2), $C = 2^p A$.

Similarly inverse inequalities hold for Banach spaces of cotype q.

Theorem 2.7.1. *Let $X_1, \ldots, X_n$ be a sequence of independent random variables with identical distributions and suppose that kernel Φ is a symmetric B-valued function, with rank m and $E\|\Phi\|^q < \infty$, $q \geq 1$. If Banach space B is of type p, $1 \leq p \leq 2$, then*

$$E\Big\| \sum_{1\leq i_1 \neq \cdots \neq i_m \leq n} \Phi(X_{i_1}, \ldots, X_{i_m})\Big\|^q \leq$$
$$\leq (6A^{1/p}m^3)^{mq}(1 \vee (q-1)^{mq/2})(n^{[m]})^{1\vee(q/p)}\, E\|\Phi(X_1, \ldots, X_m)\|^q, \quad (2.7.3)$$

where $n^{[m]} = n(n-1)\cdots(n-m+1)$, and constant $A > 0$ from (2.7.1).

Proof. We will apply the upper bound in (2.6.10). At first we see that

$$\Big(E_\varepsilon\Big\| \sum_{1\leq i_1 \neq \cdots \neq i_m \leq n} \Phi(X^1_{i_1}, \ldots, X^m_{i_m})\,\varepsilon_{i_1}\cdots\varepsilon_{i_m}\Big\|^2\Big)^{q/2} \leq$$
$$\leq 3^{qm}\Big(E_\varepsilon\Big\| \sum_{1\leq i_1 \neq \cdots \neq i_m \leq n} \Phi(X^1_{i_1}, \ldots, X^m_{i_m})\,\varepsilon_{i_1}\cdots\varepsilon_{i_m}\Big\|^p\Big)^{q/p} \leq$$

[by Theorem 2.5.4, conditioning on $\{X^j_1, \ldots, X^j_n\}$, $j = 1, \ldots, m$]

$$\leq (3m)^{qm}\Big(E_\varepsilon\Big\| \sum_{1\leq i_1 \neq \cdots \neq i_m \leq n} \Phi(X^1_{i_1}, \ldots, X^m_{i_m})\,\varepsilon^1_{i_1}\cdots\varepsilon^m_{i_m}\Big\|^p\Big)^{q/p} \leq$$

[Inequality (2.7.1) applied m times]

$$\leq (3A^{1/p}m)^{mq}\Big(\sum_{1\leq i_1 \neq \cdots \neq i_m \leq n} \|\Phi(X^1_{i_1}, \ldots, X^m_{i_m})\|^p\Big)^{q/p} \leq$$

[by using inequality (2.2.3)]

$$\leq (3A^{1/p}m)^{mq}(n^{[m]})^{(1\vee(q/p))-1} \sum_{1\leq i_1 \neq \cdots \neq i_m \leq n} \|\Phi(X^1_{i_1}, \ldots, X^m_{i_m})\|^q.$$

From here and (2.6.9), (2.7.3) follows since $X_1, \ldots, X_n$ are identically distributed.

Corollary 2.7.1. *If a Banach space is of type 2 then for any $q \geq 1$*

$$E\Big\| \sum_{1\leq i_1 \neq \cdots \neq i_m \leq n} \Phi(X_{i_1}, \ldots, X_{i_m})\Big\|^q \leq$$
$$\leq (6A^{1/2}m^3)^{mq}(1 \vee (q-1)^{mq/2})(n^{[m]})^{1\vee(q/2)} E\|\Phi(X_1, \ldots, X_m)\|^q.$$

Chapter 3

Law of large numbers

3.1 One-sample *UB*-statistics

Convergence with probability 1. Consider U-statistics (1.1.1) and let $\mathfrak{B}_n$ be an σ-algebra generated by an unordered set of random variables $(X_1, \dots, X_n)$ and by a sequence $X_{n+1}, X_{n+2}, \dots$, i.e. $\mathfrak{B}_n = \sigma\{\omega\colon (X_1, \dots, X_n), X_{n+1}, X_{n+2}, \dots\}$.

Theorem 3.1.1. *Assume that the B-valued kernel Φ is such that $E\|\Phi\| < \infty$. Then the convergence*

$$U_n \to \theta \quad a.s.$$

as $n \to \infty$, and the L^1-convergence take place, i.e.

$$E\|U_n - \theta\| \to 0 .$$

Proof. Denote $\mathfrak{B}_\infty = \bigcap_{n=m}^{\infty} \mathfrak{B}_n$. By virtue of the representation (see Lemma 1.1.3)

$$U_n = E(\Phi(X_1, \dots, X_n)|\mathfrak{B}_n)$$

and the theorem on the convergence of a B-valued reversed martingale (Chatterji (1968)), the limit

$$U_n \to E(\Phi|\mathfrak{B}_\infty)$$

as $n \to \infty$, exists with probability 1 and in L^1. The random variables $X_1, \dots, X_n$ are independent and have the same distribution. Therefore, according to the Hewitt–Savage theorem (the Hewitt–Savage law of 0 or 1), the σ-algebra $\mathfrak{B}_\infty = \{\varnothing, \Omega\}$ is trivial. Hence, the limit under consideration is equal to

$$E(\Phi|\mathfrak{B}_\infty) = E\Phi(X_1, \dots, X_m) = \theta .$$

Remark 3.1.1. By Theorem 2.3.3, for any $\varepsilon > 0$ we have

$$P(\sup_{k \geq n} \|U_k - \theta\| > \varepsilon) \leq \varepsilon^{-1} E\|U_n - \theta\| .$$

This inequality combined with the Borel–Cantelli lemma contains the relation between convergence with probability 1 and convergence in the sense of L^1 for *UB*-statistics.

In the following we will consider the various forms of the laws of large numbers for *UB*-statistics assuming some additional geometric properties of Banach space B.

Convergence in probability. Set

$$S_n = \sum_{1 \le i_1 < \ldots < i_m \le n} \Phi(X_{i_1}, \ldots , X_{i_m}) .$$

Note, that the usual *U*-statistics (1.1.1) can be obtained from S_n by letting $\Phi/\binom{n}{m}$, i.e. $U_n = \binom{n}{m}^{-1} S_n$.

From Theorem 3.1.1 it follows that

$$U_n \xrightarrow{P} \theta \quad \text{as} \quad n \to \infty$$

for any Bochner integrable kernel Φ.

We extend this statement in the following way.

Theorem 3.1.2. *Let B be a separable Banach space of type 2. Let $\Phi \colon X^m \to B$ be a symmetric Bochner integrable kernel of rank m, satisfying the condition*

$$n^{m(p-1)/p} E\|\Phi\| I(\|\Phi\| > n^{m/p}) \to 0 \tag{3.1.1}$$

for some p, $1 < p < 2$, as $n \to \infty$.

Then

$$n^{-m/p} S_n \xrightarrow{P} 0 \tag{3.1.2}$$

is valid.

Remark 3.1.2. Condition (3.1.1) is equivalent to the condition

$$t^p P(\|\Phi\| > t) \to 0 \quad \text{as} \quad t \to \infty . \tag{3.1.3}$$

Indeed, by Fubini's theorem in (3.1.1),

$$n^{m(p-1)/p} E\|\Phi\| I(\|\Phi\| > n^{m/p}) = n^m P(\|\Phi\| > n^{m/p}) + n^{m(p-1)/p} \int_{n^{m/p}}^{\infty} P(\|\Phi\| > t)\, dt .$$

Hence, if condition (3.1.1) takes place, then

$$n^m P(\|\Phi\| > n^{m/p}) \to 0$$

and

$$n^{m(p-1)/p} \int_{n^{m/p}}^{\infty} P(\|\Phi\| > t)\,dt \to 0$$

as $n \to \infty$. If we choose $n = [t]^{p/m}$ for $t \geq 1$ where $[t]$ is an integral part of t, then

$$\begin{aligned} t^p P(\|\Phi\| > t) &\leq ([t]+1)^p P(\|\Phi\| > [t]) \\ &\leq 2^p n^m P(\|\Phi\| > n^{m/p}) \\ &\to 0 \quad \text{as} \quad t \to \infty, \end{aligned}$$

that is (3.1.3) follows from (3.1.1). Also (3.1.1) follows from (3.1.3). Note that since $E\|\Phi\| < \infty$ and the formula

$$E\|\Phi\| = \int_0^{\infty} P(\|\Phi\| > t)\,dt$$

takes place, then there exists t_n such that $t_n > 2n^{m/p}$ and for $1 < p < 2$

$$n^{m(p-1)/p} \int_{t_n}^{\infty} P(\|\Phi\| > t)\,dt \to 0 \quad \text{as} \quad n \to \infty\,.$$

Further, by condition (3.1.3)

$$n^{m(p-1)/p} \int_{n^{m/p}}^{t_n} P(\|\Phi\| > t)\,dt \leq \rho_n n^{m(p-1)/p} \int_{n^{m/p}}^{t_n} t^{-p}\,dt \leq \rho_n (p-1)^{-1},$$

where $\rho_n = \max_{n^{m/p} \leq t \leq t_n} (t^p P(\|\Phi\| > t)) \to 0$, as $n \to \infty$.

It means that (3.1.1) follows from (3.1.3) for any Bochner integrable kernel Φ.

Proof of Theorem 3.1.2. At first we write with the help of (1.1.4)

$$\Phi(x_1, \dots, x_m) = \int_X \cdots \int_X \Phi(y_1, \dots, y_m) \prod_{s=1}^m \delta_{x_s}(dy_s) =$$

[by the property of degeneracy Φ]

$$= \int_X \cdots \int_X \Phi(y_1, \dots, y_m) \prod_{s=1}^m (\delta_{x_s}(dy_s) - P(dy_s)) =$$

$$= \Phi_1(x_1, \dots, x_m) + \bar{\Phi}_1(x_1, \dots, x_m),$$

where

$$\Phi_1(x_1, \dots, x_m) =$$

$$= \int_X \cdots \int_X \Phi(y_1, \dots, y_m) I(\|\Phi(y_1, \dots, y_m)\| \leq n^{m/p}) \prod_{s=1}^m (\delta_{x_s}(dy_s) - P(dy_s)),$$

$$\bar{\Phi}_1(x_1, \ldots, x_m) = \Phi(x_1, \ldots, x_m) - \Phi_1(x_1, \ldots, x_m) =$$
$$= \int_X \cdots \int_X \Phi(y_1, \ldots, y_m) I(\|\Phi(y_1, \ldots, y_m)\| > n^{m/p}) \prod_{s=1}^{m} (\delta_{x_s}(dy_s) - P(dy_s)).$$

The functions $\Phi_1(x_1, \ldots, x_m)$ and $\bar{\Phi}_1(x_1, \ldots, x_m)$ possess the property of complete degeneracy (1.1.7).

Let us represent S_n in the form of a sum, namely

$$S_n = S_n(\Phi_1) + S_n(\bar{\Phi}_1), \tag{3.1.4}$$

where

$$S_n(\Phi_1) = \sum_{1 \le i_1 < \ldots < i_m \le n} \Phi(X_{i_1}, \ldots, X_{i_m}),$$
$$S_n(\bar{\Phi}_1) = \sum_{1 \le i_1 < \ldots < i_m \le n} \bar{\Phi}_1(X_{i_1}, \ldots, X_{i_m}).$$

Next, for any fixed $t > 0$ we have

$$P(\|n^{-m/p} S_n\| > t) \le$$
$$\le P\Big(2\|n^{-m/p} S_n(\Phi_1)\| > t\Big) + P\Big(2\|n^{-m/p} S_n(\bar{\Phi}_1)\| > t\Big) \le$$

[by Chebyshev's inequality]

$$\le 2t^{-1} n^{-m/p} E\|S_n(\bar{\Phi}_1)\| + 4t^{-2} n^{-2m/p} E\|S_n(\Phi_1)\|^2.$$

It is obvious that

$$E\|S_n(\bar{\Phi}_1)\| \le \frac{n^m}{m!} E\|\bar{\Phi}_1\| \le \frac{2^m}{m!} n^m E\|\Phi\| I(\|\Phi\| > n^{m/p}).$$

Hence by hypothesis (3.1.1)

$$n^{-m/p} E\|S_n(\bar{\Phi}_1)\| \to 0 \quad \text{as } n \to \infty.$$

Then by Corollary 2.7.1 we have

$$E\|S_n(\Phi_1)\|^2 \le \frac{1}{(m!)^2} (6A^{1/2} m^3)^{2m} n^m E\|\Phi_1\|^2 \le$$
$$\le \frac{1}{m!} (6(2A)^{1/2} m^3)^{2m} n^m E\|\Phi\|^2 I(\|\Phi\| \le n^{m/p}).$$

Integration by parts gives

$$n^{-\frac{m}{p}(2-p)} E\|\Phi\|^2 I(\|\Phi\| \le n^{m/p}) =$$
$$= -n^m P(\|\Phi\| > n^{m/p}) + 2n^{-\frac{m}{p}(2-p)} \int_0^{n^{m/p}} t P(\|\Phi\| > t)\, dt.$$

By hypothesis (3.1.1) $n^m P(\|\Phi\| > n^{m/p}) \to 0$ as $n \to \infty$.

Let $\delta_n = \max\{t^p P(\|\Phi\| > t) : n^{m(2-p)/(2p)} \le t \le n^{m/p}\}$ for $1 < p < 2$. Then

$$n^{-\frac{m}{p}(2-p)} \int_0^{n^{m/p}} tP(\|\Phi\| > t)\, dt =$$

$$= n^{-\frac{m}{p}(2-p)} \int_0^{n^{m(2-p)/(2p)}} tP(\|\Phi\| > t)\, dt + n^{-\frac{m}{p}(2-p)} \int_{n^{m(2-p)/(2p)}}^{n^{m/p}} tP(\|\Phi\| > t)\, dt \le$$

$$\le n^{-\frac{m}{2p}(2-p)} \int_0^{\infty} P(\|\Phi\| > t)\, dt + \delta_n n^{-\frac{m}{p}(2-p)} \int_{n^{m(2-p)/(2p)}}^{n^{m/p}} t^{-p+1}\, dt \le$$

$$\le n^{-\frac{m}{2p}(2-p)} E\|\Phi\| + (2-p)^{-1}\delta_n \to 0 \quad \text{as} \quad n \to \infty.$$

Thus by hypothesis (3.1.1)

$$n^{-2m/p} E\|S_n(\Phi_1)\|^2 \to 0 \quad \text{as} \quad n \to \infty$$

for any $1 < p < 2$.

This completes the proof of (3.1.2).

Convergence in L^p, $1 \le p < 2$.

Theorem 3.1.3. *Let B be a separable Banach space of type 2. Suppose that the kernel Φ possesses the property of complete degeneracy and $E\|\Phi\|^p < \infty$ for some p, $1 \le p < 2$.*

Then the convergence in mean of order p is valid:

$$E\|n^{-m/p} S_n\|^p \to 0 \quad \text{as} \quad n \to \infty\,.$$

Proof. The case $p = 1$ is contained in Theorem 3.1.1. We assume in the following that $1 < p < 2$. Using formula (3.1.4), we write the following inequality

$$E\|n^{-m/p} S_n\|^p \le 2^{p-1} E\|n^{-m/p} S_n(\Phi_1)\|^p + 2^{p-1} E\|n^{-m/p} S_n(\bar{\Phi}_1)\|^p.$$

By Corollary 2.7.1

$$E\|n^{-m/p} S_n(\bar{\Phi}_1)\|^p \le \frac{(6A^{1/2}m^3)^{mp}}{(m!)^p} E\|\bar{\Phi}_1\|^p \le$$

$$\le \frac{1}{m!} 2^m (6A^{1/2}m^3)^{mp} E\|\Phi\|^p I(\|\Phi\| > n^{m/p}),$$

i.e.

$$E\|n^{-m/p} S_n(\bar{\Phi}_1)\|^p \to 0 \quad \text{as} \quad n \to \infty$$

for all $1 < p < 2$.

By Hölder's inequality

$$E\|n^{-m/p}S_n(\Phi_1)\|^p \le (E\|n^{-m/p}S_n(\Phi_1)\|^2)^{p/2} \le$$

[using Corollary 2.7.1]

$$\le (6(2A)^{1/2}m^3)^{mp}(n^{-\frac{m}{p}(2-p)}E\|\Phi\|^2 I(\|\Phi\| \le n^{m/p}))^{p/2}.$$

Here

$$n^{-\frac{m}{p}(2-p)}E\|\Phi\|^2 I(\|\Phi\| \le n^{m/p}) =$$

$$= n^{-\frac{m}{p}(2-p)}E\|\Phi\|^2 I(\|\Phi\| \le n^{\frac{m}{2p}}) + n^{-\frac{m}{p}(2-p)}E\|\Phi\|^2 I(n^{\frac{m}{2p}} < \|\Phi\| \le n^{\frac{m}{p}}) \le$$

$$\le n^{-\frac{m}{2p}(2-p)}E\|\Phi\|^p + E\|\Phi\|^p I(\|\Phi\| > n^{\frac{m}{2p}}).$$

Hence

$$E\|n^{-m/p}S_n(\Phi_1)\|^p \to 0 \quad as \quad n \to \infty$$

for all $1 < p < 2$ if $E\|\Phi\|^p < \infty$. This proves Theorem 3.1.3.

***UR*-statistics of two-parameter indexing sets.** Let

$$U_{n_1 n_2} = (n_1 n_2)^{-1} \sum_{\substack{i,j=1 \\ i \ne j}}^{n_1, n_2} \Phi(X_i, X_j), \tag{3.1.5}$$

where $\Phi : X^2 \to R$. It is natural to consider a two-parameter convergence in the strong law of large numbers for $U_{n_1 n_2}$, i.e. convergence as $n_1 \wedge n_2$ tends to infinity, where $n_1 \wedge n_2$ is the minimum of n_1 and n_2.

In (3.1.5) we assume that $X_1, X_2, \ldots$ are independent random variables with a common uniform distribution on (0,1).

Clearly,

$$EU_{n_1 n_2} = \theta,$$

where

$$\theta = \int_0^1 \int_0^1 \Phi(x,y)\, dxdy.$$

However, if $\min(n_1, n_2) \to \infty$, then the integrability of Φ does not guarantee the validity of the strong law of large numbers. This fact is reflected in the following theorem. Denote

$$\delta(\Phi) = \int_0^1 \int_0^1 |\Phi(x,y)| \left(\frac{1 + \ln^+(\|\Phi\|_1 |\Phi(x,y)|)}{\|\Phi(x,\cdot)\|_1 \|\Phi(\cdot,y)\|_1} \right) dxdy \tag{3.1.6}$$

where $\ln^+ z = \ln(\max(1,z))$.

Theorem 3.1.4. *Let $E|\Phi(X_1, X_2)| < \infty$. If $\min(n_1, n_2) \to \infty$, then*

$$U_{n_1 n_2} \to \theta \quad a.s.$$

if and only if the condition $\delta(\Phi) < \infty$ is satisfied.

For the proof of this theorem see McConnell (1987, 1988). It is easy to see that one can always find an integrable function Φ such that $\delta(\Phi) = \infty$.

3.2 Multi-sample *UB*-statistics

Almost sure convergence with multi-parameter indexing sets. We will apply the martingale Lemma 1.2.2.

Theorem 3.2.1. *Let B-valued kernel Φ satisfy the condition*

$$E\Big(\|\Phi\|(\ln^+\|\Phi\|)^{c-1}\Big) < \infty$$

where $\ln^+\|\Phi\| = 1 \vee \ln\|\Phi\|$. Then for $\min(n_1, \dots, n_c) \to \infty$

$$U_{\bar{n}} \to \theta \quad a.s.$$

takes place.

Proof. By virtue of Lemma 1.2.2 the sequence $(U_{\bar{n}}, \mathfrak{B}_{\bar{n}})$ is a regular reversed martingale. Then by Theorem 3.1 from Fazekas (1983) about a multidimensional strong law of large numbers for Banach-valued martingales we have

$$U_{\bar{n}} \to E\Big(\Phi(X_{ji},\ i = 1, \dots, m_j,\ j = 1, \dots, c)|\mathfrak{B}_\infty\Big) \quad a.s.$$

as $\min(n_1, \dots, n_c) \to \infty$, where

$$\mathfrak{B}_\infty = \bigcap_{\bar{n}=\bar{m}}^{\infty} \mathfrak{B}_{\bar{n}} = \mathfrak{B}_{1\infty} \times \mathfrak{B}_{2\infty} \times \cdots \times \mathfrak{B}_{c\infty}$$

and $\mathfrak{B}_{j\infty} = \bigcap_{n_j=m_j}^{\infty} \mathfrak{B}_{jn_j}$, $j = 1, \dots, c$. In addition, σ-algebras $\mathfrak{B}_{jn_j}$ were defined in Lemma 1.2.2. The random variables $X_{j1}, \dots, X_{jn_j}$ are independent and have the same distribution. Hence, according to the Hewitt–Savage zero-one law σ-algebra $\mathfrak{B}_{j\infty} = \{\varnothing_j, \Omega_j\}$ is trivial. Therefore the limit is equal to

$$E\Big(\Phi(X_{ji},\ i = 1, \dots, m_j,\ j = 1, \dots, c)\Big|\,\mathfrak{B}_\infty\Big) = \\ = E\Big(\Phi(X_{ji},\ i = 1, \dots, m_j,\ j = 1, \dots, c)\Big) = \theta.$$

It turns out that the statement of Theorem 3.2.1 remains valid if we somewhat weaken the condition of this theorem (see McConnell (1987)). Let $c = 2$ and $m_1 = m_2 = 1$ in (1.2.1). Then

$$U_{\bar{n}} = (n_1 n_2)^{-1} \sum_{i=1}^{n_1} \sum_{j=1}^{n_2} \Phi(X_{1i}, X_{2j}) \qquad (3.2.1),$$

where we also assume that $X_{11}, X_{12}, \ldots, X_{1n_1}$ and $X_{21}, X_{22}, \ldots, X_{2n_2}$ are two independent sequences of independent random variables identically distributed on (0,1).

Theorem 3.2.2. *Suppose that $B = R$ and $E|\Phi(X_{11}, X_{21})| < \infty$, and also the variable $\delta(\Phi)$ is defined in* (3.1.6). *For given function Φ, the following two statements are equivalent:*

$$\delta(\Phi) < \infty$$

and

$$U_{\bar{n}} \to \int_0^1 \int_0^1 \Phi(x, y)\, dx dy \quad a.s.$$

as $n_1 \wedge n_2 \to \infty$.

The proof of this theorem belongs to McConnell (1987). It is obvious that condition $\delta(\Phi) < \infty$ is weaker than the condition of Theorem 3.2.1. Thus, condition $\delta(\Phi) < \infty$ is necessary and sufficient for the validity of the strong law of large numbers for two-sample *UR*-statistics of type (3.2.1).

Convergence in probability. For $c \geq 2$ we define the sector index set $N(\lambda)$ as

$$N(\lambda) = \{\bar{n} = (n_1, \ldots, n_c) : \ n_j \geq m_j, \ \lambda n_i \leq n_j \leq \lambda^{-1} n_i, \ i, j = 1, \ldots, c\}, \quad (3.2.2)$$

$0 < \lambda < 1$.

Denote by a vector $\bar{q} = (q_1, \ldots, q_c)$ with components $1 \leq q_i < 2$, $i = 1, \ldots, c$. Let $\bar{r} = (r_1, \ldots, r_c)$ be a vector of ranks and a vector $\bar{d} = (d_1, \ldots, d_c)$ from (1.2.7). Let $r = r_1 + \cdots + r_c$, $d = d_1 + \cdots + d_c$, $p_c = \sum_{i=1}^{c}(r_i/q_i)$,

$$\gamma_{\bar{r}\bar{d}} = d/(d - r + p_c), \quad \gamma_{\bar{r}} = r/p_c, \qquad (3.2.3)$$

i.e. $\gamma_{\bar{r}} = \gamma_{\bar{r}\bar{r}}$. Similarly, we define $\gamma_{\bar{d}} = \gamma_{\bar{d}\bar{d}}$.

Theorem 3.2.3. *Let B be a separable Banach space of type* 2. *Suppose that the canonical functions $g_{\bar{d}}$ in* (1.2.9) *satisfy the condition*

$$t^{\gamma_{\bar{d}}} P(\|g_{\bar{d}}\| > t) \to 0 \quad as \quad t \to \infty. \qquad (3.2.4)$$

Then

$$\prod_{j=1}^{c} n_j^{-d_j/q_j} S_{\bar n \bar d} \xrightarrow{P} 0$$

if $\bar n \in N(\lambda)$ *and* $\min(n_1, \dots, n_c) \to \infty$.

Proof. For simplicity, we assume that $\min(n_1, \dots, n_c) = n_1$. Since $\bar n \in N(\lambda)$ then $\lambda n_1 \le n_j \le n_1 \lambda^{-1}$ for $j = 2, \dots, c$. Further, we write with the help of (1.1.4) the following formula

$$g_{\bar d} = g_{\bar d}^1 + \bar g_{\bar d}^1,$$

where

$$g_{\bar d}^1(x_{ji}, \quad i = 1, \dots, d_j, \quad 1 \le j \le c) =$$

$$= \int \cdots \int g_{\bar d}(y_{ji}, \ i = 1, \dots, d_j, \quad 1 \le j \le c) I\Big(\|g_{\bar d}\| \le \prod_{j=1}^{c} n_j^{d_j/q_j}\Big) \times$$

$$\times \prod_{j=1}^{c} \prod_{s=1}^{d_j} (\delta_{x_{js}}(dy_{js}) - P_j(dy_{js})),$$

$$\bar g_{\bar d}^1(x_{ji}, \quad i = 1, \dots, d_j, \quad 1 \le j \le c) =$$

$$= \int \cdots \int g_{\bar d}(y_{ji}, \ i = 1, \dots, d_j, \quad 1 \le j \le c) I\Big(\|g_{\bar d}\| > \prod_{j=1}^{c} n_j^{d_j/q_j}\Big) \times$$

$$\times \prod_{j=1}^{c} \prod_{s=1}^{d_j} (\delta_{x_{js}}(dy_{js}) - P_j(dy_{js})) .$$

The functions $g_{\bar d}^1$ and $\bar g_{\bar d}^1$ possess the property of complete degeneracy (1.2.3).

Hence we can represent $S_{\bar n \bar d}$ in (1.2.9) as

$$S_{\bar n \bar d} = S_{\bar n \bar d}(g_{\bar d}^1) + S_{\bar n \bar d}(\bar g_{\bar d}^1),$$

where

$$S_{\bar n \bar d}(g_{\bar d}^1) = \sum_{\substack{1 \le i_{j1} < \cdots < i_{jd_j} \le n_j \\ j=1,\dots,c}} g_{\bar d}^1(x_{j\alpha}, \quad \alpha = i_{j1}, \dots, i_{jd_j}, \ j = 1, \dots, c),$$

$$S_{\bar n \bar d}(\bar g_{\bar d}^1) = \sum_{\substack{1 \le i_{j1} < \cdots < i_{jd_j} \le n_j \\ j=1,\dots,c}} \bar g_{\bar d}^1(x_{j\alpha}, \quad \alpha = i_{j1}, \dots, i_{jd_j}, \ j = 1, \dots, c) .$$

Using Chebyshev's inequality we have for any $t > 0$

$$P\Big(\prod_{j=1}^{c} n_j^{-d_j/q_j} \|S_{\bar n \bar d}\| > t\Big) \le$$

$$\le 2t^{-1} \prod_{j=1}^{c} n^{-d_j/q_j} E\|S_{\bar n \bar d}(\bar g_{\bar d}^1)\| + 4t^{-2} \prod_{j=1}^{c} n^{-2d_j/q_j} E\|S_{\bar n \bar d}(g_{\bar d}^1)\|^2 .$$

Here

$$E\|S_{\bar{n}\bar{d}}(\bar{g}_{\bar{d}}^{1})\| \leq \prod_{j=1}^{c} \frac{2^{d_j}}{d_j!} n_j^{d_j} E\|g_{\bar{d}}\| I\Big(\|g_{\bar{d}}\| > \prod_{j=1}^{c} n_j^{d_j/q_j}\Big)$$

and under condition (3.2.4)

$$\prod_{j=1}^{c} n_j^{d_j(-\frac{1}{q_j}+1)} E\|g_{\bar{d}}\| I\Big(\|g_{\bar{d}}\| \geq \prod_{j=1}^{c} n^{d_j/q_j}\Big) \to 0 \quad \text{as } n_1 \to \infty .$$

By Corollary 2.7.1

$$E\|S_{\bar{n}\bar{d}}(g_{\bar{d}}^{1})\|^2 \leq \prod_{j=1}^{c} \frac{1}{d_j!}(6(2A)^{1/2} d_j^3)^{2d_j} n_j^{d_j} E\|g_{\bar{d}}\|^2 I\Big(\|g_{\bar{d}}\| \leq \prod_{j=1}^{c} n_j^{d_j/q_j}\Big).$$

The above relations give the assertion of this theorem under the assumption (3.2.4).

Convergence with probability 1 in a sector.

Theorem 3.2.4. *Let Banach space B be of type 2 and let kernel Φ satisfy the condition $E\|\Phi\| < \infty$. Then*

$$U_{\bar{n}} - \theta \to 0 \quad a.s.$$

if $\bar{n} \in N(\lambda)$ and $\min(n_1, \ldots, n_c) \to \infty$.

Proof. In view of (1.2.8), it is sufficient to establish that

$$U_{\bar{n}\bar{d}} \to 0 \quad a.s.$$

For any fixed vector $\bar{d} = (d_1, \ldots, d_c)$, such that $d = d_1 + \cdots + d_c \geq 1$, denote

$$\sigma(k) = \{\bar{n} \in N(\lambda) : 2^{k-1} \leq n_1^{d_1} \cdots n_c^{d_c} < 2^k\}, \quad k = 1, 2, \ldots$$

Let us show that for any $t > 0$

$$\sum_{k=1}^{\infty} P\{\omega : \sup_{\bar{n}\in\sigma(k)} \|U_{\bar{n}\bar{d}}\| > t\} < \infty.$$

Then the validity of $U_{\bar{n}\bar{d}} \to 0$ a.s. follows from the Borel–Cantelli lemma. Let us write $U_{\bar{n}\bar{d}}$ as the sum of two U-statistics : $U_{\bar{n}\bar{d}} = U_{\bar{n}\bar{d}}(\varphi_k) + U_{\bar{n}\bar{d}}(\bar{\varphi}_k)$, with the kernels $\varphi_k = g_{\bar{d}} I(\|g_{\bar{d}}\| < 2^k)$ and $\bar{\varphi}_k = g_{\bar{d}} I(\|g_{\bar{d}}\| \geq 2^k)$, $k = 1, 2, \cdots$, respectively. Further, we have

$$\{\omega : \sup_{\bar{n}\in\sigma(k)} \|U_{\bar{n}\bar{d}}\| > t\} = \bigcup_{\bar{n}\in\sigma(k)} \{\omega : \|U_{\bar{n}\bar{d}}\| > t\} \subset$$

$$\subset \Big(\bigcup_{\bar{n}\in\sigma(k)} \{\omega : \|U_{\bar{n}\bar{d}}(\varphi_k)\| > t\}\Big) \bigcup \Big(\bigcup_{\bar{n}\in\sigma(k)} \{\omega : \|U_{\bar{n}\bar{d}}(\bar{\varphi}_k)\| \neq 0\}\Big).$$

According to the definitions of $\bar{\varphi}_k$ and $\sigma(k)$, we obtain

$$P\Big(\bigcup_{\bar{n}\in\sigma(k)}\{\omega : U_{\bar{n}\bar{d}}(\bar{\varphi}_k)\neq 0\}\Big) \leq 2^k P(\|g_{\bar{d}}\| \geq 2^k);$$

moreover,

$$\sum_{k=1}^{\infty} 2^k P(\|g_{\bar{d}}\| \geq 2^k) = \sum_{k=1}^{\infty} 2^k \sum_{s=k}^{\infty} P(2^s \leq \|g_{\bar{d}}\| < 2^{s+1}) \leq$$

$$\leq \sum_{k=1}^{\infty} 2^{k+1} P(2^k \leq \|g_{\bar{d}}\| < 2^{k+1}) \leq 2E\|g_{\bar{d}}\| < \infty.$$

Using Theorem 2.3.4, after a simple calculation, we get

$$P\Big(\bigcup_{\bar{n}\in\sigma(k)}\{\omega : \|U_{\bar{n}\bar{d}}(\varphi_k)\| > t\}\Big) \leq$$

$$\leq P\Big(\max_{a_1 2^{(k-1)/d}\leq n_1\leq b_1 2^{k/d}} \cdots \max_{a_c 2^{(k-1)/d}\leq n_c\leq b_c 2^{k/d}} \|U_{\bar{n}\bar{d}}(\varphi_k)\| > t\Big) \leq$$

$$\leq c(t) 2^{-k} E(\|g_{\bar{d}}\|^2 I(\|g_{\bar{d}}\| < 2^k)),$$

where $d = d_1 + \cdots + d_c$; $a_i > 0$, $b_i > 0$, and $c(t)$ are positive and independent of $k = 1, 2, \ldots$. Note that

$$\sum_{k=1}^{\infty} 2^{-k} E(\|g_{\bar{d}}\|^2 I(\|g_{\bar{d}}\| < 2^k)) = E(\|g_{\bar{d}}\|^2 I(\|g_{\bar{d}}\| < 1)) +$$

$$+\sum_{k=1}^{\infty} 2^{-k} \sum_{s=1}^{k} E(\|g_{\bar{d}}\|^2 I(2^{s-1} \leq \|g_{\bar{d}}\| < 2^s)) =$$

$$= E(\|g_{\bar{d}}\|^2 I(\|g_{\bar{d}}\| < 1)) + \sum_{s=1}^{\infty} E(\|g_{\bar{d}}\|^2 I(2^{s-1} \leq \|g_{\bar{d}}\| < 2^s)) \sum_{k=s}^{\infty} 2^{-k} =$$

$$= E(\|g_{\bar{d}}\|^2 I(\|g_{\bar{d}}\| < 1)) + 2\sum_{s=1}^{\infty} 2^{-s} E(\|g_{\bar{d}}\|^2 I(2^{s-1} \leq \|g_{\bar{d}}\| < 2^s)) \leq$$

$$\leq 2E(\|g_{\bar{d}}\| I(\|g_{\bar{d}}\| < 1)) + 2\sum_{s=1}^{\infty} E(\|g_{\bar{d}}\| I(2^{s-1} \leq \|g_{\bar{d}}\| < 2^s)) \leq$$

$$\leq 2E\|g_{\bar{d}}\| < \infty.$$

These arguments conclude the proof of Theorem 3.2.4.

Theorem 3.2.5. *Let Banach space B be of type p, $1 < p \leq 2$. Suppose that kernel Φ with rank vector $\bar{r}$ satisfies the condition*

$$E\|g_{\bar{d}}\|^{\gamma_{\bar{r}\bar{d}}} < \infty, \quad r \leq d \leq m.$$

Then, if $p > \gamma_{\bar{r}}$,

$$\prod_{j=1}^{c} \binom{n_i}{r_i}^{-\frac{1}{q_i}+1} (U_{\bar{n}} - \theta) \to 0 \quad a.s.$$

as $\bar{n} \in N(\lambda)$ *and* $\min(n_1, \dots, n_c) \to \infty$.

Proof. It is sufficient to adapt the method of the proof of the preceding theorem.

Corollary 3.2.1. *Let Banach space B be of type p,* $1 < p \leq 2$, *and* $p > \gamma_{\bar{d}}$. *Suppose, that*

$$E\|g_{\bar{d}}\|^{\gamma_d} < \infty, \quad r \leq d.$$

Then

$$\prod_{j=1}^{c} n_j^{-d_j/q_j} S_{\bar{n}\bar{d}} \to 0 \quad a.s.$$

as $\bar{n} \in N(\lambda)$ *and* $\min(n_1, \dots, n_c) \to \infty$.

3.3 Von Mises' statistics

We shall consider von Mises' statistics (1.3.1) and use representation (1.3.3) in the following form

$$V_n = n^{-m} \sum_{c=1}^{m} S_{nc}, \tag{3.3.1}$$

where

$$S_{nc} = \sum_{1 \leq i_1 < \dots < i_c \leq n} \Phi_{mc}(X_{i_1}, \dots, X_{i_c})$$

and, for example, $\Phi_{mm}(x_1, \dots, x_m) = m!\Phi(x_1, \dots, x_m)$; $\Phi_{m1}(x_1) = \Phi(x_1, \dots, x_1)$.

Convergence in probability.

Theorem 3.3.1. *Assume that B-valued kernel* Φ *satisfies the conditions*

$$E\|\Phi_{mc}\|^{c/m} < \infty, \quad c = 1, \dots, m.$$

Then

$$V_n \xrightarrow{P} \theta \quad as \quad n \to \infty.$$

Proof. In view of (3.3.1) it is sufficient to prove that

$$n^{-m} S_{nc} \xrightarrow{P} 0$$

for all $c = 1, \dots, m$. If $c = m$ then this assertion follows from Theorem 3.1.1. Next let $1 \leq c \leq m-1$. We write S_{nc} in the form of a sum, namely $S_{nc} = S_{nc}(\varphi_1) + S_{nc}(\varphi_2)$, where $\varphi_1 = \Phi_{mc} I(\|\Phi_{mc}\| \leq \sqrt{n})$ and $\varphi_2 = \Phi_{mc} I(\|\Phi_{mc}\| > \sqrt{n})$. Furthermore, we have by Chebyshev's inequality

$$P(n^{-m}\|S_{nc}\| > t) \leq$$

$$\le 2t^{-1}n^{-m}E\|S_{nc}(\varphi_1)\| + 2^{c/m}t^{-c/m}n^{-c}E\|S_{nc}(\varphi_2)\|^{c/m}$$

for any $t > 0$. Here,

$$n^{-m}E\|S_{nc}(\varphi_1)\| \le n^{-1}E\|\varphi_1\| \le$$

$$\le n^{-1/2} \to 0, \quad \text{as } n \to \infty$$

and by inequality (2.2.3)

$$n^{-c}E\|S_{nc}(\varphi_2)\|^{c/m} \le E\|\varphi_2\|^{c/m} =$$

$$= E\|\Phi_{mc}\|^{c/m} I(\|\Phi_{mc}\| > \sqrt{n}) \to 0 \quad \text{as } n \to \infty.$$

These estimates give the assertion of the theorem.

Convergence with probability 1 and in L^1.

Theorem 3.3.2. *Suppose that B-valued kernel Φ satisfies the condition*

$$\max\{E\|\Phi(X_{i_1},\dots,X_{i_m})\| : 1 \le i_1,\dots, i_m \le m\} < \infty.$$

Then

$$V_n \to \theta$$

with probability 1 *and in* L^1 as $n \to \infty$.

Proof. We apply representation (1.3.3). According to Theorem 3.1.1, we get

$$U_{nc} \to \theta_{mc}$$

for all $c = 1,\dots,m$ with probability 1 and in L^1, where $\theta_{mc} = E\Phi_{mc}(X_1,\dots,X_c)$. Here $\theta_{mm} = m!\theta$. Taking into account that $\binom{n}{c}n^{-m} = O(n^{-1})$, for $c < m$, we obtain from (1.3.3) the statement of the theorem.

The assertion of this theorem remains valid even if we somewhat weaken its condition. For simplicity, suppose that $m = 2$. In this case, (1.3.3) has the form (1.3.5).

Theorem 3.3.3. *If*

$$E\|\Phi(X_1,X_2)\| < \infty, \quad E\|\Phi(X_1,X_1)\|^{1/2} < \infty,$$

then $V_n \to \theta$ with probability 1 *as $n \to \infty$.*

Proof. Under the first condition $E\|\Phi(X_1,\ x_2)\| < \infty$. Theorem 3.1.1 yields

$$2n^{-2} \sum_{1 \le i < j \le n} \Phi(X_i, X_j) \to \theta \quad a.s.$$

Under the second condition $E\|\Phi(X_1, X_1)\|^{1/2} < \infty$, the Marcinkiewicz strong law of large numbers implies that

$$n^{-2}\sum_{j=1}^{n} \|\Phi(X_j, X_j)\| \to 0 \quad a.s.$$

This and representation (1.3.5) together prove the theorem.

3.4 Estimates of convergence rates

UB-statistics. Let us define the variable

$$Q_n(t,q) = P\Big\{\omega : \sup_{k\geq n} \binom{k}{m}^{-\frac{1}{q}+1} \|U_k\| \geq t\Big\}, \quad 0 < q < 1,$$

and

$$P_n(t,q) = P\Big\{\omega : \sup_{k\geq n} \binom{k}{r}^{-\frac{1}{q}+1} \|U_k - \theta\| \geq t\Big\}, \quad 1 \leq q < 2$$

for UB-statistics (1.1.1), where r is the rank of kernel Φ in (1.1.10), and $t > 0$.

Theorem 3.4.1. *Suppose that $E\|\Phi\|^q < \infty$ for $0 < q < 1$. Then for $t > 0$ and $n \geq m \geq 1$ the following relation holds*

$$Q_n(t,q) \leq 2t^{-1}\binom{n}{m}^{-\frac{1}{q}+1} E\|\Phi\|I(\|\Phi\| \leq n^{1/q}) + 2^q t^{-q} E\|\Phi\|^q I(\|\Phi\| > n^{1/q}); \quad (3.4.1)$$

moreover, as $n \to \infty$,

$$\binom{n}{m}^{-\frac{1}{q}+1} E\|\Phi\|I(\|\Phi\| \leq n^{1/q}) \to 0$$

for all $m \geq 1$ and $0 < q \leq 1$.

Proof. First, we obtain the following estimate for $Q_n(t,q)$:

$$Q_n(t,q) \leq P\Big\{2\sup_{k\geq n} \binom{k}{m}^{-\frac{1}{q}+1} \|U_k(\Phi_n)\| \geq t\Big\} + P\Big\{2\sup_{k\geq n} \binom{k}{m}^{-\frac{1}{q}+1} \|U_k(\bar{\Phi}_n)\| \geq t\Big\},$$

where $U_k(\Phi_n)$ and $U_k(\bar{\Phi}_n)$ are UB-statistics of the form (1.1.1) with the kernels $\Phi_n = \Phi I(\|\Phi\| \leq n^{1/q})$ and $\bar{\Phi}_n = \Phi I(\|\Phi\| > n^{1/q})$, respectively.

Let $\mathfrak{B}_n = \sigma\{\omega : (X_1, \dots, X_n), X_{n+1}, X_{n+2}, \dots\}$ be the σ-algebra generated by a unordered collection $(X_1, \dots, X_n)$ and a sequence $X_{n+1}, X_{n+2}, \dots$. It is obvious that $\mathfrak{B}_n \supseteq \mathfrak{B}_{n+1}$, $n = 1, 2, \dots$. The stochastic sequence $(U_n, \mathfrak{B}_n)$, $n = m, m+1, \dots$ is a reversed martingale provided that $E\|\Phi\| < \infty$. Using this property and the inequality of Corollary 2.3.2, we obtain

$$P\Big\{2\sup_{k\geq n} \binom{k}{m}^{-\frac{1}{q}+1} \|U_k(\Phi_n)\| \geq t\Big\} \leq P\Big\{2\binom{n}{m}^{-\frac{1}{q}+1} \sup_{k\geq n} \|U_k(\Phi_n)\| \geq t\Big\} \leq$$

$$\leq 2t^{-1}\binom{n}{m}^{-\frac{1}{q}+1} E\|U_n(\Phi_n)\| \leq 2t^{-1}\binom{n}{m}^{-\frac{1}{q}+1} E\|\Phi\|I(\|\Phi\| \leq n^{1/q}).$$

Further, we have

$$P\Big\{2\sup_{k\geq n}\binom{k}{m}^{-\frac{1}{q}+1}\|U_k(\bar{\Phi}_n)\| \geq t\Big\} \leq P\Big\{2^q \sup_{k\geq n} U_k(\|\bar{\Phi}_n\|^q) \geq t^q\Big\} \leq$$

$$\leq 2^q t^{-q} EU_n(\|\bar{\Phi}_n\|^q) = 2^q t^{-q} E\|\Phi\|^q I(\|\Phi\| > n^{1/q}).$$

Here, the second inequality is based on (2.2.3). Uniting the estimates obtained above, we get (3.4.1).

Theorem 3.4.2. *Let Banach space B be of type p, $1 < p \leq 2$. Assume that a Bochner integrable kernel Φ has rank r.*

Then the estimate

$$P\Big\{\omega : \sup_{k\geq n}\|U_k - \theta\| \geq t\Big\} \leq 2t^{-1}\sum_{c=r}^{m} 2^c\binom{m}{c} E\|g_c\|I(\|g_c\| > n^{\delta(c)})+$$

$$+t^{-p}\sum_{c=r}^{m}\pi(c)\binom{n}{c}^{-p+1} E\|g_c\|^p I(\|g_c\| \leq n^{\delta(c)})$$

holds, where $\delta(c)$ is a number such that $0 < \delta(c) \leq c$;

$$\pi(c) = 2^{p+c}(m-r+1)^{p-1}m^{[p]}(6A^{1/p}c^3)^{cp}(c(c+1))^{p-1} \Big/ (c!)^{2p-1}$$

and a constant A is given in (2.7.1).

Proof. We employ (1.1.12). Let us represent each *UB*-statistics U_{kc} in (1.1.11) as the sum $U_{kc} = U_k(\varphi_{1c}) + U_k(\varphi_{2c})$, $k = n, n+1, \ldots$ with the kernels

$$\varphi_{sc}(x_1, \ldots, x_c) =$$

$$= \sum_{d=1}^{c}(-1)^{c-d} \sum_{1\leq j_1<\cdots<j_d\leq c} \bar{g}_{cd}^{(s)}(x_{j_1}, \ldots, x_{j_d}), \quad s = 1, 2, \tag{3.4.2}$$

respectively, where

$$\bar{g}_{cd}^{(s)}(x_1, \ldots, x_d) = g_{cd}^{(s)}(x_1, \ldots, x_d) - Eg_{cd}^{(s)}(X_1, \ldots, Xd),$$

and

$$g_{cd}^{1}(x_1, \ldots, x_d) =$$

$$= E(g_c(X_1, \ldots, X_c)I(\|g_c(X_1, \ldots, X_c)\| \leq n^{\delta(c)})\big|X_1 = x_1, \ldots, X_d = x_d),$$

$$g_{cd}^{(2)}(x_1, \ldots, x_d) =$$

$$= E(g_c(X_1,\dots,X_c)I(\|g_c(X_1,\dots,X_c)\| > n^{\delta(c)})\big|X_1 = x_1,\dots,X_d = x_d).$$

Note that $g_c = \varphi_{1c} + \varphi_{2c}$ and the functions φ_{1c} and φ_{2c} possess the property of complete degeneracy (1.1.7). Therefore, the stochastic sequences $(U_k(\varphi_{sc}), \mathfrak{B}_k)$, $k = n, n+1, \dots$, are reverse martingales. If $q \geq 1$, then one can derive the estimate

$$E\|\varphi_{1c}\|^q \leq (c(c+1))^{q-1}2^c E\|g_c\|^q I(\|g_c\| \leq n^{\delta(c)}),$$

$$E\|\varphi_{2c}\| \leq 2^c E\|g_c\| I(\|g_c\| > n^{\delta(c)}).$$

By virtue of (1.1.12), we have

$$P_n(t,1) \leq P\Big\{2\sup_{k\geq n}\Big\|\sum_{c=r}^{m}\binom{m}{c}U_k(\varphi_{1c})\Big\| \geq t\Big\} + P\Big\{2\sup_{k\geq n}\Big\|\sum_{c=r}^{m}\binom{m}{c}U_k(\varphi_{2c})\Big\| \geq t\Big\}.$$

Taking the maximal martingale inequality of Corollary 2.3.2 and the decoupling inequality (2.7.3) for $q = p$, we get

$$\begin{aligned} P\Big\{2\sup_{k\geq n}\Big\|\sum_{c=r}^{m}\binom{m}{c}U_k(\varphi_{1c})\Big\| \geq t\Big\} &\leq 2^p t^{-p} E\Big\|\sum_{c=r}^{m}\binom{m}{c}U_n(\varphi_{1c})\Big\|^p \\ &\leq 2^p t^{-p}(m-r+1)^{p-1}\sum_{c=r}^{m}\binom{m}{c}^p E\|U_n(\varphi_{1c})\|^p \\ &\leq 2^p t^{-p}(m-r+1)^{p-1}\sum_{c=r}^{m}\binom{m}{c}^p (6A^{1/p}c^3)^{cp}\binom{n}{c}^{-p}\frac{n^{[c]}}{(c!)^p}E\|\varphi_{1c}\|^p \\ &\leq t^{-p}\sum_{c=r}^{m}\pi(c)\binom{n}{c}^{-p+1} E\|g_c\|^p I(\|g_c\| \leq n^{\delta(c)}). \end{aligned}$$

If we employ Corollary 2.3.2 once more, we obtain

$$\begin{aligned} P\Big\{2\sup_{k\geq n}\Big\|\sum_{c=r}^{m}\binom{m}{c}U_k(\varphi_{2c})\Big\| \geq t\Big\} &\leq 2t^{-1}E\Big\|\sum_{c=r}^{m}\binom{m}{c}U_n(\varphi_{2c})\Big\| \\ &\leq 2t^{-1}\sum_{c=r}^{m}\binom{m}{c}E\|\varphi_{2c}\| \\ &\leq 2t^{-1}\sum_{c=r}^{m}\binom{m}{c}2^c E\|g_c\| I(\|g_c\| > n^{\delta(c)}). \end{aligned}$$

Combining these estimates, we get the assertion of the theorem.

Theorem 3.4.3. *Let $1 < q < 2$ and $\gamma_{rc} = cq/(q(c-r)+r)$ for $c = r,\dots,m$, so that $1 < \gamma_{rm} < \gamma_{r,m-1} < \dots < \gamma_{r,r+1} < \gamma_{rr} = q$. Let Banach space B be of type p, $1 < p \leq 2$, and $p > q$. Assume that kernel Φ with rank r satisfies the conditions*

$$E\|g_c\|^{\gamma_{rc}} < \infty, \quad c = r,\dots,m.$$

Then the estimate

$$P_n(t,q) \le \sum_{c=r}^{m} a(c) n^{-\frac{c}{2}(p-\gamma_{rc})/\gamma_{rc}} + \sum_{c=r}^{m} b(c) E(\|g_c\|^{\gamma_{rc}} I(\|g_c\|^{\gamma_{rc}} > 2^{-c} n^{c/2}))$$

holds for integers $n \ge m$ and $m \ge 1$, where the constants $a(c)$ and $b(c)$ are not dependent on n.

Proof. First, owing to representations (1.1.12), we can write the inequality

$$P_n(t,q) \le \sum_{c=r}^{m} P\Big\{ \sup_{k \ge n} \binom{k}{r}^{-\frac{1}{q}+1} \|U_{kc}\| \ge 2t_c \Big\}$$

where $t_c = t2^{-1}(m-c+1)^{-1} m^{[c]}/c!$. We estimate each term on the right-hand side separately. To do this, we use (3.4.2) with the following change. In definition (3.4.2), the variable $n^{\delta(c)}$ (in the functions $g_{cd}^{(1)}$ and $g_{cd}^{(2)}$) is replaced by $2^{jc/\gamma_{rc}}$, $j = 1, 2, \dots$.

Let $2^{i-1} \le n < 2^i$, $i \ge 1$. For such n, taking into account the substitution made in (3.4.2), we write

$$P\Big\{ \sup_{k \ge n} \binom{k}{r}^{-\frac{1}{q}+1} \|U_{kc}\| \ge 2t_c \Big\} = P\Big\{ \bigcup_{k=n}^{\infty} \Big(\binom{k}{r}^{-\frac{1}{q}+1} \|U_{kc}\| \ge 2t_c \Big) \Big\} \le$$

$$\le P\Big\{ \bigcup_{j=i}^{\infty} \Big(\max_{2^{j-1} \le n < 2^j} \binom{k}{r}^{-\frac{1}{q}+1} \|U_{kc}\| \ge 2t_c \Big) \Big\} \le$$

$$\le \sum_{j=i}^{\infty} P\Big\{ \max_{2^{j-1} \le k < 2^j} \binom{k}{r}^{-\frac{1}{q}+1} \|U_{kc}\| \ge 2t_c \Big\} \le$$

$$\le \sum_{j=i}^{\infty} P\Big\{ \max_{2^{j-1} \le k < 2^j} \binom{k}{r}^{-\frac{1}{q}+1} \|U_k(\varphi_{1c})\| \ge t_c \Big\} + \sum_{j=i}^{\infty} P\Big\{ \max_{2^{j-1} \le k < 2^j} \binom{k}{r}^{-\frac{1}{q}+1} \|U_k(\varphi_{2c})\| \ge t_c \Big\}.$$

Applying (2.3.2), we find

$$P\Big\{ \max_{2^{j-1} \le k < 2^j} \binom{k}{r}^{-\frac{1}{q}+1} \|U_k(\varphi_{1c})\| \ge t_c \Big\} \le$$

$$\le t_c^{-p} 2^p \Big\{ \binom{2^{j-1}}{r}^{\frac{p(q-1)}{q}} E\|U_{2^{j-1}}(\varphi_{1c})\|^p +$$

$$+ \sum_{k=2^{j-1}+1}^{2^j} \Big[\binom{k}{r}^{\frac{p(q-1)}{q}} - \binom{k-1}{r}^{\frac{p(q-1)}{q}} \Big] E\|U_k(\varphi_{1c})\|^p \Big\} \le$$

$$\le a_p^1(c,r) \binom{2^j}{r}^{\frac{p(q-1)}{q}} \binom{2^{j-1}}{c}^{-p+1} E\|g_c\|^p I(\|g_c\| \le 2^{jc/\gamma_{rc}}) \le$$

$$\leq a_p(c,r)2^{-jc(p-\gamma_{rc})/\gamma_{rc}} E\|g_c\|^p I\left(\|g_c\| \leq 2^{jc/\gamma_{rc}}\right),$$

where the constants $a_p^1(c,r)$ and $a_p(c,r)$ do not depend on j. When dividing the second inequality, we have used Corollary 2.7.1 and the fact that

$$E\|U_k(\varphi_{1c})\|^p \leq 2^c(c(c+1))^{p-1}(6A^{1/2}c^3)^{2p}\binom{2^{j-1}}{c}^{-p+1} E\|g_c\|^p I(\|g_c\| \leq 2^{jc/\gamma_{rc}}).$$

Let us apply (2.3.2) once again

$$P\Big\{\max_{2^{j-1}\leq k<2^j}\binom{k}{j}^{-\frac{1}{q}+1}\|U_k(\varphi_{2c})\| \geq t_c\Big\} \leq$$

$$\leq t_c^{-1}\Bigg\{\binom{2^{j-1}}{r}^{(q-1)/q} E\|U_{2^{j-1}}(\varphi_{2c})\| +$$

$$+\sum_{k=2^{j-1}+1}^{2^j}\Big[\binom{k}{r}^{(q-1)/q} - \binom{k-1}{r}^{(q-1)/q}\Big]E\|U_k(\varphi_{2c})\|\Bigg\} \leq$$

$$\leq b_p^1(c,r)\binom{2^j}{r}^{(q-1)/q} E\|\varphi_{2c}\| \leq$$

$$\leq b_p(c,r)2^{-jc(\frac{1}{\gamma_{rc}}-1)} E\|g_c\| I\left(\|g_c\| > 2^{jc/\gamma_{rc}}\right),$$

where the constants $b_p^1(c,r)$ and $b_p(c,r)$ are not dependent on j.

Taking into account the fact that $1 < \gamma_{rc} < p$, we obtain

$$\sum_{j=1}^{\infty} I(j \geq i)2^{-jc(p-\gamma_{rc})/\gamma_{rc}} E\|g_c\|^p I(\|g_c\| \leq 2^{jc/\gamma_{rc}}) \leq$$

$$\leq \sum_{j=1}^{\infty} I(j \geq i)2^{-jc(p-\gamma_{rc})/\gamma_{rc}}\Bigg(\sum_{k=1}^{j}\Big[2^{kc(p-\gamma_{rc})/\gamma_{rc}} \times$$

$$\times E\|g_c\|^{\gamma_{rc}} I\left(2^{c(k-1)} < \|g_c\| \leq 2^{ck}\right)\Big] + 1\Bigg) =$$

$$= \sum_{j=1}^{\infty} I(j \geq i)2^{-jc(p-\gamma_{rc})/\gamma_{rc}} + \sum_{k=1}^{\infty} 2^{kc(p-\gamma_{rc})/\gamma_{rc}} \times$$

$$\times E\|g_c\|^{\gamma_{rc}} I(2^{c(k-1)} < \|g_c\|^{\gamma_{rc}} \leq 2^{ck})\sum_{j=k}^{\infty} I(j \geq i)2^{-jc(p-\gamma_{rc})/\gamma_{rc}} =$$

$$= \sum_{j=i}^{\infty} 2^{-jc(p-\gamma_{rc})/\gamma_{rc}} + \sum_{k=1}^{i-1} 2^{kc(p-\gamma_{rc})/\gamma_{rc}} \times$$

$$\times E\|g_c\|^{\gamma_{rc}} I(2^{c(k-1)} < \|g_c\|^{\gamma_{rc}} \leq 2^{ck}) \sum_{j=i}^{\infty} 2^{-jc(p-\gamma_{rc})/\gamma_{rc}} +$$

$$+ \sum_{k=i}^{\infty} 2^{kc(p-\gamma_{rc})/\gamma_{rc}} E\|g_c\|^{\gamma_{rc}} I(2^{c(k-1)} < \|g_c\|^{\gamma_{rc}} \leq 2^{ck}) \sum_{j=k}^{\infty} 2^{-jc(p-\gamma_{rc})/\gamma_{rc}} =$$

$$= 2^{-ic(p-\gamma_{rc})/\gamma_{rc}}(1 - 2^{-c(p-\gamma_{rc})/\gamma_{rc}})^{-1} + (1 - 2^{-c(p-\gamma_{rc})/\gamma_{rc}})^{-1} \times$$

$$\times \sum_{k=1}^{i-1} 2^{-(i-k)(p-\gamma_{rc})/\gamma_{rc}} E\|g_c\|^{\gamma_{rc}} I(2^{c(k-1)} < \|g_c\|^{\gamma_{rc}} \leq 2^{ck}) +$$

$$+ (1 - 2^{-c(p-\gamma_{rc})/\gamma_{rc}})^{-1} \sum_{k=i}^{\infty} E\|g_c\|^{\gamma_{rc}} I(2^{c(k-1)} < \|g_c\|^{\gamma_{rc}} \leq 2^{ck}).$$

Since $2^{i-1} \leq n < 2^i$, this yields $2^{-ic(p-\gamma_{rc})/\gamma_{rc}} \leq n^{-c(p-\gamma_{rc})/\gamma_{rc}}$,

$$\sum_{k=1}^{i-1} 2^{-(i-k)c(p-\gamma_{rc})/\gamma_{rc}} E\|g_c\|^{\gamma_{rc}} I(2^{c(k-1)} < \|g_c\|^{\gamma_{rc}} \leq 2^{ck}) =$$

$$= \sum_{k=1}^{[i/2]} 2^{-(i-k)c(p-\gamma_{rc})/\gamma_{rc}} E\|g_c\|^{\gamma_{rc}} I(2^{c(k-1)} < \|g_c\|^{\gamma_{rc}} \leq 2^{ck}) +$$

$$+ \sum_{k=[i/2]+1}^{i-1} 2^{-(i-k)c(p-\gamma_{rc})/\gamma_{rc}} E\|g_c\|^{\gamma_{rc}} I(2^{c(k-1)} < \|g_c\|^{\gamma_{rc}} \leq 2^{ck}) \leq$$

$$\leq 2^{-\frac{ic}{2}(p-\gamma_{rc})/\gamma_{rc}} \sum_{k=1}^{\infty} E\|g_c\|^{\gamma_{rc}} I(2^{c(k-1)} < \|g_c\|^{\gamma_{rc}} \leq 2^{ck}) +$$

$$+ \sum_{k=[i/2]+1}^{\infty} E\|g_c\|^{\gamma_{rc}} I(2^{c(k-1)} < \|g_c\|^{\gamma_{rc}} \leq 2^{ck}) \leq$$

$$\leq n^{-\frac{c}{2}(p-\gamma_{rc})/\gamma_{rc}} E\|g_c\|^{\gamma_{rc}} I(\|g_c\| > 1) + E\|g_c\|^{\gamma_{rc}} I(\|g_c\|^{\gamma_{rc}} > 2^{-c} n^{c/2}),$$

$$\sum_{k=i}^{\infty} E\|g_c\|^{\gamma_{rc}} I(2^{c(k-1)} < \|g_c\|^{\gamma_{rc}} \leq 2^{ck}) =$$

$$= E\|g_c\|^{\gamma_{rc}} I(\|g_c\|^{\gamma_{rc}} > 2^{c(i-1)}) \leq E\|g_c\|^{\gamma_{rc}} I(\|g_c\|^{\gamma_{rc}} > 2^{-c} n^c).$$

Further, taking into account the fact that $1 < \gamma_{rc} < p$, we write

$$\sum_{j=1}^{\infty} I(j \geq i) 2^{-jc(\frac{1}{\gamma_{rc}}-1)} E\|g_c\| I(\|g_c\| > 2^{jc/\gamma_{rc}}) =$$

$$= \sum_{j=1}^{\infty} I(j \geq i) 2^{-jc(\frac{1}{\gamma_{rc}}-1)} \sum_{k=j}^{\infty} E\|g_c\| I(2^{kc/\gamma_{rc}} < \|g_c\| \leq 2^{(k+1)c/\gamma_{rc}}) =$$

$$= \sum_{k=1}^{\infty} E\|g_c\| I(2^{kc/\gamma_{rc}} < \|g_c\| \leq 2^{(k+1)c/\gamma_{rc}}) \sum_{j=1}^{k} I(j \geq i) 2^{-jc(\frac{1}{\gamma_{rc}}-1)} =$$

$$= \sum_{k=i}^{\infty} E\|g_c\| I(2^{kc/\gamma_{rc}} < \|g_c\| \le 2^{(k+1)c/\gamma_{rc}}) \sum_{j=i}^{k} 2^{-jc(\frac{1}{\gamma_{rc}}-1)} \le$$

$$\le \left(1 - 2^{-c(1-\frac{1}{\gamma_{rc}})}\right)^{-1} \sum_{k=i}^{\infty} 2^{kc(1-\frac{1}{\gamma_{rc}})} E\|g_c\| I(2^{kc/\gamma_{rc}} < \|g_c\| \le 2^{(k+1)c/\gamma_{rc}}) \le$$

$$\le \left(1 - 2^{-c(1-\frac{1}{\gamma_{rc}})}\right)^{-1} E\|g_c\|^{\gamma_{rc}} I(\|g_c\|^{\gamma_{rc}} > n^c).$$

Combining the above proved estimates and taking only those terms into account which give the main contribution with respect to n, we finally get the desired assertion of this theorem.

Corollary 3.4.1. *Under the condition of Theorem* 3.4.3

$$\binom{n}{r}^{-\frac{1}{q}+1} (U_n - \theta) \to 0 \quad a.s.$$

as $n \to \infty$.

Corollary 3.4.2. *Let Banach space B be of type p. If $p > q$ and $E\|g_c\|^q < \infty$, then*

$$n^{-c/q} S_{nc} \to 0 \quad a.s.$$

as $n \to \infty$.

***UH*-statistics.** Hilbert space H is a type 2 space. Therefore, the estimates of the convergence rate are better than those in Banach spaces of type p.

Theorem 3.4.4. *Suppose that H-valued kernel Φ of rank r satisfies the condition $E\|\Phi\|^q < \infty$ for some $1 \le q \le 2$. Then for arbitrary $t > 0$ the inequality*

$$P(\sup_{k \ge n} \|U_k - \theta\| > t) \le t^{-2}\alpha_n + t^{-p}\beta_n,$$

holds, where

$$\alpha_n = (m - r + 1)^2 \sum_{c=r}^{m} \binom{m}{c}^{-2} c(c+1)2^{c+2} \binom{n}{c}^{-1} E\|g_c\|^2 I(\|g_c\| \le \sqrt{n})$$

and

$$\beta_n = (m - r + 1)^q \sum_{c=r}^{m} \binom{m}{c}^{-q} \frac{(6c^3)^{cq}(c(c+1))^{q-1}2^{c+q}}{(c!)^{q-1}} \binom{n}{c}^{-q+1} E\|g_c\|^q I(\|g_c\| > \sqrt{n}).$$

Proof. First, using (1.1.12), we can write inequality

$$P(\sup_{k \ge n} \|U_k - \theta\| > t) \le \sum_{c=r}^{m} P(\sup_{k \ge n} \|U_{kc}\| > t_c),$$

where $t_c = t(m-r+1)^{-1}m^{[c]}/c!$. Next we represent the *UH*-statistics U_{kc} in the form of the sum of two *UH*-statistics, namely $U_{kc} = U_k(\varphi_{1c}) + U_k(\varphi_{2c})$, with the kernels (3.4.2), where we choose $\delta(c) = \sqrt{n}$ for all $c = r, \dots, m$.

Furthermore, we have

$$P(\sup_{k\geq n} \|U_{kc}\| > t_c) \leq P(2\sup_{k\geq n} \|U_k(\varphi_{1c})\| > t_c) + P(2\sup_{\geq n} \|U_k(\varphi_{2c})\| > t_c).$$

Of course, the functions φ_{1c} and φ_{2c} possess the property of complete degeneracy (1.1.7) and the sequences $(U_k(\varphi_{sc}), \mathfrak{B}_{ks})$, $k = n, n+1, \dots$; $s = 1, 2$, are reversed martingales, where $\mathfrak{B}_{kc} = \sigma\{\omega : U_k(\varphi_{sc}), U_{k+1}(\varphi_{sc}, \dots\}$, i.e. $U_{k+1}(\varphi_{sc}) = E(U_k(\varphi_{sc})|\mathfrak{B}_{k+1,s})$. Therefore, for any $q \geq 1$, the sequences $(\|U_k(\varphi_c)\|^q, \mathfrak{B}_{ks})$ are reversed submartingales, i.e. $\|U_{k+1}(\varphi_{sc})\|^q \leq E(\|U_k(\varphi_{sc})\|^q|\mathfrak{B}_{k+1,s})$ with probability one. Consequently, one can employ Corollary 2.3.2. For $U_k(\varphi_{1c})$, according to this corollary, we have

$$P(2\sup_{k\geq n} \|U_n(\varphi_{1c})\| > t_c) \leq 4t_c^{-2}E\|U_n(\varphi_{1c})\|^2 .$$

Here,

$$E\|U_n(\varphi_{1c})\|^2 = \binom{n}{c}^{-1} E\|\varphi_{1c}\|^2 \leq \binom{n}{c}^{-1} c(c+1)2^c E\|g_c\|^2 I(\|g_c\| \leq \sqrt{n}).$$

Thus,

$$P(2\sup_{k\geq n} \|U_k(\varphi_{1c})\| > t_c) \leq t_c^{-2}\binom{n}{c}^{-1} c(c+1)2^{c+2}E\|g_c\|^2 I(\|g_c\| \leq \sqrt{n}).$$

For $U_k(\varphi_{2c})$, according to Corollary 2.3.2 we obtain

$$P(2\sup_{k\geq n} \|U_k(\varphi_{2c})\| > t_c) \leq 2^q t_c^{-q} E\|U_n(\varphi_{2c})\|^q$$

for any $1 \leq q \leq 2$. Using Corollary 2.7.1, we get

$$E\|U_n(\varphi_{2c})\|^q \leq \frac{(6c^3)^{cq}}{(c!)^{q-1}}\binom{n}{c}^{-q+1} E\|\varphi_{2c}\|^q \leq$$

$$\leq \frac{(6c^3)^{cq}(c(c+1))^{q-1}2^c}{(c!)^{q-1}}\binom{n}{c}^{-q+1} E\|g_c\|^q I(\|g_c\| > \sqrt{n}) .$$

Hence

$$P(2\sup_{k\geq n} \|U_k(\varphi_{2c})\| > t_c) \leq$$

$$\leq t_c^{-q}\frac{(6c^3)^{cq}(c(c+1))^{q-1}2^{c+q}}{(c!)^{q-1}}\binom{n}{c}^{-q+1} E\|g_c\|^q I(\|g_c\| > \sqrt{n})$$

Combining all the estimates, we arrive at the proof of this theorem.

Chapter 4

Weak convergence

The *weak convergence* of distribution functions of *UB*-statistics and different functionals of them is studied in this chapter. Let (E, ρ) be a metric space and let $\mathfrak{E}$ be a class of Borel subsets of E, i.e. $\mathfrak{E}$ is a σ-algebra generated by open sets. Assume that probability measures Q_n and Q are given on $(E, \mathfrak{E})$ where $n = 1, 2, \ldots$. We say that the sequence of probability measures $\{Q_n\}$ converges weakly as $n \to \infty$ to the probability measure Q if the relation

$$Q_n(A) \to Q(A)$$

holds for any Borel set $A \in \mathfrak{E}$ which satisfies the condition $Q(\partial A) = 0$, where ∂A is the boundary of set A.

Assume that η_n and η_∞ have the distributions Q_n and Q, respectively. If weak convergence holds, we say that the sequence $\{\eta_n\}$ of random elements converges to the random element η_∞ *in distribution* and write

$$\eta_n \Rightarrow \eta_\infty .$$

Sometimes we shall use the notation

$$L(\eta_n) \to L(\eta_\infty),$$

where $L(\eta)$ denotes the law (distribution) of random variable η.

For *UB*-statistics, the principal questions in the case of weak convergence relate to the clarification of the conditions imposed on kernel Φ, and the description of the limit random variable η_∞.

4.1 Central limit theorem

In this section we consider one-sample *UB*-statistics

$$U_n = \binom{n}{m}^{-1} \sum_{1 \le i_1 < \ldots < i_m \le n} \Phi(X_{i_1}, \ldots, X_{i_m}),$$

where $X_1, \dots, X_n$ are independent identically distributed random variables with distribution P and symmetric kernel $\Phi : X^m \to B$ is non-degenerate.

The goal here is to give sufficient conditions which guarantee the asymptotic normality of U_n.

UR-statistics. Let $\sigma_c^2 = Eg_c^2(X_1, \dots, X_c)$, $c = 1, \dots, m$. Since

$$E(S_{nc}\, S_{nd}) = \begin{cases} \binom{n}{c}\sigma_c^2 & , \quad c = d \\ 0 & , \quad c \neq d, \end{cases}$$

then by formula (1.1.9) we have

$$\sigma^2(U_n) = \sum_{c=1}^{m} \binom{m}{c}^2 \binom{n}{c}^{-1} \sigma_c^2 .$$

Therefore, the variance of U-statistics $\sigma^2(U_n)$ as $n \to \infty$ has the form

$$\sigma^2(U_n) = m^2 \sigma_1^2 n^{-1} + O(n^{-2}) .$$

This relation suggests that $m^2\sigma_1^2 n^{-1}$ can be chosen as a normalizing factor in the limit theorem.

Theorem 4.1.1. *Suppose that a kernel Φ of rank $r = 1$ satisfies the conditions*

$$Eg_1^2(X_1) < \infty$$

and

$$t^{\gamma_c} P(|g_c| > t) \to 0 \quad as \quad t \to \infty \tag{4.1.1}$$

for each $c = 2, \dots, m$ where $\gamma_c = 2c/(2c-1)$. Then

$$\frac{\sqrt{n}}{m\sigma_1}(U_n - \theta) \Rightarrow \tau ,$$

as $n \to \infty$, where τ is a standard normal random variable.

Proof. First, we note that

$$\begin{aligned} 1 \; &< \; 1 + \frac{1}{2m-1} \\ &= \; \gamma_m < \gamma_{m-1} < \dots < \gamma_2 \\ &< \; \gamma_1 \\ &= \; 2. \end{aligned}$$

Next, according to (1.1.9), we have

$$U_n - \theta = mn^{-1}\sum_{i=1}^{n} g_1(X_j) + \sum_{c=2}^{m} \binom{m}{c}\binom{n}{c}^{-1} S_{nc}\,. \tag{4.1.2}$$

From (4.1.2) it follows that

$$\frac{\sqrt{n}}{m\sigma_1}(U_n - \theta) = \frac{1}{\sigma_1\sqrt{n}}(g_1(X_1) + \cdots + g_1(X_n)) + R_n,$$

where

$$R_n = (m\sigma_1)^{-1}\sum_{c=2}^{m} m^{[c]} \prod_{j=1}^{c-1}(1 - n^{-1}j)^{-1} n^{-c/\gamma_c} S_{nc}\,.$$

By the central limit theorem for real-valued random variables

$$(g_1(X_1) + \cdots + g_1(X_n))/(\sigma_1\sqrt{n}) \Rightarrow \tau$$

as $n \to \infty$. Applying Theorem 3.1.2 with $m = c$ and $p = \gamma_c$, we see that under condition (4.1.1)

$$n^{-c/\gamma_c} S_{nc} \xrightarrow{P} 0 \quad \text{as } n \to \infty$$

and, hence, $R_n \to 0$ in probability. This proves the theorem.

Corollary 4.1.1. *If*

$$E|g_c|^{\gamma_c} < \infty\,, \quad c = 1, \dots, m\,,$$

then

$$\left\{\frac{\sqrt{n}}{m\sigma_1}(U_n - \theta)\right\}_{n=m}^{\infty}$$

converges weakly in R *to a standard normal random variable* τ.

UH-statistics. We assume that $\Phi : X^m \to H$ where H is a separable real Hilbert space with a scalar product $(\cdot,\cdot)$ and a norm $\|\cdot\|$. Moreover, $E\|\Phi\| < \infty$ and $E\Phi = \theta$.

We consider the case $r = 1$. Assume first that $E\|g_1\|^2 < \infty$. Under this condition we define a covariance operator V by the formula

$$(Vx, y) = \int_X (x, g_1(z))(y, g_1(z))\, P(dz)$$

for all $x, y \in H$.

According to the central limit theorem in a Hilbert space, the weak convergence

$$n^{-1/2}\big(g_1(X_1) + \cdots + g_1(X_n)\big) \Rightarrow \tau\,,$$

takes place in H, where τ is a Gaussian random element in H with the characteristic functional

$$E\exp\{i(a,\tau)\} = \exp\Big\{-\frac{1}{2}(Va,a)\Big\}, \quad a \in H.$$

Theorem 4.2.2. *Suppose that kernel Φ of rank 1 satisfies the conditions*

$$E\|g_1\|^2 < \infty$$

and

$$t^{\gamma_c} P(\|g_c\| > t) \to 0 \quad \text{as } t \to \infty \tag{4.1.3}$$

for each $c = 2, \dots, m$, where $\gamma_c = 2c/(2c-1)$. Then the weak convergence

$$\frac{\sqrt{n}}{m}(U_n - \theta) \Rightarrow \tau,$$

as $n \to \infty$, takes place, where τ is a Gaussian random element with covariance operator V.

Proof. Rewrite the representation (4.1.2) as follows

$$\frac{\sqrt{n}}{m}(U_n - \theta) = \frac{1}{\sqrt{n}}\big(g_1(X_1) + \dots + g_1(X_n)\big) + R_n,$$

where

$$R_n = m^{-1} \sum_{c=2}^{m} m^{[c]} \prod_{j=1}^{c-1} (1 - n^{-1}j)^{-1} n^{-c/\gamma_c} S_{nc}.$$

A Hilbert space is a Banach space of type 2. Hence we can apply Theorem 3.1.2 with $m = c$ and $p = \gamma_c$. This gives $n^{-c/\gamma_c} S_{nc} \to 0$ in probability as $n \to \infty$, under condition (4.1.3).

***UB*-statistics.** Let B^* be a space dual to B. The value of a functional $x^* \in B^*$ on the element $x \in B$ is denoted by $x^*(x)$. Assume that $0 < E\|g_1(X_1)\|^2 < \infty$. Under this condition, we define the covariance function V:

$$V(x^*, y^*) = \int_X x^*(g_1(z)) y^*(g_1(z))\, P(dz),$$

where $x^*, y^* \in B^*$. By the central limit theorem in a Banach space B of type 2 the weak convergence

$$n^{-1/2}\big(g_1(X_1) + \dots + g_1(X_n)\big) \Rightarrow \tau$$

takes place as $n \to \infty$, where τ is a Gaussian random element in B with the characteristic functional

$$E\exp\{ix^*(\tau)\} = \exp\{-\frac{1}{2}V(x^*, x^*)\}, \quad x^* \in B^*.$$

Theorem 4.1.3. *Assume that B is a Banach space of type 2 and kernel Φ of rank $r = 1$ satisfies the conditions*

$$E\|g_1(X_1)\|^2 < \infty$$

and

$$t^{\gamma_c} P(\|g_c\| > t) \to 0 \quad as \quad t \to \infty \tag{4.1.4}$$

for each $c = 2, \dots, m$, where $\gamma_c = 2c/(2c-1)$. Then the weak convergence

$$m^{-1}\sqrt{n}(U_n - \theta) \Rightarrow \tau$$

holds as $n \to \infty$, where τ is a Gaussian random element with covariance function V.

Proof. It is sufficient to apply (4.1.2) and Theorem 3.1.2 for $m = c$ and $p = \gamma_c$, since under condition (4.1.4), $R_n \to 0$ in probability as $n \to \infty$ for Banach space B of type 2.

Remark 4.1.1. Conditions (4.1.4) can be written in the following form

$$E\|g_c\| I(\|g_c\| > n^{(2c-1)/2}) = o(n^{-1/2}) \tag{4.1.5}$$

for each $c = 2, \dots, m$, as $n \to \infty$ (see Theorem 3.1.2). In particular, (4.1.5) takes place if $E\|g_c\|^{\gamma_c} < \infty$.

Banach-valued von Mises' statistics. Representation (1.3.3) indicates that von Mises' statistics V_n is a linear combination of UB-statistics of the type (1.1.1) with kernels of the form Φ_{mc} from (1.3.4). Thus, all weak limit theorems, which hold for UB-statistics, can be generalized for von Mises' statistics.

Theorem 4.1.4. *Let B be a separable Banach space of type 2. Let $\Phi : X^m \to B$ be a symmetric measurable kernel of rank $r = 1$ which satisfies the conditions*

$$E\|g_1(X_1)\|^2 < \infty,$$

$$t^{2c/(2c-1)} P(\|g_c\| > t) \to 0 \quad as \quad t \to \infty \tag{4.1.6}$$

for each $c = 2, \dots, m$, and

$$t^{2c/(2m-1)} P(\|\Phi_{mc}\| > t) \to 0 \quad as \quad t \to \infty \tag{4.1.7}$$

for each $c = 1, \dots, m-1$.

Then the weak convergence $m^{-1}\sqrt{n}(V_n - \theta) \Rightarrow \tau$ as $n \to \infty$, takes place, where τ is a Gaussian random element in Theorem 4.1.3.

Proof. Representation (1.3.3) can be written in the form

$$V_n - \theta = n^{-m} n^{[m]}(U_n - \theta) + \theta n^{-m}(n^{[m]} - n^m) + \sum_{c=1}^{m-1} n^{-m} \sum_{1\le i_1<\ldots<i_c\le n} \Phi_{mc}(X_{i_1},\ldots,X_{i_c}).$$

From this relation we obtain

$$m^{-1}\sqrt{n}(V_n - \theta) = n^{[m]} n^{-m} n^{-1/2}\big(g_1(X_1) + \cdots + g_1(X_n)\big) +$$

$$+\theta n^{-m}(n^{[m]} - n^m) m^{-1}\sqrt{n} + \sum_{c=2}^{m}(m-1)^{[c-1]}\prod_{j=1}^{c-1}(1 - n^{-1}j)^{-1} n^{-c/\gamma_c} S_{nc} +$$

$$+m^{-1}\sum_{c=1}^{m-1} n^{-c/\gamma_{mc}} S_n(\Phi_{mc}),$$

where $\gamma_c = 2c/(2c-1)$, $c = 2,\ldots,m$ and $\gamma_{mc} = 2c/(2m-1)$, $c = 1,\ldots,m-1$; the element S_{nc} was defined in (1.1.8) and

$$S_n(\Phi_{mc}) = \sum_{1\le i_1<\ldots<i_c\le n} \Phi_{mc}(X_{i_1},\ldots,X_{i_c}).$$

Under condition (4.1.6), $n^{-c/\gamma_c} S_{nc} \to 0$ in probability as $n \to \infty$. Therefore, taking into account the proof of Theorem 4.1.3, it remains to verify that under condition (4.1.7), $n^{-c/\gamma_{mc}} S_n(\Phi_{mc}) \to 0$ in probability as $n \to \infty$. For this purpose we represent the variable $S_n(\Phi_{mc})$ as a sum $S_n(\Phi_{mc}) = S_n(\Phi^1_{mc}) + S_n(\bar\Phi^1_{mc})$ in which $\Phi^1_{mc} = \Phi_{mc} I(\|\Phi_{mc}\| \le n^{(2m-1)/2})$ and $\bar\Phi^1_{mc} = \Phi_{mc} I(\|\Phi_{mc}\| > n^{(2m-1)/2})$. Then, for any $t > 0$ we have

$$P(n^{-c/\gamma_{mc}}\|S_n(\Phi_{mc})\| > 2t) \le P(n^{-c/\gamma_{mc}}\|S_n(\Phi^1_{mc})\| > t) + P(n^{-c/\gamma_{mc}}\|S_n(\bar\Phi^1_{mc})\| > t).$$

By Chebyshev's inequality

$$\begin{aligned} P(n^{-c/\gamma_{mc}}\|S_n(\Phi^1_{mc})\| > t) &\le t^{-1} n^{-c/\gamma_{mc}} E\|S_n(\Phi^1_{mc})\| \\ &\le t^{-1} n^{-c(\frac{1}{\gamma_{mc}}-1)} E\|\Phi_{mc}\| I(\|\Phi_{mc}\| \le n^{(2m-1)/2}) \end{aligned}$$

[with the help of integration by parts]

$$\begin{aligned} &\le -t^{-1} n^{c} P(\|\Phi_{mc}\| > n^{(2m-1)/2}) + \\ &\quad + t^{-1} n^{-\frac{2(m-c)-1}{2}} \int_0^{n^{(2m-1)/2}} P(\|\Phi_{mc}\| > x)\, dx. \end{aligned}$$

Here, under condition (4.1.7)

$$n^c P(\|\Phi_{mc}\| > n^{(2m-1)/2}) \to 0 \quad \text{as } n \to \infty$$

and

$$n^{-\frac{2(m-c)-1}{2}} \int_0^{n^{(2m-1)/2}} P(\|\Phi_{mc}\| > x)\, dx \leq$$

$$\leq \rho_n\, n^{-\frac{2(m-c)-1}{2}} \int_{n^{1/4}}^{n^{(2m-1)/2}} x^{-\gamma_{mc}}\, dx + n^{-\frac{2(m-c)-1}{2}} \int_0^{n^{1/4}} P(\|\Phi_{mc}\| > x)\, dx \leq$$

$$\leq (1-\gamma_{mc})^{-1}\rho_{nc} + n^{-1/4} \to 0 \quad \text{as} \quad n \to \infty$$

where

$$\rho_{nc} = \sup\{x^{\gamma_{mc}} P(\|\Phi_{mc}\| > x) : n^{1/4} \leq x \leq n^{(2m-1)/2}\} \to 0$$

for all $c = 1, \ldots, m-1$. Therefore, $n^{-c/\gamma_{mc}} S_n(\Phi^1_{mc}) \to 0$ in probability as $n \to \infty$.

Next, using (2.2.3) and Chebyshev's inequality, we write

$$\begin{aligned} P(n^{-c/\gamma_{mc}} \|S_n(\bar{\Phi}^1_{mc})\| > t) &\leq t^{-\frac{c}{m}} n^{-\frac{c^2}{m\gamma_{mc}}} E\|S_n(\bar{\Phi}^1_{mc})\|^{\frac{c}{m}} \\ &\leq t^{-\frac{c}{m}} n^{\frac{c}{2m}} E\|\Phi_{mc}\|^{\frac{c}{m}} I(\|\Phi_{mc}\| > n^{\frac{(2m-1)}{2}}) \end{aligned}$$

[under condition (4.1.7) with the help of integration by parts]

$$\begin{aligned} &\leq t^{-\frac{c}{m}} n^c P(\|\Phi_{mc}\| > n^{\frac{(2m-1)}{2}}) \\ &\quad + t^{\frac{-c}{m}} \frac{C}{m} n^{\frac{c}{2m}} \int_{n^{(2m-1)/2}}^{\infty} x^{\frac{2}{m}-1} P(\|\Phi_{mc}\| > x)\, dx \to 0 \end{aligned}$$

as $n \to \infty$. Hence, $n^{-c/\gamma_{mc}} S_n(\bar{\Phi}^1_{nc})$ in probability, which completes the proof of this theorem.

Corollary 4.1.2. *Let Banach space B be of type 2. If $m = 2$ and the following conditions are satisfied*

$$0 < E\|g_1\|^2 < \infty,$$

$$t^{4/3} P(\|g_2\| > t) \to 0 \quad \textit{as} \quad t \to \infty,$$

$$t^{2/3} P(\|\Phi(X_1, X_2)\| > t) \to 0 \quad \textit{as} \quad t \to \infty,$$

then the weak convergence

$$2^{-1}\sqrt{n}(V_n - \theta) \Rightarrow \tau$$

holds.

4.2 Convergence to a chaos

The main goal of this section is to show that if kernel Φ is a P-canonical function with finite second moment, then the limit distribution of the sequence $\{n^{m/2}U_n\}$ is the distribution of an element of a chaos. For convenience we describe at first the real chaos space, which appears as weak limits of UR-statistics.

Let $\left(\Omega, \mathfrak{F}, P\right)$ be a probability space. Let $\{\tau_i\}_{i=1}^{\infty}$ be a set of orthogaussian random variables and let $\mathcal{H}$ be a Gaussian subspace of $L^2(\Omega, \mathfrak{F}, P)$. This means that $\mathcal{H}$ is a closed linear span generated by $\{\tau_i\}_{i=1}^{\infty}$. Let $\mathcal{P}_m := \mathcal{P}_m(\mathcal{H})$ be a closed linear span of the following set of random variables: $\{\tau_1 \tau_2 \cdots \tau_r : r \leq m, \ \tau_i \in \mathcal{H}\}$ consisting of the product of r elements from $\mathcal{H}$ for $r \leq m$. The chaos space of order m associated with $\mathcal{H}$, $\mathcal{K}_m := \mathcal{K}_m(\mathcal{H})$ is the orthogonal complement of $\mathcal{P}_{m-1}$ in $\mathcal{P}_m$, $\mathcal{P}_m \ominus \mathcal{P}_{m-1}$. It is well known that a complete orthonormal system for $\mathcal{K}_m(\mathcal{H})$ is the set of functions $H_{k_1}(\tau_{\alpha_1}), \ldots, H_{k_r}(\tau_{\alpha_r})$ where $r \leq m$, $k_1 + \cdots + k_r = m$, $1 \leq k_i \leq m$, H_k is a Hermite polynomial of degree k and leading coefficient 1, and $\{\tau_{\alpha_1}, \ldots, \tau_{\alpha_r}\}$ runs over all the subsets of size r of $\{\tau_i\}_{i=1}^{\infty}$.

Now we define $L_c^2(X^m, P^m)$ as the set of all P-canonical measurable functions from L^2:

$$L_c^2(X^m, P^m) = \left\{\Phi \in L^2(X^m, P^m) : \quad \Phi \text{ is symmetric and completely degenerate}\right\}.$$

Let $\{\varphi_i\}_{i=1}^{\infty}$ be an orthonormal set of $L^2(X, P)$ such that together with a function identically equal to 1 it forms an orthonormal basis of $L^2(X, P)$. Then $\Phi \in L^2(X^m, P^m)$ is P-canonical if Φ has an L^2-expansion

$$\Phi(x_1, \ldots, x_m) = \sum_{(i_1, \ldots, i_m)=1}^{\infty} \Phi_{i_1, \ldots, i_m} \varphi_{i_1}(x_1) \cdots \varphi_{i_m}(x_m) \tag{4.2.1}$$

with the coefficients $\Phi_{i_1, \ldots, i_m}$ invariant with respect to permutations of indices

$$\Phi_{i_1, \ldots, i_m} = E\Big(\Phi(X_1, \ldots, X_m)\varphi_{i_1}(X_1) \cdots \varphi_{i_m}(X_m)\Big).$$

That is, Φ is in the closed linear span of $\{\varphi_{i_1}(x_1) \cdots \varphi_{i_m}(x_m) : \quad i_1, \ldots, i_m \geq 1\}$. The following map is a linear isometry

$$K_m : \ L_c^2(X^m, P^m) \to \mathcal{K}_m(\mathcal{H}),$$

$$K_m \left(\sum_{(i_1, \ldots, i_m)=1}^{\infty} \Phi_{i_1, \cdots, i_m} \varphi_{i_1} \cdots \varphi_{i_m} \right) \mapsto \sum_{(i_1, \ldots, i_m)=1}^{\infty} \Phi_{i_1, \ldots, i_m} \prod_{j=1}^{\infty} H_{j(i_1, \ldots, i_m)}(\tau_j) \tag{4.2.2}$$

where $j(i_1, \ldots, i_m) = \sum_{k=1}^{m} I(i_k = j)$. Here for Gaussian space $\mathcal{H}$, system $\mathcal{K}_m(\mathcal{H})$ is a real chaos of order m.

For given complete orthogaussian system $\{\tau_i\}_{i=1}^{\infty}$ of $\mathcal{H}$ and Banach space B, a B-valued $\mathcal{H}$-chaos of order m is defined as the set of random variables of the form

$$\sum_{(i_1,\dots,i_m)=1}^{\infty} x_{i_1,\dots,i_m} \prod_{j=1}^{\infty} H_{j(i_1,\dots,i_m)}(\tau_j)$$

where $x_{i_1,\dots,i_m} \in B$ and the limit takes place a.s. or in L^p for any $0 < p < \infty$. Let $\mathcal{K}_m(\mathcal{H}, B)$ be a B-valued $\mathcal{H}$-chaos of order m.

Therefore, any element of chaos $\mathcal{K}_m(\mathcal{H}, B)$ is in the linear span of

$$\{\prod_{j=1}^{\infty} H_{j(i_1,\dots,i_m)}(\tau_j) : i_1, \dots, i_m \geq 1\}.$$

***UR*-statistics.** We will prove that the weak limit of *UR*-statistics with a canonical kernel has the distribution of an element of chaos $\mathcal{K}_m(\mathcal{H})$ and can be written in the form of multiple stochastic integrals.

Theorem 4.2.1. *Let $\Phi \in L_c^2(X^m, P^m)$. Then the weak convergence*

$$n^{m/2} U_n \Rightarrow K_m(\Phi)$$

takes place as $n \to \infty$, where $K_m(\Phi)$ is an element of chaos $\mathcal{K}_m(\mathcal{H})$, defined in (4.2.2). *That is*

$$K_m(\Phi) = \sum_{(i_1,\dots,i_m)=1}^{\infty} \Phi_{i_1,\dots,i_m} \prod_{j=1}^{\infty} H_{j(i_1,\dots,i_m)}(\tau_j).$$

Proof. By substituting (4.2.1) into (1.1.1), we obtain the following representation in the space $L^2(X^n, P^n)$,

$$n^{[m]} U_n = \sum_{(i_1,\dots,i_m)=1}^{\infty} \Phi_{i_1,\dots,i_m} \operatorname{Per} A_n^m(\varphi_{i_1}, \dots, \varphi_{i_m}), \tag{4.2.3}$$

where

$$\operatorname{Per} A_n^m(\varphi_{i_1}, \dots, \varphi_{i_m}) = \sum_{1 \leq j_1 \neq \dots \neq j_m \leq n} \prod_{s=1}^{m} \varphi_{i_s}(X_{j_s})$$

is the permanent of matrix

$$A_n^m(\varphi_{i_1}, \dots, \varphi_{i_m}) = [a_{sj} = \varphi_{i_s}(X_j); \ \ 1 \leq s \leq m, \ 1 \leq j \leq n].$$

Relation (4.2.3) shows that the weak convergence of U-statistics depends on the asymptotic behaviour of permanents $\operatorname{Per} A_n^m(\varphi_{i_1}, \dots, \varphi_{i_m})$.

Lemma 4.2.1. *Let $i_1, \dots, i_m \geq 1$. Then the weak convergence*

$$n^{-m/2}\operatorname{Per} A_n^m(\varphi_{i_1}, \dots, \varphi_{i_m}) \Rightarrow H_{j(i_1,\dots,i_m)}(\tau_j)$$

takes place as $n \to \infty$, where $H_0(x) \equiv 1$, $x \in R$.

Proof. Consider the structure of matrix $A_n^m(\varphi_{i_1}, \dots, \varphi_{i_m})$. Fix $i_1, \dots, i_m$. Denote for simplicity $m_d = j(i_1, \dots, i_m)$ as $j(i_1, \dots, i_m) \geq 1$ and $d = 1, \dots, k$. Hence, $m_1 + \dots + m_k = m$ for some $k \geq 1$, dependent on $i_1, \dots, i_m$ and $m_d = 0$ for all $d \geq k+1$. Therefore, we have a multiple matrix (see Borovskikh and Korolyuk (1994b))

$$A_n^m(\varphi_{i_1}, \dots, \varphi_{i_m}) = \begin{bmatrix} A_n^{m_1}(\varphi_1) \\ \vdots \\ A_n^{m_k}(\varphi_k) \end{bmatrix},$$

where

$$A_n^{m_d}(\varphi_d) = \left[a_{sj} = \varphi_d(X_j);\ \ 1 \leq s \leq m_d,\ 1 \leq j \leq n\right], d = 1, \dots, k$$

are sampling matrices.

Then, by Theorem 4.5 in Borovskikh and Korolyuk (1994b) the weak convergence

$$n^{-m/2}\operatorname{Per} A_n^m(\varphi_{i_1}, \dots, \varphi_{i_m}) \Rightarrow \prod_{d=1}^{k} H_{m_d}(\tau_d)$$

holds as $n \to \infty$. This relation coincides with the assertion of Lemma 4.2.1.

Let Φ_N denote the N-th partial sum of the Fourier series (4.2.1) for function Φ, i.e.

$$\Phi_N(x_1, \dots, x_m) = \sum_{(i_1,\dots,i_m)=1}^{N} \Phi_{i_1,\dots,i_m}\varphi_{i_1}(x_1)\cdots\varphi_{i_m}(x_m).$$

By Lemma 4.2.1, we have

$$n^{m/2}U_n(\Phi_N) \Rightarrow K_m(\Phi_N),$$

as $n \to \infty$, where

$$K_m(\Phi_N) = \sum_{(i_1,\dots,i_m)=1}^{N} \Phi_{i_1,\dots,i_m}\prod_{j=1}^{\infty} H_{j(i_1,\dots,i_m)}(\tau_j).$$

For $K_m(\Phi_N)$, we get

$$K_m(\Phi_N) \to K_m(\Phi)$$

as $N \to \infty$ in L^2 since $EK_m^2(\Phi - \Phi_N) \to 0$, and hence $K_m(\Phi_N) \Rightarrow K_m(\Phi)$.

Let $\mathcal{L}(X, Y)$ be the Lévy–Prokhorov distance between the random elements X and Y. Note that the distance $\mathcal{L}(X, Y)$ metrizes the weak convergence.

Consider the inequality

$$\mathcal{L}(n^{m/2}U_n(\Phi), K_m(\Phi)) \leq$$

$$\leq \mathcal{L}(n^{m/2}U_n(\Phi), n^{m/2}U_n(\Phi_N)) + \mathcal{L}(n^{m/2}U_n(\Phi_N), K_m(\Phi_N)) + \mathcal{L}(K_m(\Phi_N), K_m(\Phi)).$$

By virtue of the proof above,

$$\mathcal{L}(n^{m/2}U_n(\Phi_N), K_m(\Phi_N)) \to 0$$

as $n \to \infty$ for any fixed N and

$$\mathcal{L}(K_m(\Phi_N), K_m(\Phi)) \to 0$$

as $N \to \infty$.

According to the property of the Lévy–Prokhorov metric, we have for all $n \geq m$

$$\begin{aligned} \mathcal{L}(n^{m/2}U_n(\Phi), n^{m/2}U_n(\Phi_N)) &\leq 2\left(n^m E U_n^2(\Phi - \Phi_N)\right)^{1/3} \\ &\leq 2m\left(E(\Phi - \Phi_N)^2\right)^{1/3} \to 0 \end{aligned}$$

as $N \to \infty$ uniformly on n.

As a result we get

$$\mathcal{L}(n^{m/2}U_n(\Phi), K_m(\Phi)) \to 0 \quad \text{as} \quad N \to \infty.$$

This proves Theorem 4.2.1.

Note that in (4.2.2)

$$\prod_{j=1}^{\infty} H_{j(i_1,\dots,i_m)}(\tau_j) \;\; = \, :\tau_{i_1} \cdots \tau_{i_m}:$$

for any fixed $i_1, \dots, i_m \geq 1$. Thus,

$$K_m(\Phi) = \sum_{(i_1,\dots,i_m)=1}^{\infty} \Phi_{i_1,\dots,i_m} :\tau_{i_1} \cdots \tau_{i_m}: .$$

The Wick polynomial is represented in terms of stochastic integrals as follows

$$:\tau_{i_1} \cdots \tau_{i_m}: \stackrel{\mathrm{d}}{=} \int_X \cdots \int_X \varphi_{i_1}(x_1) \cdots \varphi_{i_m}(x_m) W(dx_1) \dots W(dx_m),$$

where $W(\cdot)$ is a Gaussian random measure.

Thus

$$K_m(\Phi) \stackrel{\mathrm{d}}{=} I_m^W(\Phi) = \int_X \cdots \int_X \Phi(x_1, \dots, x_m) W(dx_1) \dots W(dx_m) \tag{4.2.4}$$

and

$$n^{m/2} U_n \Rightarrow I_m^W(\Phi)$$

as $n \to \infty$.

The following theorem shows that the sufficient condition $\Phi \in L_c^2(X^m, P^m)$ for the weak convergence of completely degenerate *UR*-statistics is also necessary.

Theorem 4.2.2. *If the sequence* $\{n^{m/2}U_n(\Phi)\}_{n=m}^{\infty}$ *is stochastically bounded, then*

$$E\Phi^2(X_1, \dots, X_m) < \infty$$

and

$$E\Phi(x_1, \dots, x_{m-1}, X_m) = 0$$

for almost every $(x_1, \dots, x_{m-1}) \in X^{m-1}$ *(and therefore* $\{n^{m/2}U_n(\Phi)\}_{n=m}^{\infty}$ *converges in distribution).*

Proof of this theorem belongs to Giné and Zinn (1994). It was established that the decoupled, randomized U-statistics are tight; Khintchine's inequality together with the Paley–Zygmund inequality allow us to conclude that U-statistics based on Φ^2 also form a tight sequence. Therefore, by positiveness, U-statistics based on the truncations $\Phi^2 I(\Phi^2 \le c)$ are tight uniformly in n and c; this implies $E\Phi^2 < \infty$ by the law of large numbers for U-statistics with the bounded kernels $\Phi^2 I(\Phi^2 \le c)$. Strict details of this proof can be found in Giné and Zinn (1994).

Consider the general situation for a kernel Φ of arbitrary rank.

Theorem 4.2.3. *Assume that a kernel* Φ *has rank* r *and satisfies the conditions*

$$Eg_r^2(X_1, \dots, X_r) < \infty$$

and

$$t^{2c/(2c-r)} P(|g_c| > t) \to 0 \quad as \quad t \to \infty \tag{4.2.5}$$

for each $c = r+1, \dots, m$.

Then the weak convergence

$$n^{r/2} U_n \Rightarrow \binom{m}{r} K_r(g_r)$$

takes place as $n \to \infty$, *where*

$$K_r(g_r) = \sum_{(i_1,\dots,i_r)=1}^{\infty} g_{i_1,\dots,i_r} \prod_{j=1}^{\infty} H_{j(i_1,\dots,i_r)}(\tau_j),$$

with

$$g_{i_1,\dots,i_r} = E\Big(g_r(X_1, \dots, X_r)\varphi_{i_1}(X_1)\cdots\varphi_{i_r}(X_r)\Big).$$

Moreover

$$K_r(g_r) \overset{\mathrm{d}}{=} I_r^W(g_r).$$

Proof. According to the Hoeffding representation (1.1.12) with $\theta = 0$, we have

$$n^{r/2}U_n = \binom{m}{r} n^{r/2} U_{nr} + R_{nr}\,,$$

where

$$R_{nr} = \sum_{c=r+1}^{m} m^{[c]} \prod_{j=1}^{c-1} (1 - n^{-1}j)^{-1} n^{-c/\gamma_{rc}} S_{nc}\,,$$

with $\gamma_{rc} = 2c/(2c - r)$.

By Theorem 3.1.2 for $m = c$ and $p = \gamma_{rc}$, $n^{-c/\gamma_{rc}} S_{nc} \to 0$ in probability as $n \to \infty$. Hence, under condition (4.2.5), $R_{nr} \to 0$ in probability as $n \to \infty$. Since $Eg_r^2 < \infty$ then by Theorem 4.2.1, $n^{r/2}U_{nr} \Rightarrow K_r(g_r)$. This completes the proof of the theorem.

Remark 4.2.1. Suppose that an integrable kernel Φ has rank r. The general problem should be formulated in the following way,– 'Is it true that if, for some $1 \le r \le m$, the sequence $\{n^{r/2}U_n(\Phi)\}_{n=m}^{\infty}$ is stochastically bounded, then $E|\Phi| < \infty$, $g_c \equiv 0$ for $c < r$, $Eg_r^2 < \infty$ and $n^{r/2}U_{nc} \to 0$ in probability for $r < c \le m$?'

Theorem 4.2.3 answers this question in the affirmative for $r = m$. Apparently, condition (4.2.5) and the relation $n^{r/2}U_{nc} \to 0$ in probability for $r < c \le m$ are equivalent.

Corollary 4.2.1. *If $r = 1$, then*

$$n^{1/2}U_n \Rightarrow m\sigma_1\tau$$

where τ is a standard normal random variable.

Proof. In this case, we have

$$\begin{aligned} K_1(g_1) &= m\sum_{i=1}^{\infty} g_i \prod_{j=1}^{\infty} H_{j(i)}(\tau_j) \\ &= m\sum_{i=1}^{\infty} g_i\tau_i \end{aligned}$$

since $\prod_{j=1}^{\infty} H_{j(i)}(\tau_j) = \tau_i$. In addition, $\sum_{i=1}^{\infty} g_i\tau_i \stackrel{d}{=} \sigma_1\tau$, since τ_i are independent standard normal random variables where $\sigma_1^2 = \sum_{i=1}^{\infty} g_i^2$.

Corollary 4.2.2. *If $r = 2$, then*

$$nU_n \Rightarrow \frac{m(m-1)}{2} \sum_{i=1}^{\infty} \lambda_i(\tau_i^2 - 1)\,,$$

where $\{\lambda_i\}$ *are the eigenvalues of the operator*

$$S: \ f \mapsto E(g_2(X_1, X_2)f(X_2)|\ X_1),$$

acting from $L^2(X,P)$ *into* $L^2(X,P)$ *and satisfying*

$$\sum_{i=1}^{\infty} \lambda_i^2 = Eg_2^2 .$$

Proof. In fact the integral operator

$$Sf(x) = \int g_2(x,y)f(y)P(dy)$$

is a selfadjoint compact (Hilbert–Schmidt) linear operator. Therefore, S has a sequence of eigenfunctions φ_i with eigenvalues λ_i. Moreover, $\{\varphi_i\}$ is an orthonormal basis. Suppose that $\varphi_0 \equiv 1$. Since $S\varphi_0 = g_1 = 0$, we assume that $\lambda_0 = 0$. Thus,

$$\begin{aligned} g_{ij} &= E\Big(g_2(X_1,X_2)\varphi_i(X_1)\varphi_j(X_2)\Big) \\ &= E(\varphi_i(X_1)S\varphi_j(X_1)) \\ &= \lambda_j \delta_{ij} \end{aligned}$$

and

$$\begin{aligned} K_2(g_2) &= \binom{m}{2} \sum_{i=1}^{\infty} \lambda_i \ :\tau_i^2: \\ &= \frac{m(m-1)}{2} \sum_{i=1}^{\infty} \lambda_i(\tau_i^2 - 1) . \end{aligned}$$

The second equality here is due to the fact that $:\tau_i^2 := H_2(\tau_i) = \tau_i^2 - 1$.

***UH*-statistics.** We suppose that $B = H$ is a Hilbert space.

Theorem 4.2.4. *If a Hilbert valued kernel* Φ *of rank* r *is such that*

$$E\|g_r(X_1, \dots, X_r)\|^2 < \infty$$

and

$$t^{2c/(2c-r)} P(\|g_c\| > t) \to \quad as \ \ t \to \infty$$

for each $c = r+1, \dots, m$, *then the weak convergence*

$$n^{r/2}(U_n - \theta) \Rightarrow \binom{m}{r} I_r^W(g_r)$$

takes place as $n \to \infty$.

Proof. Let us rewrite (1.1.12) as follows

$$n^{r/2}(U_n - \theta) = \binom{m}{r} n^{r/2} U_{nr} + R_{nr}\,, \tag{4.2.6}$$

where

$$R_{nr} = \sum_{c=r+1}^{m} m^{[c]} \prod_{j=1}^{c-1} (1 - n^{-1}j)^{-1} n^{-c/\gamma_{rc}} S_{nc}$$

with $\gamma_{rc} = 2c/(2c - r)$.

First, we see that under the conditions of the theorem, $R_{nr} \to 0$ in probability as $n \to \infty$ (by Theorem 3.1.2 for $m = c$ and $p = \gamma_{rc}$).

Therefore, it is sufficient to prove that

$$n^{r/2} U_{nr} \Rightarrow I_r^W(g_r)\,.$$

According to the general theory (see Gikhman and Skorokhod (1971)), this relation holds if the sequence of measures $\{\mu_n\}$ generated by the random elements $\{n^{r/2}U_{nr}\}$ is weakly compact and

$$\lim_{n\to\infty} E\exp(i\eta_n(h)) = E\exp(i\eta_\infty(h))$$

for all $h \in H$, where

$$\eta_n(h) = (n^{r/2}U_{nr}, h)\,, \quad \eta_\infty(h) = (I_r^W(g_r), h)\,.$$

By Theorem 2 of Gikhman and Skorokhod (1971), p. 448, the family of measures $\{\mu_n\}$ is weakly compact if and only if

1) for any $\varepsilon > 0$, there exists $\lambda > 0$ such that

$$\mu_n\{x : \ \|x\| > \lambda\} < \varepsilon$$

for all n;

2) there exists a basis $\{e_j\}, j \geq 1$, in which the series

$$\sigma_n^2 = \sum_{j=1}^{\infty} \int_{\|x\|<\lambda} (e_j, x)^2 \mu_n(dx)$$

converges uniformly in n for any $\lambda > 0$.

In order to verify 1) under the condition $E\|g_r\|^2 < \infty$, we write

$$\begin{aligned} \mu_n\{x : \ \|x\| > \lambda\} &\leq \lambda^{-2} \int \|x\|^2 \mu_n(dx) \\ &= \lambda^{-2} n^r E\|U_{nr}\|^2 \end{aligned}$$

$$= \lambda^{-2} r! \Big(1 - \frac{1}{n}\Big)^{-1} \dots \Big(1 - \frac{r-1}{n}\Big)^{-1} E\|g_r\|^2 .$$

To verify 2), we write

$$\begin{aligned}
\sigma_n^2 &= n^r \sum_{j=1}^{\infty} \int_{\|U_{nr}\|<\lambda} \Big(e_j, U_{nr}(x_1, \dots, x_n)\Big)^2 P(dx_1) \dots P(dx_n) \\
&= n^r \int_{\|U_{nr}\|<\lambda} \|U_{nr}(x_1, \dots, x_n)\|^2 P(dx_1) \dots P(dx_n) \\
&\leq n^r E\|U_{nr}\|^2 \\
&= n^r \binom{n}{r}^{-1} E\|g_r\|^2 \\
&\leq r^r E\|g_r\|^2 .
\end{aligned}$$

Next, by definition,

$$\eta_n(h) = n^{-r/2} \sum_{1 \leq i_1 \neq \dots \neq i_r \leq n} \Big(h, g_r(X_{i_1}, \dots, X_{i_r})\Big)$$

is a normed U-statistic with degenerate kernel $\Big(h,\ g_r(x_1, \dots, x_r)\Big) \in R$. Therefore, if the condition $E\|g_r\|^2 < \infty$ is satisfied, then, according to Theorem 4.2.1, the weak convergence

$$\eta_n(h) \Rightarrow \eta_\infty(h)$$

takes place as $n \to \infty$, where

$$\eta_\infty(h) =$$

$$= \sum_{(i_1, \dots, i_r)=1}^{\infty} E\Big\{\Big(h, g_r(X_1, \dots, X_r)\Big)\varphi_{i_1}(X_1) \cdots \varphi_{i_r}(X_r)\Big\} \prod_{j=1}^{\infty} H_{j(i_1, \dots, i_r)}(\tau_j) =$$

$$= I_r^W((h, g_r)) .$$

This proves the main statement of the theorem.

UB-statistics. Here we investigate the weak convergence of U-statistics with values in a Banach space of type 2.

Theorem 4.2.5. *Let B be a separable Banach space of type 2. Let $\Phi :\ X^m \to B$ be a symmetric measurable kernel of rank r, $1 \leq r \leq m$. Suppose that*

$$E\|g_r(X_1, \dots, X_r)\|^2 < \infty$$

and

$$t^{2c/(2c-r)} P(\|g_c\| > t) \to 0 \quad as \quad t \to \infty$$

for each $c = r+1, \dots, m$. *Then there is a linear map* $K_r : L_c^2(X^r, P^r) \to \mathcal{K}_r(\mathcal{H}, B)$ *such that the weak convergence in* B

$$n^{r/2}(U_n - \theta) \Rightarrow \binom{m}{r} K_r(g_r)$$

holds and

$$K_r(g_r) \stackrel{\mathrm{d}}{=} I_r^W(g_r).$$

Proof. We apply representation (4.2.6). First, by Theorem 3.1.2, $R_{nr} \to 0$ in probability as $n \to \infty$. Next we shall prove that under the condition $E\|g_r\|^2 < \infty$ the weak convergence

$$n^{r/2} U_{nr} \Rightarrow K_r(g_r)$$

takes place. Note that by Theorem 4.2.1 for any $x^* \in B^*$

$$n^{-r/2} \sum_{1 \le i_1 \ne \dots \ne i_r \le n} x^*(g_r(X_{i_1}, \dots, X_{i_r})) \Rightarrow x^*(K_r(g_r))$$

and

$$x^*(K_r(g_r)) \stackrel{\mathrm{d}}{=} x^*(I_r^W(g_r)).$$

Hence we only need to show tightness in the form of finite-dimensional approximations (see, e.g. Theorem 4.5, Chapter 1 in Araujo and Giné (1980)). For this goal we now consider the following lemma.

Lemma 4.2.2. *Let* $\Phi : X^m \to B$. *Suppose that* $\Phi(X_1, \dots, X_m)$ *is a* B*-valued* P*-canonical random variable with finite second moment:* $E\|\Phi(X_1, \dots, X_m)\|^2 < \infty$.

Then for each $\delta > 0$ *there is a sample canonical function* f *such that*

$$\left(E\|\Phi - f\|^2\right)^{1/2} \le \delta.$$

Proof. Let $\{x_j\}$ be a dense set having the same range as Φ. We define the point set

$$B_j = \left\{\|\Phi - x_i\| > \delta 4^{-m} \text{ for } i = 1, \dots, j-1, \|\Phi - x_j\| \le \delta 4^{-m}\right\} \in X^m.$$

Then $h = \sum_{j=1}^{\infty} x_j I(B_j)$ satisfies $\|\Phi - h\|_2 \le \delta 4^{-m}$. Taking large N we can get that $h_N = \sum_{j=1}^{N} x_j I(B_j)$ satisfies $\|\Phi - h\|_2 \le \delta 3^{-m}$. Since B_j is a measurable set in X^m, $I(B_j)$ can be approximated in L^1 by functions from the linear span of $I(A_1 \times \dots \times A_m)$, where $A_i \in \mathfrak{X}$. Hence, there exists a function $g = \sum_{j=1}^{k} y_j I(A_1^{(j)} \times \dots \times A_m^{(j)})$ with y_j in B and $A_i^{(j)} \in \mathfrak{X}$ such that $\|\Phi - g\|_2 \le \delta 2^{-m}$. For given function φ we define $\pi_m^c \varphi = (\delta_{x_1} - P) \dots (\delta_{x_c} - P) P^{m-c} \varphi$ and $S_m \varphi = (m!)^{-1} \sum_\sigma \varphi(x_{\sigma(1)}, \dots, x_{\sigma(m)})$ where the sum runs over all permutations σ of $\{1, \dots, m\}$. Note, that for $\varphi \in L^2$

$$\|S_m \varphi\|_2 \le (m!)^{-1} \sum_\sigma \|\varphi(x_{\sigma(1)}, \dots, x_{\sigma(m)})\|_2 = \|\varphi\|_2$$

and

$$\begin{aligned}
\|\pi_m^m\varphi\|_2 &= \Big\| \sum_{J\in\{1,\dots,m\}} (-1)^{\#J} E_J\varphi \Big\|_2 \\
&\le \sum_{J\in\{1,\dots,m\}} \|E_J\varphi\|_2 \\
&= \sum_{J\in\{1,\dots,m\}} \left(E_{J^c}\|E_J\varphi\|^2\right)^{1/2} \\
&\le 2^m \|\varphi\|_2 ,
\end{aligned}$$

where $\#J$ denotes the cardinality of J and E_J means expectation with respect to the set of coordinates J. Since Φ is a canonical symmetric function, $S_m\pi_m^m\Phi = \Phi$. Furthermore

$$\|S_m\pi_m^m g - \Phi\|_2 = \|S_m\pi_m^m(g-\Phi)\|_2 \le 2^m\|g-\Phi\|_2.$$

Therefore function $f = S_m\pi_m^m g$ satisfies the required conditions. This proves the lemma.

Now we can prove that for all $\varepsilon > 0$ there is a finite dimensional subspace $F \subset B$ such that

$$\sup_n P\{d(n^{r/2}U_{nr}, F) > \varepsilon\} \le \varepsilon$$

where $d(x,F) = \inf\{\|x-y\| : y \in F\}$ denotes the distance of $x \in B$ to subspace F. Given $\varepsilon > 0$, by Lemma 4.2.2 with $m = r$, $\delta = A^{-r/2}(2r)^{-4r}\varepsilon^{3/2}$ there is a simple canonical function f such that $E\|g_r - f\|^2 \le A^{-r}(2r)^{-8r}\varepsilon^3$ with constant A defined by (2.7.1). According to Corollary 2.7.1

$$E\|n^{r/2}U_{nr}(g_r - f)\|^2 \le A^r(2r)^{8r}\, E\|g_r - f\|^2 \le \varepsilon^3.$$

Hence if F is a finite-dimensional space spanned by the value of f, we have

$$\begin{aligned}
P\Big(d\Big(n^{r/2}U_{nr}(g_r), F\Big) > \varepsilon\Big) &\le P\Big(n^{r/2}\|U_{nr}(g_r) - U_{nr}(f)\| > \varepsilon\Big) \\
&= P\Big(n^{r/2}\|U_{nr}(g_r - f)\| > \varepsilon\Big) \\
&\le \varepsilon^{-2} n^r E\|U_{nr}(g_r - f)\|^2 \\
&\le \varepsilon
\end{aligned}$$

with the result following.

Banach-valued von Mises' statistics. We shall apply representation (1.3.2). With the help of canonical functions g_c we define the following symmetric functions

$$g_{cd}(x_1,\dots,x_d) = \sum_{\substack{\nu_1+\dots+\nu_d=c \\ \nu_j\ge 1}} \frac{c!}{\nu_1!\dots\nu_d!} g_c(x_1^{\nu_1},\dots,x_d^{\nu_d})$$

where $x_j^{\nu_j} = (x_j, \dots, x_j)$ is a vector with ν_j coordinates.

Theorem 4.2.6. *Let B be a Banach space of type 2. Suppose that kernel $\Phi : X^m \to B$ of rank r satisfies the conditions*

$$\max\{E\|g_r(X_{i_1}, \dots, X_{i_r})\|^2 : \ 1 \le i_1 \le \dots \le i_r \le r\} < \infty$$

and

$$E\|g_{cd}(X_1, \dots, X_d)\|^{2d/(2c-r)} < \infty$$

for all $d = 1, \dots, c$, and $c = r+1, \dots, m$.

Then the weak convergence

$$n^{r/2}(V_n - \theta) \Rightarrow \binom{m}{r} I_r^Q(g_r)$$

holds as $n \to \infty$, where $Q(\cdot)$ is a centered Gaussian random measure with covariance function $EQ(A)Q(B) = P(A \cap B) - P(A)P(B)$, $A, B \in \mathfrak{X}$.

Proof. Rewrite representation (1.3.2) with the help of (1.3.4) in such a form

$$n^{r/2}(V_n - \theta) = \binom{m}{r} \int_X \cdots \int_X g_r(x_1, \dots, x_r) Q_n(dx_1) \dots Q_n(dx_r) +$$

$$+ \sum_{c=r+1}^{m} \binom{m}{c} \sum_{d=1}^{c} n^{-d/\gamma_d} \sum_{1 \le i_1 < \dots < i_d \le n} g_{cd}(X_{i_1}, \dots, X_{i_d}),$$

where $\gamma_d = 2d/(2c - r)$; $Q_n(A) = \sqrt{n}(P_n(A) - P(A))$, and $P_n(A)$ is the empirical measure, $A \in \mathfrak{X}$.

Further, under the conditions of this theorem the relations

$$\int_X \cdots \int_X g_r(x_1, \dots, x_r) \prod_{s=1}^{r} Q_n(dx_s) \Rightarrow \int_X \cdots \int_X g_r(x_1, \dots, x_r) \prod_{s=1}^{r} Q(dx_s)$$

and

$$n^{-d/\gamma_d} \sum_{1 \le i_1 < \dots < i_d \le n} g_{cd}(X_{i_1}, \dots, X_{i_d}) \xrightarrow{P} 0$$

hold as $n \to \infty$. The proof of these relations is very similar to that of the above stated theorems. Therefore it is omitted.

The statement of Theorem 4.2.6 holds even if the conditions on kernel Φ are weakened. This is verified in the following result.

Theorem 4.2.7. *Let B be a Banach space of type 2. Assume that kernel $\Phi : X^2 \to B$ of rank $r = 2$ satisfies the conditions*

$$E\|\Phi(X_1, X_2)\|^2 < \infty$$

and

$$E\|\Phi(X_1, X_1)\| < \infty .$$

Then the weak convergence

$$nV_n \Rightarrow E\,\Phi(X_1, X_1) + \int_X \int_X \Phi(x, y) W(dx) W(dy)$$

takes place as $n \to \infty$.

Proof. We represent V_n in the form

$$nV_n = \frac{1}{n} \sum_{j=1}^{n} \Phi(X_j, X_j) + \frac{1}{n} \sum_{1 \le i \ne j \le n} \Phi(X_i, X_j)$$

where by virtue of the strong law of large numbers

$$\frac{1}{n} \sum_{j=1}^{n} \Phi(X_j, X_j) \to E\,\Phi(X_1, X_1) \quad a.s.$$

and according to Theorem 4.2.5 with $m = r = 2$ we have the weak convergence

$$\frac{1}{n} \sum_{1 \le i \ne j \le n} \Phi(X_i, X_j) \Rightarrow \int_X \int_X \Phi(x, y) W(dx) W(dy) .$$

This proves the theorem.

4.3 Multi-sample *UB*-statistics

Consider the asymptotic distribution of *U*-statistics of type (1.2.1). We show that in some situations the limit distribution of multi-sample *UB*-statistics is the distribution of an element of a multiple chaos. This element is the Wiener–Itô multiple integral of a *P*-canonical function with respect to Gaussian random measures.

Asymptotic normality in a sector. Suppose, that for a vector $\bar{n} = (n_1, \dots, n_c) \in N(\lambda)$ we have the relations $n_j n_1^{-1} \to \lambda_j, \quad j = 1, \dots, c$, as $n_1 \to \infty$ where $\lambda_1 = 1$ and $0 < \lambda_j < \infty$ for $j = 2, \dots, c$. Denote $g_j(x) = g_{d_1 \dots d_c}(x)$ for $x \in X_j$ when $d_1 = \dots = d_{j-1} = d_{j+1} = \dots = d_c = 0$ and $d_j = 1$. It means that $g_j(x) = E(\Phi - \theta | X_{j1} = x), j = 1, \dots, c$. Let $\sigma_j^2 = E\|g_j(X_{j1})\|^2, \quad j = 1, \dots, c$. Under the condition $\sigma_j^2 < \infty$ we define the covariance function

$$V_j(x^*, y^*) = \int_{X_j} x^*(g_j(z)) y^*(g_j(z))\, P_j(dz) , \quad x^*, y^* \in B^*$$

and let

$$V = \sum_{j=1}^{c} m_j^2 \lambda_j^{-1} V_j .$$

According to the central limit theorem in a Banach space B of type 2, if $0 < \sigma_j^2 < \infty$ then the weak convergence

$$n^{-1/2}\Big(g_j(X_{j1}) + \cdots + g_j(X_{jn_j})\Big) \Rightarrow \tau_j$$

takes place as $n_j \to \infty$, where τ_j is a Gaussian random element in B with the characteristic functional

$$E\exp\{ix^*(\tau_j)\} = \exp\{-\frac{1}{2}V_j(x^*, x^*)\}, \quad x^* \in B^*.$$

Denote

$$\tau = \sum_{j=1}^{c} m_j \lambda_j^{-1/2} \tau_j .$$

It is obvious that τ is a Gaussian random element in B with covariance function V.

Theorem 4.3.1. *Let B be a Banach space of type 2. Assume that symmetric kernel $\Phi: X^m \to B$ satisfies the conditions*

$$0 < \max\{\sigma_1^2, \ldots, \sigma_c^2\} < \infty$$

and

$$t^{\gamma_{\bar d}} P(\|g_{\bar d}\| > t) \to 0 \quad as \quad t \to \infty$$

where $\gamma_{\bar d} = 2d/(2d-1)$, $d = 2, \ldots, m$ with $d = d_1 + \cdots + d_c$, $m = m_1 + \cdots + m_c$.

Then the weak convergence

$$n_1^{1/2}(U_{\bar n} - \theta) \Rightarrow \tau$$

takes place as $n_1 \to \infty$.

Proof. By formula (1.2.8) we have

$$n_1^{1/2}(U_{\bar n} - \theta) = \sum_{j=1}^{c} m_j (n_1/n_j)^{1/2} n_j^{-1/2} \sum_{i=1}^{n_j} g_j(X_{ji}) + R_{\bar n},$$

where

$$R_{\bar n} = \sqrt{n_1} \sum_{d=2}^{m} \sum_{d_1 + \cdots + d_c = d} \prod_{j=1}^{c} \binom{m_j}{d_j} \binom{n_j}{d_j}^{-1} S_{\bar n \bar d}.$$

By the central limit theorem in B

$$\sum_{j=1}^{c} m_j (n_1/n_j)^{1/2} n_j^{-1/2} \sum_{i=1}^{n_j} g_j(X_{ji}) \Rightarrow \tau .$$

Consider $R_{\bar n}$. Note, that

$$\sqrt{n_1} \prod_{j=1}^{c} \binom{m_j}{d_j} \binom{n_j}{d_j}^{-1} = \prod_{j=1}^{c} m_j^{[d_j]} \prod_{s=1}^{d_j - 1} (1 - s n_j^{-1}) n_j^{-d_j/q_j}$$

with

$$q_1 = \frac{2d_1}{2d_1 - 1}, \quad q_2 = \ldots = q_c = 1\,.$$

Hence,

$$R_{\bar{n}} = \sum_{d=2}^{m} \sum_{d_1+\cdots+d_c=d} \prod_{s=1}^{d_j-1} (1 - sn_j^{-1})^{-1} \prod_{j=1}^{c} n_j^{-d_j/q_j} S_{\bar{n}\bar{d}}\,.$$

Taking into account the preceding notation (3.2.3) for chosen q_i, $i = 1, \ldots, c$, we have

$$\begin{aligned} \gamma_{\bar{d}\bar{d}} &= \frac{d_1 + \cdots + d_c}{\sum_{i=1}^{c}(d_i/q_i)} \\ &= \frac{d_1 + \cdots + d_c}{d_1 + \cdots + d_c - \frac{1}{2}} \\ &= \frac{2d}{2d-1} = \gamma_{\bar{d}}\,. \end{aligned}$$

Therefore, by Theorem 3.2.3

$$R_{\bar{n}} \xrightarrow{P} 0$$

as $n_1 \to \infty$. This proves the theorem.

Asymptotic normality as $\min(n_1, \ldots, n_c) \to \infty$. For simplicity we consider *UR*-statistics. Denote

$$\hat{\sigma}^2 = \sum_{j=1}^{c} m_j^2 n_j^{-1} \sigma_j^2\,,$$

where $\sigma_j^2 = Eg_j^2(X_{j1})$.

Theorem 4.3.2. *Assume that real-valued kernel Φ satisfies the following conditions*

$$\min\{\sigma_1^2, \ldots, \sigma_c^2\} > 0\,, \quad \min\{\sigma_1^2, \ldots, \sigma_c^2\} < \infty$$

and

$$t^{\gamma_{\bar{d}}} P(|g_{\bar{d}}| > t) \to 0 \quad \text{as } t \to \infty$$

for each $\bar{d}$ with $d = 1, \ldots, m$, where $\gamma_{\bar{d}} = 2d/(2d-1)$ and $d = d_1 + \cdots + d_c$, $m = m_1 + \cdots + m_c$.

Then the weak convergence

$$\hat{\sigma}^{-1}(U_n - \theta) \Rightarrow \tau$$

holds as $\min(n_1, \ldots, n_c) \to \infty$, and τ is the standard normal random variable.

Proof. Without loss of generality we can assume that $n_1 \le n_2 \le \ldots n_c$. Note that

$$n_1^{-1/2}\sigma_1 \le \hat{\sigma} \le n_1^{-1/2}(\sigma_1^2 + \cdots + \sigma_c^2)^{1/2}\,.$$

By formula (1.2.8)

$$\hat{\sigma}^{-1}(U_n - \theta) = \sum_{j=1}^{c} m_j (n_j \hat{\sigma})^{-1} \sum_{i=1}^{n_j} g_j(X_{ji}) + \sqrt{n_1}\hat{\sigma}^{-1} R_{\bar{n}} .$$

Here by the central limit theorem for sums of independent nonidentically distributed random variables in R, the weak convergence

$$\sum_{j=1}^{c} m_j (n_j \hat{\sigma})^{-1} \sum_{i=1}^{n_j} g_j(X_{ji}) \Rightarrow \tau$$

is valid, and by Theorem 3.2.3

$$\sqrt{n_1}\hat{\sigma}^{-1} R_{\bar{n}} \xrightarrow{P} 0$$

as $\min(n_1, \dots, n_c) \to \infty$. This proves the theorem.

Asymptotic nonnormality. Consider the two-sample case, i.e. $c = 2$. We shall apply in this situation the Hoeffding representation (1.2.6). The results of this section show that the asymptotic normality of multi-sample *UB*-statistics is provided by the value of rank $r = 1$ of kernel Φ and by the way of convergence to infinity of the numbers n_i and n_j under the condition $\min(n_1, \dots, n_c) \to \infty$. Other conditions on these parameters lead to another type of limiting distribution different from the normal distribution. Here we investigate the following cases for *UR*-statistics.

Proposition 4.3.1. *Let*

$$g_{10}(x) \neq 0, \quad g_{01}(x) \neq 0$$

and $Eg_{10}^2(X_{11}) < \infty$. *Then*

$$\sigma_1^{-1} n_1^{-1/2}\Big(g_{10}(X_{11}) + \cdots + g_{10}(X_{1n_1})\Big) \Rightarrow \tau$$

as $n_1 \to \infty$ *where* τ *is a standard normal random variable. If* $1 < \alpha \leq 2$ *then for some conditions on the distribution of random variable* $g_{01}(X_{21})$, *the following relation is valid*

$$n_2^{-1/\alpha}\Big(g_{01}(X_{21}) + \cdots + g_{01}(X_{2n_2})\Big) \Rightarrow \eta_\alpha$$

as $n_2 \to \infty$ *where* η_α *is an* α*-stable random variable.*

If n_1 *and* n_2 *tend to infinity such that*

$$n_1^{1/2} n_2^{-(\alpha-1)/\alpha} \to 1$$

then the weak convergence

$$n_1^{1/2}(U_{\bar{n}} - \theta) \Rightarrow 2\sigma_1 \tau + \eta_\alpha$$

takes place under some additional moment conditions on the canonical functions $g_{\bar d}$ for $d \geq 2$.

Proof. It is sufficient to apply (1.2.6) under the conditions of this theorem.

Proposition 4.3.2. *Suppose that* $E\Phi^2 < \infty$ *and* $n_1 n_2^{-2} \to 0$ *as* $n_1 \to \infty$ *and* $n_2 \to \infty$. *If*

$$g_{10}(x) = 0, \quad g_{01}(x) = 0, \quad g_{11}(x) = 0,$$

$$g_{02}(x) = 0, \quad g_{20}(x) \neq 0,$$

then the weak convergence

$$n_1(U_{\bar n} - \theta) \Rightarrow \eta_{20},$$

takes place where

$$\eta_{20} = \sum_{j=1}^{\infty} \lambda_{1j}(\tau_{1j}^2 - 1).$$

In addition $\tau_{11}, \tau_{12}, \ldots$ *are independent standard normal random variables and* λ_{1j} *are eigenvalues of linear operator* S, *which acts from* L^2 *into* L^2 *as follows*

$$S: \quad f \to E(g_{20}(X_{11}, x) f(X_{11})).$$

Proof. In the conditions of this proposition, $U_{\bar n} - \theta$ in (1.2.6) is asymptotically equivalent to a one-sample U-statistic $U_{n_1}(g_{20})$ with kernel g_{20} that is

$$U_{n_1}(g_{20}) = \binom{n_1}{2}^{-1} \sum_{1 \leq i < j \leq n} g_{20}(X_{1i}, X_{1j}).$$

Hence, by Corollary 4.2.2, the weak convergence

$$n_1 U_{n_1}(g_{20}) \Rightarrow \eta_{20}$$

takes place as $n_1 \to \infty$.

Proposition 4.3.3. *Suppose that* $E\Phi^2 < \infty$ *and let*

$$g_{10}(x) = 0, \quad g_{01}(x) = 0, \quad g_{11}(x) = 0,$$

$$g_{20}(x, y) = 0, \quad g_{02}(x, y) \neq 0.$$

If $n_2 n_1^{-2} \to 0$ *as* $n_1 \to \infty$ *and* $n_2 \to \infty$ *then the following weak convergence takes place,*

$$n_2(U_{\bar n} - \theta) \Rightarrow \eta_{02},$$

where

$$\eta_{02} = \sum_{j=1}^{\infty} \lambda_{2j}(\tau_{2j}^2 - 1),$$

$\tau_{21}, \tau_{22}, \ldots$ *are independent standard normal random variables and* $\{\lambda_{2j}\}$ *are eigenvalues of the linear operator*

$$S : f \to E(g_{02}(X_{21}, x)\, f(X_{21})), \quad f \in L^2 .$$

Proof. In representation (1.2.6), $U_{\bar{n}} - \theta$ is asymptotically equivalent to one-sample U-statistic

$$U_{n_2}(g_{02}) = \binom{n_2}{2}^{-1} \sum_{1 \le i < j \le n_2} g_{02}(X_{2i}, X_{2j}) .$$

Thus, the assertion follows from Corollary 4.2.2.

Proposition 4.3.4. *Assume that* $E\Phi^2 < \infty$ *and let*

$$g_{10}(x) = 0, \quad g_{01}(x) = 0, \quad g_{20}(x,y) = 0,$$

$$g_{02}(x,y) = 0, \quad g_{11}(x,y) \neq 0.$$

Then the weak convergence

$$(n_1 n_2)^{1/2}(U_{\bar{n}} - \theta) \Rightarrow 4\eta_{11},$$

where

$$\eta_{11} = \sum_{j=1}^{\infty} \lambda_j \tau_j \xi_j ,$$

takes place as $\min(n_1, n_2) \to \infty$. *Here* $\{\tau_j\}$ *and* $\{\xi_j\}$ *are two independent sequences of independent standard normal random variables and* $\{\lambda_j\}$ *are some constants.*

Proof. Under the conditions of this proposition, $U_{\bar{n}} - \theta$ in (1.2.6) is asymptotically equivalent to $4U_{\bar{n}}(g_{11})$, where $U_{\bar{n}}(g_{11})$ is a two-sample U-statistic of the form

$$U_{\bar{n}}(g_{11}) = (n_1 n_2)^{-1} \sum_{i=1}^{n_1} \sum_{j=1}^{n_2} g_{11}(X_{1i}, X_{2j}) .$$

Note, that $g_{11}(x,y)$ is a degenerate function with respect to each variable, i.e.

$$E g_{11}(X_{11}, y) = 0, \quad E g_{11}(x, X_{21}) = 0.$$

Provided that $E g_{11}^2 < \infty$, there exist orthonormal systems of functions $\{\varphi_j(x)\}$ and $\{f_j(y)\}$ in space L^2 such that

$$g_{11}(x,y) = \sum_{j=1}^{\infty} \lambda_j \varphi_j(x) f_j(y) ,$$

where the series converges in L^2. Moreover, $\{\varphi_j(x)\}$ and $\{f_j(y)\}$ are eigenfunctions of kernel $g_{11}(x,y)$, and $\{\lambda_j\}$ are eigenvalues of it in the sense that

$$\int \varphi_j(x) g_{11}(x,y) P_1(dx) = \lambda_j f_j(y),$$

$$\int g_{11}(x,y) f_j(y) P_2(dy) = \lambda_j \varphi_j(x).$$

Then we get

$$(n_1 n_2)^{1/2} U_{\bar{n}}(g_{11}) = \sum_{j=1}^{\infty} \lambda_j \Big\{ n_1{}^{-1/2} \sum_{i=1}^{n_1} \varphi_j(X_{1i}) \Big\} \Big\{ n_2{}^{-1/2} \sum_{i=1}^{n_2} f_j(X_{2i}) \Big\}.$$

Since

$$E\varphi_j(X_{11}) = Ef_j(X_{21}) = 0,$$
$$E\varphi_j^2(X_{11}) = Ef_j^2(X_{21}) = 1,$$

for fixed arbitrary $j = 1, 2, \ldots,$ we have

$$n_1{}^{-1/2} \sum_{i=1}^{n_1} \varphi_j(X_{1i}) \Rightarrow \tau_j,$$

$$n_2{}^{-1/2} \sum_{i=1}^{n_2} f_j(X_{2i}) \Rightarrow \xi_j$$

by virtue of the classic central limit theorem. Taking into account the equality $E\varphi_j(X_{11}) f_j(X_{21}) = 0$ together with the Cramer–Wold method gives

$$(n_1 n_2)^{1/2} U_{\bar{n}}(g_{11}) \Rightarrow \eta_{11}$$

as $\min(n_1, n_2) \to \infty$. From here the desired result follows.

Let us calculate the characteristic function $\psi(t) = E\exp(it\eta_{11})$ of the random variable η_{11}. Due to the properties of independent sequences $\{\tau_j\}$ and $\{\xi_j\}$, we obtain

$$\begin{aligned} \psi(t) &= E\exp(it\sum_{j=1}^{\infty} \lambda_j \tau_j \xi_j) \\ &= \prod_{j=1}^{\infty} E\exp(it\lambda_j \tau_j \xi_j) \\ &= \prod_{j=1}^{\infty} E\exp\Big(-\frac{t^2}{2}\lambda_j^2 \tau_j^2\Big) \\ &= \prod_{j=1}^{\infty} (1 + t^2\lambda_j^2)^{-1/2}. \end{aligned}$$

Consequently,

$$\psi(t) = \prod_{j=1}^{\infty} (1 + t^2\lambda_j^2)^{-1/2}.$$

Proposition 4.3.5. *Suppose that $E\Phi^2 < \infty$ and let*

$$g_{10}(x) = 0\,,$$

$$g_{01}(x) \neq 0, \quad g_{20}(x,y) \neq 0\,.$$

Let $\min(n_1, n_2) \to \infty$. *If* $n_2 n_1^{-2} \to 0$ *then*

$$n_2^{1/2}(U_{\bar{n}} - \theta) \Rightarrow 2\sigma_2 \tau\,,$$

and if $n_2 n_1^{-2} \to \infty$, *then*

$$n_1(U_{\bar{n}} - \theta) \Rightarrow \eta_{20}\,.$$

Proof. Under the conditions of this proposition, U-statistic $U_{\bar{n}} - \theta$ in (1.2.6) is distributed as $n^{-1/2}\, 2\sigma_2\tau + n_1^{-1}\eta_{20}$ asymptotically.

Proposition 4.3.6. *Assume that $E\Phi^2 < \infty$ and let*

$$g_{10}(x) = 0, \quad g_{01}(x) = 0\,,$$

$$g_{20}(x,y) \neq 0, \quad g_{11}(x,y) \neq 0, \quad g_{02}(x,y) \neq 0\,.$$

Then, if $n_2/n_1 \to \beta$, $0 < \beta < \infty$, *as* $\min(n_1, n_2) \to \infty$, *the weak convergence*

$$(n_1 n_2)^{1/2}(U_{\bar{n}} - \theta) \Rightarrow \beta^{1/2}\eta_{20} + 4\eta_{11} + \beta^{-1/2}\eta_{02}$$

takes place.

Proof. Now U-statistic $U_{\bar{n}} - \theta$ in (1.2.6) is asymptotically distributed as the sum

$$n_1^{-1}\eta_{20} + 4(n_1 n_2)^{-1/2}\eta_{11} + \eta_2^{-1}\eta_{02}.$$

Convergence to a multiple stochastic integral. Suppose that in (1.2.1) $B = R$. Let $W_j(\cdot)$, $j = 1, \ldots, c$, denote the independent Gaussian random measures on the measurable spaces $(X_j, \mathfrak{X}_j)$, $j = 1, \ldots, c$, respectively, with zero mean value and covariance $EW_j(A)W_j(B) = P_j(A \cap B)$ for arbitrary sets $A, B \in \mathfrak{X}_j$. Assume the canonical functions $g_{\bar{d}}$ satisfy the condition $Eg_{\bar{d}}^2 < \infty$. Then, according to the general theory of multiple stochastic integrals, the Wiener–Itô $\bar{d}$-multiple stochastic integral $I_{\bar{d}}^W(g_{\bar{d}})$ of a given function $g_{\bar{d}}$ with respect to the vector measure $W = (W_1, \ldots, W_c)$ can be defined. This integral has the following symbolic notation

$$I_{\bar{d}}^W(g_{\bar{d}}) = \int \cdots \int g_{\bar{d}}(x_{ji},\ i = 1, \ldots, d_j,\ 1 \le j \le c) \prod_{j=1}^{c} \prod_{i=1}^{d_j} W_j(dx_{ji})\,.$$

Theorem 4.3.3. *Suppose that for some fixed vector $\bar{d} = (d_1, \dots, d_c)$ the function $g_{\bar{d}}$ satisfies the condition $0 < Eg_{\bar{d}}^2 < \infty$. Then the weak convergence*

$$\prod_{j=1}^{c} n_j^{d_j/2} U_{\bar{n}\bar{d}} \Rightarrow I_{\bar{d}}^{W}(g_{\bar{d}})$$

holds as $\min(n_1, \dots, n_c) \to \infty$.

Proof. It is sufficient to apply the method of the proof of Theorem 4.2.1.

By Lemma 1.2.1 for $B = R$ we have

$$E(S_{\bar{n}\bar{d}} S_{\bar{n}\bar{d}'}) = \begin{cases} \prod_{j=1}^{c} \binom{n_j}{d_j} Eg_{\bar{d}}^2, & \bar{d} = \bar{d}', \\ 0, & \bar{d} \neq \bar{d}'. \end{cases}$$

Taking into account this formula and representation (1.2.8) we obtain

$$E(U_{\bar{n}} - \theta)^2 = \sum_{d=r}^{m} \sum_{d_1+\dots+d_c=d} \prod_{j=1}^{c} \binom{m_j}{d_j}^2 \binom{n_j}{d_j}^{-1} Eg_{\bar{d}}^2 .$$

If here $Eg_{\bar{d}}^2 \neq 0$ and $\min(n_1, \dots, n_c) \to \infty$ then

$$E(U_{\bar{n}} - \theta)^2 = \sigma_{\bar{n}\bar{r}}^2 + o(\sigma_{\bar{n}\bar{r}}^2)$$

where

$$\sigma_{\bar{n}\bar{r}}^2 = \sum_{r_1+\dots+r_c=r} \prod_{j=1}^{c} \binom{m_j}{r_j}^2 r_j!\, n_j^{-r_j}\, Eg_{\bar{r}}^2 .$$

This ends the proof.

Theorem 4.3.4. *Assume kernel Φ has rank r and satisfies the conditions*

$$\mu = \min_{r_1+\dots+r_c=r} Eg_{\bar{r}}^2 > 0 ,$$

$$E|g_{\bar{d}}|^{\gamma_{\bar{r}\bar{d}}} < \infty, \quad \bar{d} = (d_1, \dots, d_c)$$

where $\gamma_{\bar{r}\bar{d}} = 2d/(2d - r)$, $d = r, r+1, \dots, m$. Then the weak convergence

$$\sigma_{\bar{n}\bar{r}}^{-1}(U_{\bar{n}} - \theta) \Rightarrow \sum_{d_1+\dots+d_c=r} g_{\bar{n}}(\bar{d}) \prod_{j=1}^{c} \binom{m_j}{d_j} I_{\bar{d}}^{W}(g_{\bar{d}})$$

takes place as $\min(n_1, \dots, n_c) \to \infty$.

In addition,

$$g_{\bar{n}}(\bar{d}) = \left(\sum_{r_1+\dots+r_c=r} \prod_{j=1}^{c} \binom{m_j}{r_j}^2 r_j! n_j^{d_j - r_j}\, Eg_{\bar{r}}^2 \right)^{-1/2} .$$

Proof. According to (1.2.8), we write

$$\sigma_{\bar{n}\bar{r}}^{-1}(U_{\bar{n}} - \theta) = I_{\bar{n}\bar{r}} + R_{\bar{n}}$$

where

$$I_{\bar{n}\bar{r}} = \sum_{d_1+\cdots+d_c=r} \prod_{j=1}^{c} \binom{m_j}{d_j} \left(\sigma_{\bar{n}\bar{r}} \prod_{j=1}^{c} n_j^{d_j/2}\right)^{-1} \left(\prod_{j=1}^{c} n_j^{d_j/2} U_{\bar{n}\bar{r}}\right),$$

and

$$R_{\bar{n}} = \sigma_{\bar{n}\bar{r}}^{-1} \sum_{d=r+1}^{m} \sum_{d_1+\cdots+d_c=d} \prod_{j=1}^{c} \binom{m_j}{d_j} \binom{n_j}{d_j}^{-1} S_{\bar{n}\bar{d}}.$$

Since for $d_1 + \cdots + d_c = r$ we have $\gamma_{\bar{r}\bar{d}} = 2$ and $Eg_{\bar{r}}^2 < \infty$, then by Theorem 4.3.3

$$I_{\bar{n}\bar{r}} \Rightarrow \sum_{d_1+\cdots+d_c=r} g_{\bar{n}}(\bar{d}) \prod_{j=1}^{c} \binom{m_j}{d_j} I_{\bar{d}}^{W}(g_{\bar{d}}).$$

Note that

$$\begin{aligned}\sigma_{\bar{n}\bar{r}}^2 &\geq \frac{\mu}{r!} \sum_{r_1+\cdots+r_c=r} \frac{r!}{r_1! \cdots r_c!} \prod_{j=1}^{c} n_j^{-r_j} \\ &= \frac{\mu}{r!}(n_1^{-1} + \cdots + n_c^{-1})^r.\end{aligned}$$

By analogy with this inequality we can write

$$\sigma_{\bar{n}\bar{r}}^2 \leq \frac{E\Phi^2}{r!}(m_1^2 n_1^{-1} + \cdots + m_c^2 n_c^{-1})^r.$$

Applying these estimates and also the method of the proof of Theorem 3.2.3, we obtain

$$R_{\bar{n}} \xrightarrow{P} 0 \quad \text{as} \quad \min(n_1, \ldots, n_c) \to \infty.$$

This proves the theorem.

Corollary 4.3.1. *Assume that* $n_j n_1^{-1} \to \lambda_j$, $j = 1, \ldots, c$ *as* $\min(n_1, \ldots, n_c) \to \infty$. *Then* $q_{\bar{n}}(\bar{d}) \to q(\bar{d})$, *where*

$$q(\bar{d}) = \left(\sum_{r_1+\cdots+r_c=r} \prod_{j=1}^{c} \binom{m_j}{r_j}^2 r_j! \lambda_j^{d_j - r_j} Eg_{\bar{r}}^2\right)^{-1/2}.$$

Corollary 4.3.2. *If* $r = 1$ *then the solution of the equation* $d_1 + \cdots + d_c = 1$ *are* c *identity vectors* $\bar{d} = (1, 0, \ldots, 0)$, $\bar{d} = (0, 1, \ldots, 0)$, $\ldots, \bar{d} = (0, 0, \ldots, 1)$. *Therefore, each stochastic integral* $I_{\bar{d}}^{W}(g_{\bar{d}})$ *is a Gaussian random variable with zero mean value and variance* $Eg_{\bar{d}}^2$. *Since a linear combination of independent normal random variables has a normal distribution then in this situation the statement of Theorem 4.3.4 is reduced to the one of Theorem 4.3.2.*

Corollary 4.3.3. *Let $c = 2$ and $r = 2$. The solutions of the equation $d_1 + d_2 = 2$ are three vectors $\bar{d} = (2,0)$, $\bar{d} = (1,1)$ and $\bar{d} = (0,2)$. In this situation we have*

$$\sum_{r_1+\cdots+r_c=r} q_{\bar{n}}(\bar{d}) \prod_{j=1}^{c} \binom{m_j}{d_j} I_{\bar{d}}^W(g_{\bar{d}}) = \binom{m_1}{1}\binom{m_2}{1} I_{11}^W(g_{11}) \times$$

$$\times \left(2\binom{m_1}{2}^2 n_1^{-1} n_2 Eg_{20}^2 + \binom{m_1}{1}\binom{m_2}{1} Eg_{11}^2 + 2\binom{m_2}{2}^2 n_1 n_2^{-1} Eg_{02}^2\right)^{-1/2} +$$

$$+\binom{m_1}{2} I_{20}^W(g_{20}) \left(2\binom{m_1}{2}^2 Eg_{20}^2 + \binom{m_1}{1}\binom{m_2}{1} n_1 n_2^{-1} Eg_{11}^2 + 2\binom{m_2}{2}^2 n_1^2 n_2^{-2} Eg_{02}^2\right)^{-1/2} +$$

$$+\binom{m_2}{1} I_{02}^W(g_{02}) \left(2\binom{m_1}{2}^2 n_1^{-2} n_2^2 Eg_{20}^2 + \binom{m_1}{1}\binom{m_2}{1} n_1^{-1} n_2 Eg_{11}^2 + 2\binom{m_2}{2}^2 Eg_{02}^2\right)^{-1/2}.$$

Therefore, if $n_1 n_2^{-1} \to 0$, then

$$\sigma_{\bar{n}\bar{r}}^{-1}(U_{\bar{n}} - \theta) \Rightarrow (2Eg_{20}^2)^{-1/2} I_{20}^W(g_{20})$$

and if $n_2 n_1^{-1} \to 0$, then

$$\sigma_{\bar{n}\bar{r}}^{-1}(U_{\bar{n}} - \theta) \Rightarrow (2Eg_{20}^2)^{-1/2} I_{02}^W(g_{02}).$$

The case $n_1 n_2^{-1} \to \lambda > 0$ corresponds to Corollary 4.3.1.

Corollary 4.3.4. *Let $r = m_1 + \cdots + m_c$. Vector $\bar{d} = \bar{m} = (m_1, \ldots, m_c)$ is the unique solution of the equation $d_1 + \cdots + d_c = m_1 + \cdots + m_c$, $0 \le d_i \le m_i$, $i = 1, \ldots, c$. In this situation*

$$\sum_{d_1+\cdots+d_c=r} q_{\bar{n}}(\bar{d}) \prod_{j=1}^{c} \binom{m_j}{d_j} I_{\bar{d}}^W(g_{\bar{d}}) = \left(\prod_{j=1}^{c} m_j! Eg_{\bar{m}}^2\right)^{-1/2} I_{\bar{m}}^W(g_{\bar{m}}).$$

4.4 Poisson approximation

Random permanents in the Poisson approximation scheme. Consider *UR*-statistics with the kernel $\Phi(x_1, \ldots, x_m) = x_1 \cdots x_m$ and

$$X_j = X_j^{(n)}, \quad 1 \le j \le n$$

where $X_1^{(n)}, \ldots, X_n^{(n)}$ are independent Bernoulli random variables with the distribution

$$P\{X_j^{(n)} = 1\} = 1 - P(X_j^{(n)} = 0) = \lambda / n.$$

It is well-known (see Barbour *et al.* (1992)) that the sums

$$\nu_n := \sum_{j=1}^{n} X_j^{(n)}$$

are weakly convergent to the Poisson distributed random variable ν with parameter λ, $\nu_n \Rightarrow \nu \quad \text{as} \quad n \to \infty$,

$$P\big(\nu = k\big) = e^{-\lambda}\lambda^k \,/\, k!\,, \quad k \geq 0\,.$$

We now introduce a matrix with random elements

$$A_n^m(X^{(n)}) := [a_{ij} = X_j^{(n)}; \quad 1 \leq i \leq m, \quad 1 \leq j \leq n]\,.$$

Then (see Borovskikh and Korolyuk (1994b))

$$\begin{aligned} n^{[m]}U_n(\Phi) &= \operatorname{Per} A_n^m(X^{(n)}) \\ &= \nu_n^{[m]}\,, \end{aligned}$$

where $\nu_n^{[m]} = \nu_n(\nu_n - 1)\dots(\nu_n - m + 1)$.

Hence, in the Poisson approximation scheme the following is valid

$$n^{[m]}U_n(\Phi) \Rightarrow \nu^{[m]} \quad \text{as} \quad n \to \infty\,. \tag{4.4.1}$$

Now we consider *UR*-statistics with the kernel

$$\bar{\Phi}(x_1, \dots, x_m) = (x_1 - \lambda/n)\cdots(x_m - \lambda/n)$$

and the random permanent of a matrix with centered random elements

$$A_n^m(\bar{X}^{(n)}) := [a_{ij} = X_j^{(n)} - \lambda/n; \quad 1 \leq i \leq m, \quad 1 \leq j \leq n]\,.$$

Theorem 4.4.1. *Under the conditions of Poisson approximation schemes the weak convergence*

$$n^{[m]}U_n(\bar{\Phi}) \Rightarrow C_m(\nu, \lambda) \quad \text{as} \quad n \to \infty$$

takes place. Here

$$C_m(\nu, \lambda) := \sum_{c=0}^{m} (-\lambda)^{m-c} m^{[c]} \nu^{[c]} \,/\, c!$$

is the stochastic Charlier–Poisson polynomial.

Proof. We now use the relation

$$n^{[m]}U_n(\bar{\Phi}) = \operatorname{Per} A_n^m(\bar{X}^{(n)})$$

and the shift decomposition formula for permanents to obtain

$$n^{[m]}U_n(\bar{\Phi}) = \sum_{c=0}^{m} q_n(c)(-\lambda)^{m-c} m^{[c]} \nu_n^{[c]} \,/\, c!\,,$$

where $q_n(c) = (n-c)^{[m-c]} \,/\, n^{m-c} \to 1$ as $n \to \infty$. Take the weak limit in this formula to get the assertion of the theorem.

Permanent functionals. Consider a sample in the series scheme $X_j^{(n)}$, $1 \leq j \leq n$ of independent and identically distributed random variables taking values in a measurable space $(X, \mathfrak{X})$ with distribution function $P_n(B) := P\{X_j^{(n)} \in B\} = \lambda(B)/n$, $B \in \mathfrak{X}$, where $\lambda(B)$ is some Radon measure with $\lambda(X) < \infty$. Introduce the random δ-measures on $\mathfrak{X}$, $\delta_j^{(n)}(B) := I(X_j^{(n)} \in B)$, $B \in \mathfrak{X}$. Let $\varphi : X \to R$ be real-valued continuous functions with a compact support on X. Introduce the functional matrix

$$A_n^m(\varphi) := [a_{ij} = \varphi(X_j^{(n)});\quad 1 \leq i \leq m,\ 1 \leq j \leq n]\,.$$

The permanent functional is defined as the permanent of this functional matrix and can be represented as the integral over the permanent random measure in the following form

$$\operatorname{Per} A_n^m(\varphi) = \int_{X^m} \prod_{s=1}^{m} \varphi(x_s) \operatorname{Per} A_n^m(dx_1, \ldots, dx_m)\,.$$

Consider *UR*-statistics with the kernel $\Phi(x_1, \ldots, x_m) = \varphi(x_1) \cdots \varphi(x_m)$. It is obvious that

$$n^{[m]}U_n(\Phi) = \operatorname{Per} A_n^m(\varphi)\,.$$

Theorem 4.4.2. *Under the Poisson approximation condition $P_n(B) = \lambda(B)/n$, $B \in \mathfrak{X}$, the weak convergence as $n \to \infty$*

$$n^{[m]}U_n(\Phi) \Rightarrow I_m^{\nu}(\Phi)$$

takes place, where the multiple Itô–Poisson integral $I_m^{\nu}(\Phi)$ can be represented as the Charlier–Poisson polynomial of one-dimensional stochastic integrals with respect to Poisson random measure $\nu(B)$ in the following form

$$I_m^{\nu}(\Phi) = m! \cdot \sum_{i_1+2i_2+\cdots+mi_m=m} \prod_{s=1}^{m} (-1)^{i_s-1} [I_1^{\nu}(\varphi^s)/s]^{i_s} \,/\, i_s!\,,$$

where

$$I_1^{\nu}(\varphi^s) = \int_X \varphi^s(x)\nu(dx)\,.$$

Proof. The generating function of permanent functionals has the following representation (see Borovskikh and Korolyuk (1994b))

$$\sum_{m=0}^{n} \frac{z^m}{m!} \operatorname{Per} A_n^m(\varphi) = \prod_{j=1}^{n} [1 + z\varphi(X_j^{(n)})]$$
$$= \exp\left\{ \int_X \ln(1 + z\varphi(x))\nu_n(dx) \right\}$$

where

$$\nu_n(B) := \sum_{j=1}^{n} \delta_j^{(n)}(B), \quad B \in \mathfrak{X}.$$

From this formula, due to the fact that for any function $f(z)$ analytical in some neighbourhood of $z = 0$,

$$\frac{d^m}{dz^m} e^{f(z)} = m! e^{f(z)} \sum_{i_1+2i_2+\cdots+mi_m=m} \prod_{s=1}^{m} \left[\frac{d^s f(z)}{dz^s} \right]^{i_s} / i_s!,$$

we have

$$\operatorname{Per} A_n^m(\varphi) = \frac{d^m}{dz^m} \left\{ \exp \int_X \ln[1 + z\varphi(x)]\, \nu_n \right\} \Big|_{z=0}$$
$$= m! \sum_{i_1+2i_2+\cdots+mi_m=m} \prod_{s=1}^{m} (-1)^{i_s - 1} \left[I_1^{\nu_n}(\varphi^s) / s \right]^{i_s} / i_s!,$$

where

$$I_1^{\nu_n}(\varphi^s) := \int_X \varphi^s(x)\, \nu_n(dx).$$

Under the Poisson approximation condition

$$I_1^{\nu_n}(\varphi^s) \Rightarrow I_1^{\nu}(\varphi^s) \quad \text{as} \quad n \to \infty$$

and therefore

$$\operatorname{Per} A_n^m(\varphi) \Rightarrow I_m^{\nu}(\Phi) \quad \text{as} \quad n \to \infty.$$

This proves the theorem.

Consider *UR*-statistics with the kernel

$$\bar{\Phi}(x_1, \ldots, x_m) = (\varphi(x_1) - \lambda(\varphi)/n) \cdots (\varphi(x_m) - \lambda(\varphi)/n)$$

where

$$\lambda(\varphi) := \int_X \varphi(x)\lambda(dx).$$

Theorem 4.4.3. *Under the Poisson approximation condition* $P_n(B) = \lambda(B)/n$, $B \in \mathfrak{X}$ *as* $n \to \infty$

$$n^{[m]} U_n(\bar{\Phi}) \Rightarrow I_m^{\bar{\nu}}(\Phi)$$

takes place, where $\bar{\nu}(B) = \nu(B) - \lambda(B)$, $B \in \mathfrak{X}$, *is the centered Poisson random measure with intensity* $\lambda(B) = E\nu(B)$. *Additionally*

$$I_m^{\bar{\nu}}(\Phi) = \sum_{c=0}^{m} [-\lambda(\varphi)]^{m-c} m^{[c]} I_c^{\nu}(\Phi) \,/\, c!\,,$$

where $\Phi(x_1, \ldots, x_m) = \varphi(x_1) \cdots \varphi(x_m)$.

Proof. In a similar way to the proof of Theorem 4.4.1, we apply the shift decomposition formula (1.34) with $h = -\lambda(\varphi)/n$ from Borovskikh and Korolyuk (1994b)

$$\begin{aligned} n^{[m]} U_n(\bar{\Phi}) &= \operatorname{Per} A_n^m(-\lambda(\varphi)/n + \varphi) \\ &= \sum_{c=0}^{m} q_n(c)[-\lambda(\varphi)]^{m-c} m^{[c]} \operatorname{Per} A_n^c(\varphi) \,/\, c!\,, \end{aligned}$$

where $q_n(c) = n^{[m]-m} n^{c-[c]} \to 1$ as $n \to \infty$.

Taking the weak limit and due to Theorem 4.4.2, obtain, as $n \to \infty$

$$\operatorname{Per} A_n^m(-\lambda(\varphi)/n + \varphi) \Rightarrow \sum_{c=0}^{m} [-\lambda(\varphi)]^{m-c} m^{[c]} I_c^{\nu}(\varphi) \,/\, c!\,.$$

But the right-hand side of this of this relation is the Charlier representation of the Itô–Poisson integral over the centered Poisson measure

$$I_m^{\bar{\nu}}(\Phi) = \sum_{c=0}^{m} [-\lambda(\varphi)]^{m-c} m^{[c]} I_c^{\nu}(\Phi) \,/\, c!\,.$$

Hence we get the proof of Theorem 4.4.3.

Permanent symmetric statistics. Consider symmetric statistics over a permanent δ-measure of the general form (1.5.4) under the Poisson approximation condition.

Theorem 4.4.4. *Let* $\Phi : X^m \to R$ *be a symmetric function such that there exist a compact set* $B \in X$ *and constant* $c > 0$ *such that* $\Phi(x_1, \ldots, x_m) = 0$ *if* $(x_1, \ldots, x_m) \notin B^m$ *and* $|\Phi(x_1, \ldots, x_m)| \leq c^m$ *if* $(x_1, \ldots, x_m) \in B^m$.

Then under the Poisson approximation condition $P_n(B) = \lambda(B)/n$ *the weak convergences*

$$n^{[m]}\, U_n^{\delta}(\Phi) \Rightarrow I_m^{\nu}(\Phi)$$

and

$$n^{[m]}\, U_n^{\bar{\delta}}(\Phi) \Rightarrow I_m^{\bar{\nu}}(\Phi)$$

take place as $n \to \infty$.

Proof. To prove the theorem we consider the second statement of the theorem about convergence to $I_m^{\bar{\nu}}(\Phi)$. This statement takes place if, for every fixed $t \in R$ and $n \to \infty$, the relation

$$|E \exp(itn^{[m]}U_n^{\bar{\delta}}(\Phi)) - E\exp(itI_m^{\bar{\nu}}(\Phi))| \to 0$$

is valid.

Under the Poisson approximation condition in Borovskikh and Korolyuk (1994b) it was proved by the composition method that the expression on the left-hand side of this relation does not exceed the value $c(\lambda)(|t|+t^2)n^{-1}$ with some constant $c(\lambda) > 0$. From this estimate the statement of the theorem follows.

4.5 Stable approximation

Symmetric polynomials. Consider UR-statistics with the kernel $\Phi(x_1, \ldots, x_m) = x_1 \cdots x_m$. Let $X_1, \ldots, X_m$ be independent identically distributed random variables with $EX_1 = 0$. Then

$$n^{[m]} U_n = m!\, P_n^m(X),$$

where $P_n^m(X)$ is a symmetric polynomial

$$P_n^m(X) = \sum_{1 \le i_1 < \ldots < i_m \le n} \prod_{s=1}^{m} X_s .$$

Denote by

$$\eta_{nk}(X) = n^{-k/\alpha} \sum_{j=1}^{n} X_j^k, \quad k = 1, 2, \ldots$$

where $1 < \alpha < 2$. Assume that the condition of attraction to the stable law is valid, i.e.

$$\eta_{n1}(X) \Rightarrow \eta(\alpha),$$

where $\eta(\alpha)$ is a random variable which has a stable distribution with basic parameter α. It is well-known that this condition of attraction implies the relation

$$\eta_{nk}(X) \Rightarrow \eta(\alpha/k)$$

as $n \to \infty$, where $\eta(\alpha/k)$ is a stable random variable with basic parameter α/k (other parameters are defined by k and by the parameters of distribution $\eta(\alpha)$).

Thus, as $n \to \infty$,

$$(\eta_{n1}(X), \ldots, \eta_{nk}(X)) \Rightarrow (\eta(\alpha), \ldots, \eta(\alpha/k)).$$

Here the components of the limit vector are dependent variables.

Theorem 4.5.1. *Under the condition of attraction to the stable law the weak convergence*

$$n^{m(\alpha-1)/\alpha} U_n \Rightarrow m!\, P_\infty^m(\eta)$$

takes place where

$$P_\infty^m(\eta) = \sum_{i_1+2i_2+\cdots+mi_m=m} \prod_{k=1}^{m} (-1)^{i_k-1}(k^{i_k}/i_k!)\eta^{i_k}(\alpha/k)\,.$$

Proof. The statement of this theorem follows from the formula

$$P_n^m(X/n^{1/\alpha}) = \sum_{i_1+2i_2+\cdots+mi_m=m} \prod_{k=1}^{m} (-1)^{i_k-1}(k^{i_k}/i_k!)\eta_{nk}^{i_k}(X)$$

and the relation

$$P_n^m(X/n^{1/\alpha}) \Rightarrow P_\infty^m(\eta),$$

since the condition of attraction to the stable law is valid.

Weak convergence of *UH*-statistics to stable laws. Consider *UH*-statistics from (1.1.1). Assume that rank $r = 1$ and denote by Q the probability measure on a separable real Hilbert space H which is induced by the mapping $g_1 : X \to H$. Since $r = 1$, Q is a non-degenerate distribution. We say that Q belongs to the domain of attraction of probability measure μ on H if there exist numbers $b_n > 0$ and vectors $a_n \in H$ such that the distribution of a random element

$$b_n^{-1}(g_1(X_1) + \cdots + g_1(X_n)) - a_n$$

converges weakly to the probability measure μ. A measure $\mu = \mu_\alpha$ is called stable with index α, $0 < \alpha < 2$, if its characteristic functional

$$\hat{\mu}(x) = \int_H \exp(i(x,y))\mu(dy), \quad x \in H$$

has the form

$$\hat{\mu}(x) = \exp\{i(x,a) - \int_S |(x,u)|^\alpha \Gamma(du) + iC(\alpha,x)\}\,,$$

where

$$C(\alpha,x) = \begin{cases} \operatorname{tg}(\frac{\pi\alpha}{2}) \int_S (x,u)|(x,u)|^{\alpha-1}\,\Gamma(du), & \alpha \neq 1\,, \\ \frac{2}{\pi} \int_S (x,u) \ln|(x,u)|\,\Gamma(du), & \alpha = 1\,, \end{cases}$$

Γ is a finite Borel measure on $S = \{x \in H : \|x\| = 1\}$, and vector a belongs to H.

According to Kuelbs and Mandrekar (1974), the following assertion holds: If Q belongs to the domain of attraction of a non-degenerate stable law with index $1 < \alpha < 2$ on a real separable Hilbert space H, then

$$Q(\|x\| > t) \sim t^{-\alpha} h(t) \quad \text{as} \quad t \to \infty\,,$$

where $h(t)$ is a slowly varying function; in addition, if $\mu = \mu_\alpha$ is a stable law with index $1 < \alpha < 2$ on H, then $\mu(\|x\| > t) \sim t^{-\alpha}c$. It was also shown by Kuelbs and Mandrekar (1974), that in the case of attraction to a stable law with index α the constants b_n have the form $b_n = n^{1/\alpha}h(n)$, where $h(n)$ is a slowly varying function of natural argument. It is evident that the stable law with index α belongs to the domain of its own attraction and in this case $b_n = bn^{1/\alpha}$, where b is a constant. We can say in this connection that the probability measure belongs to the domain of normal attraction of a stable law if it belongs to the domain of attraction of this law and $b_n = bn^{1/\alpha}$. For simplicity, we assume that $b = 1$.

We have $E\|g_c\| < \infty$ for all $c = 1, \dots, m$, since $E\|\Phi\| < \infty$. Consider the problem of weak convergence only for $1 < \alpha < 2$. Denote for $c = 1, \dots, m$

$$\gamma_c(\alpha) = \frac{c\alpha}{c\alpha - \alpha + 1}.$$

Note that

$$1 + \frac{\alpha - 1}{\alpha(m-1)+1} = \gamma_m(\alpha) < \gamma_{m-1}(\alpha) < \dots < \gamma_2(\alpha) = \frac{2\alpha}{\alpha+1} < \gamma_1(\alpha) = \alpha,$$

if $1 < \alpha < 2$.

Theorem 4.5.2. *Assume that Q belongs to the domain of a normal attraction of a stable law μ_α with index $1 < \alpha < 2$. Then, under conditions*

$$t^{\gamma_c(\alpha)} P(\|g_c\| > t) \to 0 \quad as \quad t \to \infty$$

for each $c = 2, \dots, m$, the weak convergence

$$m^{-1} n^{(\alpha-1)/\alpha}(U_n - \theta) \Rightarrow \tau_\alpha,$$

as $n \to \infty$, takes place where the stable H-valued random element τ_α has the distribution μ_α.

Proof. We rewrite representation (4.1.2) in the form

$$m^{-1} n^{(\alpha-1)/\alpha}(U_n - \theta) = n^{-1/\alpha}(g_1(X_1) + \dots + g_1(X_n)) +$$
$$+ \sum_{c=2}^{m} (m-1)^{[c-1]} \prod_{j=1}^{c-1} (1 - n^{-1}j)^{-1} n^{-c/\gamma_c(\alpha)} S_{nc}.$$

By virtue of the assumption of the theorem we have

$$n^{-1/\alpha}(g_1(X_1) + \dots + g_1(X_n)) \Rightarrow \tau_\alpha$$

as $n \to \infty$. Moreover, we can apply Theorem 3.1.2 for $m = c$ and $p = \gamma_c(\alpha)$. This gives $n^{-c/\gamma_c(\alpha)} S_{nc} \to 0$ in probability, as $n \to \infty$, and hence we have the proof of this theorem.

4.6 Approximation with increasing degrees

In this section the Hoeffding decomposition for U-statistics is used

$$U_n(\Phi) = \theta + \sum_{c=r}^{m} m^{[c]} U_n(g_{mc})/c! \tag{4.6.1}$$

where $r \geq 1$ is a rank of kernel $\Phi : X^m \to R$ and $g_{mc} := g_{mc}(x_1, \ldots, x_c)$ are completely degenerate kernels determined by the formula

$$g_{mc}(x_1, \ldots, x_c) := \int_{X^m} \Phi(y_1, \ldots, y_m) \prod_{s=1}^{c} (\delta_{x_s}(dy_s) - P(dy_s)) \prod_{s=c+1}^{m} P(dy_s).$$

We shall consider the situation when the degree m is increasing together with $n \to \infty$.

Gaussian approximation scheme. Let $X_1, \ldots, X_n$ be independent and identically distributed random variables which satisfy the conditions

$$EX_j = \mu \neq 0, \quad E(X_j - \mu)^2 = \sigma^2 > 0$$

and $v = \mu/\sigma > 0$.

Theorem 4.6.1. *Let $\Phi(x_1, \ldots, x_m) = x_1 \cdots x_m$ and $\gamma_n := m/\sqrt{n} \to 0$ as $n \to \infty$. Then the weak convergence*

$$(\sigma \mu^{m-1} \gamma_n)^{-1} (U_n(\Phi) - \mu^m) \Rightarrow \tau$$

takes place, where τ is a standard normal random variable.

Proof. In this case $\theta = \mu^m$ and $g_{mc} = \mu^{m-c} \prod_{s=1}^{c} (x_s - \mu)$.

Denote

$$\begin{aligned} g_c(x_1, \ldots, x_c) &= \mu^{-m} g_{mc}(x_1, \ldots, x_m) \\ &= \mu^{-c} \prod_{s=1}^{c} (x_s - \mu). \end{aligned}$$

Function g_c does not depend on m and moreover $E g_c^2 = E g_c^2(X_1, \ldots, X_c) = (\sigma/\mu)^{2c} = v^{-2c}$.

By choosing a normalized coefficient, the following representation for this U-statistic is obtained from (4.6.1) with $r = 1$

$$(\sigma \mu^{m-1} \gamma_n)^{-1} (U_n(\Phi) - \mu^m) = \frac{1}{\sigma \sqrt{n}} \sum_{j=1}^{n} (X_j - \mu) + \Delta_n,$$

where

$$\Delta_n = v\,\gamma_n^{-1} \sum_{c=2}^{m} m^{[c]}\,U_n(g_c)\,/\,c!\,.$$

Since

$$\begin{aligned} E\,U_n^2(g_c) &= \binom{n}{c}^{-1} E\,g_c^2 \\ &= c!\,n^{-[c]}\,v^{-2c}\,, \end{aligned}$$

then we obtain

$$\begin{aligned} E\,\Delta_n^2 &= \gamma_n^{-2} \sum_{c=2}^{m} v^{-2(c-1)}\,m^{2[c]}\,n^{-[c]}\,/\,c! \\ &\le \sum_{c=2}^{m} (\frac{1}{v^2})^{c-1}\,\gamma_n^{2(c-1)}\,/\,c! \\ &\le v^{-2}\,\gamma_n^2\,\exp(\gamma_n^2/v^2) \to 0 \;\text{ as }\; n \to \infty. \end{aligned}$$

By virtue of the central limit theorem and the last estimate, the proof of this theorem follows.

Theorem 4.6.2. *Let* $\Phi(x_1, \dots, x_m) = x_1 \cdots x_m$*, and assume that degree* m *increases so that*

$$\gamma_n := m\,/\,\sqrt{n} \;\to\; \gamma > 0 \text{ as } n \;\to\; \infty\,.$$

Then the weak convergence

$$\mu^{-m}\,U_n(\Phi) \;\Rightarrow\; \exp\left(a\tau - a^2/2\right)$$

takes place with $a := \gamma/v$.

Proof. By definition

$$U_n(\Phi) = n^{-[m]}\,\mathrm{Per}\,A_n^m(X)$$

with a matrix

$$A_n^m(X) = \left[\,a_{ij} = X_j;\; 1 \le i \le m,\; 1 \le j \le n\right].$$

Setting $\xi_j = (X_j - \mu)\,/\,\sigma$, so that $X_j/\mu = 1 + \xi_j/v$, we get formula

$$\mu^{-m}\,U_n(\Phi) = n^{-[m]}\,\mathrm{Per}\,A_n^m(1 + \xi/v)\,.$$

By using the shift decomposition formula (see Borovskikh and Korolyuk (1994b)), we obtain

$$\mu^{-m}\,U_n(\Phi) = \sum_{c=0}^{m} m^{[c]}\,n^{-[c]}\,v^{-c}\,\mathrm{Per}\,A_n^c(\xi)\,/\,c!\,.$$

We rewrite this sum as follows

$$\sum_{c=0}^{m} m^{[c]}\,n^{-[c]}\,v^{-c}\,\mathrm{Per}\,A_n^c(\xi)\,/\,c! =$$

$$= \sum_{c=0}^{m} (\frac{m}{nv})^c \operatorname{Per} A_n^c(\xi) \,/\, c! - \Delta_{n1} + \Delta_{n2} \,,$$

where

$$\begin{aligned}
\Delta_{n1} &= \sum_{c=m+1}^{n} (\frac{m}{nv})^c \operatorname{Per} A_n^c(\xi) \,/\, c! \,, \\
\Delta_{n2} &= \sum_{c=0}^{m} (\frac{m}{nv})^c \varepsilon_n(c) \operatorname{Per} A_n^c(\xi) \,/\, c! \,, \\
\varepsilon_n(c) &= \prod_{s=1}^{c-1} (1 - \frac{s}{m})/(1 - \frac{s}{n}) - 1 \,,
\end{aligned}$$

setting $\varepsilon_n(0) = \varepsilon_n(1) = 0$.

By definition of the generating function of permanents

$$\sum_{c=0}^{n} (\frac{m}{nv})^c \operatorname{Per} A_n^c(\xi) \,/\, c! = \prod_{j=1}^{n} (1 + \frac{b_n \xi_j}{v\sqrt{n}}) \,.$$

Here we have the following limit relation

$$\prod_{j=1}^{n} (1 + \frac{b_n \xi_j}{v\sqrt{n}}) \Rightarrow \exp(a\tau - a^2/2) \,.$$

Further, show that under the condition of this theorem the following asymptotic estimates hold

$$E\,\Delta_{nk}^2 \to 0 \text{ as } n \to \infty \,, \quad k = 1, 2 \,.$$

First note that for

$$q_s = (1 - \frac{s}{m}) \,/\, (1 - \frac{s}{n})$$

the following inequalities are true $1 - s/m \le q_s \le 1$. Then

$$\varepsilon_n^2(c) \le 1 - \prod_{s=1}^{c-1} q_s = \sum_{s=1}^{c-1} (1 - q_s) \prod_{j=s+1}^{c-1} q_j \le$$

$$\le \sum_{s=1}^{c-1} (1 - q_s) \le \sum_{s=1}^{c-1} s/m = \frac{c(c-1)}{2m}$$

and since $E\,(\operatorname{Per} A_n^c(\xi))^2 = n^{[c]}$, it follows that

$$\begin{aligned}
E\,\Delta_{n1}^2 &= \sum_{c=m+1}^{n} (\frac{m}{nv})^{2c} n^{[c]} \,/\, c! \\
&\le \sum_{c=m+1}^{n} (\frac{m}{v\sqrt{n}})^{2c} \,/\, c! \\
&\le (\frac{m}{v\sqrt{n}})^{2(m+1)} \exp((\frac{m}{v\sqrt{n}})^2) \,/\, (m+1)! \to 0,
\end{aligned}$$

$$E\,\Delta_{n2}^2 \le \frac{1}{2m} \sum_{c=2}^{m} (\frac{m}{v\sqrt{n}})^{2c} \frac{1}{(c-2)!}$$

$$\leq \frac{1}{2m}\left(\frac{m}{v\sqrt{n}}\right)^4 \exp\left(\left(\frac{m}{v\sqrt{n}}\right)^2\right) \to 0.$$

These estimates complete the proof of Theorem 4.6.2.

Consider the general case, assuming that in (4.6.1) rank r is fixed.

Theorem 4.6.3. *Suppose that the following conditions are satisfied:*

$$\gamma_n := m/\sqrt{n} \to 0 \quad as \quad n \to \infty;$$

$$\int_{X^m} \Phi^2(x_1, \dots, x_m) \prod_{s=1}^{m} P(dx_s) < \infty;$$

there exist normalizing coefficients σ_{mr} *such that*

$$\lim_{m\to\infty} \sigma_{mr}\, g_{mr}(x_1, \dots, x_r) = g_r(x_1, \dots, x_r)$$

and

$$E\left[\sigma_{mr}\, g_{mr} - g_r\right]^2 \to 0 \quad as \quad m \to \infty,$$

$$\sigma_{mr}^2 \sum_{c=r+1}^{m} \gamma_n^{2(c-r)}\, E\, g_{mc}^2 \,/\, c! \to 0 \quad as \quad n \to \infty.$$

Then the weak convergence

$$\sigma_{mr}(U_n(\Phi) - \theta)\, \gamma_n^{-r} \Rightarrow I_r^W(g_r)\,/\,r!$$

takes place.

Proof. By using the Hoeffding decomposition formula (4.6.1) , we obtain the representation

$$\sigma_{mr}(U_n(\Phi) - \theta)\, \gamma_n^{-r} = m^{[r]}\, \gamma_n^{-r}\, U_n(\sigma_{mr}\, g_{mr})\,/\,r! + \Delta_n,$$

where

$$\Delta_n = \sum_{c=r+1}^{m} m^{[c]}\, \gamma_n^{-r}\, U_n(\sigma_{mr}\, g_{mc})\,/\,c!.$$

For Δ_n we have the estimate

$$E\,\Delta_n^2 = \sigma_{mr}^2\, \gamma_n^{-2r} \sum_{c=r+1}^{m} m^{2[c]}\, n^{-[c]}\, E\, g_{mc}^2\,/\,c! \leq$$

[since $m^{2[c]}\, n^{-[c]} \leq (m/\sqrt{n})^{2c}$]

$$\leq \sigma_{mr}^2 \sum_{c=r+1}^{m} \gamma_n^{2(c-r)}\, E\, g_{mc}^2\,/\,c! \to 0 \text{ as } n \to \infty.$$

And moreover we can write

$$m^{[r]}\, \gamma_n^{-r}\, U_n(\sigma_{mr}\, g_{mr})\,/\,r! =$$

$$= n^{r/2} U_n(g_r)/r! + n^{r/2} U_n(g_r)(m^{[r]} m^{-r} - 1)/r! + \Delta_{nr} ,$$

where

$$\Delta_{nr} = m^{[r]} \gamma_n^{-r} U_{nr}(\sigma_{mr} g_{mr} - g_r) / r! .$$

By virtue of Theorem 4.2.1

$$n^{r/2} U_n(g_r) \Rightarrow I_r^W(g_r) .$$

The value Δ_{nr} allows the following estimate

$$\begin{aligned} E \Delta_{nr}^2 &= m^{2[r]} n^{-r} \gamma_n^{-2r} E (\sigma_{mr} g_{mr} - g_r)^2 / r! \\ &\le E (\sigma_{mr} g_{mr}) - g_r)^2 \to 0 \quad \text{as} \quad m \to \infty . \end{aligned}$$

This proves Theorem 4.6.3.

Theorem 4.6.4. *Suppose that the following conditions are satisfied:*

$$\gamma_n = m / \sqrt{n} \to \gamma > 0 \quad \text{as} \quad m, n \to \infty ;$$

$$\int_{X^m} \Phi^2(x_1, \dots, x_m) \prod_{s=1}^{m} P(dx_s) < \infty ;$$

there exist normalizing coefficients σ_{mr} *such that*

$$\lim_{m \to \infty} \sigma_{mr} g_{mc}(x_1, \dots, x_c) = g_c(x_1, \dots, x_c) ,$$

$$\sum_{c=r}^{m} \gamma_n^{2c} q_m^2(c) E (\sigma_{mr} g_{mc} - g_c)^2 / c! \to 0 ,$$

$$\sum_{c=r}^{\infty} \gamma_n^{2c} (1 - q_m(c))^2 E g_c^2 / c! \to 0$$

with $q_m(c) = \prod_{s=1}^{c} (1 - s/m)$;
the following series converges:

$$\sum_{c=r}^{\infty} \gamma^{2c} E g_c^2 / c! < \infty$$

Then the weak convergence

$$\sigma_{mr}(U_n(\Phi) - \theta) \Rightarrow \sum_{c=r}^{\infty} \gamma_c I_c^W(g_c) / c!$$

takes place.

Proof. First note that $m^{[c]} = \gamma_n^c n^{c/2} q_m(c)$. Consider normalized U-statistics under the Hoeffding decomposition (4.6.1)

$$\sigma_{mr}(U_n(\Phi) - \theta) = \sum_{c=r}^{m} m^{[c]} U_n(\sigma_{mr} g_{mc}) / c! .$$

Here the terms can be transformed to

$$m^{[c]}\, U_n(\sigma_{mr}\, g_{mc}) = \gamma_n^c\, n^{c/2}\, U_n(g_c) + \Delta_{nc}\,,$$

where

$$\Delta_{nc} = \gamma_n^c\,(1 - q_m(c))\, n^{c/2}\, U_n(g_c) + \gamma_n^c\, q_m(c)\, n^{c/2}\, U_n(\sigma_{mr} g_{mc} - g_c)\,.$$

By virtue of Theorem 4.2.1

$$\sum_{c=r}^{m} \gamma_n^c\, n^{c/2}\, U_n(g_c) \,/\, c! \Rightarrow \sum_{c=r}^{\infty} \gamma_c\, I_c^W(g_c) \,/\, c!,$$

and furthermore

$$E\,(\sum_{c=r}^{m} \Delta_{nc} \,/\, c!)^2 \le$$

$$\le 2 \sum_{c=r}^{m} \gamma_n^{2c}\, q_m^2(c)\, E(\sigma_{mr}\, g_{mc} - g_c)^2 \,/\, c! \,+$$

$$+2 \sum_{c=r}^{m} \gamma_n^{2c}\, (1 - q_m(c))^2\, E\, g_c^2 \,/\, c! \;\to 0 \quad \text{as} \quad m, n \to \infty\,.$$

Hence, we get the desired result for Theorem 4.6.4.

Poisson approximation schemes. Now we assume that degree m is increasing and satisfies the relation

$$b_n := m/n \to b, \quad 0 < b < 1 \quad \text{as} \quad n \to \infty\,. \tag{4.6.2}$$

Consider *UR*-statistics with the kernel

$$\Phi(x_1, \ldots, x_m) = (1 + x_1) \cdots (1 + x_m)$$

and introduce a random matrix with the shift

$$A_n^m(1 + X^{(n)}) = [a_{ij} = 1 + X_j^{(n)};\;\; 1 \le i \le m,\; 1 \le j \le m]\,,$$

constructed by means of the Bernoulli sample $X_j^{(n)},\;\; 1 \le j \le n$ in a series scheme, satisfying the Poisson approximation condition

$$\begin{aligned} P(X_j^{(n)} = 1) &= 1 - P(X_j^{(n)} = 0) \\ &= \lambda/n \end{aligned}$$

with some parameter $\lambda > 0$. Then

$$U_n(\Phi) = n^{-[m]}\, \mathrm{Per}\, A_n^m(1 + X^{(n)})\,. \tag{4.6.3}$$

Using the centered variables

$$\bar{X}_j^{(n)} = X_j^{(n)} - \lambda/n \quad \text{with} \quad E\bar{X}_j^{(n)} = 0\,,$$

we introduce a random matrix with centered shift

$$A_n^m(1+\bar{X}^{(n)}) = [a_{ij} = 1 + \bar{X}_j^{(n)};\ \ 1 \le i \le m,\ 1 \le j \le n]\,.$$

Then UR-statistics with kernel $\bar{\Phi}(x_1, \ldots, x_m) = (1 + x_1 - \lambda/n) \cdots (1 + x_m - \lambda/m)$ have the permanent form

$$U_n(\bar{\Phi}) = n^{-[m]} \operatorname{Per} A_n^m(1+\bar{X}^{(n)})\,. \tag{4.6.4}$$

Theorem 4.6.5. *Under the Poisson approximation condition with increasing degrees, the weak convergences*

$$U_n(\Phi) \Rightarrow (1+b)^{\nu}$$

and

$$U_n(\bar{\Phi}) \Rightarrow (1+b)^{\nu} e^{-\lambda b}$$

as $n \to \infty$ take place.

Proof. The permanent of matrix (4.6.3) by the shift decomposition formula has the representation

$$\begin{aligned} n^{-[m]} \operatorname{Per} A_n^m(1+X^{(n)}) &= \sum_{c=0}^{m} m^{[c]} n^{-[c]} \operatorname{Per} A_n^c(X^{(n)})/c! \\ &= \sum_{c=0}^{m} m^{[c]} n^{-[c]} \nu_n^{[c]}/c!\,. \end{aligned}$$

Using condition (4.6.2) transforms this as follows

$$n^{-[m]} \operatorname{Per} A_n^m(1+X^{(n)}) = \sum_{c=0}^{n} b^c \nu_n^{[c]}/c! - \Delta_{n1} - \Delta_{n2}, \tag{4.6.5}$$

where

$$\begin{aligned} \Delta_{n1} &:= \sum_{c=m+1}^{n} b^c \nu_n^{[c]}/c!, \\ \Delta_{n2} &:= \sum_{c=0}^{m} \varepsilon_n(c) \nu_n^{[c]}/c!, \\ \varepsilon_n(c) &:= b^c - m^{[c]}/n^{[c]}. \end{aligned}$$

The first term of (4.6.5) is the generating function of the permanent of matrices without shift $A_n^c(X^{(n)})$, i.e. the generating function of the Bernoulli sample $X_j^{(n)}, 1 \leq j \leq n$, satisfying the Poisson approximation condition. So that,

$$\begin{aligned}\sum_{c=0}^{m} b^c \nu_n^{[c]}/c! &= \prod_{j=1}^{n}(1 + bX_j^{(n)}) \\ &= (1+b)^{\nu_n} \Rightarrow (1+b)^{\nu} \quad as \quad n \to \infty.\end{aligned}$$

Now we establish the fact that, for increasing degrees (4.6.2), the second and the third terms of (4.6.5) satisfy the estimate

$$E|\Delta_{nk}| \to 0 \ as \ \ n \to \infty, \quad k = 1, 2. \tag{4.6.6}$$

Thus the proof of the first statement of the theorem is complete.

Note that considering $\nu_n^{[c]}$ as the permanent of sampling matrix $A_n^c(X^{(n)})$,

$$\begin{aligned}E\nu_n^{[c]} &= E\,\mathrm{Per}\, A_n^c(X^{(n)}) \\ &= n^{[c]} \prod_{j=1}^{c} EX_j^{(n)} \\ &= \lambda^c n^{[c]}/n^c \leq \lambda^c.\end{aligned}$$

Moreover $E\nu_n^{[c]} \to \lambda^c$, as $n \to \infty$, for fixed c. The sum in Δ_{n1} allows the following estimate

$$\begin{aligned}E\Delta_{n1} &\leq \sum_{c=m+1}^{n} (b\lambda)^c/c! \\ &\leq (b\lambda)^{m+1} e^{b\lambda}/(m+1)! \to 0 \quad as \quad m \to \infty.\end{aligned}$$

Thus, estimate (4.6.6) for $k = 1$ is true. Further, under condition (4.6.2),

$$m^{[c]}/n^{[c]} = b_n^c \prod_{s=1}^{c-1}(1 - a_s), \ \ a_s \leq s/m.$$

Using the algebraic inequality

$$\prod_{s=1}^{c-1}(1 - a_s) > 1 - \sum_{s=1}^{c-1} a_s \quad for \quad 0 < a_s < 1 \tag{4.6.7}$$

and, assuming that $m < n$,

$$\begin{aligned}b_n^c &\geq m^{[c]}/n^{[c]} \\ &\geq b_n^c \left(1 - \frac{c(c-1)}{m}\right).\end{aligned}$$

Hence the following estimate holds true, as $m \to \infty$,

$$\begin{aligned} 0 &\leq \varepsilon_{n1}(c) := b_n^c - m^{[c]}/n^{[c]} \\ &\leq b_n^c c(c-1)/m \to 0. \end{aligned}$$

To complete the estimation of sum Δ_{n2} in (4.6.5) note that

$$\begin{aligned} \varepsilon_n(c) &:= b^c - m^{[c]}/n^{[c]} \\ &= b^c - b_n^c + \varepsilon_{n1}(c). \end{aligned}$$

Applying the evident inequality

$$|b^c - b_n^c| \leq c|b - b_n|$$

we conclude that $\varepsilon_n(c)$ satisfies the following estimate

$$|\varepsilon_n(c)| \leq c|b - b_n| + c(c-1)b_n^c/m \to 0 \quad as \ n, m \to \infty.$$

So, for Δ_{n2}, estimate (4.6.7) is obtained as follows:

$$\begin{aligned} E|\Delta_{n2}| &\leq \sum_{c=0}^{m} |\varepsilon_n(c)|\lambda^c/c! \\ &\leq |b - b_n| \sum_{c=1}^{m} \lambda^c/(c-1)! + \frac{1}{m}\sum_{c=2}^{m}(b_n\lambda)^c/(c-2)! \\ &\leq |b - b_n|\lambda e^\lambda + (b_n\lambda)^2 e^{b_n\lambda}/m \to 0 \quad \text{as} \ n, m \to \infty. \end{aligned}$$

Therefore, the first statement of Theorem 4.6.5 holds.

Further, apply the shift decomposition formula with parameter $h = 1 - \lambda/n$. We write the permanent of matrix (4.6.4) in the following form

$$\begin{aligned} n^{-[m]} \operatorname{Per} A_n^m(1 + \bar{X}^{(n)}) &= \sum_{c=0}^{m}(1 - \lambda/n)^{m-c} m^{[c]} n^{-[c]} \nu_n^{[c]}/c! \\ &= (1 - \lambda/n)^m (1 + b)^{\nu_n} - \\ &\quad -(1 - \lambda/n)^m(\Delta_{n1} + \Delta_{n2}) + \Delta_{n3}, \end{aligned} \tag{4.6.8}$$

where $\Delta_{nk}, k = 1, 2$ are defined in (4.6.5) and

$$\begin{aligned} \Delta_{n3} &:= \sum_{c=0}^{m} q_n(c) m^{[c]} n^{-[c]} \nu_n^{[c]}/c!, \\ q_n(c) &:= (1 - \lambda/n)^{m-c} - (1 - \lambda/n)^m \\ &\leq \lambda c/n. \end{aligned}$$

Note that

$$\begin{aligned} E|\Delta_{n3}| &\le \sum_{c=1}^{m} \frac{\lambda^{c+1}}{(c-1)!}\frac{1}{n} \\ &\le \lambda^2 e^{\lambda}/n \to 0 \quad \text{as } n \to \infty . \end{aligned}$$

Taking the limit in (4.6.8), we get the second statement of Theorem 4.6.5.

Permanent functionals with shifts are considered as the permanents of functional sampling matrices with shifts:

$$A_n^m(1+\varphi) := [a_{ij} = 1 + \varphi(X_j^{(n)});\ \ 1 \le i \le m,\ 1 \le j \le n]$$

and with centered shifts

$$A_n^m(1+\bar{\varphi}) := [a_{ij} = 1 + \varphi(X_j^{(n)}) - \lambda(\varphi)/n;\ \ 1 \le i \le m,\ 1 \le j \le n]$$

where $\varphi(x) : X \to R$ is a real-valued bounded function with compact support on X and a sample $X_j^{(n)}$, $1 \le j \le n$ satisfies the Poisson approximation condition $P_n(B) = \lambda(B)/n$, so that

$$E\varphi(X_j^{(n)}) = \lambda(\varphi)/n ,$$

$$\lambda(\varphi) := \int_X \varphi(x)\lambda(dx) .$$

Consider UR-statistics with kernel $\Phi(x_1, \ldots, x_m) = (1+\varphi(x_1)) \cdots (1+\varphi(x_m))$. Then

$$U_n(\Phi) = n^{-[m]} \operatorname{Per} A_n^m(1+\varphi) .$$

Similarly, if $\bar{\Phi}(x_1, \ldots, x_m) = (1+\varphi(x_1) - \lambda(\varphi)/n) \cdots (1+\varphi(x_m) - \lambda(\varphi)/n)$, then

$$U_n(\bar{\Phi}) = n^{-[m]} \operatorname{Per} A_n^m(1+\bar{\varphi}) .$$

Theorem 4.6.6. *Under the Poisson approximation condition and for increasing degrees the weak convergences*

$$U_n(\Phi) \Rightarrow \exp \int_X \ln[1 + b\varphi(x)]\nu(dx)$$

and

$$U_n(\bar{\Phi}) \Rightarrow \exp\{\int_X \ln[1 + b\varphi(x)]\nu(dx) - b\lambda(\varphi)\}$$

as $n \to \infty$ take place.

Proof. Acting just as in the proof of Theorem 4.6.5, we obtain at first the following representation

$$n^{-[m]} \operatorname{Per} A_n^m(1+\varphi) = \prod_{j=1}^{n} [1 + b\varphi(X_j^{(n)})] + \Delta_n , \tag{4.6.9}$$

where Δ_n satisfies the estimate $E|\Delta_n| \to 0$ as $n \to \infty$. Setting

$$\nu_n(B) := \sum_{j=1}^{n} \delta_j^{(n)}(B)$$

we get, for the first term of (4.6.9), the following representation

$$\prod_{j=1}^{n} [1 + b\varphi(X_j^{(n)})] = \exp \int_X \ln[1 + b\varphi(x)] \nu_n(dx) . \tag{4.6.10}$$

Under the Poisson approximation condition, $\nu_n(B) \Rightarrow \nu(B)$ as $n \to \infty$ with $E\nu(B) = \lambda(B)$. Taking the limit in (4.6.9) and by virtue of (4.6.10), the first statement of the theorem is obtained.

The second statement of the theorem is a consequence of the following representation

$$n^{-[m]} \operatorname{Per} A_n^m(1 + \bar{\varphi}) =$$

$$= (1 - \lambda(\varphi)/n)^m \prod_{j=1}^{n} [1 + b\varphi(X_j^{(n)})] + \Delta_n , \tag{4.6.11}$$

where Δ_n satisfies the estimate

$$E|\Delta_n| \to 0 \quad \text{as} \quad n \to \infty .$$

Note that for increasing degrees (4.6.2)

$$(1 - \lambda(\varphi)/n)^m \to e^{-b\lambda(\varphi)} .$$

Taking the limit in (4.6.11), we obtain the proof of Theorem 4.6.6.

The general case.

Theorem 4.6.7. *Assume that the following conditions are satisfied:*

$$b_n := m \,/\, n \to b, \quad 0 < b < 1 ;$$

there exist normalizing coefficients $a_m \geq 0$ such that

$$\lim_{m \to \infty} a_m g_{mc}(x_1, \dots , x_c) = g_c(x_1, \dots , x_c)$$

where functions g_c are continuous and have compact supports, $|g_c| \leq K^c$;

$$\sum_{c=r}^{m} \frac{b_n^c}{c!} \int_X \cdots \int_X |a_m g_{mc}(x_1, \dots , x_c) - g_c(x_1, \dots , x_c)| \prod_{s=1}^{c} \lambda_n(dx_s) \to 0$$

as $n \to \infty$, where $\lambda_n(B) = nP_n(B)$, $B \in \mathfrak{X}$.

Then, in the Poisson approximation condition

$$\lim_{n\to\infty} \lambda_n(B) = \lambda(B), \quad B \in \mathfrak{X}$$

with some Radon measure $\lambda(B)$*, the weak convergence*

$$a_m\Big(U_n(\Phi) - \theta\Big) \Rightarrow \sum_{c=r}^{\infty} \frac{b^c}{c!} I_c^{\nu}(g_c)$$

with Itô–Poisson stochastic integrals

$$I_c^{\nu}(g_c) := \int_{X^c} g_c(x_1, \ldots, x_c) \prod_{s=1}^{c} \nu(dx_s)$$

takes place.

Proof. First, transform the Hoeffding decomposition of U-statistics (4.6.1) to the form

$$a_m(U_n(\Phi) - \theta) = \sum_{c=r}^{m} m^{[c]} U_n(g_c)/c! + \Delta_n\,,$$

where

$$\Delta_n = \sum_{c=r}^{m} m^{[c]} U_n(a_m g_{mc} - g_c)\,/\,c!\,.$$

Here Δ_n satisfies the estimate

$$E|\Delta_n| \le \sum_{c=r}^{m} \frac{b_n^c}{c!} \int_X \cdots \int_X |a_m g_{mc}(x_1, \ldots, x_c) - g_c(x_1, \ldots, x_c)| \prod_{s=1}^{c} \lambda_n(dx_s) \to 0$$

and hence $\Delta_n \to 0$ in probability.

Further, by Theorem 4.4.4

$$\sum_{c=r}^{m} m^{[c]} U_n(g_c)/c! \Rightarrow \sum_{c=r}^{\infty} \frac{b^c}{c!} I_c^{\nu}(g_c)\,.$$

This proves Theorem 4.6.7.

Corollary 4.6.1. *For the kernel*

$$\Phi(x_1, \ldots, x_m) = \prod_{s=1}^{m} \varphi(x_s)\,,$$

under the conditions of Theorem 4.6.7 and

$$\lambda_n(\varphi) = n \int_X \varphi(x) P_n(dx) \to \int_X \varphi(x)\lambda(dx) = \lambda(\varphi) \neq 0\,,$$

the weak convergence as $n \to \infty$

$$\lambda_n^{-m}(\varphi) U_n(\Phi) \Rightarrow \exp \int_X \ln(1 - bg(x))\nu(dx) \tag{4.6.12}$$

takes place with $g(x) = 1 - \varphi(x)/\lambda(\varphi)$.

Proof. By the Hoeffding decomposition formula (4.6.1)

$$U_n(\Phi) = \sum_{c=0}^{m} \frac{m^{[c]}}{c!} U_n(g_{mc})$$

with

$$g_{mc} = [\lambda_n(\varphi)]^{m-c} \prod_{s=1}^{c} [\varphi(x_s) - \lambda_n(\varphi)] .$$

Hence, under the condition of the Poisson approximation scheme,

$$\lambda_n^{-m}(\varphi) g_{mc} \longrightarrow (-1)^c \prod_{s=1}^{c} g(x_s) = (-1)^c g_c, m \to \infty .$$

By Theorem 4.6.7 the limiting result in (4.6.12) has the form

$$\lambda_n^{-m}(\varphi) U_n(\Phi) \Rightarrow \sum_{c=0}^{\infty} (-1)^c \frac{b^c}{c!} I_c^{\nu}(g_c) . \tag{4.6.13}$$

By the Itô–Poisson formula the right-hand side of (4.6.13) is equal to the right-hand side of (4.6.12).

Suppose that $E|\Phi(X_1^{(n)}, \dots, X_m^{(n)})| < \infty$, and define functions $\Phi_n^v : X^m \to R$ by

$$\Phi_n^v(x_1, \dots, x_m) = \int_X \cdots \int_X \Phi(y_1, \dots, y_m) \prod_{s=1}^{m} (P_n(dy_s) + v\delta_{x_s}(dy_s)) ,$$

where v is a fixed arbitrary number, $|v| < \infty$. Introduce the family of functionals (with respect to the parameter v):

$$U_n(\Phi_n^v) = n^{-[m]} \sum_{1 \le i_1 \ne \dots \ne i_m \le n} \Phi_n^v(X_{i_1}^{(n)}, \dots, X_{i_m}^{(n)}) .$$

So $U_n(\Phi_n^v)$ is the integral functional from Φ with respect to the permanent measure $\pi_n^{P_n + v\delta}(B_1, \dots, B_m)$. For $c = 0, 1, \dots, m$ denote

$$\Phi_{mc}(x_1, \dots, x_c) = \int_X \cdots \int_X \Phi(x_1, \dots, x_c, x_{c+1}, \dots, x_m) \prod_{s=c+1}^{m} P_n(dx_s)$$

and suppose that there exists the limit

$$\lim_{n \to \infty} \Phi_{mc}(x_1, \dots, x_c) = \Phi_c(x_1, \dots, x_c) .$$

Example 4.6.1. Let

$$\Phi(x_1, \dots, x_m) = \Phi(x_1 + \dots + x_m) ,$$

where $\Phi(x)$ is a continuous function with compact support. Then

$$\begin{aligned}\Phi_{mc}(x_1,\dots,x_c) &= E\Phi(x_1+\cdots+x_c+X_{c+1}^{(n)}+\cdots+X_m^{(n)}) \\ &= E\Phi(x_1+\cdots+x_c+\int_X y\nu_{m-c}(dy)).\end{aligned}$$

Since, under conditions (4.6.2),

$$\nu_{m-c}(B) \Rightarrow \nu_{b\lambda}(B)\,, \quad B \in \mathfrak{X}\,,$$

where $\nu_{b\lambda}(B)$ is the Poisson random measure with intensity $b\lambda(B)$ then, under the condition $n \to \infty$,

$$\Phi_{mc}(x_1,\dots,x_c) \longrightarrow E\Phi(x_1+\cdots+x_c+\int_X y\nu_{b\lambda}(dy)).$$

Thus, in the given example,

$$\Phi_c(x_1,\dots,x_c) = E\Phi(x_1+\cdots+x_c+\int_X y\nu_{b\lambda}(dy)).$$

Theorem 4.6.8. *Assume that condition* (4.6.2) *is satisfied and*

$$|\Phi_c(x_1,\dots,x_c)| \le \prod_{s=1}^{c} \varphi(x_s)\,, \tag{4.6.14}$$

where $\varphi(x) \ge 0$ is a continuous function with compact support. Also let, under $n \to \infty$,

$$\sum_{c=0}^{m} \frac{1}{c!}(vb_n)^c \int_X \cdots \int_X |\Phi_{mc}(x_1,\dots,x_c) - \Phi_c(x_1,\dots,x_c)| \prod_{s=1}^{c} \lambda_n(dx_s) \to 0. \tag{4.6.15}$$

Then the weak convergence

$$U_n(\Phi_n^v) \Rightarrow \sum_{c=0}^{\infty} \frac{(vb)^c}{c!} I_c^\nu(\Phi_c) \tag{4.6.16}$$

takes place.

Proof. FIrst we represent Φ_n^v as

$$\Phi_n^v(X_1^{(n)},\dots,X_m^{(n)}) = \sum_{c=0}^{m} \frac{v^c}{c!} \sum_{1\le i_1 \ne \dots \ne i_c \le m} \Phi_{mc}(X_{i_1}^{(n)},\dots,X_{i_c}^{(n)}).$$

Substituting this formula into $U_n(\Phi_n^v)$,

$$U_n(\Phi_n^v) = \sum_{c=0}^{m} \frac{v^c}{c!} m^{[c]} n^{-[c]} \sum_{1\le i_1 \ne \dots \ne i_c \le n} \Phi_{mc}(X_{i_1}^{(n)},\dots,X_{i_c}^{(n)}).$$

Write this formula as

$$U_n(\Phi_n^\delta) = \sum_{c=0}^{m} \frac{v^c}{c!} b_n^c \sum_{1\le i_1\ne\cdots\ne i_c\le n} \Phi_c(X_{i_1}^{(n)},\dots,X_{i_c}^{(n)})$$

$$+\Delta_{n1} + \Delta_{n2},$$

where

$$\Delta_{n1} = \sum_{c=0}^{m} \frac{v^c}{c!}(m^{[c]}n^{-[c]} - b_n^c) \sum_{1\le i_1\ne\cdots\ne i_c\le n} \Phi_c(X_{i_1}^{(n)},\dots,X_{i_c}^{(n)}),$$

$$\Delta_{n2} = \sum_{c=0}^{m} \frac{v^c}{c!} m^{[c]}n^{-[c]} \sum_{1\le i_1\ne\cdots\ne i_c\le n} \left(\Phi_{mc}(X_{i_1}^{(n)},\dots,X_{i_c}^{(n)}) - \Phi_c(X_{i_1}^{(n)},\dots,X_{i_c}^{(n)})\right).$$

Taking into account (4.6.2) and condition (4.6.14) we conclude that the limit of the sums with respect to c coincides with the random value from the right-hand side of (4.6.16). That is why to prove (4.6.16) it is sufficient to check that $\Delta_{ni} \to 0$ in probability. For this purpose, taking into account (4.6.14) we write

$$\begin{aligned} E|\Delta_{n1}| &\le \frac{1}{2m}\sum_{c=2}^{m} \frac{v^c}{(c-2)!} b_n^c \int_X\cdots\int_X |\Phi_c(x_1,\dots,x_c)| \prod_{s=1}^{c} \lambda_n(dx_s) \\ &\le \frac{v^2}{2}\frac{b_n}{n}\sum_{c=0}^{m-2}\frac{v^c}{c!} b_n^c \lambda_n^c(h) \\ &\le \frac{v^2}{2}\frac{b_n}{n}\exp(vb_n\lambda_n(h)) \to 0. \end{aligned}$$

Further

$$E|\Delta_{n2}| \le \sum_{c=0}^{m} \frac{1}{c!}(vb_n)^c \int_X\cdots\int_X |\Phi_{mc}(x_1,\dots,x_c)$$

$$-\Phi_c(x_1,\dots,x_c)| \prod_{s=1}^{c} \lambda_n(dx_s) \to 0$$

by virtue of condition (4.6.15). As a result statement (4.6.16) is proved.

4.7 Symmetric statistics

Consider *SB*-statistics, defined in Section 1.4. Write representation (1.4.2) in the form

$$S_n - \theta = \sum_{c=1}^{n} \frac{1}{c!} \sum_{1\le i_1\ne\dots\ne i_c\le n} g_{nc}(X_{i_1},\dots,X_{i_c}), \tag{4.7.1}$$

where $S_n = S(X_1,\dots,X_n)$ and $\theta = ES_n$.

At first we will explain sufficient conditions which guarantee the asymptotic normality of *SB*-statistics.

Central limit theorem for *SB*-statistics. Denote

$$s_n = \sum_{c=2}^{n} \frac{1}{c!} \sum_{1 \le i_1 \ne \ldots \ne i_c \le n} g_{nc}(X_{i_1}, \ldots, X_{i_c}).$$

Let $0 < E\|g_{n1}(X_1)\|^2 < \infty$, $n = 1, 2, \ldots$ and assume a function $g : X \to B$ exists such that

$$\lim_{n\to\infty} g_{n1}(x) = g(x),$$

$$E\|g_{n1}(X_1) - g(X_1)\|^2 \to 0$$

as $n \to \infty$ and $Eg(X_1) = 0, \quad 0 < E\|g(X_1)\|^2 < \infty$.

Define covariance function V:

$$V(x^*, y^*) = \int_X x^*(g_1(z)) y^*(g_1(z)) P(dz),$$

where $x^*, y^* \in B^*$.

Theorem 4.7.1. *Assume that B is a Banach space of type 2 and s_n satisfies the condition $n^{-1/2} s_n \to 0$ in probability as $n \to \infty$.*

Then the weak convergence

$$n^{-1/2}(S_n - \theta) \Rightarrow \tau$$

holds as $n \to \infty$, where τ is a Gaussian random element with covariance function V.

Proof. From (4.7.1) we have

$$n^{-1/2}(S_n - \theta) = n^{-1/2}\left(g(X_1) + \cdots + g(X_n)\right) +$$

$$+ n^{-1/2} \sum_{j=1}^{n} (g_{n1}(X_j) - g(X_j)) + n^{-1/2} s_n .$$

By the central limit theorem in a Banach space of type 2

$$n^{-1/2}\left(g(X_1) + \cdots + g(X_n)\right) \Rightarrow \tau .$$

By inequality (2.7.2)

$$E\|n^{-1/2} \sum_{j=1}^{n} (g_{n1}(X_j) - g(X_j)\|^2 \le c n^{-1} \sum_{j=1}^{n} E\|g_{n1}(X_j) - g(X_j)\|^2 =$$

$$= cE\|g_{n1}(X_1) - g(X_1)\|^2 \to 0$$

as $n \to \infty$. This completes the proof of Theorem 4.7.1.

Convergence of *SH*-statistics to multiple Wiener–Itô integrals. Now we assume that in 4.7.1, $g_{nc}(x_1,\dots,x_c) = n^{-c/2}g_c(x_1,\dots,x_c)$ and the canonical functions $g_c : X^c \to H$ satisfy the condition of complete degeneracy

$$\int_X g(x_1,\dots,x_{c-1},y)\,P(dy) = 0, c = 1,2,\dots .$$

Therefore

$$S_n - \theta = \sum_{c=1}^{n} \frac{1}{c!} n^{-c/2} \sum_{1\le i_1 \ne \dots \ne i_c \le n} g_c(X_{i_1},\dots,X_{i_c}) . \tag{4.7.2}$$

Theorem 4.7.2. *Let $X_1,\dots,X_n,\dots$ be an i.i.d. sequence with common distribution P. Then the symmetric H-valued statistic*

$$\sum_{c=1}^{n} \frac{1}{c!} n^{-c/2} \sum_{1\le i_1 \ne \dots \ne i_c \le n} g_c(X_{i_1},\dots,X_{i_c})$$

converges in distribution to the multiple Wiener–Itô integral

$$\sum_{c=1}^{\infty} \frac{1}{c!} \int_X \cdots \int_X g_c(x_1,\dots,x_c) \prod_{s=1}^{c} W(dx_s) ,$$

if the symmetric canonical functions g_c satisfy

$$\sum_{c=1}^{\infty} \frac{1}{c!} \int_X \cdots \int_X \|g_c(X_1,\dots,X_c)\|^2 \prod_{s=1}^{c} P(dx_s) < \infty .$$

Here $W(\cdot)$ is a Gaussian random measure with intensity P.

Proof. For any fixed N, $N \ge 1$, denote

$$S_{nN}^{\delta} = \sum_{c=1}^{\min(n,N)} \frac{1}{c!} n^{-c/2} \sum_{1\le i_1 \ne \dots \ne i_c \le n} g_c(X_{i_1},\dots,X_{i_c})$$

and

$$S_N^W = \sum_{c=1}^{N} \frac{1}{c!} \int_X \cdots \int_X g_c(X_1,\dots,X_c) \prod_{s=1}^{c} W(dx_s) .$$

Further, if $\mathcal{L}(X,Y)$ is the Lévy–Prokhorov distance between the random elements X and Y, then we have the inequality

$$\mathcal{L}(S_{nn}^{\delta}, S_{\infty}^{W}) \le \mathcal{L}(S_{nn}^{\delta}, S_{nN}^{\delta}) + \mathcal{L}(S_{nN}^{\delta}, S_N^W) + \mathcal{L}(S_N^W, S_{\infty}^W) .$$

By the property of the Lévy–Prokhorov metric

$$\mathcal{L}(S_{nn}^{\delta}, S_{nN}^{\delta}) \le 2(E\|S_{nn}^{\delta} - S_{nN}^{\delta}\|^2)^{1/3} \le$$

$$\le 2\Big(\sum_{c=N+1}^{\infty} \frac{1}{c!} E\|g_c(X_1,\dots,X_c)\|^2\Big)^{1/3} \to 0$$

as $N \to \infty$ uniformly on n, and similarly,

$$\mathcal{L}(S_N^W, S_\infty^W) \le 2(E\|S_N^W - S_\infty^W\|^2)^{1/3} \le$$

$$\le 2(\sum_{c=N+1}^{\infty} \frac{1}{c!} E\|g_c(X_1, \dots, X_c)\|^2)^{1/3} \to 0$$

as $N \to \infty$.

By Theorem 4.2.4

$$\mathcal{L}(S_{nN}^\delta, S_N^W) \to 0$$

as $n \to \infty$ for any fixed N.

As a result we obtain

$$\mathcal{L}(S_{nn}^\delta, S_\infty^W) \to 0$$

as $n \to \infty$. This proves Theorem 4.7.2.

Convergence of *SR*-statistics to multiple Itô–Poisson integrals. Here we assume that $X_1^{(n)}, \dots, X_n^{(n)}$ are independent, identically distributed random elements on X with distribution P_n. Denote $\nu_n(B) = \delta_{X_1^{(n)}}(B) + \dots + \delta_{X_n^{(n)}}(B)$, $B \in \mathfrak{X}$ and suppose that distribution ν_n converges weakly to a distribution of a Poisson random measure ν with intensity λ.

Theorem 4.7.3. *Assume that real continuous canonical functions g_c, $c = 1, 2, \dots$ satisfy the following conditions:*

there exists a compact set B such that for any $c \ge 1$ $g_c(x_1, \dots, x_c) = 0$ if $(x_1, \dots, x_c) \notin B^c$;

there exists a constant K such that for any $c \ge 1$ $|g_c(x_1, \dots, x_c)| \le K^c$.

Then the symmetric statistic

$$\sum_{c=1}^{n} \frac{1}{c!} \sum_{1 \le i_1 \ne \dots \ne i_c \le n} g_c(X_{i_1}^{(n)}, \dots, X_{i_c}^{(n)})$$

converges in distribution to the multiple Itô–Poisson integral

$$\sum_{c=1}^{\infty} \frac{1}{c!} \int_X \cdots \int_X g_c(x_1, \dots, x_c) \prod_{s=1}^{c} \bar{\nu}(dx_s) .$$

Here $\bar{\nu}(\cdot) = \nu(\cdot) - \lambda(\cdot)$ is a centered Poisson random measure with intensity λ.

Proof. This theorem is a consequence of Theorem 4.6.7.

***SR*-statistics with random sample sizes.** For $p > 0$, let N_p be a positive random variable taking integer values independently of i.i.d. random variables $X_1, X_2, \dots$ which take values in an arbitrary measurable space $(X, \mathfrak{X})$.

In the following, we assume that

$$N_p/p \Rightarrow Y \ \ as \ \ p \to \infty \tag{4.7.3}$$

for some positive random variable Y independent of X_i's. Under this condition we would like to study the limit behavior of statistic

$$p^{-c/2} \sum_{1 \le i_1 \ne \ldots \ne i_c \le N_p} g_c(X_{i_1}, \ldots, X_{i_c}) \ \ as \ \ p \to \infty$$

and *SR*-statistics of the form (4.7.2).

For each $T > 0$, let G_T be a space of sequences of canonical functions $g = \{g_c, c \ge 1\}$ such that

$$\|g\|_T^2 = \sum_{c=1}^{\infty} \frac{T^c}{c!} \int_X \cdots \int_X g_c^2(x_1, \ldots, x_c) \prod_{s=1}^{c} P(dx_s) < \infty \,.$$

Let $G = \bigcap_{T>0} G_T$.

Theorem 4.7.4. *Let $g \in G$ and denote by L_Y the Laplace Transform of Y. Then the finite-dimensional distributions of*

$$\sum_{c=1}^{\infty} \frac{1}{c!} p^{-c/2} \sum_{1 \le i_1 \ne \ldots \ne i_c \le N_p} g_c(X_{i_1}, \ldots, X_{i_c})$$

converge weakly to those of

$$\sum_{c=1}^{\infty} \frac{1}{c!} I_c^M(g_c)$$

as $p \to \infty$, where $I_c^M(g_c)$ are multiple integrals with respect to a random measure M on $(X, \mathfrak{X})$ such that for sets $B_1, \ldots, B_m \in \mathfrak{X}$ the vector $(M(B_1), \ldots, M(B_m))$ has the characteristic functional

$$E \exp(it_1 M(B_1) + \cdots + it_m M(B_m)) = L_Y\Big(\frac{1}{2} \sum_{i,j=1}^{m} t_i t_j \, P(B_i \cap B_j)\Big) \,.$$

The proof of this theorem is given in Feldman *et al.* (1993), where also a rigorous construction of multiple integrals is given $I_c^M(g_c)$.

In the following examples we have some random measures M which correspond to different distributions of sample size N_p.

Example 4.7.1 (Poisson sample size). Let N_p be a Poisson random variable with mean $p > 0$, then $N_p/p \to 1$ with probability 1 and M coincides with the Gaussian measure $W(\cdot)$ such that $E\,W(B) = 0$, $E\,W(A)W(B) = P(A \cap B)$, $A, B \in \mathfrak{X}$.

Example 4.7.2 (Geometric sample size). If N_p is a geometric random variable with mean p, then Y is a standard exponential and M is a random measure such that

$$E\exp(it_1\,M(B_1)+\cdots+it_m\,M(B_m))=\Big(1+\frac{1}{2}\sum_{i,j=1}^{m}t_it_j\,P(B_i\cap B_j)\Big)^{-1}.$$

Example 4.7.3 (Discrete stable sample size). Let Y be a positive stable random variable with Laplace transform $E\exp(-tY)=\exp(-t^{\,\alpha/2})$, $t>0$, where the index of Y, $\alpha/2$, is less than 1. Assume that N_p is the following discrete version of Y:

$$P(N_p=k)=P(k-1<pY\le k),\quad k\ge 1.$$

Then $N_p/p\to Y$ and the random measure M is symmetrically stable with parameter α :

$$E\exp(it_1\,M(B_1)+\cdots+it_m\,M(B_m))=\exp\Big\{-(\frac{1}{2}\sum_{i,j=1}^{m}t_it_j\,P(B_i\cap B_j))^{\alpha/2}\Big\}.$$

Now we consider SR-statistics

$$\nu_{N_n}=\sum_{c=1}^{N_n}\frac{1}{c!}n^{-c/2}\sum_{1\le i_1\ne\ldots\ne i_c\le N_n}g_c(X_{i_1}^{(n)},\ldots,X_{i_c}^{(n)}),$$

where N_n satisfies condition (4.7.3) and $X_1^{(n)},X_2^{(n)},\ldots,X_n^{(n)},\ldots$ is an i.i.d. array, satisfying the condition of the Poisson approximation.

If $n\to\infty$ then under condition (4.7.3)

$$\sum_{j=1}^{N_n}\delta_j^{(n)}\Rightarrow\nu(\cdot,Y), \tag{4.7.4}$$

where $\nu(\cdot,Y)$ is a random measure on $(X,\mathfrak{X})$ such that for set $B\in\mathfrak{X}$ random variable $\nu(B,Y)$ has the characteristic functional

$$E\,e^{is\nu(B,Y)}=E\exp\{Ys(e^{is\lambda(B)}-1)\},\ s\in R.$$

Theorem 4.7.6. *Under the conditions of Theorem* 4.7.3

$$\nu_{N_n}\Rightarrow\sum_{c=1}^{\infty}\frac{1}{c!}\int_X\cdots\int_X g_c(x_1,\ldots,x_c)\prod_{s=1}^{c}\nu(dx_s,Y)$$

as $n\to\infty$.

For the proof of this theorem see the proof of Theorem 4.6.7 and relation (4.7.4).

4.8 U-statistics with varying kernels

Consider U-statistics

$$U_n = n^{-[m]} \sum_{1 \le i_1 \ne \ldots \ne i_m \le n} \Phi_n(X_{i_1}, \ldots, X_{i_m}) \tag{4.8.1}$$

with a symmetric kernel $\Phi_n : (x_1, \ldots, x_m) \to \Phi_n(x_1, \ldots, x_m)$, $m \le n$. Kernel Φ_n is assumed to depend on n. In this case Φ_n is called a varying kernel. Note that the various functions of Φ_n such as $\Phi_c(x_1, \ldots, x_c)$, $g_c(x_1, \ldots, x_c), \ldots$ also depend on n. Therefore, in order to stress this dependence, we equip these functions with the index n: $\Phi_{nc}(x_1, \ldots, x_c)$, $g_{nc}(x_1, \ldots, x_c), \ldots$.

If $\Phi_n \in B$ and rank $r = 1$, then the asymptotic normality of U_n follows from Theorem 4.7.1. If $\Phi_n \in B$ and rank $r \ge 2$, then the asymptotic normality of such UB-statistics is guaranteed by Daffer's central limit theorem for martingales in a Banach space (see Daffer (1984)).

In what follows in this section we assume that $\Phi_n \in R$.

Martingale theorems and asymptotic normality. The behavior of the variance $\sigma^2(U_n)$ of such UR-statistics as $n \to \infty$, generally speaking, differs from that of kernel Φ independent of n. Therefore, the conditions implying the weak convergence of U-statistics with varying kernel are distinct.

Lemma 1.1.2 and the central limit theorem for sums of martingale differences (see, e.g. Hall and Heyde (1980), Borovskikh and Korolyuk (1988i), Liptser and Shiryaev (1986)) enable us to state the following assertion for such UR-statistics.

Theorem 4.8.1. *Let $E\Phi_n = 0$ and $E\Phi_n^2 < \infty$. In order that*

$$\sigma^{-1}(U_n) U_n \Rightarrow \tau, \quad \textit{as } n \to \infty,$$

where τ is a standard normal random variable, it suffices that the conditions

$$\sigma^{-2}(U_n) \sum_{k=1}^{n} E\left(\xi_{nk}^2 I(|\xi_{nk}| > \varepsilon\sigma(U_n)) | \mathfrak{F}_{k-1}\right) \xrightarrow{P} 0,$$

$$\sigma^{-2}(U_n) \sum_{k=1}^{n} E\left(\xi_{nk}^2 | \mathfrak{F}_{k-1}\right) \xrightarrow{P} 1,$$

are satisfied, where $0 < \varepsilon < 1$. Additionally, the explicit form of the stochastic sequence $(\xi_{nk}, \mathfrak{F}_k)$ is given in Lemma 1.1.2.

The conditions of this theorem can be simplified with the help of martingale moment inequalities of type (2.2.2) and (2.2.4), if we assume, in addition, that $E|\Phi_n|^{2+\delta} < \infty$ for some $\delta > 0$.

Corollary 4.8.1. *Let $m = 2$ and assume that degenerate kernel Φ_n satisfies $E|\Phi_n|^{2+\delta} < \infty$, $\delta > 0$. Denote*

$$G_n(x,y) = E(\Phi_n(X_1,x)\Phi_n(y,X_1)),$$
$$\sigma_{n2}^2 = E\Phi_n^2(X_1,X_2).$$

If the conditions

$$n^{-\delta/2}\sigma_{n2}^{-(2+\delta)}E|\Phi_n|^{2+\delta} \to 0,$$
$$\sigma_{n2}^{-(2+\delta)}E|G_n(X_1,X_2)|^{(2+\delta)/2} \to 0$$

hold as $n \to \infty$, then the weak convergence

$$\frac{n}{\sigma_{2n}\sqrt{2}}U_n \Rightarrow \tau \tag{4.8.2}$$

takes place.

Corollary 4.8.2. *If $E\Phi_n^4 < \infty$, then relation (4.8.2) holds provided that*

$$\sigma_{n2}^{-2}(EG_n^2(X_1,X_2) + n^{-1}E\Phi_n^4(X_1,X_2)) \to 0$$

as $n \to \infty$.

Corollary 4.8.3. *Assume that there exist positive constants a and b independent of n, such that*

$$\sigma_{n2} \geq a > 0, \quad E\Phi_n^4(X_1,X_2) \leq b < \infty$$

If $EG_n^2(X_1,X_2) \to 0$ as $n \to \infty$, then (4.8.2) holds.

Theorem 4.8.2. *Let kernel Φ_n satisfy the conditions $E\Phi_n(X_1,X_2) = 0$ and suppose that $E\Phi_n^2(X_1,X_2) = 1$. If*

$$\binom{n}{2}^{-1} \sum_{1\leq i<j\leq n} \Phi_n^2(X_1,X_2) \xrightarrow{P} 1$$

as $n \to \infty$, and there exists a number $\lambda > 0$ such that

$$n^{-2} \sum_{1\leq i<j<k\leq n} \Phi_n(X_i,X_j)\Phi_n(X_i,X_k) \xrightarrow{L^1} \lambda$$

then the weak convergence

$$\frac{n}{\sqrt{2+4\lambda}}U_n \Rightarrow \tau$$

takes place.

Proof. Let $\mathfrak{B}_j$ be the σ-algebra generated by an unordered set of random variables $(X_1,\dots,X_j)$ and by a sequence $X_{j+1},X_{j+2},\dots$, i.e.

$$\mathfrak{B}_j = \sigma\{\omega : (X_1,\dots,X_j), X_{j+1}, X_{j+2},\dots\}.$$

Let $E_j(\cdot) = E(\cdot|\mathfrak{B}_j)$ be a conditional expectation with respect to $\mathfrak{B}_j$. By formula (1.1.15)

$$E_j(\Phi_n(X_1, X_2)) = \binom{j}{2}^{-1} \sum_{i=2}^{j} \sum_{k=1}^{i-1} \Phi_n(X_i, X_k)$$

for all $j \geq 2$.

Denote

$$Z_{nj} = n[E_j(\Phi_n(X_1, X_2)) - E_{j+1}(\Phi_n(X_1, X_2))] =$$

$$= 2n(j-1)^{-1}[E_{j+1}(\Phi_n(X_1, X_2)) - j^{-1} \sum_{k=1}^{j} \Phi_n(X_{j+1}, X_k)].$$

Hence, if $r_n > n$, then

$$nU_n = \sum_{j=n}^{r_n - 1} Z_{nj} + n\, E_{r_n}(\Phi_n(X_1, X_2)). \tag{4.8.3}$$

For each n, we choose the integer $r_n > n$ so that $nr_n^{-1} \to 0$ as $n \to \infty$. The sequence

$$(Z_{n, r_n - j + 1}, \mathfrak{B}_{r_n - j + 1}) \quad j = 1, 2, \dots, r_n - 1,$$

forms a martingale difference. Obviously, the second condition of the theorem yields

$$nE(\Phi_n(X_1, X_2)\Phi_n(X_1, X_3)) \to \lambda.$$

Further, since $E\Phi_n^2(X_1, X_2) = 1$ and $nr_n^{-1} \to 0$ then

$$n^2 E(E_{r_n}\Phi_n(X_1, X_2))^2 = \frac{n^2}{r_n(r_n - 1)}\Big[2E\Phi_n^2(X_1, X_2) +$$

$$+ 4(r_n - 2)E(\Phi_n(X_1, X_2)\Phi_n(X_1, X_3))\Big] \to 0.$$

This relation and Chebyshev's inequality imply

$$nE_{r_n}(\Phi_n(X_1, X_2)) \xrightarrow{P} 0$$

as $n \to \infty$. Thus, due to (4.8.3), one should prove that

$$\sum_{j=n}^{r_n - 1} Z_{nj} \Rightarrow \tau\sqrt{2 + 4\lambda}$$

as $n \to \infty$. For this purpose, we exploit the martingale central limit theorem 4.2.1 from Borovskikh and Korolyuk (1988i).

According to this theorem it is sufficient to verify the conditions

$$\sum_{j=n}^{r_n - 1} EZ_{nj}^2 \to 2 + 4\lambda,$$

$$\sum_{j=n}^{r_n-1} E_{j+1}(Z_{nj})^2 \xrightarrow{P} 2+4\lambda,$$

$$\sum_{j=n}^{r_n-1} E_{j+1}\left(Z_{nj}^2 I(|Z_{nj}|\geq \varepsilon)\right) \xrightarrow{P} 0 \tag{4.8.4}$$

for any $\varepsilon > 0$. These simple arguments allow us to conclude that

$$n^2 \sum_{j=n}^{r_n-1} (j-1)^{-2} \left(E_{j+1}\left(\Phi_n(X_1,X_2)\right)\right)^2 \xrightarrow{L^1} 0 .$$

If $k \leq j$ then

$$E_{j+1}\left(\Phi_n(X_{j+1},X_k)\right) = E_{j+1}\left(\Phi_n(X_1,X_2)\right) .$$

Therefore,

$$\sum_{j=n}^{r_n-1} E_{j+1}(Z_{nj})^2 = 4n^2 \sum_{j=n}^{r_n-1} (j-1)^{-2} j^{-1} E_{j+1}(\Phi_n(X_1,X_2)) +$$

$$+4n^2 \sum_{j=n}^{r_n-1} (j-1)^{-1} j^{-1} E_{j+1}(\Phi_n(X_1,X_2)\Phi_n(X_1,X_3)) -$$

$$-4n^2 \sum_{j=n}^{r_n-1} (j-1)^{-2} (E_{j+1}(\Phi_n(X_1,X_2)))^2 .$$

From here we obtain the first relation in (4.8.4). Further, note that

$$|4n^2 \sum_{j=n}^{r_n-1} (j-1)^{-2} j^{-1} E_{j+1}\left(\Phi_n^2(X_1,X_2)\right) - 2| \leq$$

$$\leq \max_{n\leq j\leq r_n} |E_{j+1}\left(\Phi_n^2(X_1,X_2)-1\right)|\, 4n^2 \sum_{j=n}^{r_n-1} (j-1)^{-2} j^{-1} + O(n^{-1}) .$$

Set $\xi_{nj} = |E_{j+1}(\Phi_n^2(X_1,X_2)-1)|$. For every fixed $n \geq 1$, the sequence

$$(\xi_{n,r_n-j+1}, \mathfrak{B}_{r_n-j+1}), \quad j = 1, \ldots, r_n - n + 1 ,$$

is a submartingale. Therefore, by the Doob inequality for any $\varepsilon > 0$

$$P(\max_{n\leq j\leq r_n} \xi_{n_j} > \varepsilon) \leq \varepsilon^{-1} E|\xi_{nn}| =$$

$$= \varepsilon^{-1} E\left|\binom{n}{2}^{-1} \sum_{i=2}^{n}\sum_{j=1}^{i-1} \Phi_n^2(X_i,X_j) - 1\right| .$$

Since $E\Phi_n^2(X_1,X_2) = 1$, the first condition of the theorem is equivalent to

$$\binom{n}{2}^{-1} \sum_{1\leq i<j\leq n} \Phi_n^2(X_i,X_j) \xrightarrow{L^1} 1,$$

and hence,

$$4n^2 \sum_{j=n}^{r_n-1} (j-1)^{-2} j^{-1} E_{j+1}\left(\Phi_n^2(X_1,X_2)\right) \xrightarrow{P} 2.$$

By employing a similar argument, we get

$$|4n \sum_{j=n}^{r_n-1} (j-1)^{-1} j^{-1} \{nE_{j+1}(\Phi_n(X_1,X_2)\Phi_n(X_1,X_3)) - \lambda\}| \leq$$

$$\leq 8 \max_{n \leq j \leq r_n} |E_{j+1}(n\Phi_n(X_1,X_2)\Phi_n(X_1,X_3)) - \lambda| \to 0$$

provided that the second condition of the theorem holds. Hence, we obtain the second relation in (4.8.4).

To prove the third relation in (4.8.4) it suffices to establish that

$$n^2 \sum_{j=n}^{r_n-1} (j-1)^{-2} E_{j+1}\Big\{(j^{-1} \sum_{k=1}^{j} \Phi_n(X_{j+1},X_k))^2 I(|Z_{nj}| \geq \varepsilon)\Big\} \xrightarrow{P} 0$$

for any $\varepsilon > 0$ (see Borovskikh and Korolyuk (1994a), Weber (1983)).

Remark 4.8.1. If $E\Phi_n^4(X_1,X_2) < \infty$, then the first condition of Theorem 4.8.2 follows from

$$E\left(\binom{n}{2}^{-1} \sum_{1 \leq i < j \leq n} \Phi_n^2(X_i,X_j) - 1\right)^2 \to 0,$$

which, in turn, follows from the relations

$$n^{-2} E\Phi_n^4(X_1,X_2) \to 0,$$

$$n^{-1} E(\Phi_n^2(X_1,X_2)\Phi_n^2(X_1,X_3)) \to 0.$$

An example of a varying kernel. In the theorems presented above, the normal distribution appears everywhere as a limit law. If we change the conditions imposed on kernel Φ_n, then it is possible to obtain a limiting distribution governed by another law. First, we give an example which illustrates this possibility.

Theorem 4.8.3. *Let $X_1, \ldots, X_n$ be independent random variables with a uniform distribution on the interval $[-1,1]$. Define a kernel $\Phi_n(x,y) = \sin(a_n^2 xy)$, where $a_n > 0$, $n = 1,2,\ldots$.*

1. *If $a_n \to \infty$ as $n \to \infty$, then $nU_n \Rightarrow \tau$,*
2. *If $a_n \to 0$ as $n \to \infty$, then $n\sigma_{n2}^{-1} U_n \Rightarrow \tau^2 - 1$.*

Here τ is a standard normal random variable.

Proof. This kernel is degenerate because

$$E\Phi_n(X_1, y) = \frac{1}{2}\int_{-1}^{1} \Phi_n(x, y)\,dx =$$

$$= \frac{1}{2}\int_{-1}^{1} \sin(a_n^2 xy)\,dx = 0$$

for all $|y| \le 1$. To verify statement 1 we use Corollary 4.8.3. Write

$$\sigma_{n2}^2 = \frac{1}{4}\int_{-1}^{1}\int_{-1}^{1} \sin^2(a_n^2 xy)\,dx\,dy =$$

$$= \frac{1}{2} - \frac{1}{4a_n^2}\int_0^{2a_n^2} \frac{\sin u}{u}\,du\,.$$

Thus, if $a_n \to \infty$, then $\sigma_{n2}^2 \to 1/2$. Besides $E\Phi_n^4(X_1, X_2) \le 1$. Further, we have

$$EG_n^2(X_1, X_2) = \frac{1}{4}\int_{-1}^{1}\int_{-1}^{1}\left(\frac{1}{2}\int_{-1}^{1} \sin(a_n^2 xu)\sin(a_n^2 yu)du\right)dx\,dy =$$

$$= \frac{1}{16a_n^4}\int_{-1}^{1}\int_{-1}^{1}\left(\frac{\sin(a_n^2(x-y))}{x-y} - \frac{\sin(a_n^2(x+y))}{x+y}\right)^2 dx\,dy\,.$$

From this formula by simple estimations, we see that $EG_n^2(X_1, X_2) \to 0$ as $n \to \infty$, i.e. all the conditions of Corollary 4.8.3 are satisfied, and this proves statement 1.

We now establish statement 2. Under the condition $a_n \to 0$ we have

$$\sigma_{n2}^2 = \frac{1}{9}a_n^4 + O(a_n^6),$$

taking into account that $\sin x = x - x^3/6 + o(x^5)$ as $x \to 0$. Denote

$$Q_n = \frac{6}{n-1}\sum_{1\le i<j\le n} X_i X_j\,,$$

and

$$R_n = \sum_{1\le i<j\le n}\left[\frac{2}{(n-1)\sigma_{n2}}\sin(a_n^2 X_i X_j) - \frac{6}{n-1}X_i X_j\right].$$

Then

$$n\sigma_{n2}^{-1}U_n = Q_n + R_n\,,$$

where

$$ER_n^2 = \frac{2n}{n-1}E[\sigma_{n2}^{-1}\sin(a_n^2 X_1 X_2) - 3X_1 X_2]^2 = O(a_n^2)$$

as $a_n \to 0$. In addition

$$Q_n = \left(\frac{1}{\sqrt{n-1}}\sum_{j=1}^{n}(\sqrt{3}X_j)\right)^2 - \frac{1}{n-1}\sum_{j=1}^{n}(\sqrt{3}X_j)^2\,,$$

i.e. $Q_n \Rightarrow \tau^2 - 1$ by virtue of the central limit theorem and the strong law of large numbers. This proves statement 2.

Poisson convergence. Consider U-statistics (4.8.1) with varying kernels.

Theorem 4.8.4. *Let Φ_n be a sequence of zero-one functions. Assume m is fixed and $n \to \infty$. Suppose for some $\lambda > 0$*

$$\binom{n}{m} E\Phi_n(X_1, \dots, X_m) \to \lambda$$

and that

$$n^{2m-1} E\left(\Phi_n(X_1, \dots, X_m)\Phi_n(X_1, \dots, X_{m-1}, X_{m+1})\right) \to 0 .$$

Then

$$\binom{n}{m} U_n \Rightarrow \nu$$

where ν is a Poisson distributed random variable with parameter λ.

Proof. Note, that

$$\binom{n}{m} U_n = \sum_{i_m=m}^{n} S_{i_m},$$

where

$$S_{i_m} = \sum_{1 \le i_1 < \dots < i_{m-1} \le i_m - 1} \Phi_n(X_{i_1}, \dots, X_{i_{m-1}}, X_{i_m}) .$$

Let $\delta_{i_m} = I(S_{i_m} \ge 1)$. By the estimation of Brown and Silverman (1978),

$$P\left(\sum_{i_m=m}^{n} S_{i_m} \ne \sum_{i_m=m}^{n} \delta_{i_m}\right) \le$$

$$\le c n^{2m-1} E\left(\Phi_n(X_1, \dots, X_m)\Phi_n(X_1, \dots, X_{m-1}, X_{m+1})\right) \to 0 ,$$

which converges to zero by the conditions. It remains to show that

$$\sum_{i_m=m}^{n} \delta_{i_m} \Rightarrow \nu$$

as $n \to \infty$. Since δ_{i_m} are zero-one for this relation it suffices to apply Theorem 2 of Brown and Eagleson (1971), i.e. to note that

$$\max_{m \le i \le n} E(\delta_i | \mathfrak{F}_{i-1}) \xrightarrow{P} 0$$

and

$$\sum_{i=m}^{n} E(\delta_i | \mathfrak{F}_{i-1}) \xrightarrow{P} \lambda$$

where $\mathfrak{F}_i = \sigma\{\omega : X_1, \dots, X_i\}$. The truth of these conditions are verified by direct calculations.

Convergence to infinitely divisible distributions. Consider U-statistics (4.8.1) for m=2 and let $f_n(x,y) = \binom{n}{2}^{-1} \Phi_n(x,y)$.

Theorem 4.8.5. *Assume that* $E|f_n(X_1,X_2)| < \infty$ *and*

$$n^3\, E_{X_2} \left(E_{X_1}|f_n(X_1,X_2)|\right)^2 \to 0\,,$$

as $n \to \infty$.

Let $F_n(t) = P(f_n(X_1,X_2) < t)$ *be a distribution function of the random variable* $f_n(X_1,X_2)$. *If there exist real values* α *and* a_n *and a finite positive measure* dF *on* R *such that the complete convergence*

$$\frac{1}{2}n^2 \frac{t^2}{1+t^2}\, dF_n(t) \xrightarrow{c} dF(t)$$

takes place, and

$$\frac{1}{2}n^2 E \frac{f_n(X_1,X_2)}{1+f_n^2(X_1,X_2)} - a_n \to \alpha,$$

then

$$U_n - a_n \Rightarrow \eta\,,$$

as $n \to \infty$, *where* η *has an infinitely divisible characteristic function*

$$\varphi_\eta(t) = \exp\left(it\alpha + \int_{-\infty}^{\infty} \left(e^{itx} - 1 - \frac{itx}{1+x^2}\right) \frac{1+x^2}{x^2}\, dF(x)\right).$$

Proof. Let ν be a Poisson random variable with parameter n independent of $X_1, X_2, \dots$. Set

$$U_\nu = \sum_{1 \le i < j \le \nu} f_n(X_i, X_j).$$

If $\nu > n$, then

$$U_\nu - U_n = \sum_{j=n+1}^{\nu} \sum_{i=1}^{j-1} f_n(X_i, X_j)\,.$$

This yields $E(|U_\nu - U_n||\nu) \le \nu(\nu - n) E|f_n(X_1,X_2)|$.

Similarly, if $\nu < n$, then

$$E(|U_\nu - U_n||\nu) \le n(n-\nu) E|f_n(X_1,X_2)|\,.$$

Consequently,

$$E|U_\nu - U_n| \le E\left((n + |\nu - n|)|\nu - n|\right) E|f_n| =$$

$$= \left(nE|\nu - n| + E(\nu - n)^2\right) E|f_n| \le (n\sqrt{n} + n)E|f_n|.$$

According to the conditions and Hölder's inequality we can write

$$E|f_n| = E_{X_2}E_{X_1}|f_n(X_1, X_2)| \le \left(E_{X_2}(E_{X_1}|f_n(X_1, X_2)|)^2\right)^{1/2} = o(n^{-3/2}).$$

Therefore, $E|U_\nu - U_n| \to 0$ as $n \to \infty$. In Jammalamadaka and Janson (1986), it was established that $U_\nu - a_n \to \eta$ as $n \to \infty$.

Example 4.8.1. Suppose that $X_1, \ldots, X_n$ have absolutely continuous distributions in R^k with density $p(x)$. Set

$$f_n(x, y) = I(\|x - y\| < r_n), \ r_n \to 0.$$

Since $X_1 - X_2$ has the density $p * \bar{p}$, where $\bar{p}(x) = p(-x)$, then

$$\theta_n = Ef_n(X_1, X_2) = \int_{\|x\| < r_n} p * \bar{p}(x)\, dx.$$

Suppose that $p \in L^2$. Under this condition

$$r_n^{-k}\theta_n \Rightarrow c_k\, p * \bar{p}(0) = c_k \int p^2(x)\, dx,$$

where $c_k = \pi^{k/2}/\Gamma(1 + k/2)$ is the volume of a sphere with radius 1 in R^k. If $r_n \to 0$, $n^2 r_n^k \to \alpha, 0 < \alpha < \infty$, then

$$U_n \Rightarrow \nu,$$

where ν is the Poisson random variable with parameter $\lambda = \alpha\, c_k \int p^2(x)\, dx/2$.

4.9 Weighted *U*-statistics

Consider weighted *U*-statistics, which are of the following form

$$U_n = \sum_{1 \le i_1 \ne \ldots \ne i_m \le n} c_n(i_1, \ldots, i_m)\Phi(X_{i_1}, \ldots, X_{i_m}) \tag{4.9.1}$$

where kernel $\Phi : X^m \to R$ and functions $c_n : R^m \to R$ are symmetric, i.e. they are invariant under all permutations of their arguments.

We investigate the limit behavior of weighted *UR*-statistics (4.9.1) under some conditions on c_n.

Nonrandom weighting functions. Suppose that the weighting functions have the form

$$c_n(i_1, \ldots, i_m) = (\sigma^2 n)^{-m/2}\, c(i_1) \ldots c(i_m)$$

with a sequence $c(i), i = 1, 2, \ldots$, such that $c(i)$ is bounded, and the limit

$$\lim_{n\to\infty} n^{-1}\left(c^2(1) + \cdots + c^2(n)\right) = \sigma^2 > 0 \tag{4.9.2}$$

exists.

Theorem 4.9.1. *Let kernel Φ possess the property of complete degeneracy:*

$$E\Phi(x_1, \ldots, x_{m-1}, X_m) = 0 \quad a.s.$$

and

$$E\Phi^2(X_1, \ldots, X_m) < \infty.$$

Then the convergence in distribution

$$U_n \Rightarrow I_m^W(\Phi)$$

holds.

Proof. By definition,

$$U_n = \sum_{1\le i_1\ne\ldots\ne i_m\le n} \Phi(X_{i_1}, \ldots, X_{i_m}) \prod_{s=1}^{m} \left(\frac{c(i_s)}{\sigma\sqrt{n}}\right).$$

Taking account of the property of the complete degeneracy of $\Phi(x_1, \ldots, x_m)$, we can write the representation

$$U_n = \int_X \cdots \int_X \Phi(x_1, \ldots, x_m) \,\mathrm{Per}\; A_n^m(dx_1, \ldots, dx_m; c)$$

with permanent random measure

$$\mathrm{Per}\; A_n^m(B_1, \ldots, B_m; c) = \sum_{1\le i_1,\ne\ldots\ne i_m\le n} \prod_{s=1}^{m} [\left(\delta_{x_{i_s}}(B_s) - P(B_s)\right) c(i_s)/(\sigma\sqrt{n})]$$

where $B_1, \ldots, B_m \in \mathfrak{X}$.

This measure has the property of weak convergence

$$\mathrm{Per}\; A_n^m(B_1, \ldots, B_m; c) \Rightarrow \prod_{s=1}^{m} W(B_s)$$

where $W(\cdot)$ is a Gaussian random measure with zero mean value and covariance $E\,W(A)W(B) = P(A\cap B), \quad A, B \in \mathfrak{X}$.

These relations make it possible to apply Lemma 4.4.1 from Borovskikh and Korolyuk (1994b) (detailed calculations are also presented in Borovskikh and Korolyuk (1994b), Chapter 5) and hence to prove this theorem.

Now we consider the case when we do not assume that kernel Φ possesses the property of complete degeneracy. Let the limit

$$\lim_{n\to\infty} n^{-1/2}(c(1) + \cdots + c(n)) = \delta \tag{4.9.3}$$

exist. The limit behavior of U_n is different in the cases when δ is a finite nonzero limit, than when it is equal to zero or to infinity. We describe the case when δ is a finite nonzero limit. In this situation we express U_n by means of the Hoeffding decomposition (1.1.9). Then the contributions of the different terms in the Hoeffding representation have the same order and the weak limit of U_n can be represented as a sum of multiple stochastic integrals.

Theorem 4.9.2. *Let $c(i)$, $i = 1, 2, \ldots$ be a bounded sequence satisfying (4.9.2)-(4.9.3) and a square integrable kernel Φ.*

Then the sequence of weighted U-statistics U_n converges in distribution to the sum of stochastic integrals

$$d_m\theta + \sum_{c=1}^{m} d_{m-c}\binom{m}{c} I_c^W(g_c)$$

as $n \to \infty$, where $g_c(x_1, \ldots, x_c)$ is a canonical function from (1.1.6), and the sequence d_c is defined by the recurrence relation: $d_0 = 1$, $d_1 = \delta/\sigma$, and

$$d_c = \sigma^{-c}\delta^c - \sum_{s=1}^{[c/2]} \frac{c!}{2^s s!(c-2s!)} d_{c-2s}.$$

Proof. By the Hoeffding decomposition formula

$$U_n = \sum_{s=0}^{m} \binom{m}{s} \sum_{1 \le i_1 \ne \ldots \ne i_s \le n} \sigma_n(i_1, \ldots, i_s) \prod_{p=1}^{s} \left(\frac{c(i_p)}{\sigma\sqrt{n}}\right) g_s(X_{i_1}, \ldots, X_{i_s})$$

where

$$\sigma_n(i_1, \ldots, i_s) = \sum_{1 \le j_1 \ne \ldots \ne j_{m-s} \le n, j_p \in \{1, \ldots, n\} \setminus \{i_1, \ldots, i_s\}, 1 \le p \le m-s} \prod_{q=1}^{m-s} \left(\frac{c(i_q)}{\sigma\sqrt{n}}\right).$$

By Lemma 4 in Major (1994)

$$\sigma_n(i_1, \ldots, i_s) = d_{m-s} + \rho_{ns}(i_1, \ldots, i_s)$$

such that

$$\lim_{n \to \infty} \sup_{1 \le s \le m} \sup_{1 \le i_1, \ldots, i_s \le n} |\rho_{ns}(i_1, \ldots, i_s)| = 0.$$

Therefore,

$$U_n = \sum_{s=0}^{m} \binom{m}{s} d_{m-s} \sum_{1 \le i_1 \ne \ldots \ne i_s \le n} \prod_{p=1}^{s} \left(\frac{c(i_p)}{\sigma\sqrt{n}}\right) g_s(X_{i_1}, \ldots, X_{i_s}) + \rho_n,$$

where

$$\rho_n = \sum_{s=1}^{m} \binom{m}{s} \sum_{1 \le i_1 \ne \ldots \ne i_s \le n} \rho_{ns}(i_1, \ldots, i_s) \prod_{p=1}^{s} \left(\frac{c(i_p)}{\sigma\sqrt{n}}\right) g_s(X_{i_1}, \ldots, X_{i_s}).$$

Here by Theorem 4.9.1

$$\sum_{1\le i_1\ne\ldots\ne i_s\le n}\ \prod_{p=1}^{s}\left(\frac{c(i_p)}{\sigma\sqrt{n}}\right) g_s(X_{i_1},\ldots,X_{i_s}) \Rightarrow I_s^W(g_s)$$

as $n\to\infty$. The random variables $g_s(X_{j_1},\ldots,X_{j_s})$ and $g_s(X_{j_1^1},\ldots,X_{j_s^1})$ are uncorrelated if the sets $\{j_1,\ldots,j_s\}$ and $\{j_1^1,\ldots,j_s^1\}$ are different, since the functions g_s satisfy the property of complete degeneracy. Hence

$$E\,\rho_n^2\to 0 \quad \text{as} \quad n\to\infty\,.$$

This last estimate together with preceding relations imply Theorem 4.9.2.

Random weighting functions. Suppose that the weighting functions in (4.9.1) have the form

$$c_n(i_1,\ldots,i_m)=n^{-m/\alpha}\,\eta_{i_1}(\alpha)\cdots\eta_{i_m}(\alpha),$$

where $\eta_1(\alpha),\ldots,\eta_n(\alpha)$ are independent random variables belonging to the domain of normal attraction of some α-stable law, $1<\alpha<2$; X_j and $\eta_j(\alpha)$ are independent. Therefore,

$$U_n=n^{-m/\alpha}\sum_{1\le i_1\ne\ldots\ne i_m\le n}\eta_{i_1}(\alpha)\cdots\eta_{i_m}(\alpha)\,\Phi(X_{i_1},\ldots X_{i_m})\,.$$

Introduce into consideration the random matrix

$$A_n^m(B_1,\ldots,B_m;\eta)=\left[a_{ij}=n^{-1/\alpha}\,\eta_j(\alpha)\delta_{X_j}(B_i);\quad 1\le i\le m,\ 1\le j\le n\right],$$

where $B_i\in\mathfrak{X}$, $i=1,\ldots,m$ with respect to the permanent measure

$$\text{Per } A_n^m(B_1,\ldots,B_m;\eta)$$

represents this weighted U-statistic U_n in the form of permanent integral functional

$$U_n=\int_X\cdots\int_X\Phi(x_1,\ldots x_m)\text{ Per } A_n^m(dx_1,\ldots,dx_m;\eta). \tag{4.9.4}$$

Let $L_\alpha(\cdot)$ be a stable random measure with index α, $1<\alpha<2$, over the measurable space $(X,\mathfrak{X})$, i.e. a random variable $L_\alpha(B), B\in\mathfrak{X}$ has a stable distribution with given index α and a characteristic function

$$E\exp(itL_\alpha(B))=\exp(-P(B)|t|^\alpha)\,.$$

The weak limit of U_n as $n\to\infty$ proves to coincide (see Dehling *et al.* (1990)) with a multiple α-stable stochastic integral of the form

$$I_m^{L_\alpha}(\Phi)=\int_X\cdots\int_X\Phi(x_1,\ldots,x_m)L_\alpha(dx_1)\ldots L_\alpha(dx_m)\,.$$

Theorem 4.9.3. *If* $E|\Phi|^p < \infty$ *with some* $p > \alpha$ *then the weak convergence*

$$U_n \Rightarrow I_m^{L_\alpha}(\Phi) \tag{4.9.5}$$

takes place as $n \to \infty$.

Proof. The result of (4.9.5) could be deduced immediately from the permanent Lemma 4.4.1 and from Borovskikh and Korolyuk (1994b). Taking into account (4.9.4) and the Corollary 1.1 we get

$$\text{Per } A_n^m(B_1, \dots, B_m; \eta) =$$

$$= n^{-1/\alpha} \sum_{q_1=1}^{n} \eta_{q_1}(\alpha)\delta_{X_{q_1}}(B_1) \cdots n^{-1/\alpha} \sum_{q_m \in Q_1^{m-1}(q_1,\cdots,q_{m-1})} \eta_{q_m}(\alpha)\delta_{X_{q_m}}(B_m)$$

Here for each $s = 1, \dots, m$

$$n^{-1/\alpha} \sum_{q_s=1}^{n-s+1} \eta_{q_s}(\alpha)\delta_{X_{q_s}}(B_s) \Rightarrow L_\alpha(B_s)$$

because the characteristic function

$$E\exp\{itn^{-1/\alpha} \sum_{q_s=1}^{n-s+1} \eta_{q_s}(\alpha)\delta_{X_{q_s}}(B_s)\} =$$

$$= \Big(1 + P(B_s)(E\exp\{itn^{-1/\alpha}\eta_1(\alpha)\} - 1)\Big)^{n-s+1} \to \exp(-P(B_s)|t|^\alpha),$$

as $n \to \infty$. By virtue of these relations

$$\text{Per } A_n^m(B_1, \dots, B_m; \eta) \Rightarrow \prod_{s=1}^{m} L_\alpha(B_s).$$

Now convergence (4.9.5) follows from (4.9.4) and Lemma 4.4.1 in Borovskikh and Korolyuk (1994b), where, instead of Gaussian measure W, we substitute the stable measure L_α. We omit the technical estimates.

Chapter 5

Functional limit theorems

In this chapter we consider the weak convergence of probability measures, connected with U-statistics, in functional spaces. The random U-processes on the stochastic sequence $\{U_n(\Phi)\}$ can be defined in different ways. It can treat a partial sum process similar to $(U_{[nt]}(\Phi),\ 0 \le t \le 1)$ as a sequence of processes indexed by t for fixed Φ. Alternatively it can treat $(U_n(\Phi),\ \Phi \in \mathcal{F})$ as a stochastic process indexed by a family $\mathcal{F}$ of measurable symmetric kernels. The proof of the weak convergence of processes $(U_{[nt]}(\Phi),\ 0 \le t \le 1)$ is carried out according to the general scheme of the Prokhorov theorem (see Araujo and Giné (1980) and Billingsley (1968)): first it establishes the weak convergence of finite-dimensional distributions, and then it verifies the property of tightness of the family of probability measures. The study of U-processes parallels that of *empirical processes* (see Arcones and Giné (1993b), Nolan and Pollard (1987, 1988)). For example, let $\mathcal{F}$ be a class of P-canonical (i.e. completely degenerate) functions Φ. As in empirical process theory, we will say that the central limit theorem holds for U-processes $\{n^{m/2}U_n(\Phi) : \Phi \in \mathcal{F}\}$ (or for $\mathcal{F}$) if

$$n^{m/2}U_n(\Phi) \Rightarrow K_m(\Phi) \text{ in } l^\infty(\mathcal{F})$$

and if the limit chaos process $\{K_m(\Phi) : \Phi \in \mathcal{F}\}$ has a version with bounded uniformly continuous paths in $(\mathcal{F}, e_P)$, where $e_P(f,g) = \|f - g\|_{L^2(X^m,P^m)}$. Here the weak convergence $\Rightarrow$ is convergence in distribution in $l^\infty(\mathcal{F})$, the space of all bounded functions from $\mathcal{F}$ to R, in the sense of Hoffman-Jørgensen (1984). With the help of the following criterion this weak convergence reduces to an *asymptotic equicontinuity* condition as follows. If ξ_n, ξ are random elements taking values in $l^\infty(\mathcal{F})$, then the law of ξ is Radon and $\xi_n \Rightarrow \xi$ in $l^\infty(\mathcal{F})$ if and only if the finite-dimensional distributions of ξ_n converge in law to those of ξ, and there exists a pseudometric ρ on $\mathcal{F}$ such that $(\mathcal{F}, \rho)$ is totally bounded and asymptotically equicontinuous with respect to ρ on $\mathcal{F}$

$$\lim_{\delta\to 0}\limsup_{n\to\infty} \Pr{}^*_P\left\{\sup_{\rho(f,g)\le\delta} |\xi_n(f) - \xi_n(g)| > \varepsilon\right\} = 0$$

for every $\varepsilon > 0$. If this takes place, then the process ξ has a version with bounded uniformly ρ-continuous paths and conversely, for any ρ for which there is a version of ξ in $C_u(\mathcal{F}, \rho)$, the conditions $(\mathcal{F}, \rho)$ totally bounded and asymptotic equicontinuous are necessary and sufficient for the central limit theorem of $\mathcal{F}$ (see, e.g. Andersen *et al.* (1988), Giné and Zinn (1986), Sheehy and Wellner (1992)).

5.1 Non-degenerate kernels

Consider one-sample *UR*-statistics. Define a random process $\xi_n = (\xi_n(t),\ 0 \le t \le 1)$, by setting

$$\xi_n(t) = \begin{cases} 0 & \text{if } 0 \le t \le (m-1)/n, \\ \xi_n(k/n) & \text{if } k/n < t \le (k+1)/n, \qquad k = m-1, \dots, n-1 \end{cases}$$

and

$$\xi_n(k/n) = \frac{k}{m\sigma_1\sqrt{n}}(U_k - \theta), \qquad k = m, \dots, n$$

where $\sigma_1^2 = Eg_1^2(X_1) > 0$.

The process ξ_n belongs to a space $D[0,1]$ of all real-valued functions without jumps of the second kind (for example, continuous from the left and having the right limits) (see Billingsley (1968)).

Theorem 5.1.1. *Assume that the kernel Φ of rank $r = 1$ satisfies the conditions*

$$E|g_c(X_1, \dots, X_c)|^{\gamma_c} < \infty$$

for each $c = 1, 2, \dots, m$, where $\gamma_c = 2c/(2c-1)$.

Then the weak convergence

$$\xi_n \Rightarrow w$$

as $n \to \infty$ takes place in a space $D[0,1]$, where w is a standard Wiener process on $[0,1]$.

Proof. We apply the canonical Hoeffding representation (1.1.12). Define a random process $\xi_{n1} = (\xi_{n1}(t),\ 0 \le t \le 1)$, by setting

$$\xi_{n1}(t) = \begin{cases} 0 & \text{if } 0 \le t \le n^{-1}, \\ \xi_{n1}(k/n) & \text{if } k/n < t \le (k+1)/n, \qquad k = 1, \dots, n-1 \end{cases}$$

where

$$\xi_{n1}(k/n) = \frac{1}{\sigma_1\sqrt{n}}(g_1(X_1) + \dots + g_1(X_k)), \qquad k = 1, \dots, n.$$

By virtue of the invariance principle,

$$\xi_{n1} \Rightarrow w$$

holds provided that $0 < \sigma_1 < \infty$ (see, e.g. Billingsley (1968)).

According to Theorem 4.1 in Billingsley (1968) to prove the weak convergence $\xi_n \Rightarrow w$ it suffices to establish that

$$\rho(\xi_n, \xi_{n1}) = \sup_{0 \le t \le 1} |\xi_n(t) - \xi_{n1}(t)| \xrightarrow{\mathrm{P}} 0.$$

It follows from the definitions of processes ξ_n and ξ_{n1}, that the inequality

$$\rho(\xi_n, \xi_{n1}) \le \max_{1 \le k \le m-1} |\xi_{n1}(k/n)| + \frac{1}{m\sigma_1\sqrt{n}} \max_{m \le k \le n} \left(k \left| \sum_{c=2}^{m} \binom{m}{c} U_{kc} \right| \right)$$

is valid with probability 1. Therefore,

$$\rho(\xi_n, \xi_{n1}) \le \frac{1}{m\sigma_1\sqrt{n}} \max_{1 \le k \le m-1} \left| \sum_{i=1}^{k} g_1(X_i) \right| + \frac{1}{m\sigma_1\sqrt{n}} \sum_{c=2}^{m} \binom{m}{c} \max_{m \le k \le n} k|U_{kc}|.$$

It is obvious that

$$\frac{1}{m\sigma_1\sqrt{n}} \max_{1 \le k \le m-1} \left| \sum_{i=1}^{k} g_1(X_i) \right| \xrightarrow{\mathrm{P}} 0$$

as $n \to \infty$ and $Eg_1^2(X_1) < \infty$.

Now we shall prove that

$$\frac{1}{\sqrt{n}} \max_{m \le k \le n} k|U_{kc}| \xrightarrow{\mathrm{P}} 0 \tag{5.1.1}$$

for each $c = 2, \dots, m$. Let us represent the U-statistic U_{kc} in the form of the sum of two U-statistics, namely $U_{kc} = U_{kc}(\Phi_{1c}) + U_{kc}(\Phi_{2c})$, with the kernels

$$\Phi_{sc}(x_1, \dots, x_c) = \sum_{d=1}^{c} (-1)^{c-d} \sum_{1 \le i_1 < \dots < i_d \le c} (g_{cd}^{(s)}(x_{i_1}, \dots, x_{i_d}) -$$
$$- Eg_{cd}^{(s)}(X_{i_1}, \dots, X_{i_d})), \qquad s = 1, 2,$$

where

$$g_{cd}^{(1)}(x_1, \dots, x_d) =$$
$$= E\left(g_c(X_1, \dots, X_c) I(|g_c(X_1, \dots, X_c)| \le \sqrt{n}) \middle| X_1 = x_1, \dots, X_d = x_d \right),$$
$$g_{cd}^{(2)}(x_1, \dots, x_d) =$$
$$= E\left(g_c(X_1, \dots, X_c) I(|g_c(X_1, \dots, X_c)| > \sqrt{n}) \middle| X_1 = x_1, \dots, X_d = x_d \right).$$

The functions Φ_{1c} and Φ_{2c} possess the property of complete degeneracy (1.1.7), and the sequences $(U_k(\Phi_{sc}), \mathfrak{B}_k)$, $k = m, \dots, n$; $s = 1, 2$, are reversed martingales, where $\mathfrak{B}_k = \sigma\{\omega : (X_1, \dots, X_k), X_{k+1}, X_{k+2}, \dots\}$ is the σ-algebra generated by the unordered collection $(X_1, \dots, X_k)$ and by $X_{k+1}, X_{k+2}, \dots$. Therefore,

$U_{k+1}(\Phi_{sc}) = E(U_k(\Phi_{sc})|\mathfrak{B}_{k+1})$ and, for any $p \geq 1$, the sequences $(|U_k(\Phi_{sc})|^p, \mathfrak{B}_k)$ are reversed submartingales, i.e. $|U_{k+1}(\Phi_{sc})|^p \leq E(|U_k(\Phi_{sc})|^p|\mathfrak{B}_{k+1})$ with probability 1. Consequently, for the proof (5.1.1) one can employ Theorem 2.3.2. At first we have for any $\varepsilon > 0$

$$P(\max_{m\leq k\leq n} k|U_{kc}| > \varepsilon\sqrt{n}) \leq$$

$$\leq P(2\max_{m\leq k\leq n} k|U_k(\Phi_{1c})| > \varepsilon\sqrt{n}) + P(2\max_{m\leq k\leq n} k|U_k(\Phi_{2c})| > \varepsilon\sqrt{n}).$$

For $U_k(\Phi_{1c})$, according to the inequality (2.3.2) one can write

$$P(2\max_{m\leq k\leq n} k|U_k(\Phi_{1c})| > \varepsilon\sqrt{n}) \leq$$

$$\leq 4\varepsilon^{-2}n^{-1}\left(m^2EU_m^2(\Phi_{1c}) + \sum_{k=m+1}^{n}(k^2-(k-1)^2)EU_k^2(\Phi_{1c})\right).$$

Here,

$$\begin{aligned} EU_k^2(\Phi_{1c}) &= \binom{k}{c}^{-1} E\Phi_{1c}^2(X_1,\dots,X_c) \\ &\leq \binom{k}{c}^{-1} 2^{2c}E(g_c^2(X_1,\dots,X_c)I(|g_c(X_1,\dots,X_c)| \leq \sqrt{n})) \\ &\leq 2^{2c}c!\,E|g_c|^{\gamma_c}n^{(2-\gamma_c)/2}k^{-[c]}. \end{aligned}$$

Hence, for all $c = 2,\dots,m$ and for any fixed $\varepsilon > 0$

$$P(2\max_{m\leq k\leq n} k|U_k(\Phi_{1c})| > \varepsilon\sqrt{n}) \to 0$$

as $n \to \infty$.

For $U_k(\Phi_{2c})$, according to the inequality (2.3.2), we obtain

$$P(2\max_{m\leq k\leq n} k|U_k(\Phi_{2c})| > \varepsilon\sqrt{n}) \leq$$

$$\leq 2^{\gamma_c}\varepsilon^{-\gamma_c}n^{-\gamma_c/2}\left(m^{\gamma_c}E|U_m(\Phi_{2c})|^{\gamma_c} + \sum_{k=m+1}^{n}(k^{\gamma_c}-(k-1)^{\gamma_c})E|U_k(\Phi_{2c})|^{\gamma_c}\right).$$

By using Corollary 2.2.1, we get

$$\begin{aligned} E|U_k(\Phi_{2c})|^{\gamma_c} &\leq 2^{c+1}\binom{k}{c}^{-\gamma_c+1} E|\Phi_{2c}|^{\gamma_c} \\ &\leq 2^{3c+1}\binom{k}{c}^{-\gamma_c+1} E(|g_c|^{\gamma_c}I(|g_c| > \sqrt{n})). \end{aligned}$$

With the help of this estimate we can see that

$$n^{-\gamma_c/2}\sum_{k=m+1}^{n}(k^{\gamma_c}-(k-1)^{\gamma_c})E|U_k(\Phi_{2c})|^{\gamma_c} \leq a(m,c)E(|g_c|^{\gamma_c}I(|g_c| > \sqrt{n})),$$

where $a(m,c)$ is some constant, dependent only on m and c. Therefore, for all $c = 2, \dots, m$ and for any fixed $\varepsilon > 0$

$$P(2 \max_{m \le k \le n} k|U_k(\Phi_{2c})| > \varepsilon\sqrt{n}) \to 0$$

as $n \to \infty$. This has as a consequence

$$P(\max_{m \le k \le n} k|U_{kc}| > \varepsilon\sqrt{n}) \to 0$$

i.e. the relation (5.1.1). In the result we have the complete proof of Theorem 5.1.1.

Reversed invariance principle. Consider one-sample *UR*-statistics. Let us define on the tail sequence $\{U_k,\ k \ge n\}$ the random process $\zeta_n = (\zeta_n(t),\ 0 \le t \le 1)$, where

$$\zeta_n(t) = \frac{\sqrt{n}}{m\sigma_1}(U_{n(t)} - \theta),$$
$$n(t) = \min\{\, k \ge 1 : nk^{-1} \le t\,\}, \qquad 0 \le t \le 1.$$

The process $\zeta_n(t)$ belongs to the space $D[0,1]$, and $n(t) = -[-nt^{-1}]$, where $[z]$ denotes the integral part of the variable z. Similarly, we define another random process $\zeta_{n1} = (\zeta_{n1}(t),\ 0 \le t \le 1)$ by setting

$$\zeta_{n1}(t) = \frac{\sqrt{n}}{m\sigma_1 n(t)}(g_1(X_1) + \dots + g_1(X_{n(t)})).$$

Theorem 5.1.2. *Suppose that the kernel Φ of rank $r = 1$ satisfies the conditions*

$$E|g_c(X_1, \dots, X_c)|^{\gamma_c} < \infty$$

for each $c = 1, 2, \dots, m$, where $\gamma_c = 2c/(2c-1)$.

Then the weak convergence

$$\zeta_n \Rightarrow w$$

as $n \to \infty$ takes place in a space $D[0,1]$, where w is a standard Wiener process on $[0,1]$.

Proof. By the invariance principle

$$\zeta_{n1} \Rightarrow w$$

as $n \to \infty$ and $0 < \sigma_1 < \infty$ (see, e.g. Billingsley (1968)). Further we verify that under the conditions of this theorem

$$\rho(\zeta_n, \zeta_{n1}) = \sup_{0 \le t \le 1} |\zeta_n(t) - \zeta_{n1}(t)| \xrightarrow{P} 0.$$

One can see that

$$\rho(\zeta_n, \zeta_{n1}) = \frac{\sqrt{n}}{m\sigma_1} \sup_{k\geq n} \left| \sum_{c=2}^{m} \binom{m}{c} U_{kc} \right|.$$

For any $\varepsilon > 0$ and all $c = 2, \dots, m$ we have

$$P(\sqrt{n} \sup_{k\geq n} |U_{kc}| > \varepsilon)$$
$$\leq P(2\sqrt{n} \sup_{k\geq n} |U_k(\Phi_{1c})| > \varepsilon) + P(2\sqrt{n} \sup_{k\geq n} |U_k(\Phi_{2c})| > \varepsilon).$$

By corollary 2.3.2

$$\begin{aligned} P(2\sqrt{n} \sup_{k\geq n} |U_k(\Phi_{1c})| > \varepsilon) &\leq 4\varepsilon^{-2} n E U_n^2(\Phi_{1c}) \\ &= 4\varepsilon^{-2} n \binom{n}{c}^{-1} E\Phi_{1c}^2 \\ &\leq 2^{2c+2}\varepsilon^{-2} n \binom{n}{c}^{-1} E(g_c^2 I(|g_c| \leq \sqrt{n})) \\ &\leq 2^{2c+2}\varepsilon^{-2} n^2 \binom{n}{c}^{-1} n^{-\gamma_c/2} E|g_c|^{\gamma_c} \to 0 \text{ as } n \to \infty; \end{aligned}$$

$$\begin{aligned} P(2\sqrt{n} \sup_{k\geq n} |U_k(\Phi_{2c})| > \varepsilon) &\leq (2\varepsilon^{-1})^{\gamma_c} n^{\gamma_c/2} 2^{c+1} \binom{n}{c}^{-\gamma_c+1} E|\Phi_{2c}|^{\gamma_c} \\ &\leq (2\varepsilon^{-1})^{\gamma_c} 2^{3c+1} n^{\gamma_c/2} \binom{n}{c}^{-\gamma_c+1} E(|g_c|^{\gamma_c} I(|g_c)| > \sqrt{n})) \\ &= 2^{3c+1+\gamma_c} \varepsilon^{-\gamma_c} \left(\frac{1}{c!} \prod_{j=1}^{c} (1 - n^{-1}(j-1)) \right)^{-\gamma_c+1} \\ &\quad \times E(|g_c|^{\gamma_c} I(|g_c| > \sqrt{n})) \to 0 \text{ as } n \to \infty. \end{aligned}$$

Combining these estimates we find that $\rho(\zeta_n, \zeta_{n1}) \to 0$ in probability as $n \to \infty$. This proves Theorem 5.1.2.

5.2 Degenerate kernels

Consider *UR*-statistics of degree $m = 2$, i.e.

$$U_n = \binom{n}{2}^{-1} \sum_{1\leq i<j\leq n} \Phi(X_i, X_j)$$

with a symmetric degenerate kernel Φ such that $E\Phi^2(X_1, X_2) < \infty$. The random variables $X_1, \dots, X_n$ are independent and identically distributed and take values in a measurable space $(X, \mathfrak{X})$.

For $t \in [0,1]$, define a process

$$U_n(t) = \binom{n}{2}^{-1} \sum_{1 \le i < j \le [nt]} \Phi(X_i, X_j)$$

where the empty sum is assumed to be zero. The trajectories of $U_n(t)$ belong to the space $D[0,1]$. Let $\{\lambda_j\}$, $j \ge 1$, be eigenvalues of an S-operator, generated by the kernel Φ, which acts from L^2 to L^2 as follows

$$S : f \mapsto E(\Phi(X_1, x) f(X_1)).$$

This operator is a Hilbert–Schmidt operator, i.e. the l_2-norm of its eigenvalues is finite:

$$\sum_{j=1}^{\infty} \lambda_j^2 = E\Phi^2(X_1, X_2) < \infty.$$

Let $w_1(t), w_2(t), \ldots$ be a sequence of independent standard Wiener processes, $Ew_j(t) = 0$, $Ew_j^2(t) = t$, $j \ge 1$.

The process

$$\eta_\infty(t) = \sum_{j=1}^{\infty} \lambda_j (w_j^2(t) - t)$$

appears as the limiting process for $U_n(t)$ with a degenerate kernel, as $n \to \infty$. The random function $\eta_\infty(t)$ is the weak limit of the sequence of processes

$$\eta_N(t) = \sum_{j=1}^{N} \lambda_j (w_j^2(t) - t), \qquad N = 1, 2, \ldots$$

in the space $D[0,1]$, i.e.

$$\eta_N(t) \Rightarrow \eta_\infty(t)$$

as $N \to \infty$ in the topology of the space $D[0,1]$. In fact, the convergence of finite-dimensional distributions $\eta_N(t)$ to the finite-dimensional distributions $\eta_\infty(t)$ is a consequence of the fact that $\Phi \in L^2$. Theorem 15.6 in Billingsley (1968) guarantees the tightness of the sequence $\eta_N(t)$, $N \ge 1$, and hence, the validity of convergence $\eta_N \to \eta_\infty$.

Theorem 5.2.1. *Assume that a degenerate kernel Φ satisfies the conditions $E\Phi^2 < \infty$. Then the weak convergence*

$$nU_n(t) \Rightarrow \eta_\infty(t)$$

takes place as $n \to \infty$ in the topology of the space $D[0,1]$.

Proof. Denote U_{nN} the U-statistics with the kernel

$$\Phi_N(x, y) = \sum_{j=1}^{N} \lambda_j e_j(x) e_j(y),$$

where $\{e_j\}$ are the eigenfunctions of the S-operator which corresponds to the eigenvalues $\{\lambda_j\}$. We associate a random process $U_{nN}(t)$ with U_{nN} by analogy with $U_n(t)$. Let $\mathcal{L}(\xi_1, \xi_2)$ be the Lévy–Prokhorov distance between the random elements ξ_1 and ξ_2 from $D[0,1]$. Consider the inequality

$$\mathcal{L}(nU_n(t), \eta_\infty(t)) \le \mathcal{L}(nU_n(t), nU_{nN}(t)) + \mathcal{L}(nU_{nN}(t), \eta_N(t)) + \mathcal{L}(\eta_N(t), \eta_\infty(t)).$$

Since $\eta_N(t) \to \eta_\infty(t)$ as $N \to \infty$, then $\mathcal{L}(\eta_N(t), \eta_\infty(t)) \to 0$. According to the property of Lévy–Prokhorov metric, we have

$$\mathcal{L}(nU_n(t), nU_{nN}(t)) \le \inf\left\{\varepsilon > 0 : P\left(n \sup_{0\le t\le 1} |U_n(t) - U_{nN}(t)| \ge \varepsilon\right) \le \varepsilon\right\}.$$

Evidently,

$$P\left(n \sup_{0\le t\le 1} |U_n(t) - U_{nN}(t)| \ge \varepsilon\right) = P\left(\frac{2}{n-1} \max_{2\le k\le n} |S_k(\Phi - \Phi_N)| \ge \varepsilon\right),$$

where

$$S_k(\Phi - \Phi_N) = \sum_{1\le i<j\le k} (\Phi - \Phi_N)(X_i, X_j).$$

Therefore, for any $\varepsilon > 0$

$$\mathcal{L}(nU_n(t), nU_{nN}(t)) \le \varepsilon + P\left(\frac{2}{n-1} \max_{2\le k\le n} |S_k(\Phi - \Phi_N)| \ge \varepsilon\right).$$

Estimate the probability of the right-hand side of this inequality by Corollary 2.3.1 as follows

$$\begin{aligned} P\left(\frac{2}{n-1} \max_{2\le k\le n} |S_k(\Phi - \Phi_N)| \ge \varepsilon\right) &\le \frac{4}{(n-1)^2\varepsilon^2} ES_n^2(\Phi - \Phi_N) \\ &\le 4\varepsilon^{-2} E(\Phi - \Phi_N)^2. \end{aligned}$$

Choosing $\varepsilon = 2(E(\Phi - \Phi_N)^2)^{1/3}$, we obtain

$$\mathcal{L}(nU_n(t), nU_{nN}(t)) \le 3\left(E(\Phi - \Phi_N)^2\right)^{1/3}.$$

Since $E(\Phi - \Phi_N)^2 \to 0$ as $N \to \infty$, we have $\mathcal{L}(nU_n(t), nU_{nN}(t)) \to 0$ uniformly in n.

Thus, it suffices to establish the weak convergence

$$nU_{nN}(t) \Rightarrow \eta_N(t)$$

as $n \to \infty$ for any fixed N.

We write the process $nU_{nN}(t)$ in the form

$$nU_{nN}(t) = \frac{n}{n-1} \sum_{j=1}^{N} \lambda_j (S_{nj}^2(t) - t) - \frac{n}{n-1} \sum_{j=1}^{N} \lambda_j (T_{nj}(t) - t),$$

where

$$S_{nj}(t) = n^{-1/2} \sum_{k=1}^{[nt]} e_j(X_k),$$

$$T_{nj}(t) = n^{-1} \sum_{k=1}^{[nt]} e_j^2(X_k).$$

First, we show that for any fixed $\varepsilon > 0$

$$P\left(\sup_{0\le t\le 1} |T_{nj}(t) - t| \ge \varepsilon\right) \to 0$$

as $n \to \infty$ uniformly with respect to all $j = 1, \dots, N$.

Since

$$T_{nj}(t) - t = n^{-1} \sum_{k=1}^{[nt]} (e_j^2(X_k) - 1) + \frac{[nt] - nt}{n}$$

then for $\varepsilon > 1/n$ we have

$$P\left(\sup_{0\le t\le 1} |T_{nj}(t) - t| \ge \varepsilon\right) \le P\left(n^{-1} \max_{1\le m\le n} \left|\sum_{k=1}^{m} (e_j^2(X_k) - 1)\right| + \frac{1}{n} \ge \varepsilon\right)$$

[by Corollary 2.3.1]

$$\le \left(\varepsilon - \frac{1}{n}\right)^{-1} E\left|n^{-1} \sum_{k=1}^{n} (e_j^2(X_k) - 1)\right| \to 0 \text{ as } n \to \infty.$$

Here the expression on the right-hand side approaches zero according to the law of large numbers in L^1.

All elements in the sequence $S_{n1}(t), \dots, S_{nN}(t)$ have a zero expectation. They are not correlated with each other, and each is a sum of independent identically distributed random variables with finite dispersion. Therefore, the weak convergence

$$(S_{n1}(t), \dots, S_{nN}(t)) \Rightarrow (w_1(t), \dots, w_N(t))$$

as $n \to \infty$ takes place in the topology of the space $D^N[0,1]$. Taking into account Theorem 4.1 in Billingsley (1968), we obtain in the result the main statement of Theorem 5.2.1.

Side by side with $U_n(t)$, we define a process

$$U_{n(t)} = \binom{n(t)}{2}^{-1} \sum_{1\le i<j\le n(t)} \Phi(X_i, X_j),$$

where $n(t) = -[-n/t]$ is the least integer not less than n/t, $0 < t \le 1$. The trajectories of $U_{n(t)}$ belong to the space $D[0,1]$. Under the conditions of Theorem 5.2.1 the weak convergence

$$nU_{n(t)} \Rightarrow \eta_\infty(t)$$

takes place as $n \to \infty$ in the topology of the space $D[0,1]$. This statement can be proved similar to Theorem 5.2.1.

5.3 Weak convergence to a chaos process

Consider *UR*-statistics (1.1.1) of degree m and rank r. Taking into account the Hoeffding decomposition (1.1.9), we can define a random process

$$U_n(t) = \sum_{c=r}^{m} \binom{m}{c} \binom{n}{c}^{-1} S_{[nt],c}, \qquad 0 \le t \le 1,$$

where

$$S_{[nt],c} = \sum_{1 \le i_1 < \cdots < i_c \le [nt]} g_c(X_{i_1}, \ldots, X_{i_c}).$$

Evidently, $U_n(t)$ belongs to the space $D[0,1]$.

At first, we consider one interesting special case.

Permanent invariance principle. Suppose that

$$g_c(x_1, \ldots, x_c) = x_1 \cdots x_c.$$

Then

$$S_{[nt],c} = \frac{1}{c!} \operatorname{Per} A^c_{[nt]}(X),$$

where

$$\operatorname{Per} A^c_{[nt]}(X) = \sum_{1 \le i_1 \ne \cdots \ne i_c \le [nt]} X_{i_1} \cdots X_{i_c}$$

is a permanent random process (see Borovskikh and Korolyuk (1994b)).

Theorem 5.3.1. *Let $X_1, \ldots, X_n$ be a sequence of independent identically distributed random variables with mean value 0 and variance 1. Then the weak convergence*

$$n^{-c/2} S_{[nt],c} \Rightarrow \frac{1}{c!} H_c(w(t), t)$$

takes place as $n \to \infty$ in the topology of the space $D[0,1]$, where $w(t)$ is a standard Wiener process and $H_c(\cdot,\cdot)$ is the Hermite polynomial of two variables.

Proof. By formula (4.33) in Borovskikh and Korolyuk (1994b)

$$n^{-c/2} \operatorname{Per} A^c_{[nt]}(X) = H_c(n^{-1/2}\tau_{[nt],1}, n^{-1}\tau_{[nt],2}) + \\ + \sum_{k=3}^{c} P_k(n^{-3/2}\tau_{[nt],3}, \ldots, n^{-k/2}\tau_{[nt],k}) H_{c-k}(n^{-1/2}\tau_{[nt],1}, n^{-1}\tau_{[nt],2}) c^{[k]},$$

where

$$\tau_{[nt],k} = \sum_{j=1}^{[nt]} X_j^k, \qquad k = 1, 2, \ldots$$

By the Donsker invariance principle (see Billingsley (1968))

$$n^{-1/2}\tau_{[nt],1} \Rightarrow w(t)$$

and according to the strong law of large numbers

$$n^{-1}\tau_{[nt],2} \to t \qquad a.s.$$

Hence

$$H_c(n^{-1/2}\tau_{[nt],1}, n^{-1}\tau_{[nt],2}) \Rightarrow H_c(w(t), t)$$

in the space $D[0,1]$ as $n \to \infty$.

It is evident that for all $k \geq 3$

$$\sup_{0\leq t\leq 1} |\tau_{[nt],k}| \leq \sum_{j=1}^{n} |X_j|^k,$$

and according to the Marcinkiewicz–Kolmogorov law of large numbers

$$n^{-k/2}\sum_{j=1}^{n} |X_j|^k \to 0 \qquad a.s.$$

By virtue of this for $k \geq 3$

$$\sup_{0\leq t\leq 1} |P_k(n^{-3/2}\tau_{[nt],3}, \ldots, n^{-k/2}\tau_{[nt],k})| \to 0 \qquad a.s.$$

This proves Theorem 5.3.1.

Invariance principle in a general case. The following theorem is a direct extension of Theorem 5.2.1. At first, we define the limit random process. We assume, that $g_r \in L^2$. By analogy with (4.2.1), g_r has an L^2-expansion

$$g_r(x_1, \ldots, x_r) = \sum_{(i_1,\ldots,i_r)=1}^{\infty} g_{r,i_1,\ldots,i_r}\varphi_{i_1}(x_1)\cdots\varphi_{i_r}(x_r)$$

with the coefficients $g_{r,i_1,\ldots,i_r}$, defined by the formula

$$g_{r,i_1,\ldots,i_r} = E(g_r(X_1, \ldots, X_r)\varphi_{i_1}(X_1)\cdots\varphi_{i_r}(X_r)).$$

Now we define a chaos random process

$$K_r^t(g_r) = \sum_{(i_1,\ldots,i_r)=1}^{\infty} g_{r,i_1,\ldots,i_r} \prod_{j=1}^{\infty} H_{j(i_1,\ldots,i_r)}(w_j(t), t), \tag{5.3.1}$$

where $w_1(t), w_2(t), \ldots$ are independent standard Wiener processes, $Ew_j(t) = 0$, $Ew_j^2(t) = t$, $j \geq 1$. Process $K_r^t(g_r)$ is the limiting process for $n^{r/2}U_n(t)$ as $n \to$

∞. The chaos random function $K_r^t(g_r)$ is the weak limit of the sequence of chaos processes

$$K_{r,N}^t(g_r) = \sum_{(i_1,\dots,i_r)=1}^{N} g_{r,i_1,\dots,i_r} \prod_{j=1}^{\infty} H_{j(i_1,\dots,i_r)}(w_j(t),t), \qquad N = 1,2,\dots$$

in the space $D[0,1]$, i.e.

$$K_{r,N}^t(g_r) \Rightarrow K_r^t(g_r)$$

as $N \to \infty$ in the topology of the space $D[0,1]$. This relation is realized because of the condition $Eg_r^2 < \infty$ and the tightness of the sequence $K_{r,N}^t(g_r)$, $N \geq 1$. The property of the tightness is guaranteed by Theorem 15.6 in Billingsley (1968).

Theorem 5.3.2. *Suppose that the kernel $\Phi \in R$ is such that $E\Phi = 0$ and*

$$E|g_c|^{\gamma_{rc}} < \infty$$

for each $c = r,\dots,m$, where $\gamma_{rc} = 2c/(2c-r)$.

Then the weak convergence

$$n^{r/2}U_n(t) \Rightarrow \binom{m}{r} K_r^t(g_r)$$

takes place as $n \to \infty$ in the topology of the space $D[0,1]$.

Proof. By the Hoeffding decomposition (1.1.12) with $\theta = 0$, we find

$$n^{r/2}U_n(t) = \binom{m}{r} n^{-r/2} \sum_{1\leq i_1 \neq \dots \neq i_r \leq [nt]} g_r(X_{i_1},\dots,X_{i_r}) + R_{nr}(t)\,,$$

where

$$R_{nr}(t) = \sum_{c=r+1}^{m} m^{[c]} \prod_{j=1}^{c-1} (1-n^{-1}j)^{-1} n^{-c/\gamma_{rc}} S_{[nt],c}\,.$$

Here

$$\sup_{0\leq t\leq 1} |R_{nr}(t)| \xrightarrow{P} 0,$$

since for every $\varepsilon > 0$

$$P\left(\sup_{0\leq t\leq 1} \left|n^{-c/\gamma_{rc}} S_{[nt],c}\right| > \varepsilon\right) = P\left(\max_{c\leq k\leq n} \left|n^{-c/\gamma_{rc}} S_{kc}\right| > \varepsilon\right) \leq$$

[by Corollary 2.3.1]

$$\leq \varepsilon^{-\gamma_{rc}} E\left|n^{-c/\gamma_{rc}} S_{nc}\right|^{\gamma_{rc}} \to 0 \quad \text{as } n \to \infty$$

for each $c = r+1,\dots,m$ according to Theorem 3.1.3.

It still remains to establish that

$$n^{-r/2} \sum_{1\le i_1\ne\cdots\ne i_r\le[nt]} g_r(X_{i_1},\ldots,X_{i_r}) \Rightarrow K_r^t(g_r) \tag{5.3.2}$$

as $n\to\infty$ and $Eg_r^2<\infty$. If $\mathcal{L}(\xi_1,\xi_2)$ is the Lévy–Prokhorov distance between the random variables with values in $D[0,1]$, then we have the inequality

$$\mathcal{L}(S_n^t(g_r),K_r^t(g_r)) \le \mathcal{L}(S_n^t(g_r),S_{nN}^t(g_r))+\mathcal{L}(S_{nN}^t(g_r),K_{rN}^t(g_r))+\mathcal{L}(K_{rN}^t(g_r),K_r^t(g_r)),$$

where

$$S_n^t(g_r)=n^{-r/2}\sum_{1\le i_1\ne\cdots\ne i_r\le[nt]} g_r(X_{i_1},\ldots,X_{i_r})$$

and $S_{nN}^t(g_r)$ is defined just as $S_n^t(g_r)$ but with the kernel

$$g_{rN}(x_1,\ldots,x_r)=\sum_{(i_1,\ldots,i_r)=1}^{N} g_{r,i_1,\ldots,i_r}\varphi_{i_1}(x_1)\cdots\varphi_{i_r}(x_r).$$

Estimate of $\mathcal{L}(K_{rN}^t(g_r),K_r^t(g_r))$. Since $K_{rN}^t(g_r)\Rightarrow K_r^t(g_r)$ as $N\to\infty$, then

$$\mathcal{L}(K_{rN}^t(g_r),K_r^t(g_r))\to 0 \quad \text{as} \quad N\to\infty.$$

Estimate of $\mathcal{L}(S_n^t(g_r),S_{nN}^t(g_r))$. By virtue of the property of Lévy–Prokhorov metric, we get

$$\mathcal{L}(S_n^t(g_r),S_{nN}^t(g_r)) \le \inf\left\{\varepsilon>0: P\left(\sup_{0\le t\le 1}|S_n^t(g_r-g_{rN})|\ge\varepsilon\right)\le\varepsilon\right\}.$$

Further, we write

$$P\left(\sup_{0\le t\le 1}|S_n^t(g_r-g_{rN})|\ge\varepsilon\right)=P\left(n^{-r/2}\max_{r\le k\le n}|S_{kr}(g_r-g_{rN})|\ge\varepsilon\right)$$

[by Corollary 2.3.1]

$$\begin{aligned}&\le \varepsilon^{-2}n^{-r}ES_{nr}^2(g_r-g_{rN})\\ &\le \varepsilon^{-2}E(g_r-g_{rN})^2.\end{aligned}$$

Hence, for any $\varepsilon>0$

$$\mathcal{L}(S_n^t(g_r),S_{nN}^t(g_r))\le\varepsilon+\varepsilon^{-2}E(g_r-g_{rN})^2.$$

If $\varepsilon=2^{1/3}(E(g_r-g_{rN})^2)^{1/3}$, then

$$\mathcal{L}(S_n^t(g_r),S_{nN}^t(g_r))\le 2(E(g_r-g_{rN})^2)^{1/3}\to 0$$

as $N \to \infty$ uniformly in n since $E(g_r - g_{rN})^2 \to 0$.

Estimate of $\mathcal{L}(S^t_{nN}(g_r), K^t_{rN}(g_r))$. We show that for any fixed N the weak convergence

$$S^t_{nN}(g_r) \Rightarrow K^t_{rN}(g_r)$$

as $n \to \infty$ in the topology of the space $D[0,1]$. Then it follows from this relation that there exists a sequence $N = N(n)$ which tends to infinity as $n \to \infty$, and is such that

$$\mathcal{L}(S^t_{nN}(g_r), K^t_{rN}(g_r)) \to 0.$$

Now we represent the process $S^t_{nN}(g_r)$ in the form (see (4.2.3))

$$S^t_{nN}(g_r) = \sum_{(i_1,\dots,i_r)=1}^{N} g_{r,i_1,\dots,i_r} \left\{ n^{-r/2} \operatorname{Per} A^r_{[nt]}(\varphi_{i_1},\dots,\varphi_{i_r}) \right\}.$$

By Lemma 4.2.1 and Theorem 5.3.1 we have the weak convergence

$$n^{-r/2} \operatorname{Per} A^r_{[nt]}(\varphi_{i_1},\dots,\varphi_{i_r}) \Rightarrow \prod_{j=1}^{\infty} H_{j(i_1,\dots,i_r)}(w_j(t),t) \tag{5.3.3}$$

as $n \to \infty$ in the space $D[0,1]$. Further, according to the Cramer–Wold device (Billingsley (1968)) the finite-dimensional distributions of the vector

$$(n^{-r/2} \operatorname{Per} A^r_{[nt]}(\varphi_{i_1},\dots,\varphi_{i_r}),\ i_1,\dots,i_r = 1,\dots,N)$$

converge in law to those of the vector

$$\Big(\prod_{j=1}^{\infty} H_{j(i_1,\dots,i_r)}(w_j(t),t),\ i_1,\dots,i_r = 1,\dots,N\Big).$$

By virtue of (5.3.3) and the continuity of the process $\prod_{j=1}^{\infty} H_{j(i_1,\dots,i_r)}(w_j(t),t)$, we conclude that the sequence $(n^{-r/2} \operatorname{Per} A^r_{[nt]}(\varphi_{i_1},\dots,\varphi_{i_r}),\ i_1,\dots,i_r = 1,\dots,N)$ is dense in the space $D^{N^r}[0,1]$, and hence, the weak convergence

$$(n^{-r/2} \operatorname{Per} A^r_{[nt]}(\varphi_{i_1},\dots,\varphi_{i_r}),\ i_1,\dots,i_r = 1,\dots,N) \Rightarrow$$
$$\Rightarrow (\prod_{j=1}^{\infty} H_{j(i_1,\dots,i_r)}(w_j(t),t),\ i_1,\dots,i_r = 1,\dots,N)$$

takes place in the space $D^{N^r}[0,1]$. This implies the weak convergence $S^t_{nN}(g_r) \Rightarrow K^t_{rN}(g_r)$ in the space $D[0,1]$. Consequently, this proves Theorem 5.3.2.

Chaos processes and multiple stochastic integrals. Consider the chaos process defined by formula (5.3.1). Following Mandelbaum and Taqqu (1984) we present $K^t_r(g_r)$ via the multiple Wiener integral. Let

$$g^t_r = g_r(x_1,\dots,x_r) I(0 \le u_1 \le t) \cdots I(0 \le u_r \le t)$$

be a function, defined on the product of spaces $(X, \mathfrak{X}, P)^r \times ([0,T], \mathfrak{B}, \lambda)^r$, where $\mathfrak{B}$ is a σ-algebra of Borel sets and λ is a Lebesgue measure and $0 < t \le T < \infty$.

Denote $I_r^W(g_r^t)$ the multiple Wiener integral with respect to the Gaussian random measure $W(\cdot)$ on the product of spaces $(X, \mathfrak{X}, P) \times ([0,T], \mathfrak{B}, \lambda)$. Symbolically,

$$I_r^W(g_r^t) = \int \cdots \int g_r(x_1, \dots, x_r) I(0 \le u_1 \le t) \cdots I(0 \le u_r \le t) \times$$

$$\times W(dx_1, du_1) \cdots W(dx_r, du_r). \tag{5.3.4}$$

In particular, if $\varphi^t = \varphi(x) I(0 \le u \le t)$, $E\varphi(X_1) = 0$, $E\varphi^2(X_1) = 1$, then

$$I_1^W(\varphi^t) = \iint \varphi(x) I(0 \le u \le t) W(dx, du) \stackrel{\mathrm{d}}{=} w(t),$$

where $w(t)$ is a standard Wiener process on $[0,1]$; and if $\varphi_r^t = \prod_{s=1}^r \varphi(x_s) I(0 \le u_s \le t)$, then

$$I_r^W(\varphi_r^t) = \int \cdots \int \prod_{s=1}^r \varphi(x_s) I(0 \le u_s \le t) W(dx_1, du_1) \cdots W(dx_r, du_r) \stackrel{\mathrm{d}}{=} H_r(w(t), t).$$

Further we apply an L^2-expansion of function g_r. Substituting this expansion in (5.3.4), we get in L^2

$$I_r^W(g_r^t) =$$

$$= \sum_{(i_1, \dots, i_r)=1}^{\infty} g_{r, i_1, \dots, i_r} \int \cdots \int \prod_{s=1}^r \varphi_{i_s}(x_s) I(0 \le u_s \le t) W(dx_1, du_1) \cdots W(dx_r, du_r).$$

Here

$$\int \cdots \int \prod_{s=1}^r \varphi_{i_s}(x_s) I(0 \le u_s \le t) W(dx_1, du_1) \cdots W(dx_r, du_r) \stackrel{\mathrm{d}}{=} \prod_{j=1}^{\infty} H_{j(i_1, \dots, i_n)}(w_j(t), t).$$

Therefore we obtain in L^2

$$I_r^W(g_r^t) \stackrel{\mathrm{d}}{=} \sum_{(i_1, \dots, i_r)=1}^{\infty} g_{r, i_1, \dots, i_r} \prod_{j=1}^{\infty} H_{j(i_1, \dots, i_n)}(w_j(t), t).$$

Comparing this relation with (5.3.1) we conclude that

$$K_r^t(g_r) \stackrel{\mathrm{d}}{=} I_r^W(g_r^t).$$

This formula is similar to (4.2.4).

Multi-sample invariance principle. Assume, that in (1.2.1), $B = \mathrm{R}$. Let $(\mathfrak{B}, \lambda)$ be a measurable space, where $\mathfrak{B}$ are σ-algebra Borel subsets of $[0,1]$ and λ is a Lebesgue measure on $\mathfrak{B}$. Denote $(\mathfrak{B}_j, \lambda_j)$, $j = 1, \dots, c$, as copies of $(\mathfrak{B}, \lambda)$. Let $W_j(\cdot, \cdot)$, $j = 1, \dots, c$ be the independent Gaussian random measures defined on the product of spaces $(X_j, \mathfrak{X}_j, P_j) \times ([0,1], \mathfrak{B}_j, \lambda_j)$, $j = 1, \dots, c$, consequently, with mean value zero and covariances

$$EW_j(A,s)W_j(B,t) = \min(s,t)P_j(A \cap B)$$

for any sets $A, B \in \mathfrak{X}_j$ and $s, t \in [0,1]$.

According to this vector measure $W = (W_1, \dots, W_c)$ we also have multiple Wiener integrals of the form

$$I_{\bar d}^W(g_{\bar d}^{\bar t}) = \int \cdots \int g_{\bar d}^{\bar t} \prod_{j=1}^{c} \prod_{i=1}^{d_j} W_j(dx_{ji}, du_{ji}),$$

where

$$g_{\bar d}^{\bar t} = g_{\bar d}(x_{j1}, \dots, x_{jd_j}, j = 1, \dots, c) \prod_{j=1}^{c} \prod_{i=1}^{d_j} I(0 \le u_{ji} \le t_j).$$

In addition $\bar d = (d_1, \dots, d_c)$ and $\bar t = (t_1, \dots, t_c)$, $t_j \in (0,1]$.

For $\bar t \in (0,1]^c$ on multi-sample UR-statistics $U_{\bar n}$ and $U_{\bar n \bar d}$ we define the following random processes with trajectories in the space $D((0,1]^c)$. Let

$$U_{\bar n \bar d}^{\bar t} = \prod_{j=1}^{c} n_j^{d_j/2} t_j^{d_j} U_{[\bar n \bar t], d},$$

where $[\bar n \bar t] = ([n_1 t_1], \dots, [n_c t_c])$ and $[n_j t_j]$ denote the integral part of the variable $n_j t_j$, $j = 1, \dots, c$. Then, for UR-statistics $U_{\bar n}$ of the rank r we define a random process

$$U_{\bar n}^{\bar t} = \sigma_{\bar n \bar r}^{-1}(U_{[\bar n \bar t]} - \theta),$$

where

$$\sigma_{\bar n \bar r}^2 = \sum_{r_1 + \cdots + r_c = r} \prod_{j=1}^{c} m_j^{2[r_j]} / r_j! n_j^{-r_j} E g_{\bar r}^2.$$

Theorem 5.3.3. *Let $0 < Eg_{\bar d}^2 < \infty$. Then the weak convergence*

$$U_{\bar n \bar d}^{\bar t} \Rightarrow I_{\bar d}^W(g_{\bar d}^{\bar t}) \tag{5.3.5}$$

as $\min(n_1, \dots, n_c) \to \infty$ takes place in the topology of the space $D((0,1]^c)$.

Proof. The case $c = 1$, $r = m$ was considered in Theorem 5.3.2. For any $c \ge 1$ the proof can be given by a similar method.

Theorem 5.3.4. *Suppose that the kernel Φ has the rank r and satisfies the conditions*

$$\min_{r_1 + \cdots + r_c = r} E g_{\bar r}^2 > 0$$

and

$$E|g_{\bar d}|^{\gamma_{\bar r\bar d}} < \infty, \qquad d = r, r+1, \dots, m,$$

where $\gamma_{\bar r\bar d} = 2d/(2d-r)$, $d = d_1 + \cdots + d_c$, $r = r_1 + \cdots + r_c$, $m = m_1 + \cdots + m_c$.

Then the weak convergence

$$U_{\bar n}^{\bar t} \Rightarrow \sum_{d_1+\cdots+d_c=r} q_{\bar n}(\bar d) \prod_{j=1}^{c} \binom{m_j}{d_j} t_j^{-d_j} I_{\bar d}^W (g_{\bar d}^{\bar t})$$

as $\min(n_1, \dots, n_c) \to \infty$ *takes place in the topology of the space* $D((0,1]^c)$. *Additionally*

$$q_{\bar n}(\bar d) = \left(\sum_{r_1+\cdots+r_c=r} \prod_{j=1}^{c} m_j^{2[r_j]}/r_j!\, n_j^{d_j-r_j} E g_{\bar r}^2 \right)^{-1/2}.$$

Proof. By (1.2.8) we have

$$U_{\bar n}^{\bar t} = \sigma_{\bar n\bar r}^{-1} \sum_{d_1+\cdots+d_c} \prod_{j=1}^{c} \binom{m_j}{d_j} U_{[\bar n\bar t],\bar d} + R_{\bar n\bar r}(\bar t),$$

where

$$R_{\bar n\bar r}(\bar t) = \sigma_{\bar n\bar r}^{-1} \sum_{d=r+1}^{m} \prod_{j=1}^{c} \binom{m_j}{d_j} U_{[\bar n\bar t],\bar d}.$$

By virtue of the maximal martingale inequality under the conditions of the theorem,

$$\sup_{\bar t\in[0,1]^c} |R_{\bar n\bar r}(\bar t)| \xrightarrow{\mathrm{P}} 0$$

as $\min(n_1, \dots, n_c) \to \infty$. Next

$$\sigma_{\bar n\bar r}^{-1} \sum_{d=r+1}^{m} \prod_{j=1}^{c} \binom{m_j}{d_j} U_{[\bar n\bar t],\bar d} = \sum_{d_1+\cdots+d_c=r} q_{\bar n}(\bar d) \prod_{j=1}^{c} \binom{m_j}{d_j} t_j^{-d_j} U_{\bar n\bar d}^{\bar t}.$$

From here and (5.3.5) we get the proof of Theorem 5.3.4.

Corollary 5.3.1. *Suppose that* $n_j n_1^{-1} \to \lambda_j$, $0 < \lambda_j < \infty$, $j = 1, \dots, c$, *as* $\min(n_1, \dots, n_c) \to \infty$. *Then the weak convergence*

$$U_{\bar n}^{\bar t} \Rightarrow \sum_{d_1+\cdots+d_c=r} q(\bar d) \prod_{j=1}^{c} \binom{m_j}{d_j} t_j^{-d_j} I_{\bar d}^W (g_{\bar d}^{\bar t})$$

takes place in the space $D((0,1]^c)$, *where*

$$q(\bar d) = \left(\sum_{r_1+\cdots+r_c=r} \prod_{j=1}^{c} m_j^{2[r_j]}/r_j!\, \lambda_j^{d_j-r_j} E g_{\bar r}^2 \right)^{-1/2}.$$

5.4 Weak convergence in the Poisson approximation scheme

Family of Poisson random measures. Let $X_1^{(n)}, \ldots, X_j^{(n)}, \ldots$ be an independent identically distributed array on a measurable space $(X, \mathfrak{X})$ with distribution $P_n(B) = P(X_j^{(n)} \in B) = \lambda(B)/n$, $B \in \mathfrak{X}$, with some Radon measure $\lambda(B)$, $\lambda(B) < \infty$. We define a sequence of random point measures

$$\nu_n = \delta_1^{(n)} + \cdots + \delta_n^{(n)},$$

where $\delta_j^{(n)}$ is a Dirac measure which is given by $\delta_j^{(n)}(B) = I(X_j^{(n)} \in B)$ for $B \in \mathfrak{X}$. The distribution ν_n converges weakly to a distribution of a Poisson random measure ν with intensity λ, i.e. $\nu_n(B) \to \nu(B)$ for all $B \in \mathfrak{X}_\lambda$, where $\mathfrak{X}_\lambda = \{ B \in \mathfrak{X} : \lambda(\partial B) = 0 \}$. Additionally ∂B denotes the boundary of B (see Tjur (1980)).

Let $t \in (0, T]$ for some $T > 0$. Now we define a family of random point measures

$$\nu_{[nt]} = \delta_1^{(n)} + \cdots + \delta_{[nt]}^{(n)}$$

and a family of Poisson random measures $\nu(\cdot, t)$ so that for every $B \in \mathfrak{X}$ we get a Poisson random process $\nu(B, t)$ with $E\nu(B, t) = t\lambda(B)$. If we have fixed t, then $\nu(B, t)$ is a Poisson random measure on $(X, \mathfrak{X})$ with intensity $t\lambda(B)$. Under the conditions of Poisson approximation the weak convergence

$$\nu_{[nt]}(B) \Rightarrow \nu(B, t) \tag{5.4.1}$$

takes place as $n \to \infty$ in the topology of the space $D([0, T])$.

Poisson convergence of permanent random processes. Consider a permanent process

$$\text{Per } A_{[nt]}^m(\delta) = \sum_{1 \le i_1 \ne \cdots \ne i_m \le [nt]} \delta_{i_1}^{(n)}(B) \cdots \delta_{i_m}^{(n)}(B), \qquad B \in \mathfrak{X}.$$

Then by virtue of (4.4.1) and (5.4.1)

$$\text{Per } A_{[nt]}^m(\delta) \Rightarrow \nu^{[m]}(B, t).$$

If $\bar{\delta} = \delta - \lambda/n$, then for a centered permanent process the following weak convergence is valid

$$\text{Per } A_{[nt]}^m(\delta) \Rightarrow C_m(\nu(B, t), t\lambda(B)),$$

where

$$C_m(\nu(B, t), t\lambda(B)) = \sum_{c=0}^{m} (-t\lambda(B))^{m-c} m^{[c]} \nu^{[c]}(B, t)/c!$$

is the stochastic Charlier–Poisson polynomial. The relation for $t = 1$ corresponds to Theorem 4.4.1.

Next consider the permanent functional process

$$\operatorname{Per} A^m_{[nt]}(\varphi) = \sum_{1\le i_1\ne\cdots\ne i_m\le [nt]} \varphi(X^{(n)}_{i_1})\cdots\varphi(X^{(n)}_{i_m}),$$

where $\varphi : X \mapsto \mathrm{R}$ is a real-valued continuous function with compact support on X. By analogy we define a centered permanent functional process $\operatorname{Per} A^m_{[nt]}(\bar\varphi)$ with $\bar\varphi(x) = \varphi(x) - \lambda(\varphi)/n$, where

$$\lambda(\varphi) = \int_X \varphi(x)\lambda(dx).$$

Let $\bar\nu(B,t) = \nu(B,t) - t\lambda(B)$, $B \in \mathfrak{X}$.

Theorem 5.4.1. *Under the Poisson approximation scheme the weak convergences*

$$\operatorname{Per} A^m_{[nt]}(\varphi) \Rightarrow \int_{X^m} \prod_{s=1}^m \varphi(x_s)\nu(dx_1,t)\cdots\nu(dx_m,t)$$

and

$$\operatorname{Per} A^m_{[nt]}(\bar\varphi) \Rightarrow \int_{X^m} \prod_{s=1}^m \varphi(x_s)\bar\nu(dx_1,t)\cdots\bar\nu(dx_m,t)$$

take place as $n \to \infty$ in the topology of the space $D([0,T])$.

Proof. In just the same way as the proof of Theorem 4.4.2 we have the following representation

$$\operatorname{Per} A^m_{[nt]}(\varphi) = m! \sum_{i_1+2i_2+\cdots+mi_m=m} \prod_{s=1}^m (-1)^{i_s-1}\left[I_1^{\nu_{[nt]}}(\varphi^s)/s\right]^{i_s}/i_s!,$$

where

$$I_1^{\nu_{[nt]}}(\varphi^s) = \int_X \varphi^s(x)\nu_{[nt]}(dx).$$

Here under the conditions of the Poisson approximation scheme

$$I_1^{\nu_{[nt]}}(\varphi^s) \Rightarrow \int_X \varphi^s(x)\nu(dx,t)$$

as $n \to \infty$. Hence, the weak convergence

$$\operatorname{Per} A^m_{[nt]}(\varphi) \Rightarrow m! \sum_{i_1+2i_2+\cdots+mi_m=m} \prod_{s=1}^m (-1)^{i_s-1}\left[I_1^{\nu_{[nt]}}(\varphi^s)/s\right]^{i_s}/i_s!$$

is valid as $n \to \infty$ in the space $D([0,T])$. This proves the first statement of Theorem 5.4.1. The second statement of Theorem 5.4.1 follows for example from formula (2.56) in Borovskikh and Korolyuk (1994b).

Poisson convergence of permanent symmetric statistics. Let $\Phi : X^m \mapsto \mathrm{R}$ be a symmetric function. Consider a random process

$$S_{[nt]}(\Phi) := \sum_{1\le i_1\ne\cdots\ne i_m\le[nt]} \Phi(X_{i_1}^{(n)},\dots,X_{i_m}^{(n)})$$
$$= \int_X\cdots\int_X \Phi(x_1,\dots,x_m)\,\mathrm{Per}\,A_{[nt]}^m(dx_1,\dots,dx_m).$$

Theorem 5.4.2. *Suppose that the function Φ is continuous and there exist a compact set $B \subset X$ and constant $c > 0$ such that $\Phi(x_1,\dots,x_m) = 0$ if $(x_1,\dots,x_m) \notin B^m$ and $|\Phi(x_1,\dots,x_m)| \le c^m$ if $(x_1,\dots,x_m) \in B^m$.*

Then under the Poisson approximation condition the weak convergence

$$S_{[nt]}(\Phi) \Rightarrow \int_X\cdots\int_X \Phi(x_1,\dots,x_m)\nu(dx_1,t)\cdots\nu(dx_m,t) \tag{5.4.2}$$

takes place as $n\to\infty$ in the space $D([0,T])$.

Proof. Since

$$\mathrm{Per}\,A_{[nt]}^m(dx_1,\dots,dx_m) := \sum_{1\le i_1\ne\cdots\ne i_m\le[nt]} \delta_{i_1}^{(n)}(dx_1)\cdots\delta_{i_m}^{(n)}(dx_m) \Rightarrow \prod_{s=1}^m \nu(dx_s,t)$$

as $n\to\infty$, then the proof of relation (5.4.2) can develop the standard technique by using the general limit theorem (Billingsley (1968)).

5.5 Invariance principle for symmetric statistics

Nonrandom sample size. For R-valued kernel Φ we define a random process

$$S_n(t) = \sum_{c=1}^{[nt]} \frac{1}{c!} n^{-c/2} \sum_{1\le i_1\ne\cdots\ne i_c\le[nt]} g_c(X_{i_1},\dots,X_{i_c}), \qquad 0\le t\le T, \tag{5.5.1}$$

where $X_1,\dots,X_n$ is an i.i.d. sequence of random variables.

Theorem 5.5.1. *If canonical functions g_c satisfy the condition*

$$\sum_{c=1}^{\infty} \frac{T^c}{c!} Eg_c^2 < \infty$$

then the weak convergence

$$S_n(t) \Rightarrow \sum_{c=1}^{\infty} \frac{1}{c!} I_c^W(g_c^t)$$

takes place as $n\to\infty$ in the space $D([0,T])$ with multiple Wiener integrals $I_c^W(g_c^t)$ from (5.3.4).

For proof, see Mandelbaum and Taqqu (1984). Moreover, it should be noted that Theorem 5.5.1 essentially follows from Theorem 5.3.2.

Example 5.5.1. Let $g_c(x_1, \dots, x_c) = x_1 \cdots x_c$, $c \geq 1$, and $EX_1 = 0$, $EX_1^2 = 1$.

Then

$$S_n(t) \Rightarrow \sum_{c=1}^{\infty} \frac{1}{c!} H_c(w(t), t),$$

where $w(t)$ is a standard Wiener process. Taking into account the formula for the generating function of the Hermite polynomials of two variables we get the weak convergence

$$S_n(t) \Rightarrow -1 + \exp(w(t) - t/2)$$

as $n \to \infty$ in the space $D([0,T])$.

Along with process (5.5.1) we define another process

$$\nu_n(t) = \sum_{c=1}^{[nt]} \frac{1}{c!} \sum_{1 \leq i_1 \neq \cdots \neq i_c \leq [nt]} g_c(X_{i_1}^{(n)}, \dots, X_{i_c}^{(n)}), \qquad 0 \leq t \leq T, \tag{5.5.2}$$

where $X_1^{(n)}, X_2^{(n)}, \dots, X_n^{(n)}$ is an i.i.d. array. The trajectories of $\nu_n(t)$ belong to the space $D([0,T])$.

Theorem 5.5.2. *Assume that there exists a compact set $B \in \mathfrak{X}$ such that for any $c \geq 1$ $g_c(x_1, \dots, x_c) = 0$, if $(x_1, \dots, x_c) \notin B^c$, and $|g_c(x_1, \dots, x_c)| \leq K^c$ for some constant $K > 0$, if $(X_1, \dots, x_c) \in B^c$.*

Then under the Poisson approximation scheme the weak convergence

$$\nu_n(t) \Rightarrow \sum_{c=1}^{\infty} \frac{1}{c!} \int_X \cdots \int_X g_c(x_1, \dots, x_c) \nu(dx_1, t) \cdots \nu(dx_c, t)$$

is valid as $n \to \infty$ in the space $D([0,T])$.

Proof. This statement follows from Theorem 5.4.2.

Example 5.5.2. Let $g_c(x_1, \dots, x_c) = \varphi(x_1) \cdots \varphi(x_c)$, where $\varphi : X \mapsto \mathrm{R}$ is a continuous function with compact support and $\varphi(x) > -1$ for all $x \in X$.

Then by Theorem 5.5.2

$$\nu_n(t) \Rightarrow \sum_{c=1}^{\infty} \frac{1}{c!} \int_X \cdots \int_X \varphi(x_1) \cdots \varphi(x_c) \nu(dx_1, t) \cdots \nu(dx_c, t).$$

Now we represent this relation in another form. Namely, using definition (5.5.2) we write

$$\nu_n(t) + 1 = \prod_{j=1}^{[nt]} (1 + \varphi(X_j^{(n)})) =$$

$$= \exp\left\{\sum_{j=1}^{[nt]} \ln(1+\varphi(X_j^{(n)}))\right\}$$

$$= \exp\left\{\int_X \ln(1+\varphi(x))\nu_{[nt]}(dx)\right\}.$$

From here by virtue of (5.4.1) we get the weak convergence

$$\nu_n(t) \Rightarrow -1 + \exp\left\{\int_X \ln(1+\varphi(x))\nu(dx,t)\right\}$$

as $n \to \infty$ in the space of $D([0,T])$.

Random sample size. With the help of random variable N_p, satisfying condition (4.3.3), we define the random process

$$S_{N_p}(t) = \sum_{c=1}^{[N_p t]} \frac{1}{c!} p^{-c/2} \sum_{1 \le i_1 \ne \cdots \ne i_c \le [N_p t]} g_c(X_{i_1}, \dots, X_{i_c}), \qquad 0 \le t \le T.$$

The following theorem is a functional analogue of Theorem 4.7.4.

Theorem 5.5.3. *If $g \in G$ then a random process $S_{N_p}(t)$, $0 \le t \le T$, converges weakly in $D([0,T])$ to*

$$\sum_{c=1}^{\infty} \frac{1}{c!} I_c^M(g_c^t),$$

where multiple integrals $I_c^M(g_c^t)$ are taken with respect to the random measure M on the product space $(X, \mathfrak{X}, P) \times ([0,T], \mathfrak{B}, \lambda)$.

For the proof of this theorem see Feldman *et al.* (1993).

Let a sequence $\{N_n, n = 1, 2, \dots\}$ of positive random variables, taking integer values, satisfy condition (4.3.3). Define the random process

$$\nu_{N_n}(t) = \sum_{c=1}^{[N_n t]} \frac{1}{c!} n^{-c/2} \sum_{1 \le i_1 \ne \cdots \ne i_c \le [N_n t]} g_c(X_{i_1}^{(n)}, \dots, X_{i_c}^{(n)}), \qquad 0 \le t \le T,$$

where $X_1^{(n)}, X_2^{(n)}, \dots, X_n^{(n)}, \dots$ is an i.i.d. array, which satisfies the condition of the Poisson approximation. Hence, the weak convergence

$$\sum_{j=1}^{[N_n t]} \delta_j^{(n)} \Rightarrow \nu(B, Yt)$$

is valid, as $n \to \infty$, for $B \in \mathfrak{X}$ and $0 \le t \le T$.

Theorem 5.5.4. *Under the conditions of Theorem* (4.7.5),

$$\nu_{N_n}(t) \Rightarrow \sum_{c=1}^{\infty} \frac{1}{c!} \int_X \cdots \int_X g_c(x_1, \dots, x_c) \prod_{s=1}^{c} \nu(dx_s, Yt)$$

as $n \to \infty$ *in the space* $D([0,T])$.

For the proof of this theorem it is sufficient to apply the method in Feldman *et al.* (1993).

5.6 Functional limit theorems with varying kernels

Convergence to Gaussian processes. Consider UR-statistics of the form (4.8.1). For $t \in [0,1]$, we define the random process

$$U_n(t) = n^{-m/2} \sum_{1 \le i_1 < \cdots < i_m \le [nt]} \Phi_n(X_{i_1}, \dots, X_{i_m}), \tag{5.6.1}$$

where we suppose that the kernel Φ_n possesses the property of complete degeneracy.

The trajectories of $U_n(t)$ belong to the space $D[0,1]$. Since the kernel Φ depends on the parameter n, then the asymptotics, as $n \to \infty$, can change so that none of the limit theorems given in the previous sections will hold true. Fortunately, the martingale properties of U-statistics are operated in this situation. With the help of martingale limit theorems we can prove the functional limit theorems for processes of type (5.6.1).

At first, we define one limit random process for $U_n(t)$. Let $U_\infty(t)$ be a Gaussian process such that $EU_\infty(t) = 0$ and $EU_\infty^2(t) = \varphi(t), 0 \le t \le 1$, with some continuous function $\varphi(t)$. The characteristic function of process $U_\infty(t)$ is equal to

$$E \exp(isU_\infty(t)) = \exp\left(-\frac{s^2}{2}\varphi(t)\right), \qquad s \in \mathrm{R}.$$

If the function $\varphi(t)$ is differentiable, then $U_\infty(t)$ is expressed in terms of a standard Wiener process $w(t)$ in the form of a stochastic integral

$$U_\infty(t) = \int_0^t (\varphi'(u))^{1/2} \, dw(u).$$

Theorem 5.6.1. *Let*

$$\xi_{nk} = \sum_{1 \le i_1 < \cdots < i_{m-1} \le k-1} \Phi_n(X_{i_1}, \dots, X_{i_{m-1}}, X_k), \qquad k = m, \dots, n$$

and $\mathfrak{F}_k = \sigma\{\omega : X_1, \dots, X_k\}$. *If the conditions*

$$n^{-m} \sum_{k=m}^{n} E\left(\xi_{nk}^2 \, I(|\xi_{nk}| > \varepsilon n^{m/2}) \middle| \mathfrak{F}_{k-1}\right) \xrightarrow{\mathrm{P}} 0,$$

and

$$n^{-m} \sum_{k=m}^{[nt]} E\left(\xi_{nk}^2 \middle| \mathfrak{F}_{k-1}\right) \xrightarrow{\mathrm{P}} \varphi(t)$$

as $n \to \infty$ *hold for* $0 < t \leq 1$, *and* $0 < \varepsilon < 1$ *with some continuous function* $\varphi(t)$ *on* $[0,1]$, *then the weak convergence*

$$U_n(t) \to U_\infty(t)$$

takes place in the space $D[0,1]$.

Proof. This statement is a particular case of Liptser and Shiryaev (1989) because stochastic sequence $(\xi_{nk}, \mathfrak{F}_k)$, $k = m, \ldots, n$, forms a martingale difference.

Corollary 5.6.1. *If* $E\Phi_n^2 = 1$ *and*

$$n^{-1} E\Phi_n^4 \to 0 \quad as \ n \to \infty$$

then the weak convergence

$$U_n(t) \Rightarrow ((m-1)!)^{-1/2} \int_0^t u^{(m-1)/2}\, dw(u)$$

for each $m = 1, 2, \ldots$ *takes place.*

Proof. Verify the conditions of Theorem 5.6.1. Since $I(|\xi_{nk}| > \varepsilon n^{m/2}) \leq \varepsilon^{-2} n^{-m} \xi_{nk}^2$, then

$$E\left(n^{-m} \sum_{k=m}^{n} E\left(\xi_{nk}^2 I(|\xi_{nk}| > \varepsilon n^{m/2}) \middle| \mathfrak{F}_{k-1}\right)\right)$$
$$\leq n^{-2m} \varepsilon^{-2} \sum_{k=m}^{n} E\xi_{nk}^4$$

[by using Corollary 2.2.3]

$$\leq \varepsilon^{-2} \alpha_4^m E\Phi_n^4 n^{-2m} \sum_{k=m}^{n} \binom{k}{m-1} k^{m-1}$$
$$\leq \varepsilon^{-2} \alpha_4^m n^{-1} E\Phi_n^4 \to 0 \quad \text{as } n \to \infty.$$

This proves the first condition of Theorem 5.6.1. The validity of the second condition of Theorem 5.6.1 is verified in the same way with the help of Corollary 2.2.3. In addition, we have

$$E\left(n^{-m} \sum_{k=m}^{[nt]} E(\xi_{nk}^2 | \mathfrak{F}_{k-1})\right) =$$

$$= n^{-m} \sum_{k=m}^{[nt]} E\xi_{nk}^2 = n^{-m} \sum_{k=m}^{[nt]} \binom{k}{m-1}$$

$$= n^{-m} \binom{[nt]}{m} \to \frac{t^m}{m!},$$

i.e. $\varphi(t) = t^m/m!$. From here Corollary 5.6.1 follows.

Now we define the vector random process in $D^m[0,1]$

$$\mathbf{U}_n(t) = (U_{n1}(t), U_{n2}(t), \dots, U_{nm}(t)),$$

where

$$U_{nc}(t) = n^{-c/2} \sum_{1 \le i_1 < \dots < i_c \le [nt]} g_{nc}(X_{i_1}, \dots, X_{i_c}), \qquad c = 1, \dots, m$$

with canonical functions g_{nc}.

Theorem 5.6.1 and Corollary 5.6.1 show that $\mathbf{U}_n(t)$ has the following weak limit.

Corollary 5.6.2. *Let* $Eg_{nc}^2 = 1$ *and*

$$n^{-1} Eg_{nc}^4 \to 0 \quad \text{as } n \to \infty$$

for each $c = 1, \dots, m$.

Then the weak convergence

$$\mathbf{U}_n(t) \Rightarrow \mathbf{U}_\infty(t)$$

as $n \to \infty$ *takes place with*

$$\mathbf{U}_\infty(t) = (U_{\infty,1}(t), U_{\infty,2}(t), \dots, U_{\infty,m}(t)),$$

where

$$U_{\infty,c}(t) = ((c-1)!)^{-1/2} \int_0^t u^{(c-1)/2}\, dw_c(u), \qquad c = 1, \dots, m,$$

in addition $w_1(u), \dots, w_m(u)$ *are independent standard Wiener processes.*

Convergence to Poisson processes. Let Φ_n be a symmetric, zero-one function of m arguments. Define the random process in $D[0,1]$

$$\nu_n(t) = \sum_{1 \le i_1 < \dots < i_m \le [nt]} \Phi_n(X_{i_1}^{(n)}, \dots, X_{i_m}^{(n)}),$$

where $X_1^{(n)}, X_2^{(n)}, \dots, X_n^{(n)}, \dots$ is an i.i.d. array. Let $\rho_{nc}^2 = E\Phi_{nc}^2(X_1^{(n)}, \dots, X_c^{(n)})$ where

$$\Phi_{nc}(x_1, \dots, x_c) = E(\Phi_n(X_1^{(n)}, \dots, X_m^{(n)}) | X_1^{(n)} = x_1, \dots, X_c^{(n)} = x_c), \quad c = 1, \dots, m.$$

Theorem 5.6.2. *Suppose that*

$$\binom{n}{m}\rho_{nm}^2 \to \lambda$$

and

$$\max_{1\le c\le m-1}(n^{2m-c}\rho_{nc}^2) \to 0,$$

as $n \to \infty$. *Then the weak convergence*

$$\nu_n(t) \Rightarrow \nu(\lambda t^m)$$

takes place in the space $D[0,1]$, *where* $\nu(\lambda t^m)$ *is a Poisson process.*

Proof. Denote $\Phi_n(X_{\mathbf{i}}^{(n)}) = \Phi_n(X_{i_1}^{(n)}, \ldots, X_{i_m}^{(n)})$, $\mathbf{i} = (i_1, \ldots, i_m)$ and $J_n = \{\mathbf{i} = (i_1, \ldots, i_m) : 1 \le i_1 < \cdots < i_m \le n\}$. Let for all $k \ge m$

$$\xi_{nk} = \sum_{\mathbf{i}\in J_k\setminus J_{k-1}} \Phi_n(X_{\mathbf{i}}^{(n)}).$$

Now we have the formula

$$\nu_n(t) = \sum_{k=m}^{[nt]} \xi_{nk}$$

and the inequalities (see, Silverman and Brown (1978))

$$0 \le \xi_{nk} - I(\xi_{nk} \ge 1) \le \sum_{\mathbf{i}\ne\mathbf{j}\in J_k\setminus J_{k-1}} \Phi_n(X_{\mathbf{i}}^{(n)})\Phi_n(X_{\mathbf{j}}^{(n)}).$$

Introduce into the consideration the point random process

$$\nu_n'(t) = \sum_{k=m}^{[nt]} I(\xi_{nk} \ge 1).$$

Then

$$P(\nu_n(t) \ne \nu_n'(t)) \le \sum_{k=m}^{n} \sum_{\mathbf{i}\ne\mathbf{j}\in J_k\setminus J_{k-1}} E\Phi_n(X_{\mathbf{i}}^{(n)})\Phi_n(X_{\mathbf{j}}^{(n)}) =$$

$$= \sum_{c=1}^{m-1} \rho_{nc}^2(\#\{\mathbf{i},\mathbf{j} \in J_n : i_m = j_m, \#\{(i_1,\ldots,i_m)\cap(j_1,\ldots,j_m)\} = c\}),$$

where

$$\#\{\mathbf{i},\mathbf{j} \in J_n : i_m = j_m, \#\{(i_1,\ldots,i_m)\cap(j_1,\ldots,j_m)\} = c\} = O(n^{2m-c}).$$

Hence, under the conditions of this theorem $P(\nu_n(t) \ne \nu_n'(t)) \to 0$ as $n \to \infty$, and it is sufficient to prove the statement of theorem for a point process $\nu_n'(t)$.

Let $\mathfrak{F}^n_{[nt]} = \sigma\{\omega : X_1^{(n)}, \dots, X_{[nt]}^{(n)}\}$ and write the compensator of process $\nu'_n(t)$

$$A'_n(t) = \sum_{k=m}^{[nt]} E(I(\xi_{nk} \geq 1)|\mathfrak{F}^n_{k-1})$$

and the compensator of process $\nu_n(t)$

$$A_n(t) = \sum_{k=m}^{[nt]} E(\xi_{nk}|\mathfrak{F}^n_{k-1}).$$

We can see that $P(A_n(t) \neq A'_n(t)) \to 0$ as $n \to \infty$. Therefore, it is sufficient to prove (see Liptser and Shiryaev (1989)) that for fixed t the compensator $A_n(t)$ converges in probability to λt^m as $n \to \infty$. Since $\rho^2_{nm} = E\Phi_n(X_1^{(n)}, \dots, X_n^{(n)})$ then according to the conditions of this theorem, it is evident that $EA_n(t) \to \lambda t^m$ as $n \to \infty$, and

$$\begin{aligned}
&\sigma^2(A_n(t)) \leq \\
&\leq \sum_{c=1}^{m-1} \rho^2_{nc}(\#\{\mathbf{i},\mathbf{j} \in J_n : i_m = j_m, \#\{(i_1, \dots, i_m) \cap (j_1, \dots, j_m)\} = c\}) \\
&= O\left(\sum_{c=1}^{m-1} \rho^2_{nc} n^{2m-c}\right) \to 0, \text{ as } n \to \infty.
\end{aligned}$$

This means that $A_n(t) \xrightarrow{P} \lambda t^m$, and hence $A'_n(t) \xrightarrow{P} \lambda t^m$, as $n \to \infty$. This relation completes the proof of Theorem 5.6.2.

5.7 Weak convergence of *U*-processes

Empirical processes. Consider *UR*-statistics (1.1.1) for $m = 1$, i.e.

$$U_n(\Phi) = \frac{1}{n}\sum_{i=1}^{n} \Phi(X_i),$$

Let $P_n := n^{-1}(\delta_{X_1} + \dots + \delta_{X_n})$ denote the empirical measure of the first n observations $X_1, \dots, X_n$. Suppose that $\mathcal{F}$ is a collection of real-valued measurable functions defined on X, and write $P\Phi$ for $\int \Phi\, dP$. Thus, for $\Phi \in \mathcal{F}$,

$$P_n\Phi = \int \Phi\, dP_n = \frac{1}{n}\sum_{i=1}^{n} \Phi(X_i)$$

and

$$U_n(\Phi) = P_n\Phi, \qquad \Phi \in \mathcal{F}.$$

Hence, if $m = 1$, then *U*-process $(U_n(\Phi), \mathcal{F})$ can be viewed as an empirical process indexed by a class of functions $\mathcal{F}$.

Let P be a fixed probability measure on the measurable space $(X, \mathfrak{X})$. Let G_P be a P-Brownian bridge process, G_P is a Gaussian process indexed by $L^2(X, P)$ with zero mean and covariance

$$\mathrm{Cov}(G_P(f), G_P(g)) = P(fg) - PfPg \quad \text{for all} \quad f, g \in L^2(X, P).$$

We let

$$\rho_P^2(f, g) = E(G_P(f) - G_P(g))^2 = \mathrm{Var}_P(f(X) - g(X)),$$

where X has distribution P.

Of course for any finite collection $\{ \Phi_1, \ldots, \Phi_k \} \subset L^2(X, P)$ the weak convergence

$$(\sqrt{n}(U_n(\Phi_i) - P\Phi_i) : i \leq k) \Rightarrow (G_P(\Phi_i) : i \leq k)$$

takes place as $n \to \infty$. But a main concern in empirical process theory is the central limit theorem for larger collections $\mathcal{F} \subset L^2(X, P)$, for a fixed P.

In modern empirical process theory to get limit theorems for $(U_n(\Phi), \mathcal{F})$, which are uniform over all $\Phi \in \mathcal{F}$ we need to solve two potential problems. First, the functions in $\mathcal{F}$ should not be "too big pointwise". Second, there should not be "too many of them". To make these more precise, we define two more notions: the envelope function F of $\mathcal{F}$, and the entropy of $\mathcal{F}$.

Define a function f by

$$f(x) := \sup_{\Phi \in \mathcal{F}} |\Phi(x)| \quad \text{for} \quad x \in X.$$

Thus, $f(x) \geq |\Phi(x)|$ for all $x \in X$ and all $\Phi \in \mathcal{F}$. Since f is not necessarily measurable, we let F be the smallest measurable function above f, denoted by f^*. Thus

$$F(x) = f^*(x) = \left(\sup_{\Phi \in \mathcal{F}} |\Phi(x)| \right)^*$$

is the least measurable envelope of $\mathcal{F}$.

The second notion is that of an entropy number for the size of $\mathcal{F}$. Viewing $\mathcal{F}$ as a subset of a metric space (E, ρ), for any $\varepsilon > 0$ consider covering $\mathcal{F}$ with ε balls $B(\Phi_j, \varepsilon)$ with centers Φ_j in $\mathcal{F}$. Let $N(\varepsilon, \mathcal{F}, \rho)$ be the number of such balls required to cover $\mathcal{F}$. Thus $\mathcal{F} \subset \bigcup_{j=1}^m B(\Phi_j, \varepsilon)$ for some $\Phi_j \in \mathcal{F}$ with $m = N(\varepsilon, \mathcal{F}, \rho)$.

We assume that $\sup_{\Phi \in \mathcal{F}} |U_n(\Phi) - P\Phi| < \infty$ for all $x \in X$. Then we view $U_n(\Phi) - P\Phi$ as an element in $l^\infty(\mathcal{F})$, the space of all bounded functions from $\mathcal{F}$ into R. Let $C_u(\mathcal{F}, \rho_P)$ denote the set of all functions f in $l^\infty(\mathcal{F})$ that are uniformly continuous with respect to ρ_P.

Now let us give a heuristic summary of the two types of limit theorems, which are studied in empirical process theory (see, e.g. Wellner (1992)).

SLLN or Glivenko theorem. *If an "appropriate" entropy number $N(\varepsilon, \mathcal{F}, \rho)$ is finite for every $\varepsilon > 0$ and $\mathcal{F}$ has integrable envelope function F, then*

$$\sup_{\Phi \in \mathcal{F}} |U_n(\Phi) - P\Phi| \to 0 \quad a.s.$$

as $n \to \infty$.

CLT or Donsker theorem. *If an "appropriate" entropy number $N(\varepsilon, \mathcal{F}, \rho)$ satisfies the condition of the integrability*

$$\int_0^1 \{\ln N(u, \mathcal{F}, \rho)\}^{1/2} \, du < \infty$$

and the envelope function F of $\mathcal{F}$ is square integrable, then the weak convergence

$$\sqrt{n}(U_n(\Phi) - P\Phi) \Rightarrow G_P$$

is valid as $n \to \infty$, where G_P is a P-Brownian bridge process with ρ_P-continuous sample paths and G_P takes values in $C_u(\mathcal{F}, \rho_P)$, a subcollection of $l^\infty(\mathcal{F})$ consisting of all (bounded and) ρ_P-uniformly continuous functions on $\mathcal{F}$.

Here the weak convergence $\Rightarrow$ is convergence in distribution in $l^\infty(\mathcal{F})$ in the sense of Hoffman-Jørgensen (1984) (see, for example, Anderson and Dobrić (1987), Giné and Zinn (1986)):

$$E^* f(\sqrt{n}(U_n(\Phi) - P\Phi)) \to Ef(G_P) \quad \text{for all} \quad f \in C_b(l^\infty(\mathcal{F})),$$

where E^* denotes the upper (or outer) integral computed under $\Pr_P$, i.e. $E^* f = P f^*$. In addition f^* is the measurable support of f, and $C_b(l^\infty(\mathcal{F}))$ is the set of all bounded $\|\cdot\|_{\mathcal{F}}$-continuous real-valued functions on $l^\infty(\mathcal{F})$ with $\|f\|_{\mathcal{F}} = \sup_{f \in \mathcal{F}} |f|$.

Most of the developed central limit theorems for general empirical processes are for a fixed underlying probability measure P. A class $\mathcal{F}$ of measurable functions on a probability space $(X, \mathfrak{X}, P)$ is called a P-Donsker class if the empirical processes $\sqrt{n}(U_n(\Phi) - P\Phi)$ converge weakly to a P-Brownian bridge G_P having bounded uniformly continuous paths almost surely. Dudley (1987) investigated classes of functions $\mathcal{F}$ for which the central limit theorem holds for all probability measures P on $(X, \mathfrak{X})$ and he calls such classes universal Donsker classes. Giné and Zinn (1991) studied classes $\mathcal{F}$ for which the central limit theorem holds uniformly in all P on $(X, \mathfrak{X})$ and they called such classes uniform Donsker classes. But requiring convergence uniformly in all P is too restrictive. Sheehy and Wellner (1992) have analyzed some fixed subcollection $\mathcal{P}$ of all probability measures P on $(X, \mathfrak{X})$ for which the convergence holds uniformly over P in this subcollection. They call such a class of functions $\mathcal{F}$ a $\mathcal{P}$-uniform Donsker class.

For the first time Nolan and Pollard (1987,1988) investigated the U-process of degree $m = 2$. Their study parallels that of empirical processes. Further study of the limit theory of U-processes was done by Arcones and Giné (1993b), without restriction to the case $m = 2$. They also follow patterns from empirical process theory. In this new situation they used additional interesting techniques including a decoupling inequality (de la Peña (1992)) and integrability properties of Gaussian and Bernoulli chaos, based on hypercontractivity (Bonami (1970), Borell (1979)).

A further goal is to present some results of Arcones and Giné (1993b) about the U-processes of degree m with emphasis on laws of large numbers (Glivenko theorems) and central limit theorems (Donsker theorems).

Law of large numbers for U-processes. First we define some random entropies that will be applied in the following. With given pseudometric space $(\mathcal{F}, \rho)$ the ε-covering number of $(\mathcal{F}, \rho)$ is

$$N(\varepsilon, \mathcal{F}, \rho) = \min\left\{ n : \exists \Phi_1, \dots, \Phi_n \in \mathcal{F} \text{ such that } \sup_{\Phi \in \mathcal{F}} \min_{i \le n} \rho(\Phi, \Phi_i) \le \varepsilon \right\}.$$

Then we define $N_{np}(\varepsilon, \mathcal{F})$ as the random ε-covering numbers of $(\mathcal{F}, \rho)$ with

$$\rho_{np}(f, g) = (U_n(|f - g|^p))^{(1/p)\wedge 1}.$$

In addition $\mathcal{F}$ is a class of real functions on X^m and $X_1, X_2, \dots$ are i.i.d. random variables. If $X_1^j, X_2^j, \dots,\ j \ge 1$, are i.i.d. copies of $X_1, X_2, \dots$, then $N_{np}^{\text{dec}}(\varepsilon, \mathcal{F})$ is the random ε-covering number of $(\mathcal{F}, \rho_{np}^{\text{dec}})$ where

$$\rho_{np}^{\text{dec}}(f - g) = \left(n^{-[m]} \sum_{(i_1, \dots, i_m) \in I_n^m} |(f - g)(X_{i_1}^1, \dots, X_{i_m}^m)|^p \right)^{(1/p)\wedge 1}$$

We also define the distances

$$\tilde{\rho}_{np}(f, g) = \left(\frac{1}{n} \sum_{i_1 = 1}^{n} \left| (n-1)^{-[m-1]} \sum_{(i_2, \dots, i_m) : (i_1, \dots, i_m) \in I_n^m} (f - g)(X_{i_1}, \dots, X_{i_m}) \right|^p \right)^{(1/p)\wedge 1}$$

for $0 < p < \infty$ and connected covering numbers $\tilde{N}_{np}(\varepsilon, \mathcal{F})$, and the same distances for the decoupled statistics, $\tilde{\rho}_{np}^{\text{dec}}(f, g)$ and $\tilde{N}_{np}^{\text{dec}}(\varepsilon, \mathcal{F})$.

Theorem 5.7.1. *Let $\mathcal{F}$ be a measurable class of symmetric functions on X^m and let P be a probability measure on $(X, \mathfrak{X})$ such that $P^m F < \infty$. Then the following conditions are equivalent:*

(i) $\|U_n - P^m\|_{\mathcal{F}} \to 0$ *a.s.*,

(ii) $E\left\|n^{-m}\sum_{(i_1,\dots,i_m)\in I_n^m}\varepsilon_{i_1}\Phi(X_{i_1},\dots,X_{i_m})\right\|_{\mathcal{F}}\to 0$,

(iii) $n^{-1}\ln\tilde{N}_{n1}(\varepsilon,\mathcal{F})\to 0$ *in probability* * *for all* $\varepsilon>0$.

Proof. Note that $\|U_n-P^m\|_{\mathcal{F}}$ is a reverse submartingale. Hence, (i) is equivalent to $E\|U_n-P^m\|_{\mathcal{F}}\to 0$ by Doob's reversed submartingale limit theorem (e.g. Dudley (1989)). We first prove (ii) → (i). In what follows c will denote a constant that may vary from line to line. Write

$$E\|U_n-P^m\|\mathcal{F}\le$$

[by the decoupling inequality]

$$\le cn^{-[m]}E\left\|\sum_{(i_1,\dots,i_m)\in I_n^m}(\Phi(X_{i_1}^1,\dots,X_{i_m}^m)-P^m\Phi)\right\|_{\mathcal{F}}$$

[next $\{Y_i^j\}_{i=1}^n$ is an independent copy of $\{X_i^j\}_{i=1}^n$]

$$\le cn^{-[m]}E\left\|\sum_{(i_1,\dots,i_m)\in I_n^m}(\Phi(X_{i_1}^1,\dots,X_{i_m}^m)-\Phi(Y_{i_1}^1,\dots,Y_{i_m}^m))\right\|_{\mathcal{F}}$$

[by the triangle inequality]

$$\le cn^{-[m]}\sum_{j=1}^m E\Bigg\|\sum_{(i_1,\dots,i_m)\in I_n^m}(\Phi(X_{i_1}^1,\dots,X_{i_j}^j,Y_{i_{j+1}}^{j+1},\dots,Y_{i_m}^m)-$$
$$-\Phi(X_{i_1}^1,\dots,X_{i_{j-1}}^{j-1},Y_{i_j}^j,\dots,Y_{i_m}^m))\Bigg\|_{\mathcal{F}}$$

[if $\{\varepsilon_i\}$ is a Bernoulli sequence independent of $\{X_i^j\}$]

$$\le cn^{-[m]}E\left\|\sum_{(i_1,\dots,i_m)\in I_n^m}\varepsilon_{i_1}\Phi(X_{i_1}^1,\dots,X_{i_m}^m)\right\|_{\mathcal{F}}$$

[by Jensen's inequality on the ε_i^j's]

$$\le cn^{-[m]}E\Bigg\|\sum_{(i_1,\dots,i_m)\in I_n^m}\Big(\varepsilon_{i_1}^1\Phi(X_{i_1}^1,\dots,X_{i_m}^m)+$$
$$+\dots+\varepsilon_{i_m}^m\Phi(X_{i_1}^1,\dots,X_{i_m}^m)\Big)\Bigg\|_{\mathcal{F}}$$

[by the decoupling inequality]

$$\le cn^{-[m]}E\left\|\sum_{(i_1,\dots,i_m)\in I_n^m}(\varepsilon_{i_1}\Phi(X_{i_1},\dots,X_{i_m})+\dots+\varepsilon_{i_m}\Phi(X_{i_1},\dots,X_{i_m}))\right\|_{\mathcal{F}}$$

$$\leq cn^{-[m]} E \left\| \sum_{(i_1,\dots,i_m)\in I_n^m} \varepsilon_{i_1} \Phi(X_{i_1},\dots,X_{i_m}) \right\|_{\mathcal{F}} .$$

Hence, (ii) → (i). Further, we show (i) → (ii). For simplicity we consider the case $m = 2$. Taking into account the notations in (1.1.6) we write

$$\begin{aligned} E \left\| n^{-2} \sum_{1\leq i\neq j\leq n} \varepsilon_i \Phi(X_i, X_j) \right\|_{\mathcal{F}} &\leq \\ \leq E \left\| n^{-2} \sum_{1\leq i\neq j\leq n} \varepsilon_i P^2\Phi \right\|_{\mathcal{F}} &+ E \left\| n^{-2} \sum_{1\leq i\neq j\leq n} \varepsilon_i \Phi_1(X_i) \right\|_{\mathcal{F}} \\ + E \left\| n^{-2} \sum_{1\leq i\neq j\leq n} \varepsilon_i \Phi_1(X_j) \right\|_{\mathcal{F}} &+ E \left\| n^{-2} \sum_{1\leq i\neq j\leq n} \varepsilon_i g_2(X_i, X_j) \right\|_{\mathcal{F}} \\ \leq n^{-1/2} \|P^2\Phi\|_{\mathcal{F}} &+ 2E \left\| n^{-2} \sum_{1\leq i\neq j\leq n} \varepsilon_i \Phi_1(X_i) \right\|_{\mathcal{F}} \\ + n^{-1/2} E \left\| n^{-1} \sum_{j=1}^{n} \Phi_1(X_j) \right\|_{\mathcal{F}} &+ E \left\| n^{-2} \sum_{1\leq i\neq j\leq n} \varepsilon_i g_2(X_i, X_j) \right\|_{\mathcal{F}} . \end{aligned}$$

By symmetrization, convexity and decoupling, and again by decoupling, Jensen's inequality and the definition of g_2 in (1.1.6) we have

$$\begin{aligned} E \left\| n^{-2} \sum_{1\leq i\neq j\leq n} \varepsilon_i g_2(X_i, X_j) \right\|_{\mathcal{F}} &\leq \\ \leq cE \left\| n^{-2} \sum_{1\leq i\neq j\leq n} \varepsilon_i g_2(X_i, X'_j) \right\|_{\mathcal{F}} & \\ \leq cE \left\| n^{-2} \sum_{1\leq i\neq j\leq n} (\Phi(X_i, X^1_j) - P^2\Phi) \right\|_{\mathcal{F}} & \\ \leq cE \left\| n^{-2} \sum_{1\leq i\neq j\leq n} (\Phi(X_i, X_j) - P^2\Phi) \right\|_{\mathcal{F}} & \end{aligned}$$

(ii) is proved.

(ii) → (iii). We will apply Corollary 4.14 in Ledoux and Talagrand (1991) to the process $n^{1/2} n^{-[m]} \sum_{I_n^m} \varepsilon_{i_1} \Phi(X_{i_1},\dots,X_{i_m})$, $\Phi \in \mathcal{F}$, with the $\tilde{\rho}_{n2}$ distance. Then we obtain

$$\begin{aligned} \varepsilon(\ln \tilde{N}_{n2}(\varepsilon, \mathcal{F}))^{1/2} \leq n^{1/2} c \left(E_\varepsilon \left\| n^{-[m]} \sum_{I_n^m} \varepsilon_{i_1} \Phi(X_{i_1},\dots,X_{i_m}) \right\|_{\mathcal{F}} \right) \times \\ \times \left(\ln \left(2 + 1/E_\varepsilon \left\| n^{-[m]} \sum_{I_n^m} \varepsilon_{i_1} \Phi(X_{i_1},\dots,X_{i_m}) \right\|_{\mathcal{F}} \right) \right)^{1/2} . \end{aligned}$$

Since by (ii), $E_\varepsilon \left\| n^{-[m]} \sum_{I_n^m} \varepsilon_{i_1} \Phi(X_{i_1}, \ldots, X_{i_m}) \right\|_{\mathcal{F}} \to 0$ in probability, and $\tilde\rho_{n1} \le \tilde\rho_{n2}$, (iii) is proved.

(iii) $\to$ (ii). Let $\mathcal{F}^t = \{ \Phi I(F \le t) : \Phi \in \mathcal{F} \}$. Since

$$E \left\| n^{-[m]} \sum_{I_n^m} \varepsilon_{i_1} \Phi(X_{i_1}, \ldots, X_{i_m}) \right\|_{\mathcal{F}} \le$$
$$\le E \left\| n^{-[m]} \sum_{I_n^m} \varepsilon_{i_1} (\Phi I(F \le t))(X_{i_1}, \ldots, X_{i_m}) \right\|_{\mathcal{F}} + P^m F I(F \ge t)$$

and $P^m F I(F \ge t) \to 0$ as $t \to \infty$, it suffices to prove that

$$\lim_{t \to \infty} \limsup_{n \to \infty} E \left\| n^{-[m]} \sum_{I_n^m} \varepsilon_{i_1} \Phi(X_{i_1}, \ldots, X_{i_m}) \right\|_{\mathcal{F}^t} = 0 .$$

For fixed ω, given $\delta > 0$, let $\mathcal{F}^*$ be a subset of $\mathcal{F}^t$ such that $\#\mathcal{F}^* = \tilde N_{n1}(\delta, \mathcal{F}^t)$ and $\sup_{\Phi \in \mathcal{F}^t} \min_{\Phi^* \in \mathcal{F}^*} \tilde\rho_{n1}(\Phi, \Phi^*) \le \delta$. Then, by inequality (2.14) in Arcones and Giné (1993b)

$$n^{-[m]} E_\varepsilon \left\| \sum_{(i_1, \ldots, i_m) \in I^{m_n}} \varepsilon_{i_1} \Phi(X_{i_1}, \ldots, X_{i_m}) \right\|_{\mathcal{F}^t} \le$$
$$\le \delta + ct (\ln \tilde N_{n1}(\delta, \mathcal{F}^t))^{1/2} n^{-[m]} \times$$
$$\times \max_{\Phi \in \mathcal{F}^*} \left[\sum_{i_1} \left(\sum_{(i_2, \ldots, i_m):(i_1, \ldots, i_m) \in I_m^m} \Phi(X_{i_1}, \ldots, X_{i_m}) \right)^2 \right]^{1/2} \le$$
$$\le \delta + c(n^{-1} \ln \tilde N_{n1}(\delta, \mathcal{F}^t))^{1/2}.$$

Note that $\tilde\rho_{n1}(f I(F \ge t), g I(F \ge t)) \le \tilde\rho_{n1}(f, g) + 2U_n(F I(F \le t))$, and if $U_n(F I(F \ge t)) \le \delta/4$, then $\tilde N_{n1}(\delta, \mathcal{F}^t) \le \tilde N_{n1}(\delta/2, \mathcal{F})$. Hence

$$\Pr{}^* \{ n^{-1} \ln \tilde N_{n1}(\delta, \mathcal{F}^t) \ge \varepsilon \} \le$$
$$\le \Pr{}^* \{ n^{-1} \ln \tilde N_{n1}(\delta/2, \mathcal{F}^t) \ge \varepsilon \} + \Pr\{ U_n(F I(F \ge t)) > \delta/4 \} \to 0$$

if $E F I(F \ge t) < \delta/4$. Thus (ii) holds. This completes the proof of Theorem 5.7.1.

Corollary 5.7.1. *Let $\mathcal{F}$ be a measurable class and $P^m F < \infty$. Then the condition $n^{-1} \ln N_{n1}(\varepsilon, \mathcal{F}) \to 0$ in probability implies $\|U_n - P^m\|_{\mathcal{F}} \to 0$ a.s.*

Proof. It suffices to note that $\tilde\rho_{n1}(f, g) \le \rho_{n1}(f, g)$.

The condition $n^{-1} \ln N_{n1}(\varepsilon, \mathcal{F}) \to 0$ for all $\varepsilon > 0$ is stronger than the condition $n^{-1} \ln \tilde N_{n1}(\varepsilon, \mathcal{F}) \to 0$ for all $\varepsilon > 0$ (see Arcones and Giné (1993b)).

Central limit theorem for U-processes. Now we suppose that a class $\mathcal{F}$ consists of measurable functions having the rank $r = 1$. In this situation as in the case of a single function Φ, to prove the central limit theorem for $\{ U_n(\Phi) : \Phi \in \mathcal{F} \}$ with a Gaussian limit, it is necessary to prove that $\{ n^{1/2} U_n(g_1) : \Phi \in \mathcal{F} \}$ converges in $l^\infty(\mathcal{F})$ to a Gaussian process and that the value $\|n^{1/2} U_n(g_c)\|_{\mathcal{F}} \to 0$ in probability for all $2 \le c \le m$. Evidently, the first relation is equivalent to the class $\{ g_1 : \Phi \in \mathcal{F} \}$ being P-Donsker.

Denote $d_P(f,g)$ the pseudodistance on $\mathcal{F}$, given by $d_P^2(f,g) = P(P^{m-1}(f-g))^2$ and define the following class of functions $\mathcal{F}'_\delta = \{ f - g : f, g \in \mathcal{F}, d_P(f,g) \le \delta \}$ for all $\delta > 0$.

Theorem 5.7.2. *Suppose that $\mathcal{F}$ is a measurable class of symmetric functions on X^m of rank $r = 1$ such that $t^2 P(F > t) \to 0$ as $t \to \infty$. Then the following conditions are equivalent:*

(i) *$\mathcal{F}$ satisfies the central limit theorem –*

$$\{ n^{1/2}(U_n(\Phi) - P^m\Phi) : \Phi \in \mathcal{F} \} \Rightarrow \{ mG_P \circ P^{m-1}\Phi : \Phi \in \mathcal{F} \};$$

(ii) *$(\mathcal{F}, d_P)$ is totally bounded and*

$$\lim_{\delta \to 0} \limsup_{n \to \infty} E\|n^{1/2}(U_n(\Phi) - P^m\Phi)\| = 0$$

for some $0 < p < 2$;

(iii) *$(\mathcal{F}, d_P)$ is totally bounded and*

$$\lim_{\delta \to 0} \limsup_{n \to \infty} P\{ \|n^{-m+1/2} \sum_{(i_1,\dots,i_m) \in I_n^m} \varepsilon_{i_1} \Phi(X_{i_1}, \dots, X_{i_m})\|_{\mathcal{F}'_\delta} \ge \alpha \} = 0$$

for all $\alpha > 0$;

(iv) *$(\mathcal{F}, d_P)$ is totally bounded and*

$$\lim_{\delta \to 0} \limsup_{n \to \infty} E \left\| n^{-m+1/2} \sum_{(i_1,\dots,i_m) \in I_n^m} \varepsilon_{i_1} \Phi(X_{i_1}, \dots, X_{i_m}) \right\|_{\mathcal{F}'_\delta}^p = 0$$

for some $0 < p < 2$.

Proof. See Arcones and Giné (1993b).

Convergence of canonical U-processes to a chaos process. Here we assume that $\mathcal{F}$ is a class of canonical functions $\Phi : X^m \mapsto \mathrm{R}$. Given a probability measure P on $(X, \mathfrak{X})$, e_P denotes the pseudodistance on $\mathcal{F}$ given by $e_P(f,g) = \|f - g\|_{L^2(X^m, P^m)}$.

Theorem 5.7.3. *Let $\mathcal{F}$ be a measurable uniformly bounded class of canonical functions, satisfying the conditions*

$$\int_0^\infty (\ln N(\varepsilon, \mathcal{F}, e_P))^{m/2}\, d\varepsilon < \infty$$

and

$$\lim_{\delta\to 0} \limsup_{n\to\infty} P\{ \sup_{e_P(f,g)\le \delta n^{-m/2(m+1)}} |n^{-m/2} \sum_{(i_1,\ldots,i_m)\in I_n^m} (f-g)(X_{i_1},\ldots,X_{i_m})| > \alpha \} = 0$$

for all $\alpha > 0$. Then $\mathcal{F}$ satisfies the weak convergence

$$n^{m/2} U_n(\Phi) \Rightarrow K_m(\Phi) \text{ in } l^\infty(\mathcal{F})$$

as $n \to \infty$, where the limit chaos process $\{ K_m(\Phi) : \Phi \in \mathcal{F} \}$ has a version with bounded uniformly continuous paths in $(\mathcal{F}, e_P)$.

Proof. See Arcones and Giné (1993b).

The following result for *U*-processes is in Arcones (1994a).

Theorem 5.7.4. *Let $\mathcal{F}$ be a class of canonical functions, satisfying the condition*

$$\sup_{\Phi\in\mathcal{F}} |\Phi(x_1,\ldots,x_m)| < \infty \qquad P^m\text{-a.s.}$$

Then the weak convergence

$$n^{m/2} U_n(\Phi) \Rightarrow K_m(\Phi) \text{ in } l^\infty(\mathcal{F})$$

is valid, as $n \to \infty$, if and only if $(\mathcal{F}, e_P)$ is totally bounded and asymptotic equicontinuous

$$\lim_{\delta\to 0} \limsup_{n\to\infty} \Pr{}_P^*\{ \sup_{f,g\in\mathcal{F}: e_P(f,g)\le\delta} n^{m/2} |U_n(f-g)| \ge \alpha \} = 0$$

for all $\alpha > 0$.

Chapter 6

Approximation estimates

Let r be the rank of Banach-valued symmetric kernel $\Phi : X^m \to B$. We set $\eta_n = n^{r/2}(U_n - \theta)$. According to the limit theorems given in the preceding chapters, the weak convergence

$$\eta_n \Rightarrow \eta_\infty$$

as $n \to \infty$ takes place with a random element η_∞ depending on the value of r. Let $f \colon B \to R$ be a functional such that

$$\Delta_n := \sup_x |P(f(\eta_n) < x) - P(f(\eta_\infty) < x)| \to 0$$

as $n \to \infty$. Owing to weak convergence the property $\Delta_n \to 0$ takes place under fairly general conditions imposed on f. The problem of determining the rate at which the value Δ_n approaches zero with respect to n is called the problem of estimating the rate of convergence.

In this chapter we study this problem for some values of parameters B, P, Φ and f. As a function of these parameters the various approximation estimates for Δ_n are obtained.

For the analysis of the rate of Gaussian convergence we apply the *Fourier method*, or the *method of characteristic functions*, and the *method of compositions* (see, e.g. Sazonov (1981)). In the case of Poisson approximation we also use the *Chen–Stein method* (see, e.g. Arratia *et al.* (1990) and Barbour *et al.* (1992)).

In what follows we will give the necessary technical elements of these general methods.

6.1 General methods of estimation

The method of characteristic functions.

Lemma 6.1.1. *Let F and G be distribution functions, and let f and g be the corresponding characteristic functions. Suppose that $G(x)$ satisfies the Lipschitz* *condition with a constant* $c > 0$. *Then for every* $2^{-1} < u < 1$ *there exists* $v(u) > 0$ *such that*

$$\sup_x |F(x) - G(x)| \leq$$
$$\leq (2u-1)^{-1}\Big(\frac{1}{2\pi}\int_{-T}^{T} |f(t) - g(t)|\,|t|^{-1}\,dt + 2u\,v(u)\,cT^{-1}\Big) \quad (6.1.1)$$

for all $T > 0$.

Proof. See, e.g. Sazonov (1981).

Inequality (6.1.1) allows us to estimate the closeness of distribution functions by using the closeness of corresponding characteristic functions. In fact, Lemma 6.1.1 is valid in a more general situation. We formulate one of these versions which belong to Esséen.

Lemma 6.1.2. *Suppose that* $F(x)$ *is a nondecreasing function,* $G(x)$ *is a function of bounded variation,* $F(-\infty) = G(-\infty)$, *and* $F(+\infty) = G(+\infty)$. *Let* $f(t)$ *and* $g(t)$ *be the Fourier–Stieltjes transforms associated with these functions. Assume that* $\sup_x |G'(x)| \leq c$. *Then for every* $T > 0$ *we have*

$$\sup_x |F(x) - G(x)| \leq \pi^{-1}\int_{-T}^{T} |f(t) - g(t)|\,|t|^{-1}\,dt + 24\,\pi^{-1}cT^{-1}. \quad (6.1.2)$$

Inequalities (6.1.1) and (6.1.2) are uniform with respect to $x \in R$ and are called uniform inequalities. There also exist nonuniform inequalities which estimate the decrease in the value $|F(x) - G(x)|$ with respect to x.

Lemma 6.1.3. *Assume that* $F(x)$ *is a nondecreasing function,* $G(x)$ *is a function of bounded variation,* $F(-\infty) = G(-\infty)$, *and* $F(+\infty) = G(+\infty)$. *Let* $f(t)$ *and* $g(t)$ *be the Fourier–Stieltjes transform associated with these functions. Suppose also that*

$$\int_{-\infty}^{\infty} |x|^s\,|d(F(x) - G(x))| < \infty$$

for all $x \in R$, *and*

$$|G'(x)| \leq c\,(1 + |x|)^{-s}$$

for some integer $s \geq 2$ *and constant* $c > 0$.

Then, for every $x \in R$ *and* $T > 1$, *we have*

$$|F(x) - G(x)| \leq d\,(1 + |x|)^{-s}\Big(\int_{-T}^{T} |f(t) - g(t)|\,|t|^{-1}\,dt +$$
$$+ \int_{-T}^{T}\Big|\frac{d^s}{dt^s}((f(t) - g(t))\,t^{-1})\Big|\,dt + cT^{-1}\Big), \quad (6.1.3)$$

where $d > 0$ is a constant.

The method of compositions. Let F and G be arbitrary distributions (probability measures), and let F^n and G^n be their n-fold convolutions. The method of compositions is based on the identity

$$F^n - G^n = \sum_{k=1}^{n} F^{n-k}G^{k-1}(F-G), \tag{6.1.4}$$

where the products are regarded as convolutions of distributions. We represent each term on the right-hand side of (6.1.4) as a sum

$$F^{n-k}G^{k-1}(F-G) = (F^{n-k} - G^{n-k})G^{k-1}(F-G) + G^{n-1}(F-G), \qquad k = 1,\dots,n,$$

and then again apply identity (6.1.4) to the difference $F^{n-k} - G^{n-k}$, repeating these representations so far as necessary.

The "characteristic version" of the method of compositions. Let $S_n = X_1 + \ldots + X_n$ and $T_n = Y_1 + \cdots + Y_n$, where (X_i) and (Y_i) are two arbitrary sequences of random variables. Assume that $F_n(x)$ and $G_n(x)$ are the distribution functions of S_n and T_n, respectively, $f_n(t) = E\exp(itS_n)$, and $g_n(t) = E\exp(itT_n)$, $t \in R$. According to the inequalities (6.1.1)–(6.1.3), in order to estimate the variable $|F_n(x) - G_n(x)|$, it is helpful to know the estimates of the difference $f_n(t) - g_n(t)$ in the region $|t| \le T$. Good estimates can be obtained when the random variables $X_1,\dots,X_n$ ($Y_1,\dots,Y_n$, respectively) are independent, because in this situation we have

$$f_n(t) = \prod_{j=1}^{n} E\exp(itX_j),$$

and the product on the right-hand side can be estimated quite precisely. In the case when the variables $X_1,\dots,X_n$ are not independent, there is no such formula for $f_n(t)$. Therefore, in general situation the difference $f_n(t) - g_n(t)$ can be estimated with the help of the representation

$$f_n(t) - g_n(t) = \sum_{k=1}^{n} E\{e^{itz_{nk}}(e^{itX_k} - e^{itY_k})\}, \tag{6.1.5}$$

where $z_{nk} = Y_1 + \cdots + Y_{k-1} + X_{k+1} + \cdots + X_n$. Formula (6.1.5) is the "characteristic version" of (6.1.4), and it gives us a possibility to estimate the closeness of characteristic functions $f_n(t)$ and $g_n(t)$ in terms of the closeness of the moments of individual elements X_k and Y_k in the sums S_n and T_n.

The permanent variant of the composition method can be found in detail in Borovskikh and Korolyuk (1994b).

The Chen–Stein method. Stein (1970) introduced a technique for obtaining rates of convergence in the neighborhood of a normal limit. Stein's technique was free of the Fourier method. His technique is a method of estimating the discrepancy between the expectation of a function f of a random variable w and the expectation of the same function f of a standard normal random variable τ. The core idea of the technique is to find a function g such that

$$f(x) - Ef(\tau) = g'(x) - x\, g(x),$$

and then to use that structure of w which makes a normal limit plausible to estimate

$$E\{g'(w) - w\, g(w)\}$$

and hence

$$Ef(w) - Ef(\tau).$$

This technique was described in detail in Stein (1986).

Chen (1975) applied Stein's ideas in the Poisson setting. That a random variable ν has a Poisson distribution π_λ with mean λ is characterized by the recursion

$$\lambda\, P(\nu = k-1) = k\, P(\nu = k), \qquad k = 0, 1, 2, \ldots .$$

The basis of the Chen–Stein method for obtaining the Poisson approximation is to extend this recursion from indicator functions, $g(x) = I(x = k)$, to arbitrary bounded functions $g\colon Z^+ \to R$. Thus ν has distribution π_λ if and only if for all bounded g

$$E(\lambda\, g(\nu + 1) - \nu\, g(\nu)) = 0. \tag{6.1.6}$$

If $E(\lambda\, g(w+1) - w\, g(w))$ is sufficiently small for a suitable set of functions g, then the random variable w is close in distribution to π_λ. The mechanism for carrying this out is to write (6.1.6) as $Ef(w) = 0$, where f and g are related by Stein's equation,

$$f(k) = \lambda\, g(k+1) - k\, g(k), \qquad k = 1, 2, \ldots . \tag{6.1.7}$$

Given $f = (f(0), f(1), \ldots)$ there is a unique $g = (g(1), g(2), \ldots)$ which satisfies (6.1.7), that is

$$g(j) = \frac{E(f(\nu)\, I(\nu < j))}{\lambda\, P(\nu = j-1)},$$

where ν is a random variable within the distribution π_λ. For a given set A, let $g_{A,\lambda}$ denote the solution of Stein's equation

$$I(j \in A) - P(\nu \in A) = \lambda\, g_{A,\lambda}(j+1) - j\, g_{A,\lambda}(j)$$

so that

$$P(w \in A) - P(\nu \in A) = E(\lambda\, g_{A,\lambda}(w+1) - w\, g_{A,\lambda}(w)). \tag{6.1.8}$$

The total variation distance between w and ν is defined as

$$d_{TV}(w, \nu) = \sup_A |P(w \in A) - P(\nu \in A)|.$$

From (6.1.8) it follows that this distance is given by

$$d_{TV}(w, \nu) = \sup_A \Big| E(\lambda\, g_{A,\lambda}(w+1) - w\, g_{A,\lambda}(w)) \Big|. \tag{6.1.9}$$

Barbour and Eagleson (1983) proved that uniformly in A,

$$\|g_{A,\lambda}\| = \sup_{k \geq 1} |g_{A,\lambda}(k)| \leq \min(1,\, 1.4\, \lambda^{-1/2})$$

and

$$\|\Delta g_{A,\lambda}\| = \sup_{k \geq 1} |g_{A,\lambda}(k+1) - g_{A,\lambda}(k)| \leq \lambda^{-1}\,(1 - e^{-\lambda}). \tag{6.1.10}$$

Now, using (6.1.10), we can estimate the expression on the right-hand side of (6.1.9).

6.2 Rate of normal approximation of *UR*-statistics

Denote $F(x) = P(\tau < x)$, $x \in R$, where τ is a standard normal random variable.

Theorem 6.2.1. *Suppose that the kernel* $\Phi\colon X^2 \to R$ *satisfies the conditions*

$$E\,\Phi(X_1, X_2) = 0, \quad \sigma_1 > 0, \quad E\,|g_1(X_1)|^3 < \infty, \quad E\,|g_2(X_1, X_2)|^{5/3} < \infty.$$

Then for all $n \geq 2$ *the following estimate holds*

$$\sup_x \left| P\left(\frac{\sqrt{n}}{2\,\sigma_1}\, U_n < x \right) - F(x) \right| \leq (c_1 a + c_2 b)\, n^{-1/2} \tag{6.2.1}$$

where

$$a = \sigma_1^{-3}\, E\,|g_1(X_1)|^3, \qquad b = \sigma_1^{-5/3}\, E\,|g_2(X_1, X_2)|^{5/3}$$

and c_i, $i = 1, 2$, *are the absolute positive constants satisfying the inequalities*

$$c_1 \leq 8 + \frac{2^{11} \cdot 3^6}{\pi}, \qquad c_2 \leq 18 + \frac{2^{11} \cdot 3^9}{5\pi}.$$

Proof. For the sake of simplicity we assume that $n \geq a^4$. We apply martingale considerations to the random variable

$$S_{n2} = \sum_{i=1}^{n-1} \sum_{j=i+1}^{n} g_2(X_i, X_j)$$

in the following form. For $i = 1, 2, \dots, k$ and $k = 1, 2, \dots, n-1$, we define

$$\zeta_i = \sum_{j=i+1}^{n} g_2(X_i, X_j), \qquad \delta_n'(k) = \sum_{i=1}^{k} \sum_{j=i+1}^{n} g_2(X_i, X_j).$$

According to the property of degeneracy, we get

$$E\zeta_k = 0, \qquad E(\zeta_i \mid \zeta_{i+1}, \dots, \zeta_k) = 0,$$

if $i = 1, 2, \dots, k-1$ and for all $j = i+1, i+2, \dots, n$ we have

$$E\left(g_2(X_i, X_j) \mid X_i, g_2(X_i, X_{j+1}), \dots, g_2(X_i, X_n)\right) = 0.$$

If $1 \leq p \leq 2$, then by Corollary 2.2.1, we obtain

$$E\,|\delta_n'(k)|^p \leq 2^{2-p} \sum_{i=1}^{k} E\,|\zeta_i|^p.$$

By virtue of martingale properties, we can apply Corollary 2.2.1 to the mathematical expectation on the right-hand side once more. As a result, we get the estimate

$$E\,|\delta_n'(k)|^p \leq 2^{4-2p}\, k\,(n-1)\, E\,|g_2(X_1, X_2)|^p. \tag{6.2.2}$$

For $k = 1, 2, \dots, n-1$, we denote

$$\delta_n''(k) = \sum_{i=k+1}^{n-1} \sum_{j=i+1}^{n} g_2(X_i, X_j).$$

Obviously, the variable S_{n2} may be rewritten as $S_{n2} = \delta_n'(k) + \delta_n''(k)$ for all $k = 1, 2, \dots, n-1$.

Denote the characteristic function of a random variable $\sqrt{n}\,(2\sigma_1)^{-1}\,U_n$ by $\varphi_n(t) = E\exp(it\sqrt{n}\,(2\sigma_1)^{-1}U_n)$. Let $T_n = a^{-1}\sqrt{n}$. According to the Esséen inequality, we have

$$\sup_x |P((2\sigma_1)^{-1}\sqrt{n}\,U_n < x) - F(x)| \leq$$
$$\leq \frac{1}{\pi} \int_{-T_n}^{T_n} \left| \frac{\varphi_n(t) - e^{-t^2/2}}{t} \right| dt + \frac{24}{\pi\sqrt{2\pi}\,T_n}. \tag{6.2.3}$$

Let us estimate the integrand on the right-hand side of (6.2.3). Clearly,

$$|\varphi_n(t) - e^{-t^2/2}| \leq \left| \left(\varphi\left(\frac{t}{\sigma_1\sqrt{n}} \right) \right)^n - e^{-t^2/2} \right| +$$

$$+\left|E\exp\Big(\frac{it}{\sigma_1\sqrt{n}}\,S_n+\frac{it}{\sigma_1\sqrt{n}\,(n-1)}\,S_{n2}\Big)-E\exp\Big(\frac{it}{\sigma_1\sqrt{n}}\,S_n\Big)\right|, \quad (6.2.4)$$

where $\varphi(t) = E\exp(itg_1(X_1))$ is the characteristic function of a random variable $g_1(X_1)$, and $S_n = g_1(X_1) + \cdots + g_1(X_n)$. In the region $|t| \le T_n$,

$$\left|\Big(\varphi\Big(\frac{t}{\sigma_1\sqrt{n}}\Big)\Big)^n - e^{-t^2/2}\right| \le \frac{5}{12}\,\frac{a}{\sqrt{n}}\,|t|^3 e^{-t^2/12}. \quad (6.2.5)$$

With the help of the inequality

$$|e^{ix}-1-ix| \le \frac{3}{5}\,2^{1/3}\,|x|^{5/3},$$

which is valid for all real x, we obtain

$$\begin{aligned}\varphi_n(t) &= E\exp\Big(\frac{it}{\sigma_1\sqrt{n}}\,S_n+\frac{it}{\sigma_1\sqrt{n}\,(n-1)}\,S_{n2}\Big) =\\ &= E\exp\Big(\frac{it}{\sigma_1\sqrt{n}}\,S_n+\frac{it}{\sigma_1\sqrt{n}\,(n-1)}\,\delta_n''(k)\Big) +\\ &+\frac{it}{\sigma_1\sqrt{n}\,(n-1)}\,E\Big\{\delta_n'(k)\exp\Big(\frac{it}{\sigma_1\sqrt{n}}\,S_n+\frac{it}{\sigma_1\sqrt{n}\,(n-1)}\,\delta_n''(k)\Big)\Big\}+R_n.\end{aligned} \quad (6.2.6)$$

For R_n, by virtue of inequality (6.2.2) we have

$$\begin{aligned}|R_n| &\le \frac{3}{5}\,2^{1/3}\Big(\frac{|t|}{\sigma_1\sqrt{n}\,(n-1)}\Big)^{5/3} E\,|\delta_n'(k)|^{5/3}\\ &\le \frac{6}{5}\,\sigma_1^{-5/3}\,E\,|g_2(X_1,X_2)|^{5/3}\,k\,n^{-5/6}(n-1)^{-2/3}\,|t|^{5/3},\end{aligned} \quad (6.2.7)$$

here, $k = 1, 2, \ldots, n-1$, $n = 2, 3, \ldots$.

We set $k = n-1$ in (6.2.6). Then

$$\varphi_n(t) = E\exp\Big(\frac{it}{\sigma_1\sqrt{n}}\,S_n\Big)+\frac{it}{\sigma_1\sqrt{n}\,(n-1)}\,E\Big(S_{n2}\exp\Big(\frac{it}{\sigma_1\sqrt{n}}\,S_n\Big)\Big)+R_n, \quad (6.2.8)$$

where R_n can be estimated with the help of (6.2.7) as follows

$$|R_n| \le \frac{6}{5}\,\sigma_1^{-5/3}\,E\,|g_2(X_1,X_2)|^{5/3}\,n^{-1/2}\,|t|^{5/3}. \quad (6.2.9)$$

In view of the properties of degeneracy and independence of variables, we get

$$\begin{aligned}E\Big(S_{n2}\exp\Big(\frac{it}{\sigma_1\sqrt{n}}\,S_n\Big)\Big) &= \sum_{r=1}^{n-1}\sum_{j=r+1}^{n} E\Big(g_2(X_r,X_j)\exp\Big(\frac{it}{\sigma_1\sqrt{n}}\,S_n\Big)\Big)\\ &= \frac{n(n-1)}{2}\Big(\varphi\Big(\frac{t}{\sigma_1\sqrt{n}}\Big)\Big)^{n-2} E\Big(g_2(X_1,X_2)\exp\Big(\frac{it}{\sigma_1\sqrt{n}}\,(g_1(X_1)+g_1(X_2))\Big)\Big)\\ &= \frac{n(n-1)}{2}\Big(\varphi\Big(\frac{t}{\sigma_1\sqrt{n}}\Big)\Big)^{n-2} E\Big\{g_2(X_1,X_2)\prod_{r=1}^{2}\Big(\exp\Big(\frac{it}{\sigma_1\sqrt{n}}\,g_1(X_r)\Big)-1\Big)\Big\}.\end{aligned}$$

By the property of degeneracy, we obtain

$$\Big|E\Big\{g_2(X_1,X_2)\prod_{r=1}^{2}\Big(\exp\Big(\frac{it}{\sigma_1\sqrt{n}}\,g_1(X_r)\Big)-1\Big)\Big\}\Big| \le$$

$$\le \frac{t^2}{\sigma_1^2 n}\,E\,|g_2(X_1,X_2)\,g_1(X_1)\,g_1(X_2)|. \quad (6.2.10)$$

For the expectation on the right-hand side, the following inequality takes place

$$\frac{E\,|g_2(X_1,X_2)\,g_1(X_1)\,g_1(X_2)|}{\sigma_1^3} \le \frac{2}{5}\,\frac{E\,|g_1(X_1)|^3}{\sigma_1^3} + \frac{3}{5}\,\frac{E\,|g_2(X_1,X_2)|^{5/3}}{\sigma_1^{5/3}}.$$

In the region $|t| \le T_n$ we have

$$\Big|\varphi\Big(\frac{t}{\sigma_1\sqrt{n}}\Big)\Big| \le \exp\Big(-\frac{t^2}{3n}\Big). \quad (6.2.11)$$

Combining (6.2.4), (6.2.5) and (6.2.8)–(6.2.11) in the region $|t| \le T_n$ for all $n \ge 3$, we arrive at the estimate

$$|\varphi_n(t) - e^{-t^2/2}| \le \frac{1}{\sqrt{n}}\,f_1(t), \quad (6.2.12)$$

where

$$f_1(t) = \frac{5}{12}\,a\,|t|^3\,e^{-t^2/12} + \frac{6}{5}\,b\,|t|^{5/3} + \Big(\frac{1}{5}\,a + \frac{3}{10}\,b\Big)\,|t|^3\,e^{-t^2/9}.$$

Lemma 6.2.1. *In the region*

$$1 \le |t| \le T_n$$

the following estimate is valid for all $n \ge 3$

$$|\varphi_n(t) - e^{-t^2/2}| \le \frac{1}{\sqrt{n}}\,f_2(t), \quad (6.2.13)$$

where

$$f_2(t) = \frac{6\cdot 2^{2/3}}{5}\,b\,|t|^{-1/12} + \Big(\frac{1}{5}\,a + \frac{3}{10}\,b\Big)\,|t|^3\exp\Big(-\frac{t^2}{9}\Big) +$$
$$+ \frac{5}{12}a|t|^3\exp\Big(-\frac{t^2}{12}\Big) + \Big(a + \frac{27}{5}\,b\Big)\,|t|^{5/4}\exp\Big(-\frac{1}{6}\,|t|^{1/4}\Big).$$

Proof. With the help of (6.2.6), for all $k = 1, 2, \dots, n-2$, $n = 3, 4, \dots$, we obtain the following representation

$$\varphi_n(t) - e^{-t^2/2} = \Big(\Big(\varphi\Big(\frac{t}{\sigma_1\sqrt{n}}\Big)\Big)^n - e^{-t^2/2}\Big) +$$
$$+ E\exp\Big(\frac{it}{\sigma_1\sqrt{n}}\,S_n\Big)\,\Big(\exp\Big(\frac{it}{\sigma_1\sqrt{n}\,(n-1)}\,\delta_n''(k)\Big) - 1\Big) +$$

$$+\frac{it}{\sigma_1\sqrt{n}\,(n-1)}\,E\Big\{\delta_n'(k)\exp\Big(\frac{it}{\sigma_1\sqrt{n}}\,S_n+\frac{it}{\sigma_1\sqrt{n}\,(n-1)}\,\delta_n''(k)\Big)\Big\}+$$
$$+R_n. \tag{6.2.14}$$

where R_n is estimated according to (6.2.7). The first expression in the brackets on the right-hand side can be estimated according to (6.2.5). For the second one, we have

$$\begin{aligned}
&E\exp\Big(\frac{it}{\sigma_1\sqrt{n}}\,S_n\Big)\Big(\exp\Big(\frac{it}{\sigma_1\sqrt{n}\,(n-1)}\,\delta_n''(k)\Big)-1\Big)\\
&=\Big(\varphi\Big(\frac{t}{\sigma_1\sqrt{n}}\Big)\Big)^k E\exp\Big(\frac{it}{\sigma_1\sqrt{n}}\sum_{s=k+1}^{n}g_1(X_s)\Big)\Big(\exp\Big(\frac{it}{\sigma_1\sqrt{n}\,(n-1)}\,\delta_n''(k)\Big)-1\Big)\\
&=\frac{it}{\sigma_1\sqrt{n}\,(n-1)}\Big(\varphi\Big(\frac{t}{\sigma_1\sqrt{n}}\Big)\Big)^k E\Big(\delta_n''(k)\exp\Big(\frac{it}{\sigma_1\sqrt{n}}\sum_{s=k+1}^{n}g_1(X_s)\Big)\Big)\\
&\quad+\Big(\varphi\Big(\frac{t}{\sigma_1\sqrt{n}}\Big)\Big)^k R_n',
\end{aligned} \tag{6.2.15}$$

where R_n' is estimated according to (6.2.2) as follows

$$\begin{aligned}
|R_n'|&\le\frac{3}{5}\,2^{1/3}\Big(\frac{|t|}{\sigma_1\sqrt{n}\,(n-1)}\Big)^{5/3}E\,|\delta_n''(k)|^{5/3}\\
&\le\frac{3}{5}\,\sigma_1^{-5/3}E\,|g_2(X_1,X_2)|^{5/3}\,n^{-5/6}\,(n-1)^{-5/3}\,(n-k-1)(n-k)\,|t|^{5/3}.
\end{aligned} \tag{6.2.16}$$

Further, by virtue of the property of degeneracy and the independence of the variables, we get

$$\begin{aligned}
&E\Big(\delta_n''(k)\exp\Big(\frac{it}{\sigma_1\sqrt{n}}\sum_{s=k+1}^{n}g_1(X_s)\Big)\Big)\\
&=\sum_{r=k+1}^{n-1}\sum_{j=r+1}^{n}E\Big(g_2(X_r,X_j)\exp\Big(\frac{it}{\sigma_1\sqrt{n}}\sum_{s=k+1}^{n}g_1(X_s)\Big)\Big)\\
&=\frac{(n-k-1)(n-k)}{2}\Big(\varphi\Big(\frac{t}{\sigma_1\sqrt{n}}\Big)\Big)^{n-k-2}\\
&\quad\times E\Big(g_2(X_1,X_2)\exp\Big(\frac{it}{\sigma_1\sqrt{n}}\,(g_1(X_1)+g_1(X_2))\Big)\Big)\\
&=\frac{(n-k-1)(n-k)}{2}\Big(\varphi\Big(\frac{t}{\sigma_1\sqrt{n}}\Big)\Big)^{n-k-2}\\
&\quad\times E\Big\{g_2(X_1,X_2)\prod_{r=1}^{2}\Big(\exp\Big(\frac{it}{\sigma_1\sqrt{n}}\,g_1(X_r)\Big)-1\Big)\Big\}.
\end{aligned} \tag{6.2.17}$$

In the region $|t|\le T_n$, for all $n\ge 3$ and $k=1,2,\ldots,n-2$ relations (6.2.10), (6.2.11), and (6.2.15)–(6.2.17) imply the inequality

$$\Big|E\exp\Big(\frac{it}{\sigma_1\sqrt{n}}\,S_n\Big)\Big(\exp\Big(\frac{it}{\sigma_1\sqrt{n}\,(n-1)}\,\delta_n''(k)\Big)-1\Big)\Big|\le$$

$$\le \frac{3}{5}\sigma_1^{-5/3}\, E\,|g_2(X_1,X_2)|^{5/3}\, n^{-1/2}\,|t|^{5/3}\, e^{-(\frac{t^2 k}{3n})}$$

$$+\Big(\frac{1}{5}\,\frac{E\,|g_1(X_1)|^3}{\sigma_1^3} + \frac{3}{10}\,\frac{E\,|g_2(X_1,X_2)|^{5/3}}{\sigma_1^{5/3}}\Big)\, n^{-1/2}\,|t|^3\, e^{-t^2/9}. \quad (6.2.18)$$

Let us estimate the second expectation on the right-hand side of (6.2.14). Since

$$\delta_n'(k) = \sum_{r=1}^{k}\sum_{s=r+1}^{k} g_2(X_r,X_s) + \sum_{r=1}^{k}\sum_{s=k+1}^{n} g_2(X_r,X_s),$$

we have

$$E\Big\{\delta_n'(k)\exp\Big(\frac{it}{\sigma_1\sqrt{n}}\,S_n + \frac{it}{\sigma_1\sqrt{n}\,(n-1)}\,\delta_n''(k)\Big)\Big\}$$

$$= \sum_{r=1}^{k}\sum_{s=r+1}^{k} E\Big\{g_2(X_r,X_s)\exp\Big(\frac{it}{\sigma_1\sqrt{n}}\,S_n + \frac{it}{\sigma_1\sqrt{n}\,(n-1)}\,\delta_n''(k)\Big)\Big\}$$

$$+\sum_{r=1}^{k}\sum_{s=k+1}^{n} E\Big\{g_2(X_r,X_s)\exp\Big(\frac{it}{\sigma_1\sqrt{n}}\,S_n + \frac{it}{\sigma_1\sqrt{n}\,(n-1)}\,\delta_n''(k)\Big)\Big\}$$

$$= \Sigma_1 + \Sigma_2,$$

where Σ_1 and Σ_2 denote the first and second double sum on the right-hand side of this formula, respectively.

By virtue of the property of degeneracy and the independence of the variables, we find

$$\Sigma_1 = \frac{k(k-1)}{2}\Big(\varphi\Big(\frac{t}{\sigma_1\sqrt{n}}\Big)\Big)^{k-2} E\Big\{g_2(X_1,X_2)\prod_{r=1}^{2}\Big(\exp\Big(\frac{it}{\sigma_1\sqrt{n}}\,g_1(X_r)\Big)-1\Big)\Big\}\times$$

$$\times E\Big\{\exp\Big(\frac{it}{\sigma_1\sqrt{n}}\sum_{s=k+1}^{n} g_1(X_s) + \frac{it}{\sigma_1\sqrt{n}\,(n-1)}\,\delta_n''(k)\Big)\Big\}.$$

Relations (6.2.10), (6.2.11) and this formula imply that

$$\Big|\frac{it}{\sigma_1\sqrt{n}\,(n-1)}\,\Sigma_1\Big| \le \frac{k(k-1)}{2n\sqrt{n}\,(n-1)}\,\sigma_1^{-3}\,(E\,|g_2(X_1,X_2)|^{5/3})^{3/5}$$

$$\times\,(E\,|g_1(X_1)|^3)^{2/3}\,|t|^3\exp\Big(-\frac{t^2}{3n}\,(k-2)\Big) \le$$

$$\le \frac{3}{10}\,\frac{k(k-1)}{n(n-1)}\,\frac{1}{\sqrt{n}}\,\sigma_1^{-5/3}\,E\,|g_2(X_1,X_2)|^{5/3}\,|t|^3\exp\Big(-\frac{t^2}{3n}\,(k-2)\Big) +$$

$$+\frac{1}{5}\,\frac{k(k-1)}{n(n-1)}\,\frac{1}{\sqrt{n}}\,\sigma_1^{-3}\,E\,|g_1(X_1)|^3\,|t|^3\exp\Big(-\frac{t^2}{3n}\,(k-2)\Big) \quad (6.2.19)$$

in the region $|t| \le T_n$.

Consider the variable Σ_2. In view of the independence of the variables, we have

$$\Sigma_2 = (k-1)\Big(\varphi\Big(\frac{t}{\sigma_1\sqrt{n}}\Big)\Big)^{k-1}\sum_{s=k+1}^{n} E\Big\{g_2(X_1,X_s)\exp\Big(\frac{it}{\sigma_1\sqrt{n}}\,g_1(X_1) +$$

$$+\frac{it}{\sigma_1\sqrt{n}}\sum_{r=k+1}^{n} g_1(X_r) + \frac{it}{\sigma_1\sqrt{n}\,(n-1)}\,\delta_n''(k)\Big)\Big\}.$$

We represent variable $\delta_n''(k)$ in the form

$$\delta_n''(k) = \eta_n(X_s) + \eta_s,$$

for all $s = k+1,\ k+2,\ \dots,\ n$; here

$$\eta_n(X_s) = \sum_{l=s+1}^{n} g_2(X_l, X_s) + \sum_{l=k+1}^{s-1} g_2(X_l, X_s),$$

$$\eta_s = \sum_{r=s+1}^{n-1}\sum_{l=k+1}^{n} g_2(X_r, X_l) + \sum_{r=k+1}^{s-1}\sum_{l=k+1}^{s-1} g_2(X_r, X_l) + \sum_{r=k+1}^{s-1}\sum_{l=s+1}^{n} g_2(X_r, X_l)$$

and "empty" sums are assumed to be zero. By using this formula and the independence of the variables, we find

$$\begin{aligned}
&\Big|E\Big\{g_2(X_1, X_s)\exp\Big(\frac{it\,g_1(X_1)}{\sigma_1\sqrt{n}} + \frac{it}{\sigma_1\sqrt{n}}\sum_{r=k+1}^{n} g_1(X_r) + \frac{it\,\delta_n''(k)}{\sigma_1\sqrt{n}\,(n-1)}\Big)\Big\}\Big| \\
&= \Big|E\Big\{\exp\Big(\frac{it\,(S_n - g_1(X_s))}{\sigma_1\sqrt{n}} + \frac{it\,\eta_s}{\sigma_1\sqrt{n}\,(n-1)}\Big) \\
&\quad \times E\Big[g_2(X_1, X_s)\exp\Big(\frac{it}{\sigma_1\sqrt{n}}(g_1(X_1) + g_1(X_s)) + \frac{it}{\sigma_1\sqrt{n}\,(n-1)}\,\eta_n(X_s)\Big)\Big]\Big\}\Big| \\
&\le E\Big|E\Big(g_2(X_1, X_s)\exp\Big(\frac{it}{\sigma_1\sqrt{n}}(g_1(X_1) + g_1(X_s)) + \frac{it\,\eta_n(X_s)}{\sigma_1\sqrt{n}\,(n-1)}\Big)\Big)\Big| = \Sigma_3,
\end{aligned}$$

where the expectation under the modulus sign is taken only with respect to independent variables X_1 and X_s.

According to the property of degeneracy, we obtain

$$\begin{aligned}
&E\Big(g_2(X_1, X_s)\exp\Big(\frac{it}{\sigma_1\sqrt{n}}(g_1(X_1) + g_1(X_s)) + \frac{it}{\sigma_1\sqrt{n}\,(n-1)}\,\eta_n(X_s)\Big)\Big) \\
&= E\Big(g_2(X_1, X_2)\Big(\exp\Big(\frac{it}{\sigma_1\sqrt{n}}\,g_1(X_1)\Big) - 1\Big)\Big(\exp\Big(\frac{it}{\sigma_1\sqrt{n}}\,g_1(X_s)\Big) - 1\Big)\Big) \\
&\qquad + E\Big(g_2(X_1, X_s)\Big(\exp\Big(\frac{it\,g_1(X_1)}{\sigma_1\sqrt{n}}\Big) - 1\Big) \times \\
&\qquad\qquad \times \exp\Big(\frac{it\,g_1(X_s)}{\sigma_1\sqrt{n}}\Big)\Big(\exp\Big(\frac{it\,\eta_n(X_s)}{\sigma_1\sqrt{n}\,(n-1)}\Big) - 1\Big)\Big).
\end{aligned}$$

By virtue of this, we have

$$\Sigma_3 \le \Big|E\Big(g_2(X_1, X_s)\Big(\exp\Big(\frac{it}{\sigma_1\sqrt{n}}\,g_1(X_1)\Big) - 1\Big)\Big(\exp\Big(\frac{it}{\sigma_1\sqrt{n}}\,g_1(X_s)\Big) - 1\Big)\Big)\Big| + \Sigma_4,$$

where

$$\Sigma_4 = E\Big|E\Big(g_2(X_1, X_s)\Big(\exp\Big(\frac{it}{\sigma_1\sqrt{n}}\,g_1(X_1)\Big) - 1\Big) \times$$

$$\times \exp\Big(\frac{it}{\sigma_1\sqrt{n}}\, g_1(X_s)\Big) \Big(\exp\Big(\frac{it}{\sigma_1\sqrt{n}\,(n-1)}\, \eta_n(X_s)\Big) - 1\Big)\Big)\Big|.$$

Here the first term can be estimated according to (6.2.10). To estimate Σ_4, we use the inequalities

$$\Big|1 - \exp\Big(\frac{it}{\sigma_1\sqrt{n}}\, g_1(X_1)\Big)\Big| \le \frac{|t|}{\sigma_1\sqrt{n}}\, |g_1(X_1))|,$$

$$\Big|1 - \exp\Big(\frac{it}{\sigma_1\sqrt{n}\,(n-1)}\, \eta_n(X_s)\Big)\Big| \le 2\Big(\frac{|t|}{\sigma_1\sqrt{n}\,(n-1)}\Big)^{2/3} |\eta_n(X_s)|^{2/3}.$$

Then

$$\Sigma_4 \le 2\frac{|t|}{\sigma_1\sqrt{n}} \Big(\frac{|t|}{\sigma_1\sqrt{n}\,(n-1)}\Big)^{2/3} E\,\Big(|g_2(X_1, X_s)|\, |\eta_n(X_s)|^{2/3}\, |g(X_1)|\Big).$$

According to Hölder's inequality, we get

$$\begin{aligned}
&E(|g_2(X_1, X_s)|\, |\eta_n(X_s)|^{2/3}\, |g_1(X_1)|) \\
&\quad \le (E\, |g_2(X_1, X_s)|^{5/3})^{3/5}\, (E\, (|\eta_n(X_s)|^{2/3}\, |g_1(X_1)|)^{5/2})^{2/5} \\
&\quad \le (E\, |g_2(X_1, X_s)|^{5/3})^{3/5}\, (E\, |\eta_n(X_s)|^{5/3})^{2/5}\, (E\, |g_1(X_1)|^3)^{1/3}.
\end{aligned}$$

Taking into account Corollary 2.2.1, we obtain

$$E\, |\eta_n(X_s)|^{5/3} \le 2^{1/3}\, (n-k)\, E\, |g_2(X_1, X_2)|^{5/3}.$$

These inequalities yield

$$\Sigma_4 \le 4\sigma_1^{-5/3}\, E\, |g_2(X_1, X_2)|^{5/3}\, (E\, |g_1(X_1)|^3)^{1/3}\, |t|^{5/3}\, n^{-11/10}.$$

Combining all these estimates and taking into account the assumption $n \ge a^4$, we find

$$\begin{aligned}
\Big|\frac{it}{\sigma_1\sqrt{n}\,(n-1)}\, \Sigma_2\Big| &\le \frac{|t|\,(k-1)}{\sigma_1\sqrt{n}\, n} \exp\Big(-\frac{t^2}{3n}\,(k-1)\Big) \times \\
&\times \Big\{t^2\sigma_1^{-2}\, (E\, |g_2(X_1, X_2)|^{5/3})^{3/5}\, (E\, |g_1(X_1)|^3)^{2/3} + \\
&+ |t|^{5/3}\, 4\sigma_1^{-5/3}\, E\, |g_2(X_1, X_2)|^{5/3}\, n^{-1/10}\, (E\, |g_1(X_1)|^3)^{1/3}\Big\} \le \\
&\le \frac{16}{5}\, \frac{(k-1)}{n\sqrt{n}}\, \sigma_1^{-5/3}\, E\, |g_2(X_1, X_2)|^{5/3}\, |t|^3 \exp\Big(-\frac{t^2}{3n}\,(k-1)\Big) + \\
&+ \frac{2}{5}\, \frac{(k-1)}{n\sqrt{n}}\, \sigma_1^{-3}\, E\, |g_1(X_1)|^3\, |t|^3 \exp\Big(-\frac{t^2}{3n}\,(k-1)\Big)
\end{aligned} \tag{6.2.20}$$

in the region $|t| \le T_n$. With the help of (6.2.5), (6.2.7), (6.2.14) and (6.2.18)–(6.2.20), we obtain in the region $|t| \le T_n$,

$$|\varphi_n(t) - e^{-t^2/2}| \le \frac{6 \cdot 2^{2/3}}{5}\, bk\, n^{-3/2}\, |t|^{5/3} + \frac{5}{12}\, a \frac{1}{\sqrt{n}}\, |t|^3\, e^{-t^2/12} +$$

$$+\left(\frac{1}{5}a+\frac{3}{10}b\right)\frac{1}{\sqrt{n}}|t|^3e^{-t^2/9}+\frac{|t|}{\sqrt{n}}c_{nk}(t)\exp\Big(-\frac{t^2k}{3n}\Big),\quad (6.2.21)$$

where

$$c_{nk}(t)=\frac{16}{5}\frac{(k-1)}{n}bt^2\exp\Big(\frac{t^2}{3n}\Big)+\frac{3}{10}\frac{k(k-1)}{n(n-1)}bt^2\exp\Big(\frac{2t^2}{3n}\Big)+$$
$$+\frac{2}{5}\frac{(k-1)}{n}at^2\exp\Big(\frac{t^2}{3n}\Big)+\frac{1}{5}\frac{k(k-1)}{n(n-1)}at^2\exp\Big(\frac{2t^2}{3n}\Big).$$

Since $k\le n$ and $a\ge 1$, in the region $|t|\le T_n=\sqrt{n}/a$ we have

$$c_{nk}(t)\le c(t),$$

where

$$\begin{aligned}c(t)&=at^2\Big(\frac{2}{5}\exp\Big(\frac{1}{3}\Big)+\frac{1}{5}\exp\Big(\frac{2}{3}\Big)\Big)+bt^2\Big(\frac{16}{5}\exp\Big(\frac{1}{3}\Big)+\frac{3}{10}\exp\Big(\frac{2}{3}\Big)\Big)\\&\le t^2(a+\frac{27}{5}b).\end{aligned}$$

Thus, in the region $|t|\le T_n$

$$|\varphi_n(t)-e^{-t^2/2}|\le 2b\frac{k}{n}\frac{1}{\sqrt{n}}|t|^{5/3}+$$
$$+\left(\frac{2}{3}a+\frac{3}{10}b\right)\frac{1}{\sqrt{n}}|t|^3\exp\Big(-\frac{t^2}{12}\Big)+\frac{1}{\sqrt{n}}|t|\,c(t)\exp\Big(-\frac{t^2k}{3n}\Big).$$

Further, for all $n\ge 2$ and every t from region $1\le|t|\le T_n$, we choose an integer $k=k(t)$ according to the formula

$$k(t)=\left[\frac{1}{2}n\,|t|^{-7/4}\right]+1,$$

where $[x]$ denotes the integral part of a number x. Obviously, if t changes in region $1\le|t|\le T_n$, the number k satisfies the inequalities

$$\left[\frac{1}{2}n^{1/8}a^{7/4}\right]+1\le k\le\left[\frac{n}{2}\right]+1$$

for all $n\ge a^2$. By virtue of the property of the integral part of a number, we have the evident inequalities

$$\frac{1}{2}|t|^{-7/4}\le\frac{k}{n}\le|t|^{-7/4},$$

which hold for all t from region $1\le|t|\le T_n$. In this region $1\le|t|\le T_n$ we have also

$$c_{nk}(t)\le|t|^{1/4}(a+\frac{27}{5}b).$$

Therefore, according to inequality (6.2.21) in the region $1\le|t|\le T_n$

$$|\varphi_n(t) - e^{-t^2/2}| \le \frac{6 \cdot 2^{2/3}}{5} b\,|t|^{-1/12} + \Big(\frac{1}{5}a + \frac{3}{10}b\Big)\,|t|^3 \exp\Big(-\frac{t^2}{9}\Big) +$$
$$+\frac{5}{12}a|t|^3 \exp\Big(-\frac{t^2}{12}\Big) + \Big(a + \frac{27}{5}b\Big)\,|t|^{5/4} \exp\Big(-\frac{1}{6}|t|^{1/4}\Big).$$

This implies estimate (6.2.13).

Let us consider (6.2.3) again. We divide the domain of integration $|t| \le T_n$ into two parts: $|t| \le 1$ and $1 \le |t| \le T_n$. If integrating over the region $|t| \le 1$, we employ estimate (6.2.12). When integrating over the region $1 \le |t| \le T_n$, we apply Lemma 6.2.1. As a result of the calculations we obtain (6.2.1).

Von Mises' statistics. Consider a real-valued Mises' statistic

$$V_n = n^{-2} \sum_{i=1}^{n} \sum_{j=1}^{n} \Phi(X_i, X_j)$$

with a non-degenerate kernel $\Phi\colon X^2 \to R$ and $E\,\Phi(X_1, X_2) = 0$. Let us represent V_n in the form of a U-statistic. For this purpose, we denote

$$\Phi_n(x,y) = \frac{(n-1)}{n}\Phi(x,y) + \bar{\Phi}_n(x,x) + \bar{\Phi}_n(y,y),$$

where

$$\bar{\Phi}_n(x,y) = \frac{1}{2n}\big(\Phi(x,y) - E\,\Phi(X_1,X_1)\big).$$

Then under the condition $E\,|\Phi_n(X_1,X_2)| < \infty$ the following formula holds

$$V_n - EV_n = U_n$$

with U-statistic

$$U_n = n^{-[2]} \sum_{1 \le i \ne j \le n} \Phi_n(X_i, X_j).$$

Theorem 6.2.2. *Assume the following conditions are satisfied:*

$$E\,\Phi(X_1,X_2) = 0, \quad \sigma_1 > 0, \quad E\,|g_1(X_1)|^3 < \infty,$$
$$E\,|\Phi(X_1,X_1)| < \infty, \quad E\,|\Phi(X_1,X_2)|^{5/3} < \infty.$$

Then

$$\sup_x \Big| P\Big(\frac{\sqrt{n}}{2\sigma_1}(V_n - EV_n) < x\Big) - F(x)\Big| = O(n^{-1/2})$$

as $n \to \infty$.

Proof. This statement is an analogue of Theorem 6.2.1 for non-degenerate V-statistics. For the proof it is sufficient to apply the method stated above in Theorem 6.2.1.

6.3 Estimates with increasing degree

Consider UR-statistics

$$U_n = \binom{n}{m}^{-1} \sum_{1 \le i_1 < \cdots < i_m \le n} \Phi(X_{i_1}, \ldots, X_{i_m})$$

with a non-degenerate symmetric kernel $\Phi\colon X^m \to R$ and denote

$$F_n(x) = P\Big(\frac{\sqrt{n}}{m\sigma_1}(U_n - \theta) < x\Big), \qquad x \in R.$$

In this section we study a convergence rate in the conditions when degree m depends on n so that $m = m(n) \to \infty$ as $n \to \infty$. According to Theorem 4.1.1 the convergence

$$\sup_x \Big| F_n(x) - F(x) \Big| \to 0$$

takes place, if the following condition

$$\frac{n}{m^2 \sigma_1^2} \sum_{c=2}^{m} \binom{m}{c}^2 \binom{n}{c}^{-1} \sigma_c^2 \to 0$$

is valid as $n \to \infty$. Note, that σ_c^2, $c = 1, \ldots, m$ also depend on m.

General estimate.

Theorem 6.3.1. *Suppose that a non-degenerate kernel $\Phi\colon X^m \to R$ satisfies the conditions:*

$$E\,|g_1|^3 < \infty, \quad E\,g_c^2 < \infty, \qquad c = 2, \ldots, m.$$

Then for all $2 \le m \le \sqrt{n}$ the estimate is valid:

$$\sup_x \Big| F_n(x) - F(x) \Big| \le \Big(\frac{144}{\pi} + \frac{96}{\pi\sqrt{2\pi}}\Big) \frac{E\,|g_1|^3}{\sigma_1^3} \frac{1}{\sqrt{n}} + \frac{1}{\sqrt{2\pi}} \frac{m}{\sqrt{n}} +$$
$$+ 2 \sum_{c=2}^{m} \Big(\frac{m}{\sqrt{n}}\Big)^{2c-3} \Big(\frac{\sigma_c}{\sigma_1}\Big)^2 \Big/ (c-1)! + 2 \sum_{c=2}^{m} \Big(\frac{m}{\sqrt{n}}\Big)^{c-1} 3^{c/2}\, E\,|\tau|^c \sigma_c \Big/ c!,$$

where τ is a standard normal random variable.

Proof. Let us introduce the following random variables

$$S_n = \sigma_1^{-1} n^{-1/2} \sum_{j=1}^{n} g_1(X_j),$$

$$\delta = \frac{\sqrt{n}}{m\sigma_1} \sum_{c=2}^{m} \binom{m}{c} \binom{n}{c}^{-1} \sum_{1 \le i_1 < \cdots < i_c \le n} g_c(X_{i_1}, \ldots, X_{i_c}),$$

$$\delta_1 = \frac{\sqrt{n}}{m\sigma_1}\sum_{c=2}^{m}\binom{m}{c}\binom{n}{c}^{-1}\sum_{1\le i_1<\cdots<i_c\le k} g_c(X_{i_1},\ldots,X_{i_c}),$$

where $k = n - m[\sqrt{n}]$ and $[\sqrt{n}]$ is an integral part of $\sqrt{n}$. We set

$$\Delta_1 = \sup_x\Big|F(x) + F(x + mn^{-1/2})\Big|, \qquad \Delta_2 = P\Big(|\delta-\delta_1| \ge m/\sqrt{n}\Big),$$

$$\Delta_3 = \sup_x\Big|P(S_n + \delta_1 < x) - F(x)\Big|.$$

One can find that

$$\sup_x\Big|F_n(x) - F(x)\Big| \le \Delta_1 + \Delta_2 + \Delta_3.$$

We estimate each Δ_i on the right-hand side. By virtue of the properties of the normal distribution function $F(x)$, we have

$$\Delta_1 \le \frac{1}{\sqrt{2\pi}}\frac{m}{\sqrt{n}}.$$

According to Chebyshev's inequality, we get

$$P\Big(|\delta-\delta_1| \ge m/\sqrt{n}\Big) \le n\,m^{-2}\,E(\delta-\delta_1)^2.$$

Additionally,

$$\delta - \delta_1 = \frac{\sqrt{n}}{m\sigma_1}\sum_{c=2}^{m}\binom{m}{c}\binom{n}{c}^{-1}\sum_{i_c=k+1}^{n}\;\sum_{1\le i_1<\cdots<i_{c-1}\le i_c-1} g_c(X_{i_1},\ldots,X_{i_c}).$$

Taking into account the property of degeneracy, direct calculation yields

$$E\,(\delta-\delta_1)^2 = \frac{n}{m^2\,\sigma_1^2}\sum_{c=2}^{m}\binom{m}{c}^2\binom{n}{c}^{-2}\left(\binom{n}{c} - \binom{k}{c}\right)\sigma_c^2.$$

Therefore,

$$\Delta_2 \le \sum_{c=2}^{m}\left(\frac{m}{\sqrt{n}}\right)^{2c-3}\left(\frac{\sigma_c}{\sigma_1}\right)^2\frac{1}{(c-1)!}.$$

Let $T = \sqrt{n}\,\sigma_1^3\,(4\,E\,|g_1|^3)^{-1}$. According to the Esséen inequality, we have

$$\Delta_3 \le \frac{1}{\pi}\int_{-T}^{T}\Big|E\exp\Big(it\,(S_n+\delta_1)\Big) - \exp(-t^2/2)\Big|\,|t|^{-1}\,dt + \frac{24}{\pi\sqrt{2\pi}}\,T^{-1}.$$

Obviously,

$$\Big|E\exp\Big(it\,(S_n+\delta_1)\Big) - \exp(-t^2/2)\Big| \le$$
$$\le \Big|E\exp\Big(it\,(S_n+\delta_1)\Big) - E\exp(it\,S_n)\Big| + \Big|E\exp(it\,S_n) - \exp(-t^2/2)\Big|.$$

Let $\varphi(t) = E\exp(itg_1(X_1))$ be a characteristic function of a random variable $g_1(X_1)$. Since $X_1, \ldots, X_n$ are independent, we have

$$E\exp(itS_n) = \left(\varphi\Big(\frac{t}{\sigma_1\sqrt{n}}\Big)\right)^n.$$

By virtue of the well known estimate, we get

$$\left|E\exp(itS_n)-\exp(-t^2/2)\right| \le 16\,\sigma_1^{-3}\,E\,|g_1|^3\,n^{-1/2}\,|t|^3\exp(-t^2/3) \qquad (6.3.1)$$

in the region $|t|\le T$. Consequently,

$$\Delta_3 \le \frac{1}{\pi}\int_{-T}^{T}\left|E\exp\bigl(it\,(S_n+\delta_1)\bigr)-E\exp(it\,S_n)\right|\,|t|^{-1}\,dt + \\ + \left(\frac{144}{\pi}+\frac{96}{\pi\sqrt{2\pi}}\right)\frac{E\,|g_1|^3}{\sigma_1^3}\,\frac{1}{\sqrt{n}}.$$

Let us investigate the integrand. We have

$$E\exp(it\,(S_n+\delta_1)) - E\exp(it\,S_n) = E\bigl(\exp(it\,S_n)(1-\exp(it\,\delta_1))\bigr) = \\ = -it\,E\bigl(\delta_1\exp(it\,S_n)\bigr)+\frac{t^2}{2}\,E\bigl(\delta_1^2\lambda\exp(it\,S_n)\bigr),$$

where the random variable $\lambda=\lambda(X_1,\dots,X_k)$ satisfies the condition $|\lambda|\le 1$ and, in addition, does not depend on the sequence $X_{k+1},\dots,X_n$. By definition,

$$E\bigl(\delta_1\exp(itS_n)\bigr) = \\ = \frac{\sqrt{n}}{m\sigma_1}\sum_{c=2}^{m}\binom{m}{c}\binom{n}{c}^{-1}\sum_{1\le i_1<\dots<i_c\le k} E\Bigl(g_c(X_{i_1},\dots,X_{i_c})\exp\Bigl(\frac{it}{\sigma_1\sqrt{n}}\sum_{s=1}^{n}g_1(X_s)\Bigr)\Bigr).$$

In view of the independence of the variables, we have

$$E\Bigl(g_c(X_{i_1},\dots,X_{i_c})\exp\Bigl(\frac{it}{\sigma_1\sqrt{n}}\sum_{s=1}^{n}g_1(X_s)\Bigr)\Bigr) = \\ = \Bigl(\varphi\Bigl(\frac{t}{\sigma_1\sqrt{n}}\Bigr)\Bigr)^{n-c}E\Bigl(g_c(X_1,\dots,X_c)\exp\Bigl(\frac{it}{\sigma_1\sqrt{n}}\sum_{s=1}^{c}g_1(X_s)\Bigr)\Bigr).$$

Therefore,

$$E\bigl(\delta_1\exp(itS_n)\bigr) = \frac{\sqrt{n}}{m\sigma_1}\sum_{c=2}^{m}\binom{m}{c}\binom{n}{c}^{-1}\binom{k}{c}\Bigl(\varphi\Bigl(\frac{t}{\sigma_1\sqrt{n}}\Bigr)\Bigr)^{n-c}\times \\ \times E\Bigl(g_c(X_1,\dots,X_c)\exp\Bigl(\frac{it}{\sigma_1\sqrt{n}}\sum_{s=1}^{c}g_1(X_s)\Bigr)\Bigr).$$

According to the well known inequality, we have

$$|\varphi(t)|\le\exp\bigl(-t^2\sigma_1^2/3\bigr) \qquad (6.3.2)$$

if $|t\sigma_1|\le T$. Hence, in the region $|t|\le T$ we get

$$\Bigl|\varphi\Bigl(\frac{t}{\sigma_1\sqrt{n}}\Bigr)\Bigr|^{n-c}\le\exp\Bigl(-\frac{t^2}{3}\,\frac{(n-c)}{n}\Bigr)\le\exp\Bigl(-\frac{t^2}{6}\Bigr)$$

for all $2 \le c \le m$, $2 \le m \le \sqrt{n}$, $n \ge 4$. By the property of degeneracy $g_c(x_1, \ldots, x_c)$ we can write the representation

$$E\Big(g_c(X_1,\ldots,X_c)\exp\Big(\frac{it}{\sigma_1\sqrt{n}}\sum_{s=1}^{c} g_1(X_s)\Big)\Big) =$$
$$= E\Big(g_c(X_1,\ldots,X_c)\prod_{s=1}^{c}\Big(\exp\Big(\frac{it}{\sigma_1\sqrt{n}}g_1(X_s)\Big)-1\Big)\Big).$$

Further, taking into account the inequality $|\exp(ix)-1| \le |x|$, $x \in R$, we obtain

$$\Big|E\Big(g_c(X_1,\ldots,X_c)\exp\Big(\frac{it}{\sigma_1\sqrt{n}}\sum_{s=1}^{c} g_1(X_s)\Big)\Big)\Big| \le$$
$$\le \Big(\frac{|t|}{\sigma_1\sqrt{n}}\Big)^c E\left|g_c(X_1,\ldots,X_c)\,g_1(X_1)\ldots g_1(X_c)\right| \le \Big(\frac{|t|}{\sqrt{n}}\Big)^c \sigma_c.$$

Thus, in the region $|t| \le T$ the following estimate is valid

$$\Big|E\Big(\delta_1\exp(it\,S_n)\Big)\Big| \le e^{-t^2/6}\sum_{c=2}^{m}\Big(\frac{m}{\sqrt{n}}\Big)^{c-1}|t|^c\,\sigma_c/c!\,.$$

Then, taking again into account the independence of the variables, we obtain

$$E\Big(\delta_1^2\lambda\exp(it\,S_n)\Big) = \Big(\varphi\Big(\frac{t}{\sigma_1\sqrt{n}}\Big)\Big)^{n-k} E\Big(\delta_1^2\lambda\exp\Big(\frac{it}{\sigma_1\sqrt{n}}\sum_{s=1}^{k} g_1(X_s)\Big)\Big).$$

Since

$$E\,\delta_1^2 = \frac{n}{m\sigma_1^2}\sum_{c=2}^{m}\binom{m}{c}^2\binom{n}{c}^{-1}\binom{k}{c}\sigma_c^2,$$

in the region $|t| \le T$ we get

$$\Big|E\Big(\delta_1^2\lambda\exp(it\,S_n)\Big)\Big| \le \exp\Big(-\frac{t^2\,m}{6\sqrt{n}}\Big)\sum_{c=2}^{m}\Big(\frac{m}{\sqrt{n}}\Big)^{2c-2}\Big(\frac{\sigma_c}{\sigma_1}\Big)^2\Big/c!\,.$$

As a result, for Δ_3 we find the estimate

$$\Delta_3 \le \frac{3}{\pi}\sum_{c=2}^{m}\Big(\frac{m}{\sqrt{n}}\Big)^{2c-3}\Big(\frac{\sigma_c}{\sigma_1}\Big)^2\Big/c! +$$
$$+\Big(\frac{6}{\pi}\Big)^{1/2}\sum_{c=2}^{m}\Big(\frac{m}{\sqrt{n}}\Big)^{c-1}3^{c/2}\,E\,|\tau|^c\,\sigma_c\Big/c! + \Big(\frac{144}{\pi}+\frac{96}{\pi\sqrt{2\pi}}\Big)\frac{E\,|g_1|^3}{\sigma_1^3}\,\frac{1}{\sqrt{n}},$$

where τ is a standard normal random variable.

Combining the estimates for Δ_i we get the proof of Theorem 6.3.1.

Sharpening estimate. Now we consider UR-statistics with the kernel

$$\Phi(x_1,\ldots,x_m) = x_1\cdots x_m.$$

Let $\mu = EX_j$, and $\sigma^2 = E(X_j - \mu)^2 > 0$, $j = 1, \ldots, n$. In this case $\theta = \mu^m$,

$$g_c(x_1, \ldots, x_c) = \mu^{m-c}(x_1 - \mu)\cdots(x_c - \mu),$$
$$\sigma_c^2 = \mu^{2(m-c)}\sigma^{2c}, \qquad c = 1, \ldots, m.$$

Theorem 6.3.2. *Let the following conditions*

$$\mu \neq 0, \qquad \beta = E|X_1 - \mu|^3 < \infty$$

be satisfied.

Then, for all $2 \le m \le \sqrt{n}$, the following estimate is true

$$\sup_x |F_n(x) - F(x)| \le \left(\frac{48\sqrt{2}}{\pi\sqrt{\pi}} + \frac{24\sqrt{3}}{\sqrt{\pi}}\right)\frac{\beta}{\sigma^3}\frac{1}{\sqrt{n}} +$$
$$+\left(1 + \left(\frac{\sqrt{2}}{\sqrt{\pi}} + \frac{3}{2}\right)\frac{\sigma}{\mu} + \frac{9}{8}\left(\frac{\sigma}{\mu}\right)^2\right)\frac{m}{\sqrt{n}} + \frac{1}{6}\left(\frac{\sigma}{\mu}\frac{m}{\sqrt{n}}\right)^2 \exp\left(\left(\frac{\sigma}{\mu}\frac{m}{\sqrt{n}}\right)^2\right).$$

Proof. According to the Hoeffding decomposition formula

$$U_n - \mu^m = \sum_{c=1}^{m} m^{[c]} n^{-[c]} \mu^{m-c} \sum_{1 \le i_1 \neq \cdots \neq i_c \le n} (X_{i_1} - \mu)\cdots(X_{i_c} - \mu)/c!\,.$$

Denote

$$S_n = \frac{1}{\sigma\sqrt{n}}\sum_{j=1}^{n}(X_j - \mu), \quad \delta_n = \frac{m-1}{\sigma\mu(n-1)\sqrt{n}}\sum_{1 \le i < j \le n}(X_i - \mu)(X_j - \mu),$$
$$R_n = \frac{\sqrt{n}}{m\,\mu^{m-1}\sigma}\sum_{c=3}^{m} m^{[c]} n^{-[c]} \mu^{m-c} \sum_{1 \le i_1 \neq \cdots \neq i_c \le n} (X_{i_1} - \mu)\cdots(X_{i_c} - \mu)/c!\,.$$

Following these notations we rewrite this formula in the form

$$\frac{\sqrt{n}}{m\sigma\mu^{m-1}}(U_n - \mu^m) = S_n + \delta_n + R_n.$$

Hence

$$F_n(x) = P(S_n + \delta_n + R_n < x), \qquad x \in R.$$

Lemma 6.3.1. *The following inequality is true*

$$\sup_x |F_n(x) - F(x)| \le \sup_x \left|P(S_n + \delta_n < x) - F(x)\right| +$$
$$+ \frac{1}{\sqrt{2\pi}}\frac{\sigma}{\mu}\frac{m}{\sqrt{n}} + \frac{1}{6}\left(\frac{\sigma}{\mu}\frac{m}{\sqrt{n}}\right)^2 \exp\left(\left(\frac{\sigma}{\mu}\frac{m}{\sqrt{n}}\right)^2\right).$$

Proof. First check that an auxiliary inequality is true,

$$\sup_x |F_n(x) - F(x)| \le \sup_x \left|P(S_n + \delta_n < x) - F(x)\right| + \Delta_1 + \Delta_2,$$

where

$$\Delta_1 := \sup_x \Big| F(x) - F(x + m\sigma/\mu\sqrt{n}) \Big|, \quad \Delta_2 := P\Big(|R_n| \geq m\sigma/\mu\sqrt{n}\Big).$$

Indeed,

$$\begin{aligned} F_n(x) &= P(S_n + \delta_n + R_n < x) \\ &= P\Big\{(S_n + \delta_n < x - R_n) \cap (|R_n| \leq m\sigma/\mu\sqrt{n})\Big\} + \\ &\quad + P\Big\{(S_n + \delta_n < x - R_n) \cap (|R_n| > m\sigma/\mu\sqrt{n})\Big\}. \end{aligned}$$

First of all, it implies

$$\begin{aligned} F_n(x) &\geq P\Big\{(S_n + \delta_n < x - R_n) \cap (|R_n| \leq m\sigma/\mu\sqrt{n})\Big\} \\ &\geq P\Big\{(S_n + \delta_n < x - m\sigma/\mu\sqrt{n}) \cap (|R_n| \leq m\sigma/\mu\sqrt{n})\Big\} \\ &\geq P(S_n + \delta_n < x - m\sigma/\mu\sqrt{n}) - P(|R_n| > m\sigma/\mu\sqrt{n}), \end{aligned}$$

and, secondly,

$$F_n(x) \leq P(S_n + \delta_n < x + m\sigma/\mu\sqrt{n}) + P(|R_n| > m\sigma/\mu\sqrt{n}).$$

An auxiliary inequality follows from these results.

Estimate of Δ_1. According to the property of function $F(x)$,

$$\Delta_1 \leq \frac{1}{\sqrt{2\pi}} \frac{\sigma}{\mu} \frac{m}{\sqrt{n}}.$$

Estimate of Δ_2. According to the Chebyshev inequality,

$$\Delta_2 \leq \frac{\mu^2}{\sigma^2} \frac{n}{m^2} ER_n^2.$$

By definition of R_n,

$$ER_n^2 = \frac{n\mu^2}{m^2\sigma^2} \sum_{c=3}^{m} m^{2[c]} n^{-[c]} \sigma^{2c} \mu^{-2c} / c!\,.$$

Here

$$m^{2[c]}/n^{[c]} \leq m^{2c}/n^c.$$

Then

$$ER_n^2 \leq \frac{n\mu^2}{m^2\sigma^2} \sum_{c=3}^{m} \left(\frac{\sigma}{\mu} \frac{m}{\sqrt{n}}\right)^{2c} / c! = \left(\frac{\sigma}{\mu} \frac{m}{\sqrt{n}}\right)^4 \sum_{c=0}^{m-3} \left(\frac{\sigma}{\mu} \frac{m}{\sqrt{n}}\right)^{2c} \frac{1}{(c+3)!} \leq$$

$$\leq \frac{1}{6}\Big(\frac{\sigma}{\mu}\frac{m}{\sqrt{n}}\Big)^4 \sum_{c=0}^{\infty}\Big(\frac{\sigma}{\mu}\frac{m}{\sqrt{n}}\Big)^{2c} \Big/ c! = \frac{1}{6}\Big(\frac{\sigma}{\mu}\frac{m}{\sqrt{n}}\Big)^4 \exp\Big(\Big(\frac{\sigma}{\mu}\frac{m}{\sqrt{n}}\Big)^2\Big).$$

As a result we obtain

$$\Delta_2 \leq \frac{1}{6}\Big(\frac{\sigma}{\mu}\frac{m}{\sqrt{n}}\Big)^2 \exp\Big(\Big(\frac{\sigma}{\mu}\frac{m}{\sqrt{n}}\Big)^2\Big).$$

This completes the proof of Lemma 6.3.1.

Lemma 6.3.2. *For all $2 \leq m \leq \sqrt{n}$, the following inequality holds*

$$\sup_x \Big|P(S_n + \delta_n < x) - F(x)\Big| \leq$$

$$\leq \Big(\frac{48\sqrt{2}}{\pi\sqrt{\pi}} + \frac{24\sqrt{3}}{\sqrt{\pi}}\Big)\frac{\beta}{\sigma^3}\frac{1}{\sqrt{n}} + \Big(1 + \frac{1}{\sqrt{2\pi}}\frac{\sigma}{\mu} + \frac{3}{2}\frac{\sigma}{\mu} + \frac{9}{8}\Big(\frac{\sigma}{\mu}\Big)^2\Big)\frac{m}{\sqrt{n}}.$$

Proof. With the same reasoning as that for the proof of Lemma 6.3.1 we find the inequality

$$\sup_x \Big|P(S_n + \delta_n < x) - F(x)\Big| \leq \sup_x \Big|P(S_n + \delta_{nk} < x) - F(x)\Big| + \Delta_1 + \Delta_3,$$

where

$$\Delta_3 := P\Big(|\delta_n - \delta_{nk}| > m\sigma/\mu\sqrt{n}\Big),$$

$$\delta_{nk} := \frac{(m-1)}{\sigma\mu\,(n-1)\sqrt{n}} \sum_{1\leq i<j\leq k} (X_i - \mu)(X_j - \mu),$$

$k = n - m\,[\sqrt{n}]$ and $[\sqrt{n}]$ denotes the integral part of $\sqrt{n}$.

Estimate of Δ_3. According to the Chebyshev inequality,

$$\Delta_3 \leq \frac{n}{m^2}\frac{\mu^2}{\sigma^2} E(\delta_n - \delta_{nk})^2.$$

By definition

$$\delta_n - \delta_{nk} = \frac{(m-1)}{\sigma\mu\,(n-1)\sqrt{n}} \sum_{j=k+1}^{n} (X_j - \mu) \sum_{i=1}^{j-1} (X_i - \mu).$$

Applying this formula we write

$$E(\delta_n - \delta_{nk})^2 = \frac{(m-1)^2}{\sigma^2\mu^2\,(n-1)^2\,n} \sum_{j=k+1}^{n} E(X_j - \mu)^2\, E\Big(\sum_{i=1}^{j-1}(X_i - \mu)\Big)^2 =$$

$$= \frac{(m-1)^2}{\sigma^2\mu^2\,(n-1)^2\,n}\sigma^4 \sum_{j=k+1}^{n} (j-1) = \frac{\sigma^2}{\mu^2}\frac{(m-1)^2(n-k)(n+k-1)}{2(n-1)^2 n}.$$

Consequently, when $k = n - m[\sqrt{n}]$ and $\sqrt{n} \geq m \geq 2$,

$$E(\delta_n - \delta_{nk})^2 \leq \frac{\sigma^2}{\mu^2}\Big(\frac{m}{\sqrt{n}}\Big)^3.$$

Hence we obtain the estimate

$$\Delta_3 \leq \frac{m}{\sqrt{n}}.$$

Estimate of $\sup_x \big|P(S_n + \delta_{nk} < x) - F(x)\big|$. First estimate it by the Esséen inequality

$$\sup_x \big|P(S_n + \delta_{nk} < x) - F(x)\big| \leq$$
$$\leq \frac{1}{\pi}\int_{-T}^{T}\big|E\exp(it\,(S_n + \delta_{nk})) - e^{-t^2/2}\big|\,|t|^{-1}\,dt + \frac{24}{\pi\sqrt{2\pi}}\,T^{-1},$$

where $T = \sqrt{n}\,\sigma^3/4\beta$. It is obvious that

$$|Ee^{it\,(S_n+\delta_{nk})} - e^{-t^2/2}| \leq |Ee^{it\,(S_n+\delta_{nk})} - Ee^{itS_n}| + |Ee^{it\,S_n} - e^{-t^2/2}|.$$

Let $\varphi(t) = E\exp(it\,(X_1 - \mu))$, $t \in R$. It can be expressed as

$$Ee^{it\,S_n} = \Big(\varphi\Big(\frac{t}{\sigma\sqrt{n}}\Big)\Big)^n$$

because $X_1, \ldots, X_n$, are independent identically distributed random variables.

In the domain $|t| \leq T$, by the well-known estimate (6.3.2) we obtain

$$\int_{-T}^{T}\big|Ee^{it\,S_n} - e^{-t^2/2}\big|\,|t|^{-1}\,dt \leq$$
$$\leq 16\,\sigma^{-3}\beta\,n^{-1/2}\int_{-\infty}^{\infty} t^2\,e^{-t^2/3}\,dt = 24\sqrt{3\pi}\,\sigma^{-3}\beta\,n^{-1/2}.$$

Further, applying the formula

$$e^{ix} = 1 + ix - x^2\int_0^1 (1-u)\,e^{ixu}\,du, \qquad x \in R,$$

we write the representation

$$E\,e^{it\,(S_n+\delta_{nk})} - E\,e^{it\,S_n} = it\,E(\delta_{nk}\,e^{itS_n}) - t^2\int_0^1 E\Big(\delta_{nk}^2\,e^{it\,(S_n+\delta_{nk}u)}\Big)(1-u)\,du.$$

By definition

$$E(\delta_{nk}\,e^{itS_n}) = \frac{m-1}{\sigma\mu\,(n-1)\sqrt{n}}\sum_{1\leq i_1<i_2\leq k} E\Big((X_{i_1}-\mu)(X_{i_1}-\mu)\,e^{\frac{it}{\sigma\sqrt{n}}\sum_{j=1}^{n}(X_j-\mu)}\Big).$$

By virtue of independence,

$$E\Big((X_{i_1}-\mu)(X_{i_2}-\mu)\,e^{\frac{it}{\sigma\sqrt{n}}\sum_{j=1}^{n}(X_j-\mu)}\Big) =$$

$$= \left(\varphi\left(\frac{t}{\sigma\sqrt{n}}\right)\right)^{n-2} \left(E\left((X_1-\mu)\, e^{\frac{it}{\sigma\sqrt{n}}(X_1-\mu)}\right)\right)^2.$$

Consequently,

$$E(\delta_{nk}\, e^{itS_n}) = \frac{(m-1)\,k\,(k-1)}{2\sigma\mu\,(n-1)\sqrt{n}} \left(\varphi\left(\frac{t}{\sigma\sqrt{n}}\right)\right)^{n-2} \left(E\left((X_1-\mu)\, e^{\frac{it}{\sigma\sqrt{n}}(X_1-\mu)}\right)\right)^2.$$

According to (6.3.2)

$$\left|\varphi\left(\frac{t}{\sigma\sqrt{n}}\right)\right| \le e^{-\frac{t^2(n-2)}{3n}}$$

and, in addition,

$$\left|E\left((X_1-\mu)\, e^{\frac{it}{\sigma\sqrt{n}}(X_1-\mu)}\right)\right| \le \frac{\sigma}{\sqrt{n}}\,|t|.$$

Therefore we obtain the inequality

$$|E(\delta_{nk}\, e^{itS_n})| \le \frac{(m-1)\,k\,(k-1)}{2\,(n-1)\,n\sqrt{n}}\,\frac{\sigma}{\mu}\,t^2\,e^{-\frac{t^2}{3}\frac{(n-2)}{n}}.$$

In accordance with this inequality,

$$\int_{-T}^{T} |E(\delta_{nk}\, e^{itS_n})|\,dt \le \frac{(m-1)\,k\,(k-1)}{2\,(n-1)\,n\sqrt{n}}\,\frac{\sigma}{\mu}\int_{-\infty}^{\infty} dt\, t^2\, e^{-\frac{t^2}{3}\frac{n-2}{n}} \le \frac{3}{2}\,\frac{\sigma}{\mu}\,\frac{m}{\sqrt{n}}.$$

Further, because of the independence, we conclude that

$$E\left(\delta_{nk}^2\, e^{it\,(S_n+\delta_{nk}u)}\right) = \left(\varphi\left(\frac{t}{\sigma\sqrt{n}}\right)\right)^{n-k} E\left(\delta_{nk}^2\, e^{it\delta_{nk}u} \exp\left(\frac{it}{\sigma\sqrt{n}}\sum_{j=1}^{k}(X_j-\mu)\right)\right).$$

Hence, in the domain $|t| \le T$,

$$\left|E(\delta_{nk}^2\, e^{it\,(S_n+\delta_{nk}u)})\right| \le E\,\delta_{nk}^2\, e^{-\frac{t^2}{3}\frac{(n-k)}{n}}$$

and

$$E\,\delta_{nk}^2 = \frac{(m-1)^2\,k\,(k-1)}{2\,(n-1)^2\,n}\,\frac{\sigma^2}{\mu^2}.$$

Then

$$\int_0^1 du\,(u-1)\int_{-T}^{T} dt\,|t|\left|E(\delta_{nk}^2\, e^{it\,(S_n+\delta_{nk}u)})\right| \le$$

$$\le \frac{(m-1)^2\,k\,(k-1)}{4\,(n-1)^2\,n}\left(\frac{\sigma}{\mu}\right)^2 \int_{-\infty}^{\infty} dt\,|t|\,e^{-\frac{t^2}{3}\frac{(n-k)}{n}} =$$

$$= \frac{(m-1)^2\,k\,(k-1)}{4\,(n-1)^2 n}\left(\frac{\sigma}{\mu}\right)^2 \frac{3n}{n-k} \le \frac{9}{8}\left(\frac{\sigma}{\mu}\right)^2 \frac{m}{\sqrt{n}}.$$

Combining these estimates we obtain

$$\sup_x \left|P(S_n+\delta_{nk} < x) - F(x)\right| \le$$

$$\le \left(\frac{48\sqrt{2}}{\pi\sqrt{\pi}} + \frac{24\sqrt{3}}{\pi}\right)\frac{\beta}{\sigma^3}\,\frac{1}{\sqrt{n}} + \left(\frac{3}{2}\,\frac{\sigma}{\mu} + \frac{9}{8}\left(\frac{\sigma}{\mu}\right)^2\right)\frac{m}{\sqrt{n}}.$$

This completes the proof of Lemma 6.3.2.

Lemma 6.3.1 and Lemma 6.3.2 imply the validity of Theorem 6.3.2.

6.4 Nonuniform estimates

Let $m = 2$ and the kernel $\Phi\colon X^2 \to R$ such that $E\,\Phi(X_1, X_2) = 0$. Let us study in this section the behavior of variable $|F_n(x) - F(x)|$ with respect to n and x, assuming also some moment conditions about the kernel Φ. For this purpose we introduce the following notation

$$S_n = n^{-1/2}\sigma_1^{-1}\sum_{j=1}^{n} g_1(X_j), \quad S_n' = n^{-1/2}\sigma_1^{-1}\sum_{j=1}^{k} g_1(X_j),$$

$$\delta_n = \frac{1}{2} n^{1/2}\sigma_1^{-1}\binom{n}{2}^{-1} \sum_{1\le i<j\le n} g_2(X_i, X_j),$$

$$\bar{\delta}_n = \frac{1}{2} n^{1/2}\sigma_1^{-1}\binom{n}{2}^{-1} \sum_{1\le i<j\le n} \bar{g}_2(X_i, X_j),$$

$$\bar{\delta}_{n1} = \frac{1}{2} n^{1/2}\sigma_1^{-1}\binom{n}{2}^{-1} \sum_{1\le i<j\le k} \bar{g}_2(X_i, X_j),$$

$$\bar{\delta}_{n2} = \bar{\delta}_n - \bar{\delta}_{n1}, \quad \beta = E\,|\Phi(X_1, X_2)|^3,$$

where

$$\bar{g}_2(x,y) = \bar{\Phi}(x,y) - \bar{g}_1(x) - \bar{g}_1(y), \quad \bar{g}_1(x) = E(\bar{\Phi}(X_1,X_2)|X_1 = x),$$

$$\bar{\Phi}(x,y) = \Phi(x,y)I(|\Phi(x,y)| \le \sqrt{n}) - E(\Phi(X_1,X_2)I(|\Phi(X_1,X_2)| \le \sqrt{n}))$$

and $k = n - [n^{2/3}]$, and $[n^{2/3}]$ is the integral part of $n^{2/3}$.

Theorem 6.4.1. *If $\sigma_1 > 0$ and $E\,|\Phi|^3 < \infty$, then for every $n \ge 2$ and all $x \in R$ the following inequality holds*

$$\left|P\left(\frac{1}{2}n^{1/2}\sigma_1^{-1}U_n < x\right) - F(x)\right| \le c\beta\sigma_1^{-3}n^{-1/2}(1+|x|)^{-3},$$

where $c > 0$ is some absolute constant.

Proof. First, we present three lemmas.

Lemma 6.4.1. *If the conditions of the theorem are satisfied, then*

$$E\,|S_n'|^3 \le c\beta\sigma_1^{-3}, \quad E\,|\bar{\delta}_{n2}|^3 \le c\beta\sigma_1^{-3}n^{-2}, \quad E\,(S_n'\bar{\delta}_{n1})^2 \le c\beta\sigma_1^{-3}n^{-1},$$
$$E\,|\bar{\delta}_{n1}|^p \le c\beta\sigma_1^{-3}n^{-p/2}, \qquad p = 1,2,3,4.$$

Lemma 6.4.2. *Suppose that $\xi_n = \xi_{n1} + \xi_{n2}$, $n \ge 1$, is a sequence of random variables. Denote by F_n and F_{n1} the distribution functions of ξ_n and ξ_{n1}, respectively. Assume that*

$$|F_{n1}(x) - F(x)| \le d\,n^{-1/2}(1+|x|)^{-3}$$

and for $|x| \geq 1$

$$P\Big(|\xi_{n2}| \geq d\,n^{-1/2}\,|x|\Big) \leq d\,n^{-1/2}\,(1+|x|)^{-3}.$$

Then

$$|F_n(x) - F(x)| \leq d\,n^{-1/2}\,(1+|x|)^{-3},$$

for all $x \in R$, *where the constant* $d > 0$ *does not depend on* n *and* x.

This statement remains valid if $\xi_n = a_n \xi_{n1} + b_n$, where the constants a_n and b_n are such that $|a_n - 1| \leq d\,n^{-1/2}$, $|b_n| \leq d\,n^{-1/2}$.

Lemma 6.4.3. *Suppose that* $G_n(x)$ *is a nondecreasing function,* $H_n(x)$ *is a function of bounded variation, and* $g_n(t)$ *and* $h_n(t)$ *are their Fourier–Stieltjes transforms, respectively. Assume that*

$$G_n(\pm\infty) = H_n(\pm\infty), \quad \int_{-\infty}^{\infty} |x|^3 \Big|d(G_n(x) - H_n(x))\Big| < \infty,$$
$$|H_n(x) - F(x)| \leq d\,n^{-1/2}\,(1+|x|)^{-3}.$$

Then for $T_n = \varepsilon\sqrt{n}$ *and for all* $x \in R$, *the following inequality holds*

$$|G_n(x) - H_n(x)| \leq$$
$$\leq d\,(1+|x|)^{-3}\bigg(\int_{-T_n}^{T_n} \Big|g_n(t) - h_n(t)\Big|\,|t|^{-1}\,dt + \int_{-T_n}^{T_n} |\Delta_n(t)|\,|t|^{-1}\,dt + d\,n^{-1/2}\bigg),$$

where

$$\Delta_n(t) = \int_{-\infty}^{\infty} \exp(itx)\,d\Big(x^3\,(G_n(x) - H_n(x))\Big).$$

The statement of Lemmas 6.4.1 and 6.4.2 can be derived by direct calculation. Lemma 6.4.3 is a version of Lemma 6.1.3. By using Lemma 6.4.1 and Chebyshev's inequality, we obtain

$$P\bigg(\bigg|\frac{\sqrt{n}}{2\sigma_1}U_n - (S_n + \bar{\delta}_n)\bigg| \geq |x|/\sqrt{n}\bigg) \leq c\beta\sigma_1^{-3}\,n^{-1/2}\,(1+|x|)^{-3},$$
$$P\Big(\Big|(S_n + \bar{\delta}_n) - (S_n + \bar{\delta}_{n1})\Big| \geq |x|/\sqrt{n}\Big) \leq c\beta\sigma_1^{-3}\,n^{-1/2}\,(1+|x|)^{-3},$$

for $|x| \geq 1$. This and Lemma 6.4.2 show that in order to prove this theorem it suffices to establish the inequality

$$\Big|P(S_n + \bar{\delta}_{n1} \leq x) - F(x)\Big| \leq c\beta\sigma_1^{-3}\,n^{-1/2}\,(1+|x|)^{-3}.$$

For this purpose, we employ Lemma 6.4.3 with $T_n = \sigma_1^3\sqrt{n}/\beta$. Denote

$$G_n(x) = P(S_n + \bar{\delta}_{n1} \leq x),$$

$$H_n(x) = P(S_n \le x) - \frac{\alpha_2}{\sqrt{2\pi}} x \exp\left(-\frac{x^2}{2}\right) + \frac{\alpha_3}{\sqrt{2\pi}}(1 - x^2) \exp\left(-\frac{x^2}{2}\right),$$

where

$$\alpha_2 = \frac{1}{2} E\,\bar{\delta}_{n1}^2, \quad \alpha_3 = \frac{1}{6}\Big(3E(S_n^2\,\bar{\delta}_{n1}) + 3E(S_n\,\bar{\delta}_{n1}^2) + E\,\bar{\delta}_{n1}^3\Big),$$

and $|\alpha_j| \le c\beta\sigma_1^{-3} n^{-1/2}$, $j = 2, 3$. The corresponding characteristic functions are

$$g_n(t) = E\exp(it\,(S_n + \bar{\delta}_{n1})),$$
$$h_n(t) = E\exp(it\,S_n) + \alpha_2\,(it)^2 \exp\Big(-\frac{t^2}{2}\Big) + \alpha_3\,(it)^3 \exp\Big(-\frac{t^2}{2}\Big).$$

Employing the arguments presented in the previous section, we find the following estimates

$$\int_{-T_n}^{T_n} \Big|g_n(t) - h_n(t)\Big|\,|t|^{-1}\,dt \le c\beta\sigma_1^{-3} n^{-1/2},$$
$$\int_{-T_n}^{T_n} |t|^{s-4}\Big|\frac{d^s}{dt^s}(g_n(t) - h_n(t))\Big|\,dt \le c\beta\sigma_1^{-3} n^{-1/2}, \qquad s = 0, 1, 2, 3.$$

Since by Lemma 6.4.3

$$|G_n(x) - H_n(x)| \le c\,(1 + |x|)^{-3}\Bigg(\int_{-T_n}^{T_n} |t|^{-1}\Big|g_n(t) - h_n(t)\Big|\,dt + \\ + \sum_{s=0}^{3} \int_{-T_n}^{T_n} |t|^{s-4}\Big|\frac{d^s}{dt^s}(g_n(t) - h_n(t))\Big|\,dt + c\beta\sigma_1^{-3} n^{-1/2}\Bigg),$$

we have

$$|G_n(x) - H_n(x)| \le c\beta\sigma_1^{-3}\,(1 + |x|)^{-3}\,n^{-1/2}$$

for all $x \in R$. For given distribution function $H_n(x)$ the conditions of this theorem are satisfied. Since $|G_n(x) - F(x)| \le |H_n(x) - F(x)| + |G_n(x) - H_n(x)|$ the statement of Theorem 6.4.1 is valid.

6.5 Rate of chaos approximation

Consider *UR*-statistics of degree m with a canonical kernel Φ. By Theorem 4.2.1 the weak limit of such *U*-statistics has the distribution of an element of chaos $\mathcal{K}_2(\mathcal{H})$. At first we assume that $m = 2$, i.e.

$$U_n = (n\,(n-1))^{-1} \sum_{1 \le i_1 \ne i_2 \le n} \Phi(X_{i_1}, X_{i_2}),$$

where a symmetric kernel $\Phi\colon X^2 \to R$ satisfies the condition of degeneracy. Let $F_n(x) = P((n-1)\,U_n < x)$ and $F_\infty(x) = \lim F_n(x)$ as $n \to \infty$. If $\Phi \in L^2(X, P)$, then for the limiting distribution function by Corollary 4.2.2, we have $F_\infty(x) = P(K_2(\Phi) < x)$, where the random variable $K_2(\Phi)$ is defined as

$$K_2(\Phi) = \sum_{j=1}^{\infty} \lambda_j \, (\tau_j^2 - 1).$$

In addition $\tau_1, \tau_2, \dots$, are independent standard normal random variables and $\{\lambda_j\}$ are the eigenvalues of the operator

$$S\colon f \to E\Big(\Phi(X_1, X_2)\, f(X_2) \,\Big|\, X_1 = x\Big).$$

In what follows, we assume that there exist infinitely many indices j for which the eigenvalues of this S-operator λ_j are nonzero. If this condition is violated, then the problem of estimating the quantity $|F_n(x) - F_\infty(x)|$ can be reduced to applying the well-known estimates of the convergence rate in the central limit theorem in finite-dimensional spaces.

Theorem 6.5.1. *Suppose that the canonical symmetric kernel $\Phi\colon X^2 \to R$ satisfies the condition*

$$E\,|\Phi(X_1, X_2)|^3 < \infty.$$

Then the following estimate holds

$$\sup |F_n(x) - F_\infty(x)| = o(n^{-1/2})$$

as $n \to \infty$.

Proof. Let us show that in the conditions of the theorem,

$$\sup_x |F_n(x) - F_\infty(x)| \le c_1\, \rho_n n^{-1/2} \tag{6.5.1}$$

where $\rho_n \to 0$ as $n \to \infty$, and the positive constant c_1 (below it may have other indices) depends on the kernel Φ only via eigenvalues $\{\lambda_j\}$ and the expectation $E|\Phi|^3$. In what follows, we shall show that the nature of variable ρ_n in (6.5.1) with the property $\rho_n \to 0$ is connected with a "truncated" kernel Φ_n which is defined as follows. For $x, y \in X$, we let

$$\begin{aligned}\Phi_n(x,y) = \Phi(x,y)\, I\Big(|\Phi(x,y)| \le \sqrt{n}\Big) - E\,\Phi(x,X_2)\, I\Big(|\Phi(x,X_2)| \le \sqrt{n}\Big) - \\ - E\,\Phi(X_1,y)\, I\Big(|\Phi(X_1,y)| \le \sqrt{n}\Big) + E\,\Phi(X_1,X_2)\, I\Big(|\Phi(X_1,X_2)| \le \sqrt{n}\Big),\end{aligned}$$

where $I(A)$ is the indicator function of the set A.

Kernel $\Phi_n(x,y)$ possesses the property of degeneracy. Further, if $x \ne y$, then

$$\Big|\Phi(x,y) - \Phi(x,y)\, I(|\Phi(x,y)| \le \sqrt{n})\Big| \le n^{-1/2}\, |\Phi(x,y)|^2\, I\Big(|\Phi(x,y)| > \sqrt{n}\Big).$$

This yields, for $x \neq y$,

$$\begin{aligned}|\Phi - \Phi_n| \le n^{-1/2} &|\Phi(x,y)|^2 I\big(|\Phi(x,y)| > \sqrt{n}\big) \\ &+ n^{-1/2} E\Big(|\Phi(x,X_1)|^2 I\big(|\Phi(x,X_1)| > \sqrt{n}\big)\Big) \\ &+ n^{-1/2} E\Big(|\Phi(X_2,y)|^2 I\big(|\Phi(X_2,y)| > \sqrt{n}\big)\Big) \\ &+ n^{-1/2} E\Big(|\Phi(X_1,X_2)|^2 I\big(|\Phi(X_1,X_2)| > \sqrt{n}\big)\Big).\end{aligned}$$

Let S_n be a linear operator that corresponds to kernel Φ_n. It is defined the same as operator S. By $\{\lambda_{nj}\}$, we denote the eigenvalues of the S_n-operator, and

$$K_2(\Phi_n) = \sum_{j=1}^{\infty} \lambda_{nj}\,(\tau_j^2 - 1).$$

For $t \in R$, we set

$$\varphi(t,\Phi) = E\exp(itK_2(\Phi)) = \prod_{j=1}^{\infty} e^{-it\lambda_j}(1 - 2it\,\lambda_j)^{-1/2},$$

$$\varphi(t,\Phi_n) = E\exp(itK_2(\Phi_n)) = \prod_{j=1}^{\infty} e^{-it\lambda_{nj}}(1 - 2it\,\lambda_{nj})^{-1/2}.$$

For $x, y \in X$, we set

$$R(t,\Phi)(x,y) = (2it)\Big\{\Phi(x,y) + 2it\int \Phi(x,z)\,\big[(I - 2it\,S)^{-1}\,\Phi(\cdot,y)\big](z)\,P(dz)\Big\},$$

where $(I - 2it\,S)^{-1}$ is the resolvent of operator S.

Lemma 6.5.1. *In the region $|t| \le n^{1/10}$, the following inequality holds*

$$|\varphi(t,\Phi) - \varphi(t,\Phi_n)| \le c_2\, n^{-1/2}\,\beta_n^{2/3}\,|t|\,|(1+t^4)\,|\varphi(t,\Phi)|,$$

where

$$\beta_n = E\,\Big(|\Phi(X_1,X_2)|^3\, I(|\Phi(X_1,X_2)| > \sqrt{n})\Big).$$

Proof. Since, for every complex z, $|1 - \exp(z)| \le |z|\exp(|z|)$ is valid, we obtain

$$|\varphi(t,\Phi) - \varphi(t,\Phi_n)| \le |a_n|\exp(|a_n|)\,|\varphi(t,\Phi)|,$$

where

$$\begin{aligned}a_n &= \frac{1}{2}\sum_{j=1}^{\infty}\Big(\ln(1 - 2it\,\lambda_j) - \ln(1 - 2it\,\lambda_{nj})\Big) \\ &= (-i)\int_0^t \sum_{j=1}^{\infty}\Big(\lambda_j\,(1 - 2iu\,\lambda_j)^{-1} - \lambda_{nj}\,(1 - 2iu\,\lambda_{nj})^{-1}\Big)\,du \\ &= -\frac{1}{2}\int_0^t \int \Big(R(u,\Phi) - R(u,\Phi_n)\Big)\,(x,x)\,P(dx)\,u^{-1}\,du.\end{aligned}$$

By virtue of the definition of $R(t,\Phi)(\cdot,\cdot)$, we get

$$R(t,\Phi)(\cdot,\cdot) = 2it\,(\Phi(\cdot,\cdot)) + \int \Phi(\cdot,z)\,R(t,\Phi)(\cdot,z)\,P(dz).$$

This formula implies that

$$\begin{aligned}\big(R(t,\Phi) - R(t,\Phi_n)\big)(x,y) &= 2it\,(\Phi-\Phi_n)(x,y) + \\ &+ \int \Big(R(t,\Phi)(x,z)\,(\Phi-\Phi_n)(z,y) + R(t,\Phi_n)(z,y)\,(\Phi-\Phi_n)(z,x) + \\ &\quad + R(t,\Phi)(x,z)\,(\Phi-\Phi_n)(z,w)\,R(t,\Phi_n)(w,y)\Big)\,P(dz)\,P(dw).\end{aligned}$$

This yields

$$\begin{aligned}&\Big|\int \big(R(t,\Phi) - R(t,\Phi_n)\big)(x,x)\,P(dx)\Big| \le \\ &\le 2\,|t|\,n^{-1/2} \int |R(t,\Phi)(x,z)|\,\Phi^2(z,x)\,I\big(|\Phi(z,x)| > \sqrt{n}\big)\,P(dx)\,P(dz) \\ &+ 2\,|t|\,n^{-1/2} \int |R(t,\Phi_n)(z,x)|\,\Phi^2(z,x)\,I\big(|\Phi(z,x)| > \sqrt{n}\big)\,P(dx)\,P(dz) \\ &+ 2\,|t|\,n^{-1/2} \int |R(t,\Phi)(x,z)|\,|R(t,\Phi_n)(w,x)| \times \\ &\qquad \times \Phi^2(z,w)\,I\big(|\Phi(z,w)| > \sqrt{n}\big)\,P(dz)\,P(dw).\end{aligned}$$

Since $\|(I-2it\,S)^{-1}\| \le 1$ for any symmetric operator S, then

$$|R(t,\Phi)(x,y)| \le 2\,|t|\,|\Phi(x,y)| + 4\,t^2\,(E\,|\Phi(X_1,x)|^3)^{1/3}\,(E\,|\Phi(X_1,y)|^3)^{1/3}.$$

With the help of this inequality and by using Hölder's inequality, we obtain

$$\begin{aligned}&\int |R(t,\Phi)(x,z)|\,\Phi^2(z,x)\,I\big(|\Phi(z,x)| > \sqrt{n}\big)\,P(dx)\,P(dz) \le \\ &\quad \le 2\,|t|\,E\big(|\Phi(X_1,X_2)|^3\,I(|\Phi(X_1,X_2)| > \sqrt{n})\big) + \\ &\qquad + 4\,t^2 \int \big(E\,|\Phi(X_1,x)|^3\big)^{1/3} \big(E\,|\Phi(X_1,y)|^3\big)^{1/3} \times \\ &\qquad\qquad \times \Phi^2(y,x)\,I(|\Phi(y,x)| > \sqrt{n})\,P(dx)\,P(dy) \le \\ &\quad \le 2\,|t|\,\beta_n + 4\,t^2\,\big(E\,|\Phi(X_1,X_2)|^3\big)^{2/3}\,\beta_n^{2/3}.\end{aligned}$$

By virtue of Hölder's inequality, we also have

$$\begin{aligned}&\int |R(t,\Phi)(x,z)|\,|R(t,\Phi_n)(w,x)|\,\Phi^2(z,w)\,I(|\Phi(z,w)| \ge \sqrt{n})\,P(dx)\,P(dz)\,P(dw) \le \\ &\le \beta_n^{2/3} \int \Big(\int |R(t,\Phi)(x,z)|^3\,P(dz)\Big)^{1/3} \Big(\int |R(t,\Phi_n)(w,x)|^3\,P(dw)\Big)^{1/3} P(dx) \le \\ &\qquad \le c_3\,t^2\,(1+|t|)^2\,\beta_n^{2/3}.\end{aligned}$$

Combining these estimates we obtain the following estimate for a_n

$$|a_n| \le c_4\,|t|\,(1+t^4)\,n^{-1/2}\,\beta_n^{2/3}.$$

This yields Lemma 6.5.1.

Note that under the conditions of the theorem $\beta_n \to 0$ as $n \to \infty$. Lemma 6.5.1 implies the following inequality

$$|\varphi(t, \Phi_n)| \leq c_5 \, |\varphi(t, \Phi)|,$$

for $|t| \leq n^{1/10}$. We set $\varphi_n(t, \Phi) = E \exp(it\,(n-1)\,U_n)$. For $\varphi_n(t, \Phi)$, it was proved by Borovskikh and Korolyuk (1984c), that the following inequality holds in the region $|t| \leq n^{1-\varepsilon}$ for any $\varepsilon > 0$

$$|\varphi_n(t, \Phi)| \leq c_6 \left((1 + |t|)^{-q_1} + n^{-q_2}\right) \tag{6.5.2}$$

with large arbitrary numbers $q_i \geq 1$.

Lemma 6.5.2. *In the region $|t| \leq n^{1-\varepsilon}$, the following inequality holds,*

$$|\varphi_n(t, \Phi) - \varphi_n(t, \Phi_n)| \leq c_7 \, |t| \, (1 + t^4)\left((1 + |t|)^{-q_1} + n^{-q_2}\right) \rho_n n^{-1/2},$$

where the value $\rho_n \to 0$ as $n \to \infty$.

Proof. Following Götze (1979), we can write the formula

$$\varphi_n(t, \Phi) - \varphi_n(t, \Phi_n) = \sum_{\substack{j,r=1 \\ j \neq r}}^{n} \int \varphi_{jr}(x_j, x_r) \left(1 - \exp(itn^{-1}\,(\Phi_n - \Phi)(x_j, x_r))\right) P(dx_j, dx_r),$$

where

$$\varphi_{jr}(x_j, x_r) = \int \exp\left(itn^{-1} \sum_{\substack{m,p=1 \\ m \neq p}}^{n} \Phi_{mp}^{jr}(x_m, x_p)\right) \prod_{m \neq j,r} P(dx_m),$$

$$\Phi_{mp}^{jr} = \begin{cases} \Phi, & \text{if } (m,p) \geq (j,r), \\ \Phi_n & \text{otherwise.} \end{cases}$$

Each term on the right-hand side of this formula can be estimated by

$$\left|\int \varphi_{jr}(x,y) \left(1 + itn^{-1}\,(\Phi - \Phi_n)(x,y) - \exp(itn^{-1}\,(\Phi_n - \Phi)(x,y))\right) P(dx)\,P(dy)\right| +$$
$$+ |t|\,n^{-1} \left|\int \varphi_{jr}(x,y)\,(\Phi_n - \Phi)(x,y)\,P(dx)\,P(dy)\right|. \tag{6.5.3}$$

Since $|1 + iu - \exp(iu)| \leq u^2/2$, for $u \in R$, the first term in (6.5.3) does not exceed the following expression,

$$2\,t^2 n^{-2} \int |\varphi_{jr}(x,y)| \left\{\Phi^2(x,y)\, I\left(|\Phi(x,y)| > \sqrt{n}\right) + \right.$$
$$+ \left(E\left(\Phi(x, X_1)\, I(|\Phi(x, X_1)| > \sqrt{n})\right)\right)^2 +$$

$$+\left(E\left(\Phi(X_2,y)\,I(|\Phi(X_2,y)|>\sqrt{n})\right)\right)^2+$$
$$+\left(E\left(\Phi(X_1,X_2)\,I(|\Phi(X_1,X_2)|>\sqrt{n})\right)\right)^2\Big\}\,P(dx)\,P(dy). \tag{6.5.4}$$

It follows from the proof of Lemma 2.5 in Borovskikh and Korolyuk (1984c) that for $\varphi_{jr}(x,y)$ the following estimate is valid

$$|\varphi_{jr}(x,y)|\le c_8\left((1+|t|)^{-q_1}+n^{-q_2}\right) \tag{6.5.5}$$

in the region $|t|\le n^{1-\varepsilon}$ with large arbitrary $q_i\ge 1$. Therefore, in the region $|t|\le n^{1-\varepsilon}$, the integral in (6.5.4) does not exceed the following expression

$$c_9\,n^{-1/2}\,\beta_n\left((1+|t|)^{-q_1}+n^{-q_2}\right).$$

Let us estimate the second term in (6.5.3). For simplicity, we assume that $j<r$. The case $j>r$ can be considered quite similarly. Denote

$$r_n(t)=\int\varphi_{jr}(x_r,x_j)\,(\Phi_n-\Phi)(x_r,x_j)\,P(dx_r)\,P(dx_j).$$

We represent $r_n(t)$ in a more convenient form by taking into account the definition of $\varphi_{jr}(x,y)$. At first, we write the representation

$$\sum_{m,p=1}^{n}\Phi^{jr}_{mp}(x_m,x_p)=H+a(x_r)+b(x_j)+h(x_r,x_j) \tag{6.5.6}$$

where

$$H=\sum_{m=j+1}^{n}{}^{(r)}\sum_{p=r+1}^{n}\Phi(x_m,x_p)+\sum_{m=1}^{j-1}\sum_{p=r+1}^{n}\Phi_n(x_m,x_p)$$
$$+\sum_{m=1}^{j-1}\sum_{p=1}^{r-1}{}^{(j)}\Phi_n(x_m,x_p)+\sum_{m=j+1}^{n}{}^{(r)}\sum_{p=1}^{r-1}{}^{(j)}\Phi_n(x_m,x_p),$$

$$a(x_r)=\sum_{p=r+1}^{n}\Phi(x_p,x_r)+\sum_{m=j+1}^{n}{}^{(r)}\Phi(x_m,x_r)$$
$$+\sum_{m=1}^{j-1}\Phi_n(x_m,x_r)+\sum_{p=1}^{r-1}{}^{(j)}\Phi_n(x_p,x_r),$$

$$b(x_j)=\sum_{p=r+1}^{n}\Phi(x_p,x_j)+\sum_{m=1}^{j-1}\Phi_n(x_m,x_j)$$
$$+\sum_{m=j+1}^{n}{}^{(r)}\Phi_n(x_m,x_j)+\sum_{p=1}^{r-1}{}^{(j)}\Phi_n(x_p,x_j),$$

$$h(x_r,x_j)=\Phi(x_r,x_j)-\Phi_n(x_r,x_j);$$

here $\sum^{(r)}$ denotes the summation over all corresponding indices except r. On the left and on the right-hand side of (6.5.6) we assume, that all terms corresponding to $m = p$ are equal to zero. With the help of (6.5.6), we can write $r_n(t)$ in the form

$$r_n(t) = \int P(dx_r)\, P(dx_j)\, (\Phi_n - \Phi)(x_r, x_j) \exp(itn^{-1}\, h(x_r, x_j)) \times \\ \times E \exp(itn^{-1}\, H) \exp(itn^{-1}\, a(x_r) + itn^{-1}\, b(x_j)).$$

To transform this relation, we employ the formula

$$\exp(x) = 1 + x \int_0^1 \exp(xu)\, du$$

for $x = itn^{-1}\, h(x_r, x_j)$, $x = itn^{-1}\, a(x_r)$, and $x = itn^{-1}\, b(x_j)$, successively. As a result, taking the fact that the function $(\Phi_n - \Phi)(x, y)$ is degenerate into account, we get

$$r_n(t) = r_{n1}(t) + r_{n2}(t),$$

where

$$r_{n1}(t) = it\, n^{-1} \int_0^1 du_1 \int P(dx_r)\, P(dx_j)\, (\Phi_n - \Phi)(x_r, x_j)\, h(x_r, x_j) \times \\ \times \exp(itu_1 n^{-1}\, h(x_r, x_j))\, E \exp(itn^{-1}\, H) \exp(itn^{-1}\, a(x_r)) + itn^{-1}\, b(x_j)),$$

$$r_{n2}(t) = -t^2\, n^{-1} \int_0^1 du_2 \int_0^1 du_3 \int P(dx_r)\, P(dx_j)\, (\Phi_n - \Phi)(x_r, x_j) \times \\ \times E \left\{ a_n(x_r)\, b_n(x_j) \exp(itn^{-1}\, H) \exp(itu_2 n^{-1}\, a(x_r) + itu_3 n^{-1}\, b(x_j)) \right\},$$

where $a_n(x_r) = n^{-1/2}\, a(x_r)$ and $b_n(x_j) = n^{-1/2}\, b(x_j)$. First, we estimate $r_{n1}(t)$ as follows

$$|r_{n1}(t)| \le |t|\, n^{-1} \int P(dx_r)\, P(dx_j) \left| (\Phi_n - \Phi)(x_r, x_j)\, h(x_r, x_j) \right| \times \\ \times \left| E \exp(itn^{-1}\, H) \exp(itn^{-1}\, a(x_r) + itn^{-1}\, b(x_j)) \right|.$$

Here we estimate the expectation under the modulus sign according to the symmetrization inequality (2.3.2) in Borovskikh and Korolyuk (1994a). As a result, we find that in the region $|t| \le n^{1-\varepsilon}$

$$|r_{n1}(t)| \le c_{10}\, |t|\, n^{-3/2}\, \rho_n\, ((1 + |t|)^{-q_1} + n^{-q_2}). \tag{6.5.7}$$

Consider now $r_{n2}(t)$. To estimate the expectation with the help of the symmetrization inequality (see Lemma 2.1 in Borovskikh and Korolyuk (1984c)), one should first divide $b_n(x_j)$ into the sum of two terms $b_{n1}(x_j)$ and $b_{n2}(x_j)$, where $b_{n1}(x_j)$ depends only on random variables $X_1, \ldots, X_k$ (except X_r and X_j), and $b_{n2}(x_j)$ depends only on $X_{k+1}, \ldots, X_n$ (except X_r and X_j); here, k is an arbitrary integer such that $1 \le k \le n$.

As in the case of $b_n(x_j)$, we represent $a_n(x_r)$ as a sum of two terms $a_{n1}(x_r)$ and $a_{n2}(x_r)$. Now

$$a_n(x_r)\, b_n(x_j) = (a_{n1}(x_r) + a_{n2}(x_r))\, (b_{n1}(x_j) + b_{n2}(x_j)) =$$

$$= a_{n1}(x_r)\, b_{n1}(x_j) + a_{n2}(x_r)\, b_{n1}(x_j) + a_{n1}(x_r)\, b_{n2}(x_j) + a_{n2}(x_r)\, b_{n2}(x_j).$$

Therefore, $r_{n2}(t)$ divides into the sum of four integral expressions, and in each we have one of the products $a_{np}(x_r)\, b_{ns}(x_j)$, p, $s = 1, 2$, instead of $a_n(x_r)\, b_n(x_j)$. In the case where $p \neq s$, we estimate these integral expressions with the help of Hölder's inequality by the following value

$$c_{11}\, t^2\, n^{-3/2}\, \rho_n \int_0^1 du_2 \int_0^1 du_3 \int \Big(P(dx_r)\, P(dx_j) \times$$

$$\times \Big| E\Big\{ a_{np}(x_r)\, b_{ns}(x_j) \exp(itn^{-1}\, H) \exp(itu_2 n^{-1}\, a(x_r) + itu_3 n^{-1}\, b(x_j))\Big\}\Big|^3\Big)^{1/3}.$$

Employing the symmetrization inequality again, we obtain the estimate of the form (6.5.7).

In the case of $p = s$, we employ the following argument. For simplicity, let $p = s = 1$. By definition, we have

$$a_{n1}(x_r) = n^{-1/2} \sum_{m=1}^{k} f_1(X_m, x_r),$$

$$b_{n1}(x_j) = n^{-1/2} \sum_{l=1}^{k} f_2(X_l, x_j),$$

where f_1 and f_2 can be written in an explicit form according to (6.5.6). Therefore, $a_{n1}(x_r)\, b_{n1}(x_j) = A_n(x_r, x_j) + n^{-1/2}\, B_n(x_r, x_j)$, where

$$A_n(x_r, x_j) = n^{-1} \sum_{m \neq l} f_1(X_m, x_r)\, f_3(X_l, x_j),$$

$$B_n(x_r, x_j) = n^{-1/2} \sum_{m=1}^{k} f_1(X_m, x_r)\, f_2(X_m, x_j).$$

We estimate the integral expression which contains $A_n(x_r, x_j)$ with the help of Hölder's inequality as follows

$$c_{12}\, t^2 n^{-3/2}\, \rho_n \int_0^1 du_2 \int_0^1 du_3 \Big(\int P(dx_r)\, P(dx_j) \times$$

$$\times \Big| E\Big\{ A_n(x_r, x_j) \exp(itn^{-1}\, H) \exp(itu_2 n^{-1}\, a(x_r) + it\, u_2 n^{-1} b(x_j))\Big\}\Big|^3\Big)^{1/3}.$$

Then, by using the symmetrization inequality, we obtain the estimate of type (6.5.7). Finally, let us estimate the integral expression which contains $B_n(x_r, x_j)$ with the help of Hölder's inequality. We have

$$t^2\, n^{-3/2} \int_0^1 du_2 \int_0^1 du_3 \Big(\int P(dx_r)\, P(dx_j)\, |(\Phi_n - \Phi)(x_r, x_j)|^3\Big)^{1/3}$$

$$\times \Big(\int P(dx_r)\, P(dx_j) \Big| E\Big\{ B_n(x_r, x_j) \exp(itn^{-1}\, H) \times$$

$$\times \exp(itu_2 n^{-1} a(x_r) + itu_3 n^{-1} b(x_j))\}\Big|^{3/2}\Big)^{2/3};$$

then, by using the symmetrization inequality, we derive the estimate of type (6.5.7). Thus, under the conditions of the theorem, we obtain

$$|r_n(t)| \le c_{13}\, n^{-3/2}\, \rho_n\, |t|\,(1+|t|)\,(1+|t|^{-q_1}+n^{-q_2})$$

in the region $|t| \le n^{1-\epsilon}$. This completes the proof of Lemma 6.5.2.

Lemma 6.5.3. *The following inequality holds in the region* $|t| \le n^{1/10}$

$$|\varphi_n(t,\Phi_n) - \varphi(t,\Phi_n)| \le c_{14}\,|t|\,(1+t^4)\,((1+|t|)^{-q_1}+n^{-q_2})\,\rho_n\, n^{-1/2} + c_{15}\, t^2\, n^{-1},$$

where $\rho_n \to 0$ *an* $n \to \infty$.

Proof. By virtue of formula (4.2.4)

$$\varphi(t,\Phi_n) = E\exp\Big(it\int_X\int_X \Phi_n(x,y)\, W(dx)\, W(dy)\Big).$$

Further, for $k = 1, \ldots, n$ we denote

$$\zeta_{nk} = n^{-1}\sum_{1\le i_1\ne i_2\le k}\Phi_n(X_{i_1},X_{i_2}) +$$
$$+\, n^{-1}\sum_{i_1=1}^{k}\sum_{i_2=k+1}^{n}\Big(\int_X \Phi_n(x,X_{i_2})\, W_{i_1}(dx) + \int_X \Phi_n(X_{i_1},y)\, W_{i_2}(dy)\Big) +$$
$$+\, n^{-1}\sum_{k+1\le i_1\ne i_2\le n}\int_X\int_X \Phi_n(x,y)\, W_{i_1}(dx)\, W_{i_2}(dy),$$

where $W_1(\cdot)$, $W_2(\cdot)$, ... are the independent copies of a Gaussian random measure $W(\cdot)$. For $t \in R$, we let

$$\varphi_{nk}(t,\Phi_n) = E\exp(it\,\zeta_{nk}).$$

It is obvious that $\varphi_{nn}(t,\Phi_n) = \varphi_n(t,\Phi_n)$. In this notation, we have the formula

$$\varphi_n(t,\Phi_n) - \varphi(t,\Phi_n) = (\varphi_{n0}(t,\Phi_n) - \varphi(t,\Phi_n)) + \\ + \sum_{k=1}^{n}(\varphi_{nk}(t,\Phi_n) - \varphi_{n\,k-1}(t,\Phi_n)). \qquad (6.5.8)$$

It can be seen that (6.5.8) is the analogue of formula (6.1.5).

Consider in (6.5.8) $\varphi_{n0}(t,\Phi_n) - \varphi(t,\Phi_n)$ and $\varphi_{nk}(t,\Phi_n) - \varphi_{n\,k-1}(t,\Phi_n)$, $k = 1, \ldots, n$. By the infinite divisibility property of a Gaussian measure,

$$W(B) \overset{d}{=} n^{-1/2}\sum_{i=1}^{n} W_i(B), \qquad B \in \mathfrak{X}$$

and for the disjoint sets B_1, B_2

$$W(B_1)\,W(B_2) \stackrel{\mathrm{d}}{=} n^{-1}\Big(\sum_{i_1=1}^{n} W_{i_1}(B_1)\Big)\Big(\sum_{i_2=1}^{n} W_{i_2}(B_2)\Big) =$$

$$= n^{-1} \sum_{1\le i_1\ne i_2\le n} W_{i_1}(B_1)\,W_{i_2}(B_2) + n^{-1}\sum_{i=1}^{n} W_i(B_1)\,W_i(B_2).$$

Hence,

$$I_2^W(\Phi_n) \stackrel{\mathrm{d}}{=} \zeta_{n0} + \alpha_n$$

with

$$\alpha_n = n^{-1}\sum_{i=1}^{n} I_2^{W_i}(\Phi_n).$$

Then, we get the representation

$$\varphi_{n0}(t,\Phi_n) - \varphi(t,\Phi_n) = E\,e^{it\zeta_{n0}}\,(1-e^{it\alpha_n}).$$

Further, for all $k = 1, \ldots, n$ we introduce the following notations

$$a_{nk} = 2n^{-1}\sum_{i_1=1}^{k-1} \Phi_n(X_{i_1},X_k) +$$

$$+ n^{-1}\sum_{i_2=k+1}^{n}\Big(\int_X \Phi_n(x,X_{i_2})\,W_k(dx) + \int_X \Phi_n(X_k,y)\,W_{i_2}(dy)\Big),$$

$$b_{nk} = 2n^{-1}\sum_{i_2=k+1}^{n}\int_X\int_X \Phi_n(x,y)\,W_k(dx)\,W_{i_2}(dy) +$$

$$+ n^{-1}\sum_{i_1=1}^{k-1}\Big(\int_X \Phi_n(x,X_k)\,W_{i_1}(dx) + \int_X \Phi_n(X_{i_1},y)\,W_k(dy)\Big),$$

$$z_{nk} = n^{-1}\sum_{1\le i_1\ne i_2\le k-1} \Phi_n(X_{i_1},X_{i_2}) +$$

$$+ n^{-1}\sum_{i_1=1}^{k-1}\sum_{i_2=k+1}^{n}\Big(\int_X \Phi_n(x,X_{i_2})\,W_{i_1}(dx) + \int_X \Phi_n(X_{i_1},y)\,W_{i_2}(dy)\Big) +$$

$$+ n^{-1}\sum_{k+1\le i_1\ne i_2\le n}\int_X\int_X \Phi_n(x,y)\,W_{i_1}(dx)\,W_{i_2}(dy).$$

Note that

$$\zeta_{nk} = z_{nk} + a_{nk}, \quad \zeta_{n\,k-1} = z_{nk} + b_{nk}.$$

In these notations we get the representation

$$\varphi_{nk}(t,\Phi_n) - \varphi_{n\,k-1}(t,\Phi_n) = E\,e^{it\,z_{nk}}\,(e^{it\,a_{nk}} - e^{it\,b_{nk}})$$

for all $k = 1, \ldots, n$.

Therefore, formula (6.5.8) reduces to the form

$$\varphi_n(t,\Phi_n) - \varphi(t,\Phi_n) = E\,e^{it\zeta_{n0}}\,(1-e^{it\,\alpha_n}) + \sum_{k=1}^{n} E\,e^{it\,z_{nk}}\,(e^{it\,a_{nk}} - e^{it\,b_{nk}}). \tag{6.5.9}$$

To estimate the expectations in (6.5.9), it is sufficient to apply the method of the proof of Lemma 6.6.3 in Borovskikh and Korolyuk (1994a). Here the detailed proof is omitted.

Now, let us show that the lemmas presented above yield (6.5.1). In fact, according to the Esséen inequality, we have

$$\sup_x |F_n(x) - F_\infty(x)| \le \frac{1}{\pi} \int_{-n^{1-\varepsilon}}^{n^{1-\varepsilon}} |\varphi_n(t,\Phi) - \varphi(t,\Phi)|\, |t|^{-1}\, dt + c\, n^{-1+\varepsilon}, \tag{6.5.10}$$

with some constant $c > 0$. For (6.5.10), we note that the distribution function $F_\infty(x)$ has a bounded density.

Consider the integral in (6.5.10). By virtue of (6.5.2) and the assumption that the number of nonzero eigenvalues λ_j is infinite, we get

$$\int_{n^{1/10} \le |t| \le n^{1-\varepsilon}} |\varphi_n(t,\Phi) - \varphi(t,\Phi)|\, |t|^{-1}\, dt = O(n^{-1})$$

as $n \to \infty$. In the region $|t| \le n^{1/10}$ we have the inequality

$$|\varphi_n(t,\Phi) - \varphi(t,\Phi)| \le |\varphi(t,\Phi) - \varphi(t,\Phi_n)| + |\varphi_n(t,\Phi) - \varphi_n(t,\Phi_n)| + |\varphi_n(t,\Phi_n) - \varphi(t,\Phi_n)|,$$

and apply Lemmas 6.5.1, 6.5.2 and 6.5.3, respectively. As a result we obtain (6.5.1) and hence the proof of Theorem 6.5.1.

Now we consider U-statistics of degree $m \ge 3$. Let $F_{nr}(x) = P(n^{r/2} U_n < x)$ and $F_{\infty r}(x) = \lim F_{nr}(x)$ as $n \to \infty$, where r is a rank of kernel $\Phi\colon X^m \to R$. Under the conditions of Theorem 4.2.3, we have

$$\sup_x |F_{nr}(x) - F_{\infty r}(x)| \to 0.$$

The permanent variant of the composition method in Borovskikh and Korolyuk (1994b) and the method presented in the proof of Theorem 6.5.1 allow us to obtain the following estimates of the convergence rate in this situation.

Theorem 6.5.2. *Let $m \ge 2$ and*

$$\eta_\infty = \int_X \cdots \int_X g_r(x_1, \ldots, x_r)\, W(dx_1) \ldots W_r(dx_r).$$

Suppose that

$$|E \exp(it\, \eta_\infty)| = O(|t|^{-\gamma})$$

as $|t| \to \infty$, where $\gamma > 0$ is some sufficiently large number. Then we have

$$\sup_x |F_{nr}(x) - F_{\infty r}(x)| = O(n^{-1/2})$$

for $E\,|\Phi|^3 < \infty$ and odd rank $r \ge 1$, and

$$\sup_x |F_{nr}(x) - F_{\infty r}(x)| = o(n^{-1/2})$$

for $E\,|\Phi|^4 < \infty$ and even rank $r \ge 2$.

6.6 Normal approximation of *UH*-statistics

Let us consider the rate of convergence on balls in the central limit theorem for *UH*-statistics (Theorem 4.1.2).

Estimates of order $n^{-1/2}$ with the explicit dependence on the covariance operator. Let $X_1, \ldots, X_n$ be independent random variables with values in a measurable space $(X, \mathfrak{X})$ with the same distribution P in it. Let H be a real separable Hilbert space with inner product $(\cdot,\cdot)$ and norm $\|\cdot\|$. Consider the *UH*-statistics

$$U_n = \binom{n}{2}^{-1} \sum_{1\le i<j\le n} \Phi(X_i, X_j)$$

with a symmetric kernel $\Phi\colon X^2 \to H$ such that $E\,\Phi(X_1,X_2) = 0$, $E\,\|\Phi(X_1,X_2)\| < \infty$.

By the Hoeffding decomposition

$$U_n = 2n^{-1}\sum_{j=1}^{n} g_1(X_j) + 2n^{-1}(n-1)^{-1} \sum_{1\le i<j\le n} g_2(X_i,X_j),$$

where

$$g_1(x) = \int_X \Phi(x,y)\,P(dy) = E\,(\Phi(X_1,X_2) \mid X_1 = x),$$
$$g_2(x,y) = \Phi(x,y) - g_1(x) - g_1(y).$$

Assume that here $\sigma^2 := E\,\|g_1(X_1)\|^2 > 0$. Assuming also that $\sigma < \infty$ we can define the covariance operator V of $g_1(X_1)$ in the usual way:

$$(Vx,y) = \int_X (x, g_1(z))\,(y, g_1(z))\,P(dz), \qquad x,y \in H.$$

Let τ be a Gaussian H-valued random variable with mean zero and covariance operator $\sigma^{-2}V$. Denote $\sigma_1^2 \ge \sigma_2^2 \ge \cdots$ the eigenvalues of V. For any $r \ge 0$, $a \in H$, let $B_r(a) = \{x \in H\colon \|x-a\| < r\}$ and $B_r = B_r(0)$.

By the central limit theorem for *UH*-statistics (see Theorem 4.1.2) if

$$E\,\|g_1(X_1)\|^2 < \infty, \quad E\,\|g_2(X_1,X_2)\|^{4/3} < \infty,$$

then

$$\Delta_n(a) = \sup_r \left| P(2^{-1}n^{1/2}\sigma^{-1}U_n \in B_r(a)) - P(\tau \in B_r(a)) \right| \to 0, \qquad n \to \infty.$$

In the following, we study the dependence of $\Delta_n(a)$ on n, the covariance operator V and $\beta = E\,\|\Phi(X_1,X_2)\|^3$. When $\beta < \infty$, an estimate of order $n^{-1/2}$ is obtained, the dependence of which on Φ involves only β, the first nine eigenvalues of V and $\sigma^2 = \operatorname{tr} V$.

Theorem 6.6.1. *There exists an absolute constant c such that for all $n \geq 2$ and $a \in H$*

$$\Delta_n(a) \leq c\,(1 + \|a\|^3)\,\sigma^4 \left(\prod_{j=1}^{9} \sigma_j^{-1} \right) \sigma_9^{-1}\, \beta^2\, n^{-1/2}. \tag{6.6.1}$$

Remark 6.6.1. In what follows we will assume that $\sigma = 1$. The general case is reduced to this one if we replace Φ by $\sigma^{-1}\,\Phi$. We assume also that $\beta < \infty$ and $\sigma_9 > 0$ since otherwise (6.6.1) is obvious. In the proof of this theorem the upper estimates for $\Delta_n(a)$ are calculated more precisely; except for n and V, which depend only on the moments

$$E\,\|g_1\|^3 = E\,\|g_1(X_1)\|^3, \quad E\,\|g_2\|^3 = E\,\|g_2(X_1, X_2)\|^3,$$

which are related to β by the inequalities

$$E\,\|g_1\|^3 \leq \beta, \quad E\,\|g_2\|^3 \leq 27\beta.$$

We formulate this theorem using only β for the sake of simplicity.

Proof. Below, c, c_1, c_2, ... denote absolute constants; the same symbol may stand for different constants. The proof is divided into several lemmas.

Lemma 6.6.1. *We have*

$$\Delta_n(a) \leq \Delta_{n1}(a) + n\,P(\|g_1(X_1)\| > n^{1/2}),$$

where

$$\Delta_{n1}(a) = \sup_r \left| P(\|\bar{S}_n - a + v_n\| < r) - P(\|\tau - a\| < r) \right|,$$

$$\bar{S}_n = n^{-1/2} \sum_{j=1}^{n} g_1(X_j)\, I(\|g_1(X_j)\| \leq n^{1/2}),$$

$I(A)$ is the indicator function of set A and

$$v_n = (n-1)^{-1}\, n^{-1/2} \sum_{j=1}^{n} \sum_{i=1}^{j-1} g_2(X_i, X_j).$$

Proof. According to the Hoeffding decomposition

$$2^{-1}\, n^{1/2}\, U_n - a = S_n - a + v_n,$$

where

$$S_n = n^{-1/2} \sum_{j=1}^{n} g_1(X_j).$$

Furthermore, for any $r \geq 0$ we can write

$$P(\|S_n - a + v_n\| < r) - P(\|\bar{S}_n - a + v_n\| < r) =$$
$$= \sum_{k=1}^{n} \Big(P(\|z_{nk} + n^{-1/2} g_1(X_k) + v_n\| < r) -$$
$$- P(\|z_{nk} + n^{-1/2} g_1(X_k) I(\|g_1(X_k)\| \leq n^{1/2}) + v_n\| < r) \Big),$$

where

$$z_{nk} = n^{-1/2} \sum_{j=1}^{k-1} g_1(X_j) I(\|g_1(X_j)\| \leq n^{1/2}) + n^{-1/2} \sum_{j=k+1}^{n} g_1(X_j) - a.$$

Moreover

$$P(\|z_{nk} + n^{-1/2} g_1(X_k) + v_n\| < r) =$$
$$= \int_X P(\|z_{nk} + n^{-1/2} g_1(x_k) + \tilde{v}_n\| < r) P(dx_k)$$
$$= \int_B P(\|z_{nk} + n^{-1/2} g_1(x_k) + \tilde{v}_n\| < r) P(dx_k)$$
$$+ \int_{\bar{B}} P(\|z_{nk} + n^{-1/2} g_1(x_k) + \tilde{v}_n\| < r) P(dx_k),$$

and

$$P\Big(\Big\|z_{nk} + n^{-1/2} g_1(X_k) I(\|g_1(X_k)\| \leq n^{1/2}) + v_n\Big\| < r\Big) =$$
$$= \int_X P\Big(\Big\|z_{nk} + n^{-1/2} g_1(x_k) I(\|g_1(x_k)\| \leq n^{1/2}) + \tilde{v}_n\Big\| < r\Big) P(dx_k)$$
$$= \int_B P(\|z_{nk} + n^{-1/2} g_1(x_k) + \tilde{v}_n\| < r) P(dx_k)$$
$$+ \int_{\bar{B}} P(\|z_{nk} + \tilde{v}_n\| < r) P(dx_k),$$

where $\bar{B} = X \backslash B$, $B = \{x \in X : \|g_1(x)\| \leq n^{1/2}\}$, $\tilde{v}_n = v_n$ with $X_k = x_k$.

Since $\Big|P(\|z_{nk} + n^{-1/2} g_1(x_k) + v_n\| < r) - P(\|z_{nk} + v_n\| < r)\Big| \leq 1$, the above observations imply Lemma 6.6.1.

Let $k = n - [n^{2/3}]$, where $[n^{2/3}]$ is the integral part of $n^{2/3}$ and denote

$$\alpha_k = \sum_{1 \leq i < j \leq k} g_2(X_i, X_j), \qquad \beta_k = \sum_{i=1}^{k} \sum_{j=k+1}^{n} g_2(X_i, X_j),$$

$$\delta_k = \sum_{k+1 \leq i < j \leq n} g_2(X_i, X_j), \qquad \bar{S}_n^k = n^{-1/2} \sum_{j=1}^{k} \bar{g}_1(X_j),$$

$$\bar{g}_1(X_j) = g_1(X_j)\, I(\|g_1(X_j)\| \le n^{1/2}), \qquad j = 1, \dots, n,$$

$$z_n = 2\varepsilon_n\, (\bar{S}_n^k - a, \alpha_k), \qquad \varepsilon_n = (n-1)^{-1}\, n^{-1/2}.$$

Lemma 6.6.2. *For $\Delta_{n1}(a)$ we have*

$$\Delta_{n1}(a) \le \Delta_{n2}(a) + c_1\,(1 + \|a\|^3)\,(\beta^2 + \sigma_1^{-1}\sigma_2^{-1})\, n^{-1/2},$$

where

$$\Delta_{n2}(a) = \sup_r \Big| P(\|\bar{S}_n - a\|^2 + z_n < r) - P(\|\tau - a\|^2 < r) \Big|.$$

Proof. First observe that

$$\begin{aligned} \|\bar{S}_n - a + v_n\|^2 &= (\bar{S}_n - a + v_n, \bar{S}_n - a + v_n) \\ &= \|\bar{S}_n - a\|^2 + 2\,(\bar{S}_n - a, v_n) + \|v_n\|^2. \end{aligned}$$

In our notation

$$\|v_n\|^2 = (n-1)^{-2}\, n^{-1} \Big\| \sum_{j=2}^{n} \sum_{i=1}^{j-1} g_2(X_i, X_j) \Big\|^2,$$

$$2\,(\bar{S}_n - a, v_n) = z_n + 2\varepsilon_n(\bar{S}_n^k - a, \beta_k + \delta_k) + 2\,(\bar{S}_n - \bar{S}_n^k, v_n),$$

and

$$(\bar{S}_n^k - a, \beta_k + \delta_k) = (\bar{S}_n^k, \beta_k) - (a, \beta_k) + (\bar{S}_n^k, \delta_k) - (a, \delta_k),$$

$$(\bar{S}_n - \bar{S}_n^k, v_n) = \varepsilon_n\,(\bar{S}_n - \bar{S}_n^k, \alpha_k) + \varepsilon_n\,(\bar{S}_n - \bar{S}_n^k, \beta_k) + \varepsilon_n\,(\bar{S}_n - \bar{S}_n^k, \delta_k).$$

Further we will need the following inequalities for any $p \ge 2$:

$$E\,\|\bar{S}_n^k\|^p \le c_p, \quad \|E\,\bar{g}_1(X_1)\| = \Big\| \int_{\|g_1(x)\| > n^{1/2}} g_1(x)\, P(dx) \Big\| \le n^{-1/2} \tag{6.6.2}$$

(see Sazonov (1981)) and

$$\begin{aligned} &E\,\|\alpha_k\|^p \le c_p\, E\,\|g_2\|^p\, k^p, \\ &E\,\|\beta_k\|^p \le c_p\, E\,\|g_2\|^p\, k^{p/2}\,(n-k)^{p/2}, \\ &E\,\|\delta_k\|^p \le c_p\, E\,\|g_2\|^p\,(n-k)^p \end{aligned} \tag{6.6.3}$$

(see Chapter 2).

The Chebyshev inequality and (6.6.3) imply

$$P(\|v_n\|^2 \ge n^{-1/2}) \le n^{1/2}\, E\,\|v_n\|^2 \le c_1\, E\,\|g_2\|^2\, n^{-1/2}.$$

Furthermore

$$P\Big(|(\bar{S}_n^k, \varepsilon_n \beta_k)| > n^{-1/2}\Big) \le p_1 + p_2,$$

where

$$p_1 = P\Big(2k\varepsilon_n \Big|\Big(E\,\bar g_1(X_1), \sum_{i=1}^{k} h(X_i)\Big)\Big| > 1\Big),$$

$$p_2 = P\Big(2\varepsilon_n \Big|\Big(\sum_{i=1}^{k} \bar g(X_i), \sum_{i=1}^{k} h(X_i)\Big)\Big| > 1\Big),$$

$$\bar g(X_i) = \bar g_1(X_i) - E\,\bar g_1(X_1), \qquad i = 1, \dots, k,$$

$$h(X_i) = \sum_{j=k+1}^{n} g_2(X_i, X_j), \qquad i = 1, \dots, k.$$

For p_1 by the Chebyshev inequality and (6.6.2)-(6.6.3)

$$p_1 \le (2k\varepsilon_n)^3\, \|E\,\bar g_1(X_1)\|^3\, E\,\Big\|\sum_{i=1}^{k} h(X_i)\Big\|^3 \le c_1\, E\,\|g_2\|^3\, n^{-1/2}.$$

To estimate p_2 we observe first that

$$\Big(\sum_{i=1}^{k} \bar g_1(X_i), \sum_{i=1}^{k} h(X_i)\Big) = \sum_{i=1}^{4} A_i,$$

where

$$A_1 = k\, E_1(\bar g(X_1), h(X_1)),$$

$$A_2 = \sum_{i=1}^{k}\Big((\bar g(X_i), h(X_i)) - E_1(\bar g(X_1), h(X_i))\Big),$$

$$A_3 = \sum_{i_1=2}^{k}\sum_{i_2=1}^{i_1-1} (\bar g(X_{i_1}), h(X_{i_2})),$$

$$A_4 = \sum_{i_1=1}^{k-1}\sum_{i_2=i_1+1}^{k} (\bar g(X_{i_1}), h(X_{i_2})),$$

and obviously,

$$E_k|A_1|^3 \le k^3\, |E_1(\bar g(X_1), h(X_1))|^3,$$

by Corollary 2.7.1

$$E_k|A_2|^3 \le c_1 k^{3/2}\, E_1\, |(\bar g(X_1), h(X_1))|^3,$$

and by (6.6.3)

$$E_k|A_3|^3 \le c_1 k^3\, E_2\, |(\bar g(X_1), h(X_2)|^3,$$

$$E_k|A_4|^3 \le c_1 k^3\, E_2\, |(\bar g(X_1), h(X_2))|^3,$$

where $E_k(\cdot)$ denotes the expectation with respect to $X_1, \dots, X_k$ with fixed $X_{k+1}, \dots, X_n$, $k = 1, \dots, n$.

Moreover, since $E\,\|g_1(X_1)\|^2 = 1$, we have

$$|E\,(\bar g(X_1), h(X_1))|^3 \le E_1\, \|h(X_1)\|^3.$$

Obviously,

$$E_1\,|(\bar g(X_1),h(X_1))|^3 \le 8\,n^{3/2}\,E_1\,\|h(X_1)\|^3,$$
$$E_2\,|(\bar g(X_1),h(X_2))|^3 \le c_1\,E_1\,\|\bar g(X_1)\|^3\,E_1\,\|h(X_1)\|^3,$$

and by Corollary 2.7.1

$$\bar E_k\,\|h(X_1)\|^3 \le c_1\,(n-k)^{3/2}\,\bar E_k\,\|g_2(X_1,X_{k+1})\|^3,$$

where $\bar E_k(\cdot)$ denotes the expectation with respect to $X_{k+1}, \ldots, X_n$ with fixed $X_1, \ldots, X_k$, $k=1,\ldots,n$.

The above implies

$$p_2 \le c_1\,E\,\|g_1\|^3\,E\,\|g_2\|^3\,n^{-1/2},$$

and hence we obtain an estimate for the probability $P(|(\bar S_n^k, \varepsilon_n\beta_k)| > n^{-1/2})$.

Similarly, using (6.6.2)–(6.6.3) we get

$$\begin{aligned}
P(|c\,(\bar S_n^k,\varepsilon_n\delta_k)|>n^{-1/2}) &\le P(c\,\|\bar S_n^k\|>n^{1/6})+P(n^{1/6}\varepsilon_n\,\|\delta_k\|>n^{-1/2})\\
&\le c^3\,E\,\|\bar S_n^k\|^3\,n^{-1/2}+n^{1/2}\,n^{3/2}\,\varepsilon_n^3\,E\|\delta_k\|^3\\
&\le c_1\,n^{-1/2}+c_2\,E\,\|g_2\|^3\,n^{-1/2},
\end{aligned}$$

$$\begin{aligned}
P(|c\,(a,\varepsilon_n\beta_k)|>n^{-1/2}) &\le c^3\,\|a\|^3\,\varepsilon_n^3\,n^{3/2}\,E\|\beta_k\|^3\\
&\le c_1\,\|a\|^3\,E\|g_2\|^3\,n^{-1/2},
\end{aligned}$$

$$\begin{aligned}
P(|c\,(a,\varepsilon_n\delta_k)|>n^{-1/2}) &\le c^3\,\|a\|^3\,\varepsilon_n^3\,n^{3/2}\,E\|\delta_k\|^3\\
&\le c_1\,\|a\|^3\,E\|g_2\|^3\,n^{-1/2},
\end{aligned}$$

$$\begin{aligned}
P(|c\,(\bar S_n-\bar S_n^k,\varepsilon_n\alpha_k)|>n^{-1/2}) &\le c^3\,n^{3/2}\,E\|\bar S_n-\bar S_n^k\|^3\,\varepsilon_n^3\,E\|\alpha_k\|^3\\
&\le c_1\,E\|g_1\|^3\,E\|g_2\|^3\,n^{-1/2},
\end{aligned}$$

$$\begin{aligned}
P(|c\,(\bar S_n-\bar S_n^k,\varepsilon_n\beta_k)|>n^{-1/2}) &\le P(c\,\|\bar S_n-S_n^k\|>1)+P(\varepsilon_n\,\|\beta_k\|>n^{-1/2})\\
&\le c^3\,E\|\bar S_n-\bar S_n^k\|^3+n^{3/2}\,\varepsilon_n^3\,E\|\beta_k\|^3\\
&\le c_1\,E\|g_1\|^3\,n^{-1/2}+c_2\,E\|g_2\|^3\,n^{-1/2},
\end{aligned}$$

$$\begin{aligned}
P(c\,|(\bar S_n-\bar S_n^k,\varepsilon_n\delta_k)|>n^{-1/2}) &\le P(c\,\|\bar S_n-S_n^k\|>1)+P(\varepsilon_n\,\|\delta_k\|>n^{-1/2})\\
&\le c^3\,E\|\bar S_n-\bar S_n^k\|^3+n^{3/2}\,\varepsilon_n^3\,E\|\delta_k\|^3\\
&\le c_1\,E\|g_1\|^3\,n^{-1/2}+c_2\,E\|g_2\|^3\,n^{-1/2}.
\end{aligned}$$

Denote now

$$y_n = 2\varepsilon_n\,(\bar S_n^k - a, \beta_k+\delta_k) + 2\,(\bar S_n - \bar S_n^k, v_n) + \|v_n\|^2.$$

Then

$$\|\bar S_n - a + v_n\|^2 = \|\bar S_n - a\|^2 + z_n + y_n$$

and the estimates above imply

$$P(|y_n| > n^{-1/2}) \le c_1\,(1 + \|a\|^3)\,E\|g_1\|^3\,E\|g_2\|^3\,n^{-1/2}.$$

Since for all $r \ge 0$

$$\begin{aligned} &P(\|\bar{S}_n - a\|^2 + z_n < r - n^{-1/2}) - P(|y_n| > n^{-1/2}) \\ &\quad \le P(\|\bar{S}_n - a\|^2 + z_n + y_n < r) \\ &\quad \le P(\|\bar{S}_n - a\|^2 + z_n < r + n^{-1/2}) + P(|y_n| > n^{-1/2}), \end{aligned}$$

we have

$$\Delta_{n1}(a) \le \Delta_{n2}(a) + P(|y_n| > n^{-1/2}) + \sup_r P(r - n^{-1/2} \le \|\tau - a\|^2 \le r + n^{-1/2}).$$

In Paulauskas and Rackauskas (1989) it is proved that

$$\sup_r P(r - n^{-1/2} \le \|\tau - a\|^2 \le r + n^{-1/2}) \le c\,\sigma_1^{-1}\sigma_2^{-1}\,n^{-1/2}.$$

These estimates imply Lemma 6.6.2.

Denote now

$$\begin{aligned} \varphi_n(t) &= E\exp(it\,\|\bar{S}_n - a\|^2 + it\,z_n), \\ \varphi(t) &= E\exp(it\,\|\tau - a\|^2), \\ \psi_n(t) &= E\exp(it\,\|\bar{S}_n - a\|^2). \end{aligned}$$

The density of the random variable $\|\tau - a\|^2$ is not greater then $c\,(\sigma_1\sigma_2)^{-1}$ (see Paulauskas and Rackauskas (1989)). Thus for $\Delta_{n2}(a)$, applying Esséen's inequality for any $T > 0$ we have

$$\Delta_{n2}(a) \le c_1 \int_{-T}^{T} |\varphi_n(t) - \varphi(t)|\,|t|\,dt + c_2\,(\sigma_1\sigma_2)^{-1}T^{-1}. \tag{6.6.4}$$

We choose in (6.6.4)

$$T = c n^{1/2}/L, \quad L = 3\,\sigma_9^{-2}\,E\|g_1\|^3, \quad \sigma_9 > 0$$

where c is a sufficiently small absolute constant, and then apply Lemma 12 from Zalesskii *et al.* (1988). Note that with such L the following conditions of Lemma 12 in Zalesskii *et al.* (1988)

$$\frac{2}{3} \le L^2 \le n/(2e), \quad P(B_L) \ge \frac{1}{2}$$

are fulfilled (for the proof see Zalesskii *et al.* (1988), (1989)). When this condition is satisfied the distribution of $\bar{g}_1 = g_1(X_1)\,I(\|g_1(X_1)\| \le n^{1/2})$ may be represented as $(Q + Q_1)/2$ where

$$Q(A) = P(A \cap B_L)/P(B_L), \quad A \in H$$

and Q_1 is a probability measure (see Sazonov (1981)).

Let V_1 be the covariance operator of Q and let $\sigma_{11}^2 \geq \sigma_{12}^2 \geq \cdots$ be its eigenvalues. By our assumption $\operatorname{tr} V = 1$. Hence (see Zalesskii *et al.* (1988)) $\operatorname{tr} V_1 \leq 2$ and

$$2\,\sigma_j^2/3 \leq \sigma_{1j}^2 \leq 2\,\sigma_j^2, \quad j = 1, \ldots, 9.$$

To estimate the integral in (6.6.4) we will need the following generalization of Lemma 11 in Zalesskii *et al.* (1988) (see also Lemma 4 in Zalesskii *et al.* (1989)).

Lemma 6.6.3. *Denote*

$$Z = n^{-1/2} \sum_{j=1}^{l+m} \bar{g}_1(X_j), \quad Z_1(f) = \sum_{j=1}^{l+m} f(X_j),$$

where $\bar{g}_1 = g_1(X_1)\, I(\|g_1(X_1)\| \leq n^{1/2})$ *and* $f\colon X \to R$ *is a measurable function,* l, m *are positive integers,* $l \leq m$, $l + m \leq n$. *Then for any* $A > 0$, *even* $k \geq 0$, *integer* $r \geq 0$, $k_j \geq 0$, $j = 1, \ldots, r$ *and any* x, $x_j \in H$, $j = 1, \ldots, r$, *for* $|t| \leq c(A)\, nL^{-1}\, (l \ln(l/L^2))^{-1/2}$ *if* $l > L^2$, *and for any* t *if* $l \leq L^2$

$$\Big|E \exp(it\,\|Z + x\|^2 + it\, Z_1(f))\, \|Z\|^k \prod_{j=1}^{r} (x_j, Z)^{k_j}\Big| \leq$$

$$\leq c_1 \prod_{j=1}^{r} \|x_j\|^{k_j} \Big(\exp(-c_2 l) + c_3\, (L^2/l)^A + h^{1/2}\, (c_4 T_1^2 lm/n^2, V_1)\Big),$$

where c_1, c_2 *are functions of* $\bar{k} = k + k_1 + \cdots + k_r$, c_3, c_4 *are functions of* A *and* $\bar{k}$

$$T_1 = \min\Big\{|t|, nL^{-1}(m \ln(m/L^2))^{-1/2}\Big\},$$
$$h(s, V_1) = \prod_{j=1}^{\infty} (1 + 2s\sigma_{1j}^4)^{-1/2}.$$

Proof. This lemma is reduced to Lemma 11 in Zalesskii *et al.* (1988) if $Z_1(f) = 0$. The proof of the present lemma is essentially the same as the proof of Lemma 11 in Zalesskii *et al.* (1988).

Let us estimate the integral in (6.6.4). Obviously

$$\int_{-T}^{T} |\varphi_n(t) - \varphi(t)|\, |t|^{-1}\, dt \leq I_1 + I_2,$$

where

$$I_1 = \int_{-T}^{T} |\psi_n(t) - \varphi(t)|\, |t|^{-1}\, dt,$$
$$I_2 = \int_{-T}^{T} |\varphi_n(t) - \psi_n(t)|\, |t|^{-1}\, dt.$$

By Lemma 12 in Zalesskii *et al.* (1988)

$$I_1 \le c_1 J,$$

where

$$\begin{aligned} J &= c_7(V)\,(E(\|g_1\|^2 + (a, g_1)^2)\,I(\|g_1\| > n^{1/2})) \\ &\quad + \left(E(\|g_1\|^3 + |(a, g_1)|^3)\,I(\|g_1\| \le n^{1/2}) + (Va, a)^{3/2}\right) n^{-1/2} \\ &\quad + c_3(V)\,\sigma_9^{-2}\,E\|g_1\|^3\,n^{-1/2}, \\ c_s(V) &= \prod_{j=1}^{s} \sigma_j^{-1} \end{aligned}$$

and for J we have

$$\begin{aligned} J &\le (1 + \|a\|^3)\,c_7(V)\,E\|g_1\|^3\,n^{-1/2} + c_3(V)\,\sigma_9^{-2}\,E\|g_1\|^3\,n^{-1/2} \\ &\le 2\,(1 + \|a\|^3)\,\beta c_9(V)\,\sigma_9^{-1}\,n^{-1/2}. \end{aligned}$$

Consider now I_2. Obviously we can write,

$$\varphi_n(t) - \psi_n(t) = E\,e^{it\|\bar{S}_n - a\|^2}\,(e^{itz_n} - 1).$$

Further we will use the representation

$$e^z - 1 = \sum_{s=1}^{r-1} z^s/s! + z^r \int_0^1 (1-u)^{r-1}\,e^{uz}\,du/(r-1)!\,, \tag{6.6.5}$$

valid for all complex z and positive integers r, with $r = 3$, $z = itz_n$. Thus we have

$$\varphi_n(t) - \psi_n(t) = it\,J_1(t) - t^2\,J_2(t)/2 + (it)^3 J_3(t)/2, \tag{6.6.6}$$

where

$$\begin{aligned} J_1(t) &= E\,(z_n\,e^{it\|\bar{S}_n - a\|^2}), \quad J_2(t) = E\,(z_n^2\,e^{it\|\bar{S}_n - a\|^2}), \\ J_3(t) &= \int_0^1 du\,(1-u)^2\,E\,(z_n^3\,e^{it\|\bar{S}_n - a\|^2 + ituz_n}). \end{aligned}$$

Lemma 6.6.4. *If* $|t| \le T = cn^{1/2}L^{-1}$ *then*

$$|J_3(t)| \le c_1\Big(L^{3/2}n^{-1/2} + \prod_{j=1}^{7}(1 + c_2\,\sigma_j^4\,t^2/n^{2/3})^{-1/4}\Big)\,(E\|g_1\|^3 + \|a\|^3)\,E\|g_2\|^3\,n^{-3/2}.$$

Proof. The random variable z_n does not depend on $X_{k+1}, \ldots, X_n$. Hence

$$\left|E(z_n^3\,e^{it\|\bar{S}_n - a\|^2 + ituz_n})\right| \le E_k|z_n|^3\,|\bar{E}_k\,e^{it\|\bar{S}_n - a\|^2}|,$$

where as above $\bar{E}_k(\cdot)$ denotes expectation with respect to $X_{k+1}, \dots, X_n$ with fixed $X_1, \dots, X_k$. Representing $\bar{S}_n = (\bar{S}_n - \bar{S}_r^k) + \bar{S}^k$ we apply Lemma 6.6.3 with $Z = \bar{S}_n - \bar{S}_n^k$, $l \sim n^{2/3}/4$, $m \sim 3n^{2/3}/4$ and $A = 3/4$. Noting here that $T_1 = |t|$, we obtain

$$|\bar{E}_k\, e^{it\|\bar{S}_n - a\|^2}| \le c_1 \Big(\exp(-c_2 n^{2/3}) + c_3\, L^{3/2} n^{-1/2} + \prod_{j=1}^{7} (1 + c_4 \sigma_j^4 t^2/n^{2/3})^{-1/4}\Big).$$

To estimate $E|z_n|^3$ we first write

$$E|z_n|^3 \le 32\, \varepsilon_n^2 \Big(E\,|(\bar{S}_n^k, \alpha_k)|^3 + \|a\|^3\, E\|\alpha_k\|^3\Big).$$

Then represent $(\bar{S}_n^k, \alpha_k)$ as

$$(\bar{S}_n^k, \alpha_k) = n^{-1/2} \Big(\sum_{j=1}^{k} \bar{g}(X_j), \sum_{1 \le i_1 < i_2 \le k} g_2(X_{i_1}, X_{i_2})\Big) +$$
$$+ n^{-1/2} k \Big(E\, \bar{g}_1(X_1), \sum_{1 \le i_1 < i_2 \le k} g_2(X_{i_1}, X_{i_2})\Big).$$

Using (6.6.2)–(6.6.3) we have

$$E\Big|\Big(E\,\bar{g}_1(X_1), \sum_{1 \le i_1 < i_2 \le k} g_2(X_{i_1}, X_{i_2})\Big)\Big|^3 \le$$
$$\le \|E\,\bar{g}_1(X_1)\|^3\, E\Big\|\sum_{1 \le i_1 < i_2 \le k} g_2(X_{i_1}, X_{i_2})\Big\|^3 \le c_1 n^{-3/2} k^3\, E\|g_2\|^3.$$

To estimate the first term on the right-hand side we will use the representation

$$\Big(\sum_{j=1}^{k} \bar{g}(X_j), \sum_{1 \le i_1 < i_2 \le k} g_2(X_{i_1}, X_{i_2})\Big) =$$
$$= \sum_{j=1}^{k} \sum_{1 \le i_1 < i_2 \le k} (\bar{g}(X_j), g_2(X_{i_1}, X_{i_2})) = \sum_{i=1}^{5} G_i,$$

where

$$G_1 = \sum_{j=3}^{k} \sum_{i_1=1}^{i_2-1} \sum_{i_2=2}^{j-1} (\bar{g}(X_j), g_2(X_{i_1}, X_{i_2})),$$

$$G_2 = \sum_{j=2}^{k} \sum_{i_1=1}^{j-1} (\bar{g}(X_j), g_2(X_{i_1}, X_j)),$$

$$G_3 = \sum_{j=2}^{k-1} \sum_{i_1=1}^{j-1} \sum_{i_2=j+1}^{k} (\bar{g}(X_j), g_2(X_{i_1}, X_{i_2})),$$

$$G_4 = \sum_{j=1}^{k} \sum_{i_2=j+1}^{k} (\bar{g}(X_j), g_2(X_j, X_{i_2})),$$

$$G_5 = \sum_{j=1}^{k} \sum_{i_1=j+1}^{i_2-1} \sum_{i_2=j+2}^{k} (\bar{g}(X_j), g_2(X_{i_1}, X_{i_2})).$$

By (6.6.3) we have

$$E\,|G_i|^3 \le c_1\, E\,\|g_1\|^3\, E\,\|g_2\|^3\, n^{9/2}, \qquad i = 1,3,5,$$
$$E\,|G_i|^3 \le c_2\, E\,\|g_2\|^3\, n^{9/2}, \qquad i = 2,4.$$

Combining these estimates we obtain

$$E\,|z_n|^3 \le c_1\,(E\,\|g_1\|^3 + \|a\|^3)\, E\,\|g_2\|^3\, n^{-3/2}.$$

This implies the proof of Lemma 6.6.4.

With $m \sim n$, partition the interval $|t| \le T = c n^{1/2} L^{-1}$ into two subsets

$$|t| \le T(m) = nL^{-1}(m \ln(m/L^2))^{-1/2} \tag{6.6.7}$$

and

$$T(m) \le |t| \le T. \tag{6.6.8}$$

Lemma 6.6.5. *For t satisfying* (6.6.7)

$$|J_1(t)| \le c_1\,(1+\|a\|^3)\, E\|g_1\|^3\,(E\|g_2\|^3)^{1/3}\Big(L^3 n^{-1/2} +$$
$$+(|t|+t^2)\prod_{j=1}^{7}(1+c_2\sigma_j^4 t^2)^{-1/4} + |t|^3\prod_{j=1}^{9}(1+c_2\sigma_j^4 t^2)^{-1/4}\Big)\, n^{-1/2},$$

and for t satisfying (6.6.8)

$$|J_1(t)| \le c_1\,(1+\|a\|^3)\, E\|g_1\|^3\,(E\|g_2\|^3)^{1/3}\, c_9(V)\, L^2 n^{-1}.$$

Proof. Let us first show that

$$J_1(t) = 4n^{-1/2}\varepsilon_n \binom{k}{2} h_1(t) + 2\,(k-2)\, n^{-1/2}\varepsilon_n \binom{k}{2} h_2(t) - 2\varepsilon_n \binom{k}{2} h_3(t), \tag{6.6.9}$$

where

$$h_1(t) = E\left((\bar g_1(X_1), g_2(X_1,X_2))\, e^{it\|\bar g_{1n}(X_1)+\bar g_{1n}(X_2)+\bar S_{n2}\|^2}\right),$$
$$h_2(t) = E\left((\bar g_1(X_1), g_2(X_2,X_3))\, e^{it\|\bar g_{1n}(X_1)+\bar g_{1n}(X_2)+\bar g_{1n}(X_3)+\bar S_{n3}\|^2}\right),$$
$$h_3(t) = E\left((a, g_2(X_1,X_2))\, e^{it\|\bar g_{1n}(X_1)+\bar g_{1n}(X_2)+\bar S_{n2}\|^2}\right),$$
$$\bar g_{1n}(X_j) = \bar g_1(X_j)\, n^{-1/2}, \quad \bar S_{nr} = n^{-1/2}\sum_{j=r+1}^{n} \bar g_1(X_j) - a.$$

Indeed, according to the definition

$$J_1(t) = 2\,n^{-1/2}\varepsilon_n \sum_{i=1}^{5} H_i(t) - 2\varepsilon_n\, H_6(t),$$

$$H_1(t) = \sum_{j=3}^{k} \sum_{i_1=1}^{i_2-1} \sum_{i_2=2}^{j-1} E\left((\bar{g}_1(X_j), g_2(X_{i_1}, X_{i_2}))\, e^{it\|\bar{S}_n - a\|^2}\right)$$
$$= \binom{k}{3} h_2(t),$$
$$H_2(t) = \sum_{j=2}^{k} \sum_{i_2=1}^{j-1} \left((\bar{g}_1(X_j), g_2(X_{i_2}, X_j))\, e^{it\|\bar{S}_n - a\|^2}\right)$$
$$= \binom{k}{2} h_1(t),$$
$$H_3(t) = \sum_{j=2}^{k-1} \sum_{i_1=1}^{j-1} \sum_{i_2=j+1}^{k} E\left((\bar{g}_1(X_j), g_2(X_{i_1}, X_{i_2}))\, e^{it\|\bar{S}_n - a\|^2}\right)$$
$$= \binom{k}{3} h_2(t),$$
$$H_4(t) = \sum_{j=1}^{k-1} \sum_{i_1=j+1}^{k} E\left((\bar{g}_1(X_j), g_2(X_j, X_{i_1}))\, e^{it\|\bar{S}_n - a\|^2}\right)$$
$$= \binom{k}{2} h_1(t),$$
$$H_5(t) = \sum_{j=1}^{k-2} \sum_{i_1=j+1}^{i_2-1} \sum_{i_2=j+2}^{k} E\left((\bar{g}_1(X_j), g_2(X_{i_1}, X_{i_2}))\, e^{it\|\bar{S}_n - a\|^2}\right)$$
$$= \binom{k}{3} h_2(t),$$
$$H_6(t) = \sum_{i_1=1}^{i_2-1} \sum_{i_2=2}^{k} E\left((a, g_2(X_{i_1}, X_{i_2}))\, e^{it\|\bar{S}_n - a\|^2}\right)$$
$$= \binom{k}{2} h_3(t).$$

These representations imply (6.6.9). Next we estimate functions $h_i(t)$ in (6.6.9). Obviously

$$\|\bar{g}_{1n}(X_1) + \bar{g}_{1n}(X_2) + \bar{S}_{n2}\|^2 = \|\bar{S}_{n2}\|^2 + u_1 + u_2,$$

where

$$u_1 = \|\bar{g}_{1n}(X_1) + \bar{g}_{1n}(X_2)\|^2, \quad u_2 = 2\,(\bar{g}_{1n}(X_1) + \bar{g}_{1n}(X_2), \bar{S}_{n2}).$$

By (6.6.5) with $r = 1$ and $z = it\,(u_1 + u_2)$, taking into consideration the property of degeneracy, we have

$$h_1(t) = it \int_0^1 du\, E\left((\bar{g}_1(X_1), g_2(X_1, X_2))\, u_1 e^{ituu_1}\, \bar{E}_2\, e^{it\|\bar{S}_{n2}\|^2 + ituu_2}\right) +$$
$$+ it \int_0^1 du\, E\left((\bar{g}_1(X_1), g_2(X_1, X_2))\, e^{ituu_1}\, \bar{E}_2(e^{it\|\bar{S}_{n2}\|^2 + ituu_2} u_2)\right).$$

We will estimate the expectation $\bar{E}_2$ with respect to variables $X_3, \ldots, X_n$ by using Lemma 6.6.3. To estimate $h_1(t)$ in the interval (6.6.7) we take in Lemma 6.6.3,

$l \sim n/2, \ \ m = n - l, \ \ A = 1$. Then $T_1 = |t|$ and

$$\left|\bar{E}_2\, e^{it\|\bar{S}_{n2}\|^2 + ituu_2}\right| \le c_1 \left(\exp(-c_2 n) + c_3\, L^2 n^{-1} + \prod_{j=1}^{5} (1 + c_2\sigma_j^4 t^2)^{-1/4}\right),$$

$$\left|\bar{E}_2(e^{it\|S_{n2}\|^2 + ituu_2}\, u_2)\right| \le c_1\,(1 + \|a\|)\,\|\bar{g}_{1n}(X_1) + \bar{g}_{n1}(X_2)\| \times$$
$$\times \left(\exp(-c_2 n) + c_3 L^2 n^{-1} + \prod_{j=1}^{5} (1 + c_2\sigma_j^4 t^2)^{-1/4}\right).$$

Moreover

$$\begin{aligned}
E\,&\Big(|(\bar{g}_1(X_1), g_2(X_1, X_2))|\,\|\bar{g}_{1n}(X_1) + \bar{g}_{1n}(X_2)\|\Big)\\
&\le n^{-1/2}\Big(E(\|g_1(X_1)\|^2\,\|g_2(X_1, X_2)\|)\\
&\quad + E(\|g_1(X_1)\|\,\|g_1(X_2)\|\,\|g_2(X_1, X_2)\|)\Big)\\
&\le 2n^{-1/2}\,(E\|g_1\|^3)^{2/3}\,(E\|g_2\|^3)^{1/3},\\
E\,&\Big(|(\bar{g}_1(X_1), g_2(X_1, X_2))|\,\|\bar{g}_{1n}(X_1) + \bar{g}_{1n}(X_2)\|^2\Big)\\
&\le 2n^{-1/2}\,(E\|g_1\|^3)^{2/3}\,(E\|g_2\|^3)^{1/3}.
\end{aligned}$$

Thus in the interval (6.6.7)

$$|h_1(t)| \le c_1\,(1 + \|a\|)\,(E\|g_1\|^3)^{2/3}\,(E\|g_2\|^3)^{1/3} \times$$
$$\times \left(L\, n^{-1/2} + |t| \prod_{j=1}^{7} (1 + c_2\sigma_j^4 t^2)^{-1/4}\right) n^{-1/2}.$$

To estimate $h_1(t)$ for t satisfying (6.6.8) take in Lemma 6.6.3, $l \sim n/\ln n, \ \ m = n - l, \ \ A = 5/4$. Then $T_1 = T(m)$ and similar reasoning gives for T satisfying (6.6.8),

$$\begin{aligned}
|h_1(t)| &\le c_1\,|t|\,(1 + \|a\|)\,(E\|g_1\|^3)^{2/3}\,(E\|g_2\|^3)^{1/3}\\
&\times \Big(\exp(-c_2 n/\ln n) + c_3 L^{5/2} n^{-5/4}\,(\ln n)^{5/4}\\
&+ \prod_{j=1}^{5} (1 + c_4\sigma_j^4 n/L^2(\ln n)(\ln(m/L^2)))^{-1/4}\Big)\, n^{-1/2}\\
&\le c_5\,(1 + \|a\|)\,(E\|g_1\|^3)^{2/3}\,(E\|g_2\|^3)^{1/3}\, c_5(V)\, L^{3/2} n^{-1}.
\end{aligned}$$

Consider now $h_2(t)$ in (6.6.9). Denote

$$\xi = \sum_{j=1}^{3} \xi_j, \quad \bar{\xi} = \sum_{j=1}^{3} E\,\bar{g}_{1n}(X_j),$$
$$\xi_j = \bar{g}_{1n}(X_j) - E\,\bar{g}_{1n}(X_j), \qquad j = 1, 2, 3.$$

Obviously

$$\|\bar{g}_{1n}(X_1) + \bar{g}_{1n}(X_2) + \bar{g}_{1n}(X_3) + \bar{S}_{n3}\| = \|\xi + \bar{S}_{n3} + \bar{\xi}\|^2 =$$

$$= \|\bar{S}_{n3} + \bar{\xi}\|^2 + u_3 + u_4 + u_5,$$

where

$$u_3 = \sum_{j=1}^{3} \|\xi_j\|^2, \quad u_4 = 2\sum_{j=1}^{3}(\xi_j, \bar{S}_{n3} + \bar{\xi}), \quad u_5 = 2 \sum_{1\le i<j\le 3} (\xi_i, \xi_j).$$

Applying (6.6.5) with $r = 2$ and $z = it\, u_5$ we can write

$$h_2(t) = h_{21}(t) + h_{22}(t) + h_{23}(t),$$

where

$$h_{21}(t) = E\left((\bar{g}_1(X_1), g_2(X_2, X_3))\, e^{it\,\|\bar{S}_{n3}+\bar{\xi}\|^2 + it(u_3+u_4)}\right),$$
$$h_{22}(t) = it\, E\left((\bar{g}_1(X_1), g_2(X_2, X_3))\, u_5\, e^{it\,\|\bar{S}_{n3}+\bar{\xi}\|^2 + it(u_3+u_4)}\right),$$
$$h_{23}(t) = -t^2 \int_0^1 du\, (1-u)\, E\left((\bar{g}_1(X_1), g_2(X_2, X_3))\, u_5^2\, e^{it\,\|\bar{S}_{n3}+\bar{\xi}\|^2 + it(u_3+u_4+uu_5)}\right).$$

Furthermore, observe that

$$|h_{23}(t)| \le t^2\, E\left(|(\bar{g}_1(X_1), g_2(X_2, X_3))|\, u_5^2\, |\bar{E}_3\, e^{it\,\|\bar{S}_{n3}+\bar{\xi}\|^2 + itu_4}|\right).$$

In the interval (6.6.7) by Lemma 6.6.3 with $l \sim n/2,\ \ m = n - l,\ \ A = 3/2$

$$|\bar{E}_3\, e^{it\,\|\bar{S}_{n3}+\bar{\xi}\|^2+itu_4}| \le c_1 \left(\exp(-c_2 n) + c_3 L^3 n^{-3/2} + \prod_{j=1}^{7} (1 + c_4 \sigma_j^4 t^2)^{-1/4}\right),$$

and for t satisfying (6.6.8) Lemma 6.6.3 with $l \sim n/\ln n,\ \ m = n - l,\ \ A = 7/4$ gives

$$|\bar{E}_3\, e^{it\,\|\bar{S}_{n3}+\bar{\xi}\|^2+itu_4}| \le c_1 \Big(\exp(-c_2 n/\ln n) + c_3 L^{7/2} n^{-7/4} (\ln n)^{7/4}$$
$$+ \prod_{j=1}^{7} \left(1 + c_4 \sigma_j^4 n/L^2 (\ln n)\, (\ln(m/L^2))\right)^{-1/4}\Big).$$

Moreover

$$E\left(|(\bar{g}_1(X_1), g_2(X_2, X_3))|\, u_5^2\right) \le c\, (E\|g_1(X_1)\|^3)^{1/3}\, (E\|g_2\|^3)^{1/3}\, n^{-3/2}.$$

Combining these inequalities we obtain an estimate for $h_{23}(t)$.

By (6.6.5) with $r = 1$, $z = it(u_3 + u_4)$ we have

$$h_{22}(t) = f_1(t) + f_2(t),$$

where

$$f_1(t) = it\, E\left((E\, \bar{g}_1(X_1), g_2(X_2, X_3))\, u_5\, e^{it\,\|\bar{S}_{n3}+\bar{\xi}\|^2 + it(u_3+u_4)}\right),$$
$$f_2(t) = -t^2 \int_0^1 du\, E\Big((\bar{g}_1(X_1) - E\, \bar{g}_1(X_1), g_2(X_2, X_3)) \times$$

$$\times\, u_5(u_3+u_4)\, e^{it\,\|\bar S_{n3}+\bar\xi\|^2+itu(u_3+u_4)}\Big).$$

Here

$$|f_1(t)| \le |t|\, E\left(|(E\,\bar g_1(X_1), g_2(X_2,X_3))\,u_5|\, |\bar E_3\, e^{it\,\|\bar S_{n3}+\bar\xi\|^2+itu_4}|\right),$$

and

$$E\left(|(E\,\bar g_1(X_1), g_2(X_2,X_3))\,u_5|\right) \le 24\,(E\|g_2\|^2)^{1/2}\, n^{-3/2}.$$

Thus for t satisfying (6.6.7) by Lemma 6.6.3 with $l \sim n/2$, $m = n-l$, $A = 1$,

$$|f_1(t)| \le c_1\,|t|\,(E\|g_2\|^2)^{1/2} \times$$
$$\times\left(\exp(-c_2 n) + c_3 L^2 n^{-1} + \prod_{j=1}^{5}(1+c_4\sigma_j^4 t^2)^{-1/4}\right) n^{-3/2},$$

and for t satisfying (6.6.8) by Lemma 6.6.3 with $l \sim n/\ln n$, $m = n-l$, $A = 5/4$

$$|f_1(t)| \le c_1\,|t|\,(E\|g_2\|^2)^{1/2}\left(\exp(-c_2 n/\ln n) + c_3 L^{5/2} n^{-5/4}(\ln n)^{5/4} + \right.$$
$$\left. + \prod_{j=1}^{5}\left(1 + c_4\sigma_j^4 n/L^2(\ln n)(\ln(m/L^2))\right)^{-1/4}\right) n^{-3/2}.$$

To estimate $f_2(t)$ we first observe that

$$|f_2(t)| \le t^2 \int_0^1 du\, E\left(|(\bar g_1(X_1) - E\,\bar g_1(X_1), g_2(X_2,X_3))\,u_5| \times\right.$$
$$\left. \times |\bar E_3(u_3+u_4)\, e^{it\,\|\bar S_{n3}+\bar\xi\|^2+ituu_4}|\right)$$

and

$$E\left(\left|(\bar g_1(X_1) - E\,\bar g_1(X_1), g_2(X_2,X_3))\,u_5\right| \left(u_3 + \sum_{j=1}^{3}\|\xi_j\|\,(1+\|a\|)\right)\right) \le$$
$$\le c\,E\|g_1\|^3\,(E\|g_2\|^3)^{1/3}\,(1+\|a\|)\,n^{-3/2}.$$

Applying now Lemma 6.6.3 for t satisfying (6.6.7) with $l \sim n/2$, $m = n-l$, $A = 3/2$, we have

$$|f_2(t)| \le c_1 t^2\,(1+\|a\|)\,E\|g_1\|^3\,(E\|g_2\|^3)^{1/3} \times$$
$$\times\left(\exp(-c_2 n) + c_3 L^3 n^{-3/2} + \prod_{j=1}^{7}(1+c_4\sigma_j^4 t^2)^{-1/4}\right) n^{-3/2},$$

and for t satisfying (6.6.8), using Lemma 6.6.3 with $l \sim n/\ln n$, $m = n-l$, $A = 7/4$, we obtain

$$|f_2(t)| \le c_1 t^2\,(1+\|a\|)\,E\|g_1\|^3\,(E\|g_2\|^3)^{1/3}$$
$$\times\left(\exp(-c_2 n/\ln n) + c_3 L^{7/2} n^{-7/4}(\ln n)^{7/4} + \right.$$

$$+\prod_{j=1}^{7}\left(1+c_4\sigma_j^4 n/L^2(\ln n)(\ln(m/L^2))\right)^{-1/4}\Big) n^{-3/2}.$$

It remains only to estimate $h_{21}(t)$. Represent it as

$$h_{21}(t)=f_3(t)+f_4(t),$$

where

$$f_3(t)=E\left((E\,\bar g_1(X_1),g_2(X_2,X_3))\,e^{it\|\bar S_{n3}+\bar\xi\|^2+it(u_3+u_4)}\right),$$
$$f_4(t)=E\left((\bar g_1(X_1)-E\,\bar g_1(X_1),g_2(X_2,X_3))\,e^{it\|\bar S_{n3}+\bar\xi\|^2+it(u_3+u_4)}\right).$$

In view of the property of degeneracy we can also write

$$f_3(t)=E\Big((E\,\bar g_1(X_1),g_2(X_2,X_3))\,e^{it\|\bar S_{n3}+\bar\xi\|^2+it\|\xi_1\|^2+2it(\xi_1,\bar S_{n3}+\bar\xi)}$$
$$\times\prod_{j=2}^{3}(e^{it\|\xi_j\|^2+2it(\xi_j,\bar S_{n3}+\bar\xi)}-1)\Big),$$
$$f_4(t)=E\Big((\bar g_1(X_1)-E\,\bar g_1(X_1),g_2(X_2,X_3))\,e^{it\|\bar S_{n3}+\bar\xi\|^2}$$
$$\times\prod_{j=1}^{3}(e^{it\|\xi_j\|^2+2it(\xi_j,\bar S_{n3}+\bar\xi)}-1)\Big).$$

Furthermore, applying (6.6.5) with $r=1$ in $f_3(t)$ and $f_4(t)$, we have

$$f_3(t)=-\,t^2\,E\Big((E\,\bar g_1(X_1),g_2(X_2,X_3))\,e^{it\|\bar S_{n3}+\bar\xi\|^2+it\|\xi_1\|^2+2it(\xi_1,\bar S_{n3}+\bar\xi)}$$
$$\times\prod_{j=2}^{3}(\|\xi_j\|^2+2(\xi_j,\bar S_{n3}+\bar\xi))\int_0^1 du\,e^{itu\|\xi_j\|^2+2itu(\xi_j,\bar S_{n3}+\bar\xi)}\Big), \tag{6.6.10}$$

$$f_4(t)=(it)^3\,E\Big((\bar g_1(X_1)-E\,\bar g_1(X_1),g_2(X_2,X_3))\,e^{it\|\bar S_{n3}+\bar\xi\|^2}$$
$$\times\prod_{j=1}^{3}(\|\xi_j\|^2+2(\xi_j,\bar S_{n3}+\bar\xi))\int_0^1 du\,e^{itu\|\xi_j\|^2+2itu(\xi_j,\bar S_{n3}+\bar\xi)}\Big). \tag{6.6.11}$$

Denote $b_j=\|\xi_j\|^2+2(\xi_j,\xi)-2(\xi_j,a)$. Then

$$\prod_{j=2}^{3}(\|\xi_j\|^2+2(\xi_j,\bar S_{n3}+\bar\xi))=\prod_{j=2}^{3}(b_j+2(\xi_j,\bar S_{n3}+a))=$$
$$=b_2b_3+2b_2(\xi_3,\bar S_{n3}+a)+2b_3(\xi_2,\bar S_{n3}+a)+4(\xi_2,\bar S_{n3}+a)(\xi_3,\bar S_{n3}+a), \tag{6.6.12}$$

where according to the notation introduced above

$$\bar S_{n3}+a=n^{-1/2}(\bar g_1(X_4)+\cdots+\bar g_1(X_n)).$$

Note that

$$\|\xi_j\| \le 2, \qquad \|\xi_j\| \le n^{-1/2}(1 + \|g_1(X_j)\|),$$

$$|b_j| \le 4\,(1 + \|a\|)\,(1 + \|g_1(X_j)\|)\,n^{-1/2}.$$

Substituting now (6.6.12) into (6.6.10) we obtain four expectations. Then we apply Lemma 6.6.3 to each of them, taking into account the estimate $\|E\,\bar{g}_1(X_1)\| \le n^{-1/2}$. For t satisfying (6.6.7) we take in Lemma 6.6.3, $l \sim n/2$, $m = n - l$, $A = 3/2$ and obtain

$$|f_3(t)| \le c_1 t^2 (1 + \|a\|)^2 (E\|g_2\|^2)^{1/2} \times$$
$$\times \Big(\exp(-c_2 n) + c_3 L^3 n^{-3/2} + \prod_{j=1}^{7} (1 + c_4 \sigma_j^4 t^2)^{-1/4}\Big)\, n^{-3/2},$$

and for t satisfying (6.6.8) we take $l \sim n/\ln n$, $m = n - l$, $A = 7/4$ and obtain

$$\begin{aligned}|f_3(t)| \le\, & c_1 t^2 (1 + \|a\|)^2 (E\|g_2\|^2)^{1/2} \\ & \times \Big(\exp(-c_2 n/\ln n) + c_3 L^{7/2} n^{-7/4} (\ln n)^{7/4} \\ & + \prod_{j=1}^{7}\big(1 + c_4 \sigma_j^4 n/L^2 (\ln n)(\ln(n/L^2))\big)^{-1/4}\Big)\, n^{-3/2}.\end{aligned}$$

To estimate $f_4(t)$ in (6.6.11) we use the formula

$$\begin{aligned}\prod_{j=1}^{3}(\|\xi_j\|^2 + 2\,(\xi_j, \bar{S}_{n3} + \bar{\xi})) = {} & b_1 b_2 b_3 + \\ & + 2b_1 b_2\,(\xi_3, \bar{S}_{n3} + a) + 2b_1 b_3\,(\xi_2, \bar{S}_{n3} + a) + 2b_2 b_3\,(\xi_1, \bar{S}_{n3} + a) \\ & + 4b_1\,(\xi_2, \bar{S}_{n3} + a)\,(\xi_3, \bar{S}_{n3} + a) + 4b_2\,(\xi_1, \bar{S}_{n3} + a)\,(\xi_3, \bar{S}_{n3} + a) \\ & + 4b_3\,(\xi_1, \bar{S}_{n3} + a)\,(\xi_2, \bar{S}_{n3} + a) \\ & + 8\,(\xi_1, \bar{S}_{n3} + a)\,(\xi_2, \bar{S}_{n3} + a)\,(\xi_3, \bar{S}_{n3} + a).\end{aligned}$$

Substituting it into (6.6.11) we obtain eight similar expectations. To estimate these expectations we again apply Lemma 6.6.3. For t satisfying (6.6.7) we take $l \sim n/2$, $m = n - l$, $A = 2$ and obtain

$$|f_4(t)| \le c_1\,|t|^3 (1 + \|a\|)^3 (E\|g_2\|^2)^{1/2} \times$$
$$\times \Big(\exp(-c_2 n) + c_3 L^4 n^{-2} + \prod_{j=1}^{9} (1 + c_4 \sigma_j^4 t^2)^{-1/4}\Big)\, n^{-3/2},$$

and for t satisfying (6.6.8) we take $l \sim n/\ln n$, $m = n - l$, $A = 9/4$ and obtain

$$|f_4(t)| \le c_1\,|t|^3 (1 + \|a\|)^3 (E\|g_2\|^2)^{1/2} \times$$

$$\times \left(\exp(-c_2 n/\ln n) + c_3 L^{9/2} n^{-9/4} (\ln n)^{9/4} + \prod_{j=1}^{9}\left(1 + c_4 \sigma_j^4 n/L^2(\ln n)(\ln(m/L^2))\right)^{-1/4}\right) n^{-3/2}.$$

Estimates for t satisfying (6.6.7) imply

$$|h_2(t)| \le c_1 (1 + \|a\|)^3 (E\|g_1\|^3) (E\|g_2\|^3)^{1/3} \times$$
$$\times \left(L n^{-1/2} + (|t| + t^2) \prod_{j=1}^{7} (1 + c_4 \sigma_j^4 t^2)^{-1/4} + |t|^3 \prod_{j=1}^{9} (1 + c_4 \sigma_j^4 t^2)^{-1/4}\right) n^{-3/2},$$

and estimates for t satisfying (6.6.8) imply

$$|h_2(t)| \le c_1 (1 + \|a\|)^3 (E\|g_1\|^3) (E\|g_2\|^3)^{1/3} c_9(V) L^{3/2} n^{-2}.$$

It remains to estimate $h_3(t)$ in (6.6.9). Applying (6.6.5) with $r = 1$ and $z = it(u_1 + u_2)$, $z = itu(u_1 + u_2)$ twice we have

$$h_3(t) = it \int_0^1 du\, E\left((a, g_2(X_1, X_2))\, u_1\, e^{ituu_1}\, \bar{E}_2\, e^{it\|\bar{S}_{n2}\|^2 + ituu_2}\right) -$$
$$-t^2 \int_0^1 dv \int_0^1 du\, u E\left((a, g_2(X_1, X_2))\, e^{ituvu_1}\, \bar{E}_2 (e^{it\|\bar{S}_{n2}\|^2 + ituvu_2} (u_1 + u_2))\right).$$

Arguing in the same way as in the proof of the estimates for $h_1(t)$, we obtain for t satisfying (6.6.7)

$$|h_3(t)| \le c_1 (1 + \|a\|)^3 (E\|g_1\|^3)^{2/3} (E\|g_2\|^3)^{1/3} \times$$
$$\times \left(L n^{-1/2} + (|t| + t^2) \prod_{j=1}^{7} (1 + c_2 \sigma_j^4 t^2)^{-1/4}\right) n^{-1},$$

and for t satisfying (6.6.8)

$$|h_3(t)| \le c_1 (1 + \|a\|^3) (E\|g_1\|^3)^{2/3} (E\|g_2\|^3)^{1/3} c_7(V) L^{3/2} n^{-3/2}.$$

Now all the estimates for $h_i(t)$, $i = 1, 2, 3$ imply Lemma 6.6.5.

Lemma 6.6.6. *If t belongs to the interval* (6.6.7) *then*

$$|J_2(t)| \le c_1 (1 + \|a\|^2) (E\|g_1\|^3)^{2/3} \left(L^2 n^{-1/2} + n^{1/6} \prod_{j=1}^{9} (1 + c_2 \sigma_j^4 t^2 / n^{1/3})^{-1/6}\right) n^{-1},$$

and if t satisfies (6.6.8)

$$|J_2(t)| \le c_1 (1 + \|a\|^2) (E\|g_1\|^3)^{2/3} c_9(V) L^3 n^{-3/2}.$$

Proof. Noting that

$$J_2(t) = E\left(z_n^2\, \bar{E}_k\, e^{it\|\bar{S}_n - a\|^2}\right),$$

we have

$$|J_2(t)| \le 4\varepsilon_n\, E\left(\|\bar S_n^k\|^2\,\|\alpha_k\|^2\,|\bar E_k\, e^{it\|\bar S_n - a\|^2}|\right) + 4\varepsilon_n^2\,\|a\|^2\, E\left(\|\alpha_k\|^2\,|\bar E_k\, e^{it\|\bar S_n - a\|^2}|\right).$$

By Hölder's inequality

$$E\left(\|\bar S_n^k\|^2\,\|\alpha_k\|^2\,|\bar E_k\, e^{it\|\bar S_n - a\|^2}|\right) \le (E\,\|\alpha_k\|^3)^{2/3}\left(E\left(\|\bar S_n^k\|^6\,|\bar E_k\, e^{it\|\bar S_n - a\|^2}|^3\right)\right)^{1/3}.$$

Since for any $x > 0$ and $p > 0$

$$\|\bar S_n^k\|^6 \le x^6 + x^{-p}\,\|\bar S_n^k\|^{p+6},$$

we have

$$E\left(\|\bar S_n^k\|^6\,|\bar E_k\, e^{it\|\bar S_n - a\|^2}|^3\right) \le x^6\, E\,|\bar E_k\, e^{it\|\bar S_n - a\|^2}|^2 + x^{-p}\, E\|\bar S_n^k\|^{p+6}.$$

Further we also use Hölder's inequality and obtain

$$E\left(\|\alpha_k\|^2\,|\bar E_k\, e^{it\|\bar S_n - a\|^2}|\right) \le (E\,\|\alpha_k\|^3)^{2/3}\,(E\,|\bar E_k\, e^{it\|\bar S_n - a\|^2}|^3)^{1/3} \le$$

$$\le (E\,\|\alpha_k\|^3)^{2/3}\,(E\,|\bar E_k\, e^{it\|\bar S_n - a\|^2}|^2)^{1/3}.$$

Then by using the elementary inequality $(\alpha+\beta)^{1/3} \le \alpha^{1/3} + \beta^{1/3}$ we get

$$|J_2(t)| \le 4\varepsilon_n^2\,(E\|\alpha_k\|^3)^{2/3}\left(x^{-p/3}\,(E\,\|\bar S_n^k\|^{p+6})^{1/3} + (x^2 + \|a\|^2)\,(E\,|\bar E_k\, e^{it\|\bar S_n - a\|^2}|^2)^{1/3}\right).$$

Applying the same argument as in the proof of Lemma 10 in Zalesskii *et al.* (1988) with $l \sim n^{2/3}$, $m = n - l$, $A = 3/2$ for $|t| \le c\,n^{1/2}L^{-1}$ we have

$$E\,|\bar E_k\, e^{it\|\bar S_n - a\|^2}|^2 \le c_1\left(L^3 n^{-1} + \prod_{j=1}^{9}(1 + c_2\sigma_j^4 T_1^2/n^{1/3})^{-1/2}\right),$$

where T_1 is defined in Lemma 6.6.3. Furthermore taking $p = 36$, $x = n^{1/12}$ and by using (6.6.2) and (6.6.3) we obtain the estimate

$$|J_2(t)| \le c_1\,(1 + \|a\|^2)\,(E\|g_2\|^3)^{2/3}\left(L^2 n^{-1/2} + n^{1/6}\prod_{j=1}^{9}(1 + c_2\sigma_j^4 T_1^2/n^{1/3})^{-1/6}\right) n^{-1},$$

which implies Lemma 6.6.6.

Lemma 6.6.7. *I_2 satisfies the inequality*

$$I_2 \le c_1\,(1 + \|a\|^3)\,\beta^2\sigma_9^{-1}\,c_9(V)\,n^{-1/2}.$$

Proof. (6.6.6) implies

$$|\varphi_n(t) - \psi_n(t)|\,|t|^{-1} \le |J_1(t)| + |t|\,|J_2(t)| + t^2\,|J_3(t)|.$$

By Lemma 6.6.4

$$\int_{-T}^{T} t^2\,|J_3(t)|\,dt \le c_1\,(1+\|a\|^3)\,E\|g_1\|^3\,E\|g_2\|^3\,c_7(V)\,n^{-1/2}.$$

By Lemma 6.6.5

$$\int_{-T}^{T} |J_1(t)|\,dt \le c_1\,(1+\|a\|^3)\,(E\|g_1\|^3)\,(E\|g_2\|^3)^{1/3}\,c_9(V)\,L^{1/2}n^{-1/2}.$$

By Lemma 6.6.6

$$\int_{-T}^{T} |t|\,|J_2(t)|\,dt \le c_1\,(1+\|a\|^2)\,(E\|g_1\|^3)^{2/3}\,c_9(V)\,L\,n^{-1/2}.$$

From here we obtain Lemma 6.6.7.

From (6.6.4) and Lemma 6.6.7 it follows that

$$\Delta_{n2}(a) \le c_1\,(1+\|a\|^3)\,\beta^2\sigma_9^{-1}\,c_9(V)\,n^{-1/2}.$$

This estimate and Lemmas 6.6.1, 6.6.2 imply (6.6.1). Theorem 6.6.1 is proved.

Estimates of order $o\,(n^{-1/2})$. Now we consider *UH*-statistics of degree $m \ge 2$ and rank $r = 1$.

Denote

$$\Delta_n = \sup_x \Big| P(\|m^{-1}\sqrt{n}\,(U_n-\theta)\|^2 < x) - P(\|\tau\|^2 < x)\Big|.$$

Theorem 6.6.2. *Assume that kernel Φ satisfies the conditions*

$$\sigma^2 := E\,\|g_1(X_1)\|^2 > 0, \quad \mu_2 = E\,\|\Phi(X_1,\dots,X_m)\|^2 < \infty.$$

Let V-operator have infinitely many nonzero eigenvalues σ_j^2, $j = 1, 2, \dots$.

Then, for all $n \ge m$, the estimate

$$\Delta_n \le c(\Phi)\Big(E\,(\|g_1\|^2\,I(\|g_1\| > \sqrt{n})) + n^{-1}\,E(\|g_1\|^4\,I(\|g_1\| \le \sqrt{n}))\Big) + \rho_n n^{-1/2},$$

holds, where $\rho_n \to 0$ as $n \to \infty$. Here, the positive constant $c(\Phi)$ depends on kernel Φ only via the eigenvalues of the V-operator σ_j^2 and μ_2.

Corollary 6.6.1. *Let $E\|g_1\|^3 < \infty$ and $\mu_2 < \infty$. Then (see Borovskikh and Korolyuk* (1994*a*))

$$\Delta_n = o(n^{-1/2}).$$

Proof. In what follows, $c_1, c_2, \ldots$ always denote positive constants, which possibly depend on kernel Φ only via the eigenvalues of the V-operator and the variables σ^2 and μ_2.

According to (1.1.9), we have

$$m^{-1}\sqrt{n}\,(U_n - \theta) = S_n + v_n + R_n, \tag{6.6.13}$$

where

$$S_n = n^{-1/2}\,(g_1(X_1) + \cdots + g_1(X_n)),$$
$$v_n = \frac{(m-1)}{(n-1)\sqrt{n}} \sum_{1\le i<j\le n} g_2(X_i, X_j),$$
$$R_n = m^{-1}\sqrt{n}\sum_{c=3}^{m}\binom{m}{c}U_{nc};$$

moreover, $R_n \equiv 0$ if $m = 1, 2$. First, we estimate $E\|R_n\|^2$. If $E\|\Phi\|^2 < \infty$, then

$$E\,\|R_n\|^2 = m^{-2}n\sum_{c=3}^{m}\binom{m}{c}^2\binom{n}{c}^{-1}E\|g_c\|^2. \tag{6.6.14}$$

Lemma 6.6.8. *Let $\gamma_n = S_n + v_n$. Under condition $\mu_2 < \infty$, the following relation holds*

$$\Delta_n = \sup_x\left|P(\|\gamma_n\|^2 < x) - P(\|\tau\|^2 < x)\right| +$$
$$+E(\|g_1\|^2\, I(\|g_1\| > \sqrt{n})) + O((n\ln n)^{-1/2}).$$

Proof. To use equality (6.6.13), we write

$$\|\gamma_n + R_n\|^2 = \|\gamma_n\|^2 + 2\,(S_n, R_n) + 2\,(v_n, R_n) + \|R_n\|^2.$$

Taking this formula into account, after some computations, we find

$$\begin{aligned}\Delta_n \le\ & \sup_x\left|P(\|\gamma_n\|^2 < x) - P(\|\tau\|^2 < x)\right|\\ &+ \sup_x\left|P(\|\tau\|^2 < x + (n\ln n)^{-1/2}) - P(\|\tau\| < x)\right|\\ &+ P(6\,|(S_n, R_n)| > (n\ln n)^{-1/2})\\ &+ P(6\,|(v_n, R_n)| > (n\ln n)^{-1/2})\\ &+ P(3\,\|R_n\|^2 > (n\ln n)^{-1/2}).\end{aligned}$$

According to (6.6.14) and the Chebyshev inequality, under the condition $\mu_2 < \infty$, we get

$$P(3\,\|R_n\|^2 > (n\ln n)^{-1/2}) \le 3\,(n\ln n)^{1/2}\,E\|R_n\|^2 = O(n^{-1}),$$

$$P(6\,|(v_n, R_n)| > (n\ln n)^{-1/2}) \le 6\,(n\ln n)^{1/2}\,E|(v_n, R_n)| \le$$
$$\le 6\,(n\ln n)^{1/2}\,(E\|v_n\|^2)^{1/2}\,(E\|R_n\|^2)^{1/2} = O(n^{-1}(\ln n)^{1/2}).$$

Further, we can write

$$P(6\,|(S_n, R_n)| > (n\ln n)^{-1/2}) \le P(6\,\|S_n\|\,\|R_n\| > (n\ln n)^{-1/2}) \le$$
$$\le P(\|S_n\| > (\ln n)^{3/2}) + P(6\,\|R_n\| > n^{-1/2}\,(\ln n)^{-2}).$$

To estimate the probability $P(\|S_n\| > (\ln n)^{3/2})$, we consider "truncated" elements $\bar{g}_j = g_1(X_j)\,I(\|g_1(X_j)\| \le \sqrt{n})$, $j = 1, \ldots, n$, and denote $S_n' = \bar{g}_1 + \bar{g}_2 + \cdots + \bar{g}_n$. Then, under condition $\mu_2 < \infty$, we obtain

$$P(\|S_n\| > (\ln n)^{3/2}) \le$$
$$\le P(\|S_n'\| > (\ln n)^{3/2}) + E(\|g_1(X_1)\|^2\,I(\|g_1(X_1)\| > \sqrt{n})),$$

where, according to the exponential inequality in Yurinskii (1976), we get $P(\|S_n'\| > (\ln n)^{3/2}) = O(n^{-1})$. Hence,

$$P(6\,|(S_n, R_n)| > (n\ln n)^{-1/2}) =$$
$$= E(\|g_1\|^2\,I(\|g_1\| > \sqrt{n})) + O(n^{-1}(\ln n)^4).$$

Let $F(x) = P(\|\tau\|^2 < x)$, $x \in R$. If $\varphi(t) = E\exp(it\,\|\tau\|^2)$, $t \in R$, then

$$\varphi(t) = \prod_{j=1}^{\infty}(1 - 2it\sigma_j^2)^{-1/2}.$$

Clearly, there exists a constant $c > 0$ dependent only on the eigenvalues $\{\sigma_j^2\}$ such that $F'(x) \le c$ for all $x \in R$. Therefore, we have

$$\sup_x \left| P(\|\tau\|^2 < x + (n\ln n)^{-1/2}) - P(\|\tau\|^2 < x) \right| \le O((n\ln n)^{-1/2}).$$

By combining these estimates, we obtain Lemma 6.6.8.

Let $B = \{x\colon \|g_1(x)\| \le \sqrt{n}\} \in \mathfrak{X}$. For any $A \in \mathfrak{X}$ we define here a probability measure Q by setting $Q(A) = P(A \cap B)/P(B)$.

Let $\xi_1, \xi_2 \ldots, \xi_n$ be independent random variables with the same distribution Q. We also assume that $(X_1, X_2, \ldots, X_n)$ and $(\xi_1, \xi_2, \ldots, \xi_n)$ are independent. Note that, under condition $\mu_2 < \infty$, we have

$$P(\bar{B}) = P(\|g_1(X_1)\| > \sqrt{n}) \le n^{-1}\,E(\|g_1\|^2\,I(\|g_1\| > \sqrt{n}))$$

and if $n \to \infty$, then $1/P(B) = 1 + P(\bar{B}) + (P(\bar{B}))^2 + O(n^{-3})$.

Lemma 6.6.9. *Let $\eta_n = S_n + v_n$, where S_n and v_n are defined by* (6.6.13) *with X_j replaced by ξ_j, respectively, for all $j = 1, 2, \ldots, n$. Then the inequality*

$$\sup_x \Big| P(\|\gamma_n\|^2 < x) - P(\|\tau\|^2 < x) \Big| \le$$
$$\le \sup_x \Big| P(\|\eta_n\|^2 < x) - P(\|\tau\|^2 < x) \Big| + n\, P(\bar{B})/P(B)$$

holds, where under condition $\mu_2 < \infty$

$$n\, P(\bar{B})/P(B) = E(\|g_1\|^2\, I(\|g_1\| > \sqrt{n})) + O(n^{-1}).$$

Proof. It suffices to show that

$$\sup_x \Big| P(\|\gamma_n\|^2 < x) - P(\|\eta_n\|^2 < x) \Big| \le n\, P(\bar{B})/P(B).$$

For this purpose, we write

$$\begin{aligned} P(\|\gamma_n\|^2 < x) &- P(\|\eta_n\|^2 < x) \\ &= E\, I(\|\gamma_n(X_1, \ldots, X_n)\|^2 < x) - E\, I(\|\eta_n(\xi_1, \ldots, \xi_n)\|^2 < x) \\ &= \sum_{k=1}^{n} \int_X \cdots \int_X I(\|\gamma_n(y_1, \ldots, y_n)\|^2 < x) \\ &\quad \times \prod_{j=1}^{k-1} P(dy_j)\, (P(dy_k) - Q(dy_k)) \prod_{j=k+1}^{n} Q(dy_j). \end{aligned}$$

Here, for any fixed $x \in R$, $y_j \in X$, $j = 1, \ldots, k-1, k+1, \ldots, n$, we have

$$\begin{aligned} \int_X & I(\|\gamma_n(y_1, \ldots, y_n)\|^2 < x)\, (P(dy_k) - Q(dy_k)) \\ &= \int_X I\Big(\|\gamma_n(y_1, \ldots, y_{k-1}, y, y_{k+1}, \ldots, y_n)\|^2 < x\Big)\, P(dy) \\ &\quad - \frac{1}{P(B)} \int_B I\Big(\|\gamma_n(y_1, \ldots, y_{k-1}, y, y_{k+1}, \ldots, y_n)\|^2 < x\Big)\, P(dy) \\ &= -\frac{P(\bar{B})}{P(B)} \int_X I\Big(\|\gamma_n(y_1, \ldots, y_{k-1}, y, y_{k+1}, \ldots, y_n)\|^2 < x\Big)\, P(dy) \\ &\quad + \frac{1}{P(B)} \int_{\bar{B}} I\Big(\|\gamma_n(y_1, \ldots, y_{k-1}, y, y_{k+1}, \ldots, y_n)\|^2 < x\Big)\, P(dy). \end{aligned}$$

Since $I\Big(\|\gamma_n(y_1, \ldots, y_{k-1}, y, y_{k+1}, \ldots, y_n)\|^2 < x\Big) \le 1$, from these estimates follows Lemma 6.6.9.

Further, we use the Esséen inequality

$$\sup_x \Big| P(\|\eta_n\|^2 < x) - P(\|\tau\|^2 < x) \Big| \le c_1\, I_n + c_2\, (n \ln n)^{-1/2}, \qquad (6.6.15)$$

where

$$I_n = \int_{-t_n}^{t_n} |\varphi_n(t) - \varphi(t)|\,|t|^{-1}\,dt$$

with $\varphi_n(t) = E\exp(it\,\|\eta_n\|^2)$, $t_n = (n\ln n)^{1/2}$.

Consider the characteristic function $\varphi_n(t)$. Let $k = k(t) = n - l$ and $l = [n/(|t|^{1-\varepsilon} + 2)]$ for all t from the region $|t| \le t_n$, where $\varepsilon > 0$ is a sufficiently small fixed number. Denote here

$$a_k = \sum_{j=2}^{k}\sum_{i=1}^{j-1} g_2(\xi_i, \xi_j),$$

$$b_k = \sum_{j=k+1}^{n}\sum_{i=1}^{k} g_2(\xi_i, \xi_j),$$

$$d_k = \sum_{j=k+1}^{n}\sum_{i=k+1}^{j-1} g_2(\xi_i, \xi_j),$$

$$\bar{S}_n = n^{-1/2}\sum_{j=1}^{k} g_1(\xi_j), \quad \tilde{S}_n = n^{-1/2}\sum_{j=k+1}^{n} g_1(\xi_j),$$

$$S_{n2}(g_2) = \sum_{1\le i<j\le n} g_2(\xi_i, \xi_j), \quad \varepsilon_n = (m-1)(n-1)^{-1}n^{-1/2}.$$

Note that $S_{n2}(g_2) = a_k + b_k + d_k$ and $S_n = \bar{S}_n + \tilde{S}_n$.

Lemma 6.6.10. *Assume that condition $\mu_2 < \infty$ is satisfied and $0 < \varepsilon < 1/32$. If $n^\varepsilon \le |t| \le (n\ln n)^{1/2}$, then*

$$|\varphi_n(t)| \le c_1(\Phi)\,(n^{-2} + |t|^{2\varepsilon}\,n^{-1}(\ln n)^2).$$

If $|t| \le n^\varepsilon$, then

$$\varphi_n(t) = E\exp(it\,\|S_n\|^2) + 2it\varepsilon_n\,E\{(S_n, S_{n2}(g_2))\exp(it\,\|S_n\|^2)\} + r_n(t),$$

where

$$|r_n(t)| \le c_2(\Phi)\,|t|\,(1+|t|)\,n^{-1+\varepsilon},$$

and the positive constants $c_i(\Phi)$ do not depend on n and depend on the kernel Φ only via the eigenvalues of the V-operator and the variables σ^2 and μ_2.

Proof. By using accepted notations, we can write

$$\eta_n = S_n + \varepsilon_n a_k + \varepsilon_n b_k + \varepsilon_n d_k;$$

therefore,

$$\|\eta_n\|^2 = \|S_n\|^2 + 2\varepsilon_n\,(S_n, a_k) + 2\varepsilon_n\,(S_n, b_k) + 2\varepsilon_n\,(S_n, d_k) +$$

$$+ \varepsilon_n^2 \|a_k\|^2 + \varepsilon_n^2 \|b_k\|^2 + \varepsilon_n^2 \|d_k\|^2$$
$$+ 2\varepsilon_n^2 (a_k, b_k) + 2\varepsilon_n^2 (a_k, d_k) + 2\varepsilon_n^2 (b_k, d_k).$$

Further, by using (6.6.5) with $r = 1$, we obtain

$$E \exp(it \|\eta_n\|^2) = E \exp(it \|S_n\|^2 + it\, u_1) + \varrho_{n1}(t), \tag{6.6.16}$$

where

$$u_1 = 2\varepsilon_n (S_n, a_k) + 2\varepsilon_n (S_n, b_k) + 2\varepsilon_n (S_n, d_k) + 2\varepsilon_n^2 (a_k, b_k) + \varepsilon_n^2 \|a_k\|^2,$$
$$\varrho_{n1}(t) = it \int_0^1 du\, E \Big(u_2 \exp(it \|S_n\|^2 + it\, u_1 + itu\, u_2)\Big),$$

and

$$u_2 = \varepsilon_n^2 \|b_k\|^2 + \varepsilon_n^2 \|d_k\|^2 + 2\varepsilon_n^2 (a_k, d_k) + 2\varepsilon_n^2 (b_k, d_k).$$

It is obvious that

$$|\varrho_{n1}(t)| \le |t|\, E\, |u_2|,$$

where

$$E\, |u_2| \le \varepsilon_n^2 \Big(E\|b_k\|^2 + E\|d_k\|^2 + 2\, E\, |(a_k, d_k)| + 2\, E\, |(b_k, d_k)|\Big).$$

Let us estimate here the expectations on the right-hand side. For this purpose, we define the functions

$$h_1(x) = E\, g_2(x, \xi_2) - E\, g_2(\xi_1, \xi_2),$$
$$h_2(x, y) = g_2(x, y) - h_1(x) - h_1(y) - E\, g_2(\xi_1, \xi_2).$$

Note that $E\, h_1(\xi_1) = 0$, and also that the symmetric function $h_2(x, y)$ possesses the property of degeneracy. It follows from the definition that

$$g_2(x, y) = h_2(x, y) + h_1(x) + h_1(y) + E\, g_2(\xi_1, \xi_2).$$

By using this relation, we can write

$$E\, \|\beta_k\|^2 \le 4 \Bigg(E \Big\| \sum_{j=k+1}^{n} \sum_{i=1}^{k} h_2(\xi_i, \xi_j)\Big\|^2 + k^2 E \Big\| \sum_{j=k+1}^{n} h_1(\xi_j)\Big\|^2$$
$$+ l^2 E \Big\|\sum_{i=1}^{k} h_1(\xi_i)\Big\|^2 + k^2 l^2 \|E\, g_2(\xi_1, \xi_2)\|^2\Bigg).$$

Here, by virtue of the property of degeneracy, we have

$$E \Big\| \sum_{j=k+1}^{n} \sum_{i=1}^{k} h_2(\xi_i, \xi_j)\Big\|^2 = kl\, E\, \|h_2(\xi_1, \xi_2)\|^2$$

and, moreover,

$$E \Big\| \sum_{j=k+1}^{n} h_1(\xi_j)\Big\|^2 = l\, E\, \|h_1(\xi_1)\|^2,$$

$$E\left\|\sum_{i=1}^{k} h_1(\xi_i)\right\|^2 = k\,E\,\|h_1(\xi_1)\|^2.$$

By virtue of the definition and the Hölder inequality, we obtain

$$E\,\|h_2(\xi_1,\xi_2)\|^2 \le 12\,E\,\|g_2(\xi_1,\xi_2)\|^2.$$

Further,

$$E\,\|h_1(\xi_1)\|^2 = E_{\xi_1}\,\|E_{\xi_2}\,g_2(\xi_1,\xi_2)\|^2 - \|E\,g_2(\xi_1,\xi_2)\|^2,$$

where the symbol E_ξ denotes the expectation with respect to the random variable ξ. Hence, $E\,\|h_1(\xi_1)\|^2 \le E_{\xi_1}\,\|E_{\xi_2}\,g_2(\xi_1,\xi_2)\|^2$. Therefore, we find

$$E\,\|\beta_k\|^2 \le 4\left(12kl\,E\,\|g_2(\xi_1,\xi_2)\|^2 + k^2l^2\,\|E\,g_2(\xi_1,\xi_2)\|^2 + \right.$$
$$\left. +2k^2l\,E_{\xi_1}\,\|E_{\xi_2}\,g_2(\xi_1,\xi_2)\|^2\right),$$

whence, we first get the estimate

$$E\,\|g_2(\xi_1,\xi_2)\|^2 \le (P(B))^{-2}\,E\,\|g_2(X_1,X_2)\|^2.$$

By using the property of degeneracy of function g_2, we can write

$$E\,g_2(\xi_1,\xi_2) = (P(B))^{-2}\int_{\bar{B}}\int_{B} g_2(x,y)\,P(dx)\,P(dy).$$

Therefore, according to the Hólder inequality, we have

$$\begin{aligned}\|E\,g_2(\xi_1,\xi_2)\|^2 &\le (P(B))^{-4}\,E\,\|g_2(X_1,X_2)\|^2\,(E\,I(\|g_1(X_1)\| > \sqrt{n}))^2\\ &\le (P(B))^{-4}\,E\,\|g_2(X_1,X_2)\|^2\\ &\quad\times \left(E\,(\|g_1(X_1)\|^2\,I(\|g_1(X_1)\| > \sqrt{n}))\right)^2 n^{-2}.\end{aligned}$$

By using a similar argument, we get

$$E_{\xi_1}\,\|E_{\xi_2}\,g_2(\xi_1,\xi_2)\|^2 = (P(B))^{-2}\,E_{\xi_1}\left\|\int_B g_2(\xi_1,x)\,P(dx)\right\|^2 \le$$

$$\le (P(B))^{-3}\,E\,\|g_2(X_1,X_2)\|^2 E\left(\|g_1(X_1)\|^2\,I(\|g_1(X_1)\| > \sqrt{n})\right) n^{-1}.$$

As a result, we obtain

$$\begin{aligned}E\,\|b_k\|^2 &\le 48\,(P(B))^{-4}\,E\,\|g_2(X_1,X_2)\|^2\\ &\quad\times\left(nl + l^2\left(E(\|g_1(X_1)\|^2\,I(\|g_1(X_1)\| > \sqrt{n}))\right)^2\right.\\ &\quad\left.+\,nl\,E\left(\|g_1(X_1)\|^2\,I(\|g_1(X_1)\| > \sqrt{n})\right)\right).\end{aligned}$$

By using the same argument, we can write

$$E\,\|d_k\|^2 = \binom{l}{2}^2 4l^{-1}\,E\,\|h_1(\xi_1)\|^2 + \binom{l}{2}^{-1} E\,\|h_2(\xi_1,\xi_2)\|^2 + \|E\,g_2(\xi_1,\xi_2)\|^2.$$

It follows from here that $E\,\|d_k\|^2 \le c_3\, E\,\|g_2(X_1,X_2)\|^2\, l^2$ with some constant $c_3 > 0$ independent on n.

Further, according to the Hölder inequality, we have

$$E\,|(a_k,d_k)| \le (E\|a_k\|^2)^{1/2}\,(E\|d_k\|^2)^{1/2},$$
$$E\,|(b_k,d_k)| \le (E\|b_k\|^2)^{1/2}\,(E\|d_k\|^2)^{1/2}.$$

Here,

$$E\,\|a_k\|^2 = \binom{k}{2}^2 \left[4k^{-1}\,E\,\|h_1(\xi_1)\|^2 + \binom{k}{2}^{-1} E\,\|h_2(\xi_1,\xi_2)\|^2 + \|E\,g_2(\xi_1,\xi_2)\|^2\right],$$

i.e. $E\,\|a_k\|^2 \le c_4\, E\,\|g_2(X_1,X_2)\|^2\, n^2$.

It follows from the inequalities stated above, that in (6.6.16)

$$|\varrho_{n1}(t)| \le c_4\, E\,\|g_2(X_1,X_2)\|^2\,|t|\,(1+|t|^{1-\varepsilon})^{-1}\, n^{-1}.$$

Consider the expectation on the right-hand side of (6.6.16) and rewrite it by using (6.6.5) with $r=2$. We get

$$E\exp(it\,\|S_n\|^2 + it\,u_1) = E\exp(it\,\|S_n\|^2 + it\,u_3) +$$
$$+2it\varepsilon_n\, E\left\{(v_{n1} + (S_n,d_k))\,\exp(it\,\|S_n\|^2 + it\,u_3)\right\} + \varrho_{n2}(t), \qquad (6.6.17)$$

where

$$u_3 = 2\varepsilon_n\,(S_n,a_k) + 2\varepsilon_n^2\,(a_k,b_k) + \varepsilon_n^2\,\|a_k\|^2 + 2\varepsilon_n\,(\bar S_n,b_k) + 2\varepsilon_n v_{n2};$$

$$v_{n1} = n^{-1/2}\sum_{i=1}^{k}\omega_n(\xi_i),$$

$$v_{n2} = n^{-1/2}\sum_{j=k+1}^{n}\sum_{i=1}^{k}(g_1(\xi_j), g_2(\xi_i,\xi_j)),$$

$$\omega_n(\xi_i) = \sum_{k+1\le j<s\le n} f(\xi_i;\xi_j,\xi_s),$$

$$f(\xi_i;\xi_j,\xi_s) = (g_1(\xi_j), g_2(\xi_i,\xi_s)) + (g_1(\xi_s), g_2(\xi_i,\xi_j)),$$

$$v_1 + v_2 = (\tilde S_n, b_k);$$

$$\varrho_{n2}(t) = -4t^4\varepsilon_n^2\int_0^1 du\,(1-u)\times$$
$$\times\, E\left\{(v_{n1} + (S_n,d_k))^2\exp\!\left(it\,\|S_n\|^2 + it\,u_3 + itu\,(v_{n1} + (S_n,d_k))\right)\right\}.$$

Clearly,

$$|\varrho_{n2}(t)| \le 8\,t^2\varepsilon_n^2\,(Ev_{n1}^2 + E(S_n,d_k)^2).$$

To estimate Ev_{n1}^2, we first write

$$Ev_{n1}^2 \le E\,\omega_n^2(\xi_1) + n\,E\,(E_{\xi_1}\omega_n(\xi_1))^2,$$

owing to independence. Denote

$$f_1(z;x) = E\,f(z;x,\xi_2) - E\,f(z;\xi_1,\xi_2),$$
$$f_2(z;x,y) = f(z;x,y) - f_1(z;x) - f_1(z;y) - E\,f(z;\xi_1,\xi_2),$$

i.e.

$$f(z;x,y) = f_2(z;x,y) + f_1(z;x) + f_1(z;y) + E\,f(z;\xi_1,\xi_2).$$

Then

$$E\,\omega_n^2(\xi_1) = \binom{l}{2}^2 \Bigg[4l^{-1}\,E\,f_1^2(\xi_1;\xi_2) + \\ + \binom{l}{2}^{-1} E\,f_2^2(\xi_1;\xi_2,\xi_3)\,E_{\xi_1}(E\,f(\xi_1;\xi_2,\xi_3))^2\Bigg],$$

$$E\,(E_{\xi_1}\,\omega_n(\xi_1))^2 = \binom{l}{2}^2 \Bigg[4l^{-1}\,E\,(E_{\xi_1}\,f_1(\xi_1;\xi_2))^2 + \\ + \binom{l}{2}^{-1} E\,(E_{\xi_1}\,f_2(\xi_1;\xi_2,\xi_3))^2 + (E\,f(\xi_1;\xi_2,\xi_3))^2\Bigg].$$

By estimating these expectations we finally get $E\,v_{n1}^2 \le c_5\,l^2$.

Further, to estimate $E(S_n,d_k)^2$ we write

$$E\,(S_n,d_k)^2 \le 2\left(E\,(\bar{S}_n,d_k)^2 + E\,(\tilde{S}_n,d_k)^2\right),$$

where we have $E\,(\bar{S}_n,d_k)^2 \le E\,\|\bar{S}_n\|^2\,E\,\|d_k\|^2 \le c_6\,l^2$ and

$$E\,(\tilde{S}_n,d_k)^2 = E\left(I(\|\tilde{S}_n\| \le \ln n)\,(\tilde{S}_n,d_k)^2\right) + E\left(I(\|\tilde{S}_n\| > \ln n)\,(\tilde{S}_n,d_k)^2\right).$$

Moreover,

$$E\left(I(\|\tilde{S}_n\| \le \ln n)\,(\tilde{S}_n,d_k)^2\right) \le (\ln n)^2\,E\|d_k\|^2 \le c_7\,(\ln n)^2\,l^2,$$
$$E\left(I(\|\tilde{S}_n\| > \ln n)\,(\tilde{S}_n,d_k)^2\right) \le n^2\,E\left(I(\|\tilde{S}_n\| > \ln n)\,\|d_k\|^2\right) =$$
$$= n^2 \sum_{c=0}^{2} \sum_{(c)} E\left\{I(\|\tilde{S}_n\| > \ln n)\,g_2(\xi_{i_1},\xi_{i_2})\,g_2(\xi_{j_1},\xi_{j_2})\right\},$$

where $\sum_{(c)}$ denotes the summation over all indices $k+1 \le i_1 < i_2 \le n$ and $k+1 \le j_1 < j_2 \le n$ such that exactly c equalities $i_\alpha = j_\beta$ are satisfied. The number of terms in the sum $\sum_{(c)}$ is equal to

$$\binom{2}{c}\binom{l-2}{2-c}\binom{l}{2}.$$

Further, the sum $\tilde{S}_n$ can be represented as follows:

$$\tilde{S}_n = n^{-1/2}\left(g_1(\xi_{i_1}) + g_1(\xi_{i_2}) + g_1(\xi_{j_1}) + g_1(\xi_{j_2})\right) + \tilde{S}_{n1},$$

where $\tilde{S}_{n1}$ does not contain terms with the indicated indices i_α and j_β. Note that $\|\tilde{S}_n\| \le 4 + \|\tilde{S}_{n1}\|$. Thus, taking into account the exponential estimate in Yurinskii (1976), we obtain $P(\|\tilde{S}_n\| > \ln n) \le P(\|\tilde{S}_{n1}\| > \ln n - 4) = O(n^{-q_1})$ with a large, arbitrarily fixed $q_1 > 0$. As a result, we get the following estimate $E\Big(I(\|\tilde{S}_n\| > \ln n)\,(\tilde{S}_n, d_k)^2\Big) \le c_8\, l^2$, i.e. $E\,(\tilde{S}_n, d_k)^2 \le (c_7(\ln n)^2 + c_8)\, l^2$. Hence, $E\,(S_n, d_k)^2 \le 2\,(c_7\,(\ln n)^2 + c_8)\, l^2$. The combination of these estimates gives

$$|\varrho_{n2}(t)| \le c_9\, t^2\,(1 + |t|^{1-\varepsilon})^{-2}\,(\ln n)^2\, n^{-1}.$$

Further, we estimate the expectation on the right-hand side of (6.6.17) in the region $n^\varepsilon \le |t| \le (n \ln n)^{1/2}$ with small arbitrarily fixed $\varepsilon > 0$. For this purpose, we apply the symmetrization inequality (see, e.g. Borovskikh and Korolyuk (1994a), Götze (1984) and Zalesskii *et al.* (1988, 1989)). Denote $l_1 = [l/2]$, and represent S_n as

$$S_n = n^{-1/2}\sum_{j=1}^{l_1} g_1(\xi_{k+j}) + n^{-1/2}\sum_{j=l_1+1}^{l} g_1(\xi_{k+j}) + \bar{S}_n.$$

Note that in (6.6.17), u_3 linearly depends on $\xi_{k+1}, \dots, \xi_n$. In this case, for the first expectation in (6.6.17), according to the symmetrization inequality, we have

$$\Big|E\exp(it\,\|S_n\|^2 + it\,u_3)\Big|^4 \le E\,\exp\Big\{2itn^{-1}\Big(\sum_{j=1}^{l_1} z_j,\ \sum_{j=l_1+1}^{l} z_j\Big)\Big\}, \tag{6.6.18}$$

where $z_j = g_1(\xi_{k+j}) - g_1(\xi^1_{k+j})$ and ξ^1_{k+j} are independent copies of ξ_{k+j}. By estimating the expectation on the right-hand side of (6.6.18) with chosen l and l_1 just as in Borovskikh and Korolyuk (1984c), we get the following estimate in the region $|t| \le (n \ln n)^{1/2}$:

$$E\exp\Big\{2itn^{-1}\Big(\sum_{j=1}^{l_1} z_j,\ \sum_{j=l_1+1}^{l} z_j\Big)\Big\} \le c_1\,(1 + |t|)^{-q} \tag{6.6.19}$$

with large arbitrarily fixed $q > 0$. Therefore, in the region $n^\varepsilon \le |t| \le (n \ln n)^{1/2}$, we have

$$\Big|E\exp(it\,\|S_n\|^2 + it\,u_3)\Big| = O(n^{-8}). \tag{6.6.20}$$

As far as relations (6.6.19) and (6.6.20) are concerned, it is worth noting that the number of nonzero eigenvalues of the V-operator is infinite and, hence, for any small arbitrarily fixed number $\varepsilon > 0$, there always exists a sufficiently large number $q > 0$ for which (6.6.18) and (6.6.19) imply (6.6.20).

To get the estimate of the second expectation on the right-hand side of (6.6.17) in the form of the inequality of type (6.6.18), we write the following representation:

$$E\Big\{(v_{n1} + (S_n, d_k))\,\exp(it\,\|S_n\|^2 + it\,u_3)\Big\} =$$

$$= n^{-1/2}\sum_{i_1=1}^{k}\ \sum_{k+1 \le i_2 < i_3 \le n} E\,\Big\{f(\xi_{i_1}; \xi_{i_2}, \xi_{i_3})\,\exp(it\,\|S_n\|^2 + it\,u_3)\Big\} +$$

$$+\, n^{-1/2} \sum_{i_1=1}^{n} \sum_{k+1 \le i_2 < i_3 \le n} E\Big\{(g_1(\xi_{i_1}), g_2(\xi_{i_2}, \xi_{i_3}))\, \exp(it\,\|S_n\|^2 + it\,u_3)\Big\}.$$

By symmetrization inequality (see Borovskikh and Korolyuk (1984c), Lemma 2.1),

$$\Big|E\Big\{f(\xi_{i_1}; \xi_{i_2}, \xi_{i_3}) \exp(it\,\|S_n\|^2 + it\,u_3)\Big\}\Big|^4 \le$$
$$\le (E\,f^2(\xi_1; \xi_2, \xi_3))^2\, E\exp\Big\{2it\,n^{-1}\Big(\sum_{j=4}^{l_1} z_j, \sum_{j=l_1+1}^{l} z_j\Big)\Big\},$$
$$\Big|E\Big\{(g_1(\xi_{i_1}), g_2(\xi_{i_2}, \xi_{i_3})) \exp(it\,\|S_n\|^2 + it\,u_3)\Big\}\Big|^4 \le$$
$$\le n^2\,(E\,\|g_2(\xi_1, \xi_2)\|^2)^2\, E\exp\Big\{2it\,n^{-1}\Big(\sum_{j=4}^{l_1} z_j, \sum_{j=l_1+1}^{l} z_j\Big)\Big\}.$$

Together with relations (6.6.19)–(6.6.20) in the region $n^\varepsilon \le |t| \le (n \ln n)^{1/2}$, this yields

$$\Big|E\Big\{(v_{n1} + (S_n, d_k)) \exp(it\,\|S_n\|^2 + it\,u_3)\Big\}\Big| = O(n^{-2}).$$

Thus, in what follows, when analysing the right-hand side of (6.6.17), it is sufficient to assume that $|t| \le n^\varepsilon$. In this analysis, we use relation (6.6.5) and the following inequalities

$$|e^{ix} - 1| \le 2\,|x|^{1-\delta}, \qquad 0 \le \delta \le 1,$$
$$|e^{ix} - 1 - ix| \le 2\,|x|^{2-\varepsilon}, \qquad 0 \le \varepsilon \le 1, \tag{6.6.21}$$

which hold for all $x \in R$. The first term on the right-hand side of (6.6.17) can be represented as

$$E\exp(it\,\|S_n\|^2 + it\,u_3) = E\exp(it\,\|S_n\|^2) +$$
$$+2\,it\varepsilon_n\, E\Big\{((S_n, a_k) + (\bar S_n, b_k) + v_{n2})\, \exp(it\,\|S_n\|^2)\Big\} + \varrho_{n3}(t) \tag{6.6.22}$$

where

$$|\varrho_{n3}(t)| \le 2\,|t|\,\varepsilon_n^2\,(E\,|(a_k, b_k)| + E\,\|a_k\|^2) +$$
$$+12\,|t|^{2-\varepsilon}\,\varepsilon_n^{2-\varepsilon}\Big(E\,|(S_n, a_k)|^{2-\varepsilon} + E\,|(\bar S_n, b_k)|^{2-\varepsilon} + E\,|v_{n2}|^{2-\varepsilon}\Big).$$

Here, the first term in the brackets appears when we apply (6.6.5) with $r = 1$, while the second term is a result of inequality (6.6.21). For the second term on the right-hand side of (6.6.17), we can write

$$2\,it\varepsilon_n\, E\Big\{(v_{n1} + (S_n, d_k)) \exp(it\,\|S_n\|^2 + it\,u_3)\Big\} =$$
$$= 2\,it\varepsilon_n\, E\Big\{(v_{n1} + (S_n, d_k)) \exp(it\,\|S_n\|^2)\Big\} + \varrho_{n4}(t)$$

where

$$|\varrho_{n4}(t)| \le 8\,|t|^{3/2}\,\varepsilon_n^2\, E\Big\{(|v_{n1}| + |(S_n, d_k)|)\,(\,|(a_k, b_k)|^{1/2} + \|a_k\|)\Big\} +$$

$$+4\,|t|^{2-\varepsilon}\,\varepsilon_n^{2-\varepsilon}\,E\Big\{(|v_{n1}|+|(S_n,d_k)|)\,\Big(|(S_n,a_k)|^{1-\varepsilon}+|(\bar{S}_n,b_k)|^{1-\varepsilon}+|v_{n2}|^{1-\varepsilon}\Big)\Big\}.$$

Here, the first expectation on the right-hand side appears if we apply the inequality (6.6.21) with $\delta = 1/2$, while the second one is a result of the application of the same inequality with $\delta = \varepsilon$.

Let us make the estimates more precise here. By the Hölder inequality, we have

$$\begin{aligned} E\,|(a_k,b_k)| &\le \left(E\,\|a_k\|^2\right)^{1/2}\left(E\,\|b_k\|^2\right)^{1/2},\\ E\,|(S_n,a_k)|^{2-\varepsilon} &\le \left(E\,\|S_n\|^{2(2-\varepsilon)/\varepsilon}\right)^{\varepsilon/2}\left(E\,\|a_k\|^2\right)^{2-\varepsilon/2},\\ E\,|(\bar{S}_n,b_k)|^{2-\varepsilon} &\le \left(E\,\|\bar{S}_n\|^{2(2-\varepsilon)/\varepsilon}\right)^{\varepsilon/2}\left(E\,\|b_k\|^2\right)^{2-\varepsilon/2},\\ E\,|v_{n2}|^{2-\varepsilon} &\le \left(E\,v_{n2}^2\right)^{2-\varepsilon/2}. \end{aligned}$$

A simple calculation shows that

$$\begin{aligned} E\,v_{n2}^2 = n^{-1}kl\,E(g_1(\xi_2),g_2(\xi_1,\xi_2))^2 &\\ +\,n^{-1}kl\,(l-1)\,E(g_1(\xi_2),g_2(\xi_1,\xi_2))\,(g_1(\xi_3),g_2(\xi_1,\xi_3))&\\ +\,n^{-1}k(k-1)\,l\,E(g_1(\xi_3),g_2(\xi_1,\xi_3))\,(g_1(\xi_3),g_2(\xi_2,\xi_3))&\\ +\,n^{-1}k(k-1)\,l(l-1)\,E(g_1(\xi_3),g_2(\xi_1,\xi_3))\,(g_1(\xi_4),g_2(\xi_2,\xi_4)).& \end{aligned}$$

Under condition $\mu_2 < \infty$, it follows from here that

$$|\varrho_{n3}(t)| \le c_1\,|t|\,(1+|t|^{1-\varepsilon})\,n^{-1+\varepsilon/2}.$$

To make the estimate for $\varrho_{n4}(t)$ more precise, we write

$$\begin{aligned} &E\left\{(|v_{n1}|+|(S_n,d_k)|)\,(|(a_k,b_k)|^{1/2}+\|a_k\|)\right\}\\ &\quad\le \Big(E(|v_{n1}|+|S_n,d_k)|)^2\Big)^{1/2}\times\Big(E(|(a_k,b_k)|^{1/2}+\|a_k\|^2\Big)^{1/2}\\ &\quad\le 2\left(E\,v_{n1}^2+E\,(S_n,d_k)^2\right)^{1/2}\times\Big((E\,\|a_k\|^2)^{1/2}(E\,\|b_k\|^2)^{1/2}+E\,\|a_k\|^2\Big)^{1/2},\\ &E\left\{(|v_{n1}|+|(S_n,d_k)|)\,\Big(|(S_n,a_k)|^{1-\varepsilon}+|(\bar{S}_n,b_k)|^{1-\varepsilon}+|v_{n2}|^{1-\varepsilon}\Big)\right\}\\ &\quad\le \Big(E\,(|v_{n1}|+|(S_n,d_k)|)^2\Big)^{1/2}\times\Big(E\Big(|(S_n,a_k)|^{1-\varepsilon}+|(\bar{S}_n,b_k)|^{1-\varepsilon}+|v_{n2}|^{1-\varepsilon}\Big)^2\Big)^{1/2}\\ &\quad\le 2\left(E\,v_{n1}^2+E\,(S_n,d_k)^2\right)^{1/2}\\ &\quad\times\Big(E\,|(S_n,a_k)|^{2(1-\varepsilon)}+E\,|(\bar{S}_n,b_k)|^{2(1-\varepsilon)}+E\,|v_{n2}|^{2(1-\varepsilon)}\Big)^{1/2}, \end{aligned}$$

according to the Hölder inequality.

Under condition $\mu_2 < \infty$, by using these inequalities, we get the following estimate

$$|\varrho_{n4}(t)| \le c_2\,|t|^{3/2}\,(1+|t|)^{1/2-\varepsilon})\,(\ln n)\,n^{-1+\varepsilon/2}.$$

Combining these estimates we find that

$$E\exp(it\,\|S_n\|^2+it\,u_1) = E\exp(it\,\|S_n\|^2)\,+$$

$$+2it\,\varepsilon_n\,E\Big\{\Big((S_n,a_k)+(S_n,d_k)+(\bar S_n,b_k)+v_{n1}+v_{n2}\Big)\,\exp(it\,\|S_n\|^2)\Big\} =: \varrho_{n5}(t)$$

in the region $|t|\le n^{\varepsilon}$, where $|\varrho_{n5}(t)|\le c_3\,|t|\,(1+|t|)\,n^{-1+\varepsilon}$.

Note that $(S_n,a_k)+(S_n,d_k)+(\bar S_n,b_k)+v_{n1}+v_{n2}=(S_n,S_{n2}(g_2))$.

Lemma 6.6.10 is proved.

Lemma 6.6.11. *The relation*

$$E\left\{(S_n,S_{n2}(g_2))\exp(it\,\|S_n\|^2)\right\}=2\,n^{-1/2}\binom{n}{2}\psi_{n1}(t)+3\,n^{-1/2}\binom{n}{3}\psi_{n2}(t)$$

holds, where

$$\begin{aligned}
\psi_{n1}(t) &= E\left\{(g_1(\xi_1),g_2(\xi_1,\xi_2))\exp\left(it\left\|n^{-1/2}(g_1(\xi_1)+g_1(\xi_2))+S_{n2}\right\|^2\right)\right\},\\
\psi_{n2}(t) &= E\left\{(g_1(\xi_1),g_2(\xi_2,\xi_3))\,\times\right.\\
&\qquad\left.\times\exp\left(it\left\|n^{-1/2}(g_1(\xi_1)+g_1(\xi_2)+g_1(\xi_3))+S_{n3}\right\|^2\right)\right\},
\end{aligned}$$

and

$$S_{nr}=n^{-1/2}\sum_{j=r+1}^{n}g_1(\xi_j),\qquad r=0,1,2,\dots .$$

Proof. By definition we have

$$\begin{aligned}
&E\left\{(S_n,S_{n2}(g_2))\exp(it\,\|S_n\|^2)\right\}\\
&\quad=-n^{-1/2}\sum_{j=1}^{n}\sum_{1\le i_1<i_2\le n}E\left\{(g_1(\xi_j),g_2(\xi_{i_1},\xi_{i_2}))\exp(it\,\|S_n\|^2)\right\}\\
&\quad=-n^{-1/2}\sum_{j=1}^{n}\sum_{i_1=2}^{j}\sum_{i_2=1}^{i_1-1}E\left\{(g_1(\xi_j),g_2(\xi_{i_1},\xi_{i_2}))\exp(it\,\|S_n\|^2)\right\}\\
&\qquad+n^{-1/2}\sum_{j=1}^{n}\sum_{i_1=j+1}^{n}\sum_{i_2=1}^{j}E\left\{(g_1(\xi_j),g_2(\xi_{i_1},\xi_{i_2}))\exp(it\,\|S_n\|^2)\right\}\\
&\qquad+n^{-1/2}\sum_{j=1}^{n}\sum_{i_1=j+1}^{n}\sum_{i_2=j+1}^{i_1-1}E\left\{(g_1(\xi_j),g_2(\xi_{i_1},\xi_{i_2}))\exp(it\,\|S_n\|^2)\right\}.
\end{aligned}$$

In this formula, we have

$$\begin{aligned}
&\sum_{j=1}^{n}\sum_{i_1=2}^{j}\sum_{i_2=1}^{i_1-1}E\left\{(g_1(\xi_j),g_2(\xi_{i_1},\xi_{i_2}))\exp(it\,\|S_n\|^2)\right\}\\
&\qquad=\sum_{j=1}^{n}\sum_{i_1=2}^{j-1}\sum_{i_2=1}^{i_1-1}E\left\{(g_1(\xi_j),g_2(\xi_{i_1},\xi_{i_2}))\exp(it\,\|S_n\|^2)\right\}\\
&\qquad\quad+\sum_{j=1}^{n}\sum_{i_2=1}^{j-1}E\left\{(g_1(\xi_j),g_2(\xi_j,\xi_{i_2}))\exp(it\,\|S_n\|^2)\right\}=
\end{aligned}$$

$$= \binom{n}{3}\psi_{n2}(t) + \binom{n}{2}\psi_{n1}(t),$$

$$\sum_{j=1}^{n}\sum_{i_1=j+1}^{n}\sum_{i_2=1}^{j} E\left\{(g_1(\xi_j), g_2(\xi_{i_1}, \xi_{i_2}))\exp(it\,\|S_n\|^2)\right\}$$

$$= \sum_{j=1}^{n}\sum_{i_1=j+1}^{n}\sum_{i_2=1}^{j-1} E\left\{(g_1(\xi_j), g_2(\xi_{i_1}, \xi_{i_2}))\exp(it\,\|S_n\|^2)\right\}$$

$$+ \sum_{j=1}^{n}\sum_{i_1=j+1}^{n} E\left\{(g_1(\xi_j), g_2(\xi_{i_1}, \xi_j))\exp(it\,\|S_n\|^2)\right\}$$

$$= \binom{n}{3}\psi_{n2}(t) + \binom{n}{2}\psi_{n1}(t),$$

$$\sum_{j=1}^{n}\sum_{i_1=j+1}^{n}\sum_{i_2=j+1}^{i_1-1} E\left\{(g_1(\xi_j), g_2(\xi_{i_1}, \xi_{i_2}))\exp(it\,\|S_n\|^2)\right\}$$

$$= \binom{n}{3}\psi_{n2}(t).$$

This proves Lemma 6.6.11.

Lemma 6.6.12. *Under condition $\mu_2 < \infty$ the inequality*

$$|\psi_{n1}(t)| \le c_3(\Phi)\,(1+|t|)^{-q} \times$$
$$\times\left(n^{-1/2}\left(\int_B \|g_1(x)\|^2\,P(dx)\right)^{1/2} + n^{-1}\,E\,\|g_1(\xi_1)\|^4\right),$$

holds in the region $|t| \le n^{1-\epsilon}$, where $q > 0$ is a large arbitrarily fixed number and $c_3(\Phi)$ depends on the kernel Φ only via eigenvalues of the V-operator and the variables σ^2 and μ_2.

Proof. It is clear that

$$\left\|n^{-1/2}(g_1(\xi_1)+g_1(\xi_2)) + S_{n2}\right\|^2 = n^{-1}\,\|g_1(\xi_1)\|^2 + n^{-1}\,\|g_1(\xi_2)\|^2 +$$
$$+2n^{-1}\,(g_1(\xi_1), g_1(\xi_2)) + 2n^{-1/2}\,(g_1(\xi_1)+g_1(\xi_2), S_{n2}) + \|S_{n2}\|^2.$$

By using (6.6.5) with $r = 1$, we can rewrite $\psi_{n1}(t)$ as follows

$$\psi_{n1}(t) = E\left\{(g_1(\xi_1), g_2(\xi_1, \xi_2))\exp(it\,\|S_{n2}\|^2 + it\,u_4)\right\} + \varrho_{n6}(t)\,, \qquad (6.6.23)$$

where

$$u_4 = n^{-1}\,\|g_1(\xi_1)\|^2 + n^{-1}\,\|g_1(\xi_2)\|^2 + 2n^{-1/2}\,(g_1(\xi_1)+g_1(\xi_2), S_{n2}),$$

$$\varrho_{n6}(t) = 2it\,n^{-1}\int_0^1 du\,E\left\{(g_1(\xi_1), g_2(\xi_1,\xi_2))\,(g_1(\xi_1), g_1(\xi_2))\,\times\right.$$
$$\left.\times\exp\left(it\,\|S_{n2}\|^2 + it\,u_4 + 2itu\,n^{-1}\,(g_1(\xi_1), g_1(\xi_2))\right)\right\}.$$

For ϱ_{n6}, we have the estimate

$$|\varrho_{n6}(t)| \leq 2\,|t|\,n^{-1}\,E\Big\{\|g_1(\xi_1)\|^2\,\|g_1(\xi_2)\|\,\|g_2(\xi_1,\xi_2)\| \times \\ \times |\bar{E}_2\,\exp(it\,\|S_{n2}\|^2 + it\,u_4)|\Big\}.$$

The expectation $\bar{E}_2$ with respect to the variables $\xi_3,\ldots,\xi_n$, can be estimated by using the symmetrization inequality (see (6.6.18)–(6.6.19)). We have

$$|\bar{E}_2\exp(it\,\|S_{n2}\|^2 + it\,u_4)| \leq c_{14}\,(1+|t|)^{-q_1}$$

with large arbitrary $q_1 > 0$ and a constant c_{14} independent of ξ_1 and ξ_2. Therefore,

$$|\varrho_{n6}(t)| \leq 2c_4\,|t|\,(1+|t|)^{-q_1}\,n^{-1}\,E\left(\|g_1(\xi_1)\|^2\,\|g_1(\xi_2)\|\,\|g_2(\xi_1,\xi_2)\|\right).$$

According to the Hölder inequality, we have, firstly,

$$E\left(\|g_1(\xi_1)\|^2\,\|g_1(\xi_2)\|\,\|g_2(\xi_1,\xi_2)\|\right) \leq \\ \leq \left(E\,(\|g_1(\xi_1)\|^2\,\|g_1(\xi_2)\|)^2\right)^{1/2}\left(E\,\|g_2(\xi_1,\xi_2)\|^2\right)^{1/2}$$

and, secondly, $E\,\|g_1(\xi_1)\|^2 \leq (E\,\|g_1(\xi_1)\|^4)^{1/2}$. Hence,

$$|\varrho_{n6}(t)| \leq 2c_4\,|t|\,(1+|t|)^{-q_1}\,n^{-1}\,E\,\|g_1(\xi_1)\|^4 \times \\ \times (E\,\|g_1(\xi_1)\|^2)^{1/2}\,(E\,\|g_2(\xi_1,\xi_2)\|^2)^{1/2}.$$

To estimate the expectation in (6.6.23), we apply the following formula

$$e^{i(x+y)} = 1 + (e^{ix}-1) + (e^{iy}-1) + (e^{ix}-1)\,(e^{iy}-1), \tag{6.6.24}$$

valid for all x, $y \in R$. Here, we set

$$x = x_1 = 2n^{-1/2}\,(g_1(\xi_1), S_{n2}) \quad \text{and} \quad y = y_1 = 2n^{-1/2}\,(g_1(\xi_2), S_{n2}).$$

Then

$$E\Big\{(g_1(\xi_1), g_2(\xi_1,\xi_2))\,\exp(it\,\|S_{n2}\|^2 + it\,u_4)\Big\} = \sum_{j=1}^{3} e_{nj}(t) + \varrho_{n7}(t), \tag{6.6.25}$$

where

$$e_{n1}(t) = E\Big\{(g_1(\xi_1), g_2(\xi_1,\xi_2))\,\exp(it\,\|S_{n2}\|^2 + it\,u_5)\Big\},$$
$$e_{n2}(t) = E\Big\{(g_1(\xi_1), g_2(\xi_1,\xi_2))\,\exp(it\,\|S_{n2}\|^2 + it\,u_5)\,(e^{itx_1}-1)\Big\},$$
$$e_{n3}(t) = E\Big\{(g_1(\xi_1), g_2(\xi_1,\xi_2))\,\exp(it\,\|S_{n2}\|^2 + it\,u_5)\,(e^{ity_1}-1)\Big\},$$
$$u_5 = n^{-1}\,\|g_1(\xi_1)\|^2 + n^{-1}\,\|g_1(\xi_2)\|^2;$$

$$\varrho_{n7} = -t^2 \int_0^1 du_1 \int_0^1 du_2\, E\Big\{x_1 y_1\, (g_1(\xi_1), g_2(\xi_1, \xi_2))$$
$$\times \exp(it\, \|S_{n2}\|^2 + it\, u_5 + it\, u_1 x_1 + it\, u_2 y_1)\Big\}.$$

First, we estimate $\varrho_{n7}(t)$ as follows

$$|\varrho_{n7}(t)| \le 2\, t^2 n^{-1} \int_0^1 du_1 \int_0^1 du_2\, E\Big\{\|g_1(\xi_1)\|\, \|g_2(\xi_1, \xi_2)\| \times$$
$$\times \Big|\bar{E}_2 \Big((g_1(\xi_1), S_{n2})\, (g_1(\xi_2), S_{n2}) \exp(it\, \|S_{n2}\|^2 + it\, u_1 x_1 + it\, u_2 y_1)\Big)\Big|\Big\}.$$

It follows from Borovskikh and Korolyuk (1984c), Lemma 2.11, p. 112 that, for any integer $r_1 \ge 0$ and $r_2 \ge 0$, the inequality

$$\Big|\bar{E}_2 \Big((g_1(\xi_1), S_{n2})^{r_1}\, (g_1(\xi_2), S_{n2})^{r_2} \exp(it\, \|S_{n2}\|^2 + it\, u_1 x_1 + it\, u_2 y_1)\Big)\Big| \le$$
$$\le c_5\, (1 + |t|)^{-q_2}\, \|g_1(\xi_1)\|^{r_1}\, \|g_1(\xi_2)\|^{r_2} \tag{6.6.26}$$

holds in the region $|t| \le n^{1-\varepsilon}$ with large arbitrary $q_2 > 0$. Finally, by using (6.6.26) with $r_1 = r_2 = 1$, we get the following estimate for $\varrho_{n7}(t)$

$$|\varrho_{n7}(t)| \le 2 c_{15}\, t^2\, (1 + |t|)^{-q_2}\, n^{-1}\, E\, \|g_1(\xi_1)\|^4 \times$$
$$\times (E\, \|g_1(\xi_1)\|^2)^{-1/2}\, (E\, \|g_2(\xi_1, \xi_2)\|^2)^{1/2}.$$

To analyse the functions $e_{nj}(t)$, $j = 1, 2, 3$, we use (6.6.24) with $x_1 = x_2 = n^{-1}\, \|g_1(\xi_1)\|^2$ and $y = y_2 = n^{-1}\, \|g_1(\xi_2)\|^2$. For $e_{n1}(t)$, we can write

$$e_{n1}(t) = e^1_{n1}(t) + e^2_{n1}(t) + e^3_{n1}(t) + e^4_{n1}(t),$$

where

$$e^1_{n1}(t) = E\, \{(g_1(\xi_1), g_2(\xi_1, \xi_2))\}\, E \exp(it\, \|S_{n2}\|^2),$$
$$e^2_{n1}(t) = E\, \{(g_1(\xi_1), g_2(\xi_1, \xi_2))\, (e^{itx_2} - 1)\}\, E \exp(it\, \|S_{n2}\|^2),$$
$$e^3_{n1}(t) = E\, \{(g_1(\xi_1), g_2(\xi_1, \xi_2))\, (e^{ity_2} - 1)\}\, E \exp(it\, \|S_{n2}\|^2),$$
$$e^4_{n1}(t) = E\, \{(g_1(\xi_1), g_2(\xi_1, \xi_2))\, (e^{itx_2} - 1)\, (e^{ity_2} - 1)\}\, E \exp(it\, \|S_{n2}\|^2).$$

According to the property of degeneracy of the function g_2, we have

$$E\, (g_1(\xi_1), g_2(\xi_1, \xi_2)) = (P(B))^{-1} \int_B E_{\xi_1}\, (g_1(\xi_1), g_2(\xi_1, x))\, P(dx). \tag{6.6.27}$$

This yields

$$\Big|E\, (g_1(\xi_1), g_2(\xi_1, \xi_2))\Big| \le n^{-1/2}\, (P(B))^{-1} \times$$
$$\times E\, \Big(\|g_1(\xi_1)\|\, \|g_2(\xi_1, X_1)\|\, \|g_1(X_1)\|\, I(\|g_1(X_1)\| > \sqrt{n})\Big).$$

To estimate $e^2_{n2}(t)$, we use (6.6.27) and inequality $|\exp(itx_2)-1| \le 2$, whereas $e^3_{n1}(t)$ and $e^4_{n1}(t)$ are estimated by using inequality $|\exp(ity_2)-1| \le |t|\,y_2$. With the help of (6.6.26) with $r_1 = r_2 = 0$ and $u_1 = u_2 = 0$, we can estimate the expectation $E\exp(it\,\|S_{n2}\|^2)$. As a result, we get

$$|e_{n1}(t)| \le c_6\,(1+|t|)^{-q_2+1} \times$$
$$\times\Big(n^{-1/2}\left(E(\|g_1(X_1)\|^2\, I(\|g_1(X_1)\| > \sqrt{n}))\right)^{1/2} + n^{-1}\, E\,\|g_1(\xi_1)\|^4\Big).$$

For $e_{n2}(t)$ in (6.6.25), we can write

$$e_{n2}(t) = e^1_{n2}(t) + e^2_{n2}(t),$$

where

$$\begin{aligned} e^1_{n2}(t) &= E\left\{(g_1(\xi_1), g_2(\xi_1,\xi_2))\, e^{itx_2}\,(e^{ity_2}-1)\right.\\ &\qquad \left.\times\, \bar{E}_2\,(e^{itx_1}-1)\exp(it\,\|S_{n2}\|^2)\right\},\\ e^2_{n1}(t) &= E\left\{(g_1(\xi_1), g_2(\xi_1,\xi_2))\, e^{itx_2}\right.\\ &\qquad \left.\times\, \bar{E}_2\,(e^{itx_1}-1)\exp(it\,\|S_{n2}\|^2)\right\}. \end{aligned}$$

The variable $e^2_{n2}(t)$ can be estimated by using (6.6.27) because x_2 and S_{n2} do not depend on ξ_2. The modulus of the expectation $\bar{E}_2$ with respect to the variables $\xi_3, \dots, \xi_n$ can be estimated with the help of (6.6.26). As a result, we get

$$|e_{n2}(t)| \le c_7\,(1+|t|)^{-q_2+1}\, n^{-1}\, E\,\|g_1(\xi_1)\|^4.$$

Consider $e_{n3}(t)$ in (6.6.25). First, by using (6.6.5), we write

$$\begin{aligned} &e^{it\,u_5}\,(e^{ity_1}-1) =\\ &\quad = (e^{it\,u_5}-1)\,(e^{ity_1}-1) + (e^{ity_1}-1) =\\ &\quad = it\,y_1\int_0^1 du\, e^{it\,uy_1}\,(e^{it\,u_5}-1) - t^2y_1^2\int_0^1 du\,(1-u)\,e^{ity_1} + ity_1. \end{aligned}$$

By inserting this relation into $e_{n3}(t)$, we obtain

$$e_{n3}(t) = e^1_{n3}(t) + e^2_{n3}(t) + e^3_{n3}(t),$$

where

$$\begin{aligned} e^1_{n3}(t) &= it\int_0^1 du\, E\left\{(g_1(\xi_1), g_2(\xi_1,\xi_2))\,(e^{it\,u_5}-1)\right.\\ &\qquad \left.\times\, \bar{E}_2\,(y_1\exp(it\,\|S_{n2}\|^2 + it\,uy_1))\right\},\\ e^2_{n3}(t) &= -t^2\int_0^1 du\,(1-u) \end{aligned}$$

$$\times\, E\Big\{(g_1(\xi_1), g_2(\xi_1,\xi_2))\, \bar{E}_2\,(y_1^2 \exp(it\,\|S_{n2}\|^2 + it\,uy_1))\Big\},$$
$$e_{n3}^3 = it\,E\Big\{(g_1(\xi_1), g_2(\xi_1,\xi_2))\, \bar{E}_2(y_1 \exp(it\,\|S_{n2}\|^2))\Big\}.$$

In $e_{n3}^1(t)$, we have $|\exp(it\,u_5) - 1| \le 2\,|t|^{1/2}\,|u_5|^{1/2}$; the expectation $\bar{E}_2$ with respect to the variables $\xi_3, \dots, \xi_n$ can be estimated by using (6.6.26). This gives

$$|e_{n3}^1(t)| \le c_8\,|t|^{3/2}\,(1+|t|)^{-q_2}\,n^{-1}\,E\,\|g_1(\xi_1)\|^4.$$

For $e_{n3}^2(t)$, by virtue of (6.6.26), we obtain

$$|e_{n2}^2(t)| \le c_9 t^2\,(1+|t|)^{-q_2}\,n^{-1}\,E\,\|g_1(\xi_1)\|^4.$$

Further, to estimate $e_{n3}^3(t)$, we write

$$\bar{E}_2\Big(y_1 \exp(it\,\|S_{n2}\|^2)\Big) = 2n^{-1/2}\,E\Big((g_1(\xi_2), S_2)\,\exp(it\,\|S_{n2}\|^2)\Big)$$

for fixed ξ_2. We also use the following formula

$$E\,\Big((g_1(\xi_2), S_{n2})\,\exp(it\,\|S_{n2}\|^2)\Big) - E\,\Big((g_1(\xi_2), \tau)\,\exp(it\,\|\tau\|^2)\Big) =$$
$$= \sum_{j=0}^{n-4} \iint (g_1(\xi_2), x+y)\,\exp(it\,\|x+y\|^2)\,F_j(dx)\,H(dy), \tag{6.6.28}$$

where $F_j = G^j \bar{N}^{n-j-4}$, $H = G - \bar{N}$, G is the distribution $g_1(\xi_1)\,n^{-1/2}$, and $\bar{N}$ is the distribution $\tau n^{-1/2}$. In (6.6.28), according to the Taylor formula, we have

$$e^{it\,\|x+y\|^2} = e^{it\,\|x\|^2} + 2it\,(x,y)\,e^{it\,\|x\|^2} +$$
$$+ \Big((2it)^2\,(x,y)^2 + 2it\,\|y\|^2\Big)\,e^{it\,\|x\|^2} + R(x,y),$$

where

$$R(x,y) = \frac{1}{2}\int_0^1 du\,(1-u)^2\,\times$$
$$\times \Big((2it)^3\,(x+uy,y)^3 + 3\,(2it)^2\,\|y\|^2\,(x+uy,y)\Big)\,\exp(it\,\|x+uy\|^2).$$

If we insert this relation into (6.6.28) and note that, by the property of the distribution N of the Gaussian random variable τ, we have

$$E\,\Big((g_1(\xi_2)\tau)\,\exp(it\,\|\tau\|^2)\Big) = 0$$

for fixed ξ_2, then we get

$$E\,\Big((g_1(\xi_2), S_{n2})\,\exp(it\,\|S_{n2}\|^2)\Big) = \sum_{j=0}^{n-4}\sum_{s=0}^{3} v_{js}(t), \tag{6.6.29}$$

where

$$v_{j0}(t) = \iint (g_1(\xi_2), x+y)\,\exp(it\,\|x\|^2)\,F_j(dx)\,H(dy),$$

$$v_{j1}(t) = 2it \iint (g_1(\xi_2), x+y)\,(x,y) \exp(it\,\|x\|^2)\, F_j(dx)\, H(dy),$$
$$v_{j2}(t) = \iint (g_1(\xi_2), x+y) \left((2it)^2 (x,y)^2 + 2it\,\|y\|^2\right) \times \exp(it\,\|x\|^2)\, F_j(dx)\, H(dy),$$
$$v_{j3}(t) = \iint (g_1(\xi_2), x+y)\, R(x+y)\, F_j(dx)\, H(dy).$$

Since

$$\int H(dy) = 0, \quad \int y\, H(dy) = n^{-1/2} (P(B))^{-1} \int_{\bar{B}} g_1(z)\, P(dz), \tag{6.6.30}$$

with the help of (6.6.26), we obtain

$$|v_{jo}(t)| \le c_1\, \|g_1(\xi_2)\|\, (1+|t|)^{-q_2}\, n^{-1} \int_{\bar{B}} \|g_1(z)\|^2\, P(dz).$$

To estimate $v_{j1}(t)$, we first write

$$\int (g_1(\xi_2), x+y)\,(x,y)\, H(dy) = $$
$$= (g_1(\xi_2), x) \int (x,y)\, H(dy) + \int (g_1(\xi_2), y)\,(x,y)\, H(dy),$$

where the first term on the right-hand side can be estimated by using (6.6.30). For the second term, in view of the fact that the measures P and N have the same covariance operator V, we get

$$\int (g_1(\xi_2), y)\,(x,y)\, H(dy) = n^{-1}\, P(\bar{B})\, (P(B))^{-1}\, (Sx, g_1(\xi_2)) -$$
$$-n^{-1} \int_{\bar{B}} (g_1(\xi_2), g_1(z))\, (x, g_1(z))\, P(dz). \tag{6.6.31}$$

In $v_{j1}(t)$, we first take (6.6.30) and (6.6.31) into account and then estimate the integral over the measure F_j by using the inequalities (6.6.2) and (6.6.26). As a result, we get

$$|v_{j1}(t)| \le c_2\, \|g_1(\xi_2)\|\, |t|\, (1+|t|)^{-q_2}\, n^{-1} \int_{\bar{B}} \|g_1(z)\|^2\, P(dz).$$

To estimate $v_{j2}(t)$, we apply relation (6.6.31) and

$$\int \|y\|^2\, H(dy) =$$
$$= n^{-1}\, P(\bar{B})\, (P(B))^{-1} \int_B \|y\|^2\, P(dy) - n^{-1} \int_{\bar{B}} \|y\|^2\, P(dy).$$

This yields

$$|v_{j2}(t)| \le c_3\, \|g_1(\xi_2)\|\, |t|\, (1+|t|)^{-q_2+1} \times$$
$$\times \left(n^{-1} \int_{\bar{B}} \|g_1(z)\|^2\, P(dz) + n^{-3/2}\, E\, \|g_1(\xi_1)\|^3\right),$$

where, according to the Hölder inequality,

$$E\,\|g_1(\xi_1)\|^3 \le \left(E\,\|g_1(\xi_1)\|^2\right)^{-1/2} E\,\|g_1(\xi_1)\|^4.$$

The integral $v_{j3}(t)$ can be estimated, by using (6.6.26), as follows

$$|v_{j3}(t)| \le c_4\,\|g_1(\xi_2)\|\,t^2\,(1+|t|)^{-q_2+1}\,n^{-3/2}\,E\,\|g_1(\xi_1)\|^3.$$

By inserting the estimates for $v_{js}(t)$ into (6.6.29), we obtain

$$|e^3_{n3}(t)| \le c_5\,|t|\,(1+|t|)^{-q_2+2} \times$$
$$\times\left(n^{-1/2}\int_{\bar B}\|g_1(z)\|^2\,P(dz) + n^{-1}\,E\,\|g_1(\xi_1)\|^4\right).$$

Thus, under condition $\mu_2 < \infty$, we have

$$|e_{n3}(t)| \le c_6\,|t|^{1/2}\,(1+|t|)^{-q_2+2} \times$$
$$\times\left(n^{-1/2}\int_{\bar B}\|g_1(z)\|^2\,P(dz) + n^{-1}\,E\,\|g_1(\xi_1)\|^4\right).$$

Lemma 6.6.12 is proved.

Lemma 6.6.13. *Under condition $\mu_2 < \infty$, the inequality*

$$|\psi_{n2}(t)| \le c_4(\Phi)\,(1+|t|)^{-q} \times$$
$$\times\left(n^{-3/2}\int_{\bar B}\|g_1(x)\|^2\,P(dx) + n^{-2}\,E\,\|g_1(\xi_1)\|^4\right),$$

holds in the region $|t| \le n^{1-\varepsilon}$; here, $q > 0$ is a large arbitrarily fixed number, $c_4(\Phi)$ depends on the kernel Φ only via the eigenvalues of the V-operator and variables σ^2 and μ_2.

Proof. First, we write the representation

$$\left\|n^{-1/2}\left(g_1(\xi_1) + g_1(\xi_2) + g_1(\xi_3)\right) + S_{n3}\right\|^2 = u_6 + u_7 + u_8 + \|S_{n3}\|^2,$$

where

$$u_6 = n^{-1}\left(\|g_1(\xi_1)\|^2 + \|g_1(\xi_2)\|^2 + \|g_1(\xi_3)\|^2\right),$$
$$u_7 = 2n^{-1}\left((g_1(\xi_1), g_1(\xi_2)) + (g_1(\xi_1), g_1(\xi_3)) + (g_1(\xi_2), g_2(\xi_3))\right),$$
$$u_8 = 2n^{-1/2}\left(g_1(\xi_1) + g_1(\xi_2) + g_1(\xi_3), S_{n3}\right).$$

Further, according to the definition of $\psi_{n2}(t)$, by relation (6.6.5) with $r = 2$, we can write

$$\psi_{n2}(t) = E\left\{(g_1(\xi_1), g_2(\xi_2,\xi_3))\,\exp(it\,\|S_{n3}\|^2 + it\,u_6 + it\,u_8)\right\} +$$

$$+ it\, E\left\{(g_1(\xi_1), g_2(\xi_2,\xi_3))\, u_7 \exp(it\,\|S_{n3}\|^2 + it\,u_6 + it\,u_8)\right\}$$
$$+ \varrho_{n8}(t), \tag{6.6.32}$$

where

$$\varrho_{n8}(t) = t^2 \int_0^1 du\,(1-u)\,\times$$
$$\times E\left\{(g_1(\xi_1), g_2(\xi_2,\xi_3))\, u_7^2 \exp(it\,\|S_{n3}\|^2 + it\,u_6 + it\,u_8 + itu\,u_7)\right\}.$$

By using (6.6.26), we get the following estimate for $\varrho_{n8}(t)$:

$$|\varrho_{n8}(t)| \le c_7 t^2 (1+|t|)^{-q_2}\, n^{-2}\, E\,\|g_1(\xi_1)\|^4.$$

To analyse the expectations in (6.6.32), we shall use relation (6.6.24) and the formulas given below. Let $x,\ y,\ z \in R$. Then

$$e^{i(x+y)} = 1 + (e^{ix}-1) + (e^{iy}-1) + xy + iy\,(e^{ix}-1) +$$
$$+ix\,(e^{iy}-1) + (1+ix-e^{ix})\,(1+iy-e^{iy}).$$

Since $e^{iz} = 1 + (e^{iz}-1)$, further we obtain

$$e^{i(x+y+z)} =$$
$$= 1 + (e^{ix}-1) + (e^{iy}-1) + (e^{iz}-1) + (e^{ix}-1)(e^{iz}-1)$$
$$+ (e^{iy}-1)(e^{iz}-1) + xy + xy\,(e^{iz}-1) + iy\,(e^{ix}-1)$$
$$+ iy(e^{ix}-1)(e^{iz}-1) + ix\,(e^{iy}-1) + ix\,(e^{iy}-1)(e^{iz}-1)$$
$$+ (1+ix-e^{ix})(1+iy-e^{iy})\,e^{iz}. \tag{6.6.33}$$

In (6.6.33), we set $x = 2n^{-1/2}\,(g_1(\xi_1), S_{n3})$, $y = 2n^{-1/2}\,(g_1(\xi_2), S_{n3})$, and $z = 2n^{-1/2}\,(g_1(\xi_3), S_{n3})$. Then we substitute the expression on the right-hand side of (6.6.33) for $\exp(it\,u_8)$ in (6.6.32). As a result, we get the representations for the expectations in (6.6.32), which are similar to (6.6.25) and convenient for analysis. By applying (6.6.5) and the relations

$$\int (g_1(x), g_2(\xi_2,\xi_3))\, Q(dx) = -(P(B))^{-1} \int_{\bar B} (g_1(x), g_2(\xi_2,\xi_3))\, P(dx),$$
$$\int (g_1(\xi_1), g_2(x,\xi_3))\, Q(dx) = -(P(B))^{-1} \int_{\bar B} (g_1(\xi_1), g_2(x,\xi_3))\, P(dx),$$
$$\int (g_1(\xi_1), g_2(\xi_2,x))\, Q(dx) = -(P(B))^{-1} \int_{\bar B} (g_1(\xi_1), g_2(\xi_2,x))\, P(dx),$$

which hold for all $\xi_i \in B$, we obtain the proof of Lemma 6.6.13.

Let us return to inequality (6.6.15) and consider I_n. Owing to $\varphi(t)$ and Lemma 6.6.10, we have

$$\int_{n^\varepsilon \le |t| \le (n \ln n)^{1/2}} |\varphi_n(t) - \varphi(t)|\, |t|^{-1}\, dt = O(n^{-1+3\varepsilon}).$$

Therefore,

$$I_n = \int_{-n^\varepsilon}^{n^\varepsilon} |\varphi_n(t) - \varphi(t)|\, |t|^{-1}\, dt + O(n^{-1+3\varepsilon}).$$

According to Lemma 6.6.10

$$I_n = \int_{-n^\varepsilon}^{n^\varepsilon} \left| E \exp(it\, \|S_n\|^2) - E \exp(it\, \|\tau\|^2) \right| |t|^{-1}\, dt +$$
$$+2\varepsilon_n \int_{-n^\varepsilon}^{n^\varepsilon} \left| E\left\{ (S_n, S_{n2}(g_2)) \exp(it\, \|S_n\|^2) \right\} \right| dt + O(n^{-1+3\varepsilon}). \qquad (6.6.34)$$

Under condition $\mu_2 < \infty$, we have in (6.6.34) (see Borovskikh and Korolyuk (1984c))

$$\left| E \exp(it\, \|S_n\|^2) - E \exp(it\, \|\tau\|^2) \right| \leq$$
$$\leq c_8\, |t|\, (1 + |t|)^{-q} \left(\int_{\bar{B}} \|g_1(x)\|^2\, P(dx) + n^{-1}\, E\, \|g_1(\xi_1)\|^4 \right).$$

By using this inequality, we can estimate the first integral in (6.6.34). The second integral can be estimated with the help of Lemmas 6.6.11, 6.6.12 and 6.6.13. Thus, we have

$$I_n \leq c_9 \left(\int_{\bar{B}} \|g_1(x)\|^2\, P(dx) + n^{-1}\, E\, \|g_1(\xi_1)\|^4 + n^{-1+3\varepsilon} \right).$$

This relation, together with Lemmas 6.6.8, 6.6.9 and inequality (6.6.15) proves Theorem 6.6.2 under condition $\mu_2 < \infty$, because $0 < \varepsilon < 1/32$.

Theorem 6.6.3. *Assume that the kernel $\Phi\colon X^m \to H$ of rank $r = 1$ satisfies the following conditions*

$$E\, \|g_1\|^3 < \infty \quad \textit{and} \quad E\, \|\Phi\|^{5/3} < \infty$$

and that the V-operator has infinitely many nonzero eigenvalues σ_j^2, $j = 1, 2, \ldots$.

Then the following estimate

$$\sup_x \left| P\left(\left\| m^{-1} \sqrt{n}\, (U_n - \theta) \right\| < x \right) - P\left(\|\tau\| < x \right) \right| = o(n^{-1/2})$$

takes place as $n \to \infty$.

Proof. It is sufficient to combine the arguments which are given in the proofs of Theorems 6.6.1 and 6.6.2. Here the detailed proof is omitted.

6.7 Multi-sample *UH*-statistics

Let us consider the convergence rate in Theorem 4.3.1 for $B = H$. We assume that $n_j = n$, $j = 1, \ldots, c$, and that the V-operator of a limit Gaussian random element $\tau \in H$ in Theorem 4.3.1 has infinitely many nonzero eigenvalues σ_j^2, $j = 1, 2, \ldots$.

Theorem 6.7.1. *Suppose that the kernel* Φ *satisfies the following conditions* $0 \leq E\|g_j\|^3 < \infty$, $\quad j = 1, \dots, c$, *and* $E\|\Phi\|^2 < \infty$.

Then the following estimate holds

$$\sup_x \Big| P(\|\sqrt{n}(U_{\bar{n}} - \theta)\| < x) - P(\|\tau\| < x) \Big| = o(n^{-1/2})$$

as $n \to \infty$.

Proof. By formula (1.2.8) we get

$$\sqrt{n}\,(U_{\bar{n}} - \theta) = S_{\bar{n}} + v_{\bar{n}} + R_{\bar{n}},$$

where $\bar{n} = (n, n, \dots, n)$,

$$S_{\bar{n}} = n^{-1/2} \sum_{i=1}^{n} \Big(\sum_{j=1}^{c} m_j\, g_j(X_{ji}) \Big),$$

$$v_{\bar{n}} = \sqrt{n} \sum_{d_1 = \cdots + d_c = 2} \prod_{j=1}^{c} \binom{m_j}{d_j} \binom{n}{d_j}^{-1} S_{\bar{n}\bar{d}},$$

$$R_{\bar{n}} = \sqrt{n} \sum_{d=3}^{m} \sum_{d_1 + \cdots + d_c = d} \prod_{j=1}^{c} \binom{m_j}{d_j} \binom{n}{d_j}^{-1} S_{\bar{n}\bar{d}}.$$

By Lemma 1.2.1

$$E\|R_{\bar{n}}\|^2 = n \sum_{d=3}^{m} \sum_{d_1 + \cdots + d_c = d} \prod_{j=1}^{c} \binom{m_j}{d_j}^2 \binom{n}{d_j}^{-1} E\|g_{\bar{d}}\|^2.$$

From this formula we find that

$$E\|R_{\bar{n}}\|^2 \leq n^{-2} \Big(\sum_{i=1}^{c} m_i^2 \Big)^3 \exp\Big(\sum_{i=1}^{c} m_i^2 / n \Big) E\|\Phi\|^2.$$

Hence (see the proof of Theorem 6.6.2) we get further the following estimate

$$\sup \Big| P(\|\sqrt{n}\,(U_{\bar{n}} - \theta)\| < x) - P(\|\tau\| < x) \Big| \leq \Delta_{\bar{n}} + n^{-1/2} \rho_n,$$

where

$$\Delta_{\bar{n}} := \sup_x \Big| P(\|S_{\bar{n}} + v_{\bar{n}}\|^2 < x) - P(\|\tau\|^2 < x) \Big|$$

and $\rho_n \to 0$ as $n \to \infty$.

Here $S_{\bar{n}}$ is a sum of independent identically distributed elements in H, and $v_{\bar{n}}$ is a linear combination of *UH*-statistics of the type (1.2.1) with degenerate kernels $g_{\bar{d}}$. In addition, $d_1 + \cdots + d_c = 2$. Therefore we can apply for the estimation of $\Delta_{\bar{n}}$ the method of the proof of Theorem 6.6.2. As a result we obtain the estimate $\Delta_{\bar{n}} \leq n^{-1/2} \rho_n$. This proves Theorem 6.7.1. The tedious technical details of the proof have been omitted.

6.8 Estimates in central limit theorem in l_{2p}

Banach space l_{2p}. Let $X = l_{2p}$ be a Banach space of infinite sequences $x = (x_1, x_2, \dots)$ with the norm

$$\|x\| = \left(\sum_{j=1}^{\infty} x_j^{2p}\right)^{1/(2p)},$$

where p is a natural number, and with a σ-algebra $\mathfrak{X}$ of Borel sets. Assume also that $X_1, \dots, X_n$ are independent X-valued random variables which have the same distribution P on the measurable space $(X, \mathfrak{X})$, and $EX_1 = 0$. Let S denote the covariance operator of the distribution P. We set $S_n = n^{-1/2}(X_1 + \dots + X_n)$ and assume that P_n is a probability measure on $(X, \mathfrak{X})$ induced by the sum S_n.

The central limit theorem in X states that if the condition $E\|X_1\|^2 < \infty$ is satisfied, then for a broad class of sets we have $P_n(A) \to N(A)$, where N is a normal distribution on $(X, \mathfrak{X})$ with zero mean value and the same covariance operator S as that of P. As a set A, we choose the sphere $A = \{x \in X \colon \|x\| < z\}$, $z \in R^+ = [0, \infty)$, and note that the random variable $\|S_n\|^{2p}$ can be written in the form of Von Mises' statistics

$$\|S_n\|^{2p} = n^{-p} \sum_{(i_1, \dots, i_{2p})=1}^{n} \Phi(X_{i_1}, \dots, X_{i_{2p}}) \tag{6.8.1}$$

with the kernel

$$\Phi(x_1, \dots, x_{2p}) = \sum_{j=1}^{\infty} x_{1j} \dots x_{2p,j},$$

where $x_i = (x_{i1}, x_{i2}, \dots, x_{ij}, \dots) \in X$, $i = 1, \dots, 2p$.

Uniform estimate. Under condition $E\|X_1\|^2 < \infty$, we introduce the following notation

$$\alpha_n = \int_{\bar{B}} \|y\|^2 P(dy), \quad \gamma_n = \frac{1}{n}\int_B \|y\|^4 P(dy),$$
$$B = \{y \in X \colon \|y\| \le \sqrt{n}\}, \quad \bar{B} = X \setminus B.$$

Below, the symbols $c, c_1, \dots$ denote some positive constants, $\varepsilon, \varepsilon_1, \dots$ are sufficiently small positive numbers, and $q, q_1, \dots$ denote large, arbitrary positive numbers. In what follows, we assume that the distribution N is not concentrated on any finite-dimensional subspace of X.

Theorem 6.8.1. *Under the condition $E\|X_1\|^2 < \infty$, the following inequality holds*

$$\sup_x \left| P_n(\|y\| < x) - N(\|y\| < x) \right| \le c(\alpha_n + \gamma_n + n^{-p+\varepsilon}),$$

for any $\varepsilon > 0$, where $(\|y\| < x) = \{y \in X \colon \|y\| < x,\ x \in R^+\}$.

Proof. Assume that $S'_n = n^{-1/2}(X'_1 + \cdots + X'_n)$, $X'_j = X_j\, I(\|X_j\| \le \sqrt{n})$, $j = 1, \ldots, n$, where $I(A)$ is the indicator function of the set $A \in \mathfrak{X}$. Then

$$\sup_x \Big| P_n(\|y\| < x) - N(\|y\| < x)\Big| \le \sup_x \Big| P(\|S'_n\|^{2p} < x) - P(\|\tau\|^{2p} < x)\Big| + \alpha_n,$$

where τ is a Gaussian element in X with the distribution N. Since the probability $P(\|\tau\|^{2p} < x)$ has a bounded density, we have, according to the Esséen inequality,

$$\sup_x \Big| P(\|S'_n\|^{2p} < x) - P(\|\tau\|^{2p} < x)\Big| \le \frac{1}{\pi}\int_{-T}^{T} |t|^{-1}\, |\varphi_n(t) - \varphi(t)|\, dt + cT^{-1}, \tag{6.8.2}$$

where

$$T = n^{p-\varepsilon}, \quad \varphi(t) = E\exp(it\,\|\tau\|^{2p}), \quad \varphi_n(t) = E\exp(it\,\|S'_n\|^{2p}).$$

To estimate the integral on the right-hand side of (6.8.2), we prove the following two lemmas.

Lemma 6.8.1. *Under the condition* $E\,\|X_1\|^2 < \infty$, *the following inequality*

$$|\varphi_n(t)| \le c_1\,(1 + |t|)^{-q_1}$$

holds in the region $|t| \le n^{p-\varepsilon}$.

Proof. By definition, we get

$$\|S'_n\|^{2p} = \sum_{j=1}^{\infty} S_{nj}^{2p},$$
$$S_{nj} = n^{-1/2}(x_{1j} + \cdots + x_{nj}),$$

where x_{ij} denotes the coordinate of the vector X'_i, $i = 1, \ldots, n$. We set

$$y_{sj} = \sum_{r=m_{s-1}+1}^{m_s} x_{rj}, \qquad s = 1, \ldots, 2p;$$

here, the choice of integers m_s is arbitrary but the following condition should be satisfied: $m_{s-1} < m_s$, $m_0 = 0$, and $m_{2p} = n$. By using (6.8.1) and the symmetrization inequality (2.3.6) in Borovskikh and Korolyuk (1994a), we find

$$|\varphi_n(t)|^{2^{2p}} \le E\exp\Big\{itn^{-p}\,(2p)!\,\Phi(\bar{y}_1, \ldots, \bar{y}_{2p})\Big\}, \tag{6.8.3}$$

where $\bar{y}_{ij} = y_{ij} - y^1_{ij}$, y^1_{ij} are independent copies of y_{ij}, and $x_i = (x_{i1}, \ldots, x_{ij}, \ldots) \in X$, $i = 1, \ldots, 2p$. Let F denote the distribution of X'_j. Side by side with measure P, we define the measure

$$P_1(A) = P(X_1 \in A \mid \|X_1\| \le L),$$

where $A \in \mathfrak{X}$, $L = m\,E\|X_1\|$, and m is a fixed number such that $m \geq 2$. We assume that $L < \sqrt{n}$ because $n \to \infty$. Then, for any $A \in \mathfrak{X}$, we have

$$F(A) = \frac{1}{2}\,(P_1(A) + P_2(A)),$$

where P_2 is a probability measure concentrated in the sphere $\{x \in X : \|x\| < \sqrt{n}\}$. Assume that S_1 is a covariance operator which corresponds to the distribution P_1. Let us show that

$$\|S - S_1\| \to 0\,, \tag{6.8.4}$$

as $m \to \infty$. In fact, by definition, we get

$$\|S - S_1\| = \sup_{\|h_1\|\leq 1, \|h_2\|\leq 1} |(Sh_1, h_2) - (S_1 h_1, h_2)|\,,$$

here, $h_i \in X^* = l_{p'}$, $l_{p'}$ is the space dual to l_{2p}, $p' = 1 + 1/(2p-1)$, and

$$(Sh_1, h_2) = E\,(X_1, h_1)\,(X_1, h_2)\,,$$
$$(S_1 h_1, h_2) = \int_X (x - \mu_1, h_1)\,(x - \mu_1, h_2)\,P(dx)\,,$$

where $\mu_1 = \int x\,P_1(dx)$, and the symbol (x, h) is defined by

$$(x, h) = x_1 h_1 + \cdots + x_j h_j + \cdots\,, \qquad x \in X,\ h \in X^*.$$

According to this, we have

$$(Sh_1, h_2) - (S_1 h_1, h_2) = \int_X (x, h_1)\,(x, h_2)\,P(dx) -$$
$$-(P(\|X_1\| \leq L))^{-1} \int_{\|x\|\leq L} (x - \mu_1, h_1)\,(x - \mu_1, h_2)\,P(dx).$$

In view of the estimate

$$\|\mu_1\| \leq L^{-1} \int_{\|x\|>L} \|x\|^2\,P(dx),$$

the last relation follows from (6.8.4) provided that $m \to \infty$ and the condition $E\,\|X_1\|^2 < \infty$ is satisfied. In what follows, the number m is fixed so that the left-hand side of (6.8.4) is as small as desired. For $t \in R$, we denote

$$\Psi(t) = E\exp(it\,\Phi(\tau_1', \ldots, \tau_{2p}')),$$

where τ_1', ..., τ_{2p}' are independent Gaussian X-valued random variables with the same covariance operator S_1. By virtue of (6.8.4) and Lemma 3.1 in Yurinskii (1983), the following inequality holds for all $t \in R$

$$|\Psi(t)| \leq c\,(1 + |t|)^{-q}, \tag{6.8.5}$$

where q is a large, arbitrary positive number, and the constant c depends on q and on the structure of the covariance operator S.

Estimate (6.8.5) plays the principal role in the subsequent analysis because the expression on the right-hand side of (6.8.3) is estimated through the function $\Psi(t)$ and only under condition $E\,\|X_1\|^2 < \infty$. Note that, according to the symmetrization inequality (2.3.6) in Borovskikh and Korolyuk (1994a) we have

$$|\Psi(t)|^{2^{2p}} \le E\exp(it\,(2p)!\,(2p)^{-p}\,\Phi(\tau_1,\dots,\tau_{2p})),$$

where τ_j are independent copies of τ, and the expectation on the right-hand side is estimated just as $\Psi(t)$ in (6.8.5).

We now return to inequality (6.8.3) and transform its right-hand side according to the argument presented in Borovskikh (1983e). First, we estimate the right-hand side for all $t \in R$ by use of the decomposition $F(A) = (P_1(A) + P_2(A))/2$, and get

$$\begin{aligned} E\exp\Big(it\,n^{-p}\,(2p)!\,\Phi(\bar y_1,\dots,\bar y_{2p})\Big) \le \\ \le c_1 \int \exp(it\,(2p)!\,\Phi(y_1,\dots,y_{2p}))\,\bar P_1^{n_1}(\sqrt{n}\,dy_1)\dots\bar P_1^{n_{2p}}(\sqrt{n}\,dy_{2p}) \\ + c_2\exp(-c_3\min(n_1,\dots,n_{2p})), \end{aligned} \tag{6.8.6}$$

where $\bar P_1^{n_j}$ denotes the n_j-fold convolution of the symmetrized measure P_1 and $n_j = [(m_j - m_{j-1})/4]$, $j = 1, \dots, 2p$.

For each $j = 1, \dots, 2p$, we choose the solutions $(l_1, \dots, l_j)$ and $(k_{j+1}, \dots, k_{2p})$ of the Diophantine equation $l_1 + \dots + l_j + k_{j+1} + \dots + k_{2p} = n$, such that $l_j \sim a_j n$, $k_{j+1} \sim b_j n^{6\varepsilon}$ as $n \to \infty$, where a_j and b_j are some suitable positive constants. Note that such integral solutions of this equation always exist. Then we set in (6.8.6) $n_s = l_s$, $s = 1, \dots, j$, and $n_s = k_s$, $s = j+1, \dots, 2p$ for all $j = 1, \dots, 2p$.

For these values of parameters, the integral in (6.8.6) can be represented in the form

$$\begin{aligned} \int \exp(it\,(2p)!\,\Phi(y_1,\dots,y_{2p}))\,\bar P_1^{n_1}(\sqrt{n}\,dy_1)\dots\bar P_1^{n_{2p}}(\sqrt{n}\,dy_{2p}) = \\ = \int \Big|\int_X \exp(i\delta_j l_1^{-1/2}\,\Phi(y, y_2,\dots,y_{2p}))\,P_1(dy)\Big|^{2l_1} \times \\ \times \bar P_1^{l_2}(\sqrt{l_2}\,dy_2)\dots\bar P_1^{k_{2p}}(\sqrt{k_{2p}}\,dy_{2p}), \end{aligned} \tag{6.8.7}$$

where $j = 1, \dots, 2p$, and

$$\delta_j = tn^{-p}\,(l_1\dots l_j k_{j+1}\dots k_{2p})^{1/2}\,(2p)!\,,$$
$$l_2\dots l_j = 1 \quad \text{for } j = 1, \quad \text{and} \quad k_{j+1}\dots k_{2p} = 1 \quad \text{for } j = 2p.$$

In (6.8.7), we estimate the integral with respect to y on the right-hand side in a standard way, by using Taylor's formula. Then, for all $j = 1, \dots, 2p$ in the region

$$|t| \le c\,n^{p-\varepsilon}\,(l_1\dots l_j)^{-1/2}\,(k_{j+1}\dots k_{2p})^{-1/2}, \tag{6.8.8}$$

the integral expression on the right-hand side of (6.8.7) does not exceed the value

$$\int \exp(-c_j\delta_j^2\, f(y_2,\dots,y_{2p}))\, \bar{P}_1^{l_2}(\sqrt{l_2}\, dy_2)\dots \bar{P}_1^{k_{2p}}(\sqrt{k_{2p}}\, dy_{2p}) + O(n^{-q_2}), \quad (6.8.9)$$

where

$$f(y_2,\dots,y_{2p}) = \int_X \Phi^2(y-\mu_1, y_2,\dots,y_{2p})\, P_1(dy).$$

Consider the regions

$$|t| \le c_{2p} n^{p+\varepsilon}\,(l_1\dots l_{2p})^{-1/2}, \quad (6.8.10)$$

$$c_j n^{p+\varepsilon}\,(l_1\dots l_j)^{-1/2}\,(k_{j+1}\dots k_{2p})^{-1/2} \le |t| \le c_{j-1} n^{p+\varepsilon}\,(l_1\dots l_{j-1})^{-1/2}\,(k_j\dots k_{2p})^{-1/2},$$
$$j = 2,3,\dots,2p, \quad (6.8.11)$$

and

$$c_1 n^{p+\varepsilon} l_1^{-1/2}\,(k_2\dots k_{2p})^{-1/2} \le |t| \le c n^{p-\varepsilon}\,(k_2\dots k_{2p})^{-1/2}. \quad (6.8.12)$$

Clearly, in order to establish Lemma 6.8.1, due to the fact that arbitrary $\varepsilon > 0$ is small, it suffices to obtain the corresponding estimate of the integral in (6.8.9) for the region (6.8.8) with $j = 1$. It is also obvious that the constants in (6.8.10)–(6.8.12) can be chosen so that the union of regions (6.8.10)–(6.8.12) gives the region (6.8.8) with $j = 1$.

For the region (6.8.10), we set $j = 2p$ in (6.8.8). Note that the integral in (6.8.9) decreases with respect to the variable δ_j monotonically. Therefore, in this situation the integral in (6.8.9) does not exceed the integral

$$\int \exp(-ct^2\, f(y_2,\dots,y_{2p}))\, \bar{P}_1^{l_2}(\sqrt{l_2}\, dy_2)\dots \bar{P}_1^{l_{2p}}(\sqrt{l_{2p}}\, dy_{2p}) \quad (6.8.13)$$

for all t in the region (6.8.10). If we denote a centered distribution of a Gaussian random variable τ_1' by N_1, then, by virtue of the properties of N_1, the integral (6.8.13) is equal to

$$\int \exp(it\sqrt{2c}\,(y_1,y_2,\dots,y_{2p}))\, N_1(dy_1)\, \bar{P}_1^{l_2}(\sqrt{l_2}\, dy_2)\dots \bar{P}_1^{l_{2p}}(\sqrt{l_{2p}}\, dy_{2p}). \quad (6.8.14)$$

In (6.8.14), we first write the integral with respect to the variable y_2, according to scheme (6.8.7), and then transform it by use of Taylor's formula. As a result, we obtain the integral of type (6.8.14) in which the measure $\bar{P}_1^{l_2}$ is replaced by the Gaussian measure N_1. Continuation of this process with respect to each variable y_j, $j = 3, \dots, 2p$, finally implies that integral (6.8.13) does not exceed the value

$$|\Psi(c_1 t)| + O(n^{-q_3})$$

in the region (6.8.10).

Assume now that t changes in the region (6.8.11), $j = 2, 3, \ldots, 2p$. The property of monotonicity with respect to δ_j ensures that, for each t, the integral in (6.8.9) does not exceed the integral

$$\int \exp(-c_j n^{2\varepsilon} f(y_2, \ldots, y_{2p}))\, \bar{P}_1^{l_2}(\sqrt{l_2}\, dy_2) \ldots \bar{P}_1^{k_{2p}}(\sqrt{k_{2p}}\, dy_{2p}), \qquad (6.8.15)$$

$$j = 2, 3, \ldots, 2p.$$

We transform the expression (6.8.15) once again, according to scheme (6.8.13)–(6.8.14). As a result, we find that the integral in (6.8.9) does not exceed the value

$$|\Psi(c_2 n^{\varepsilon})| + O(n^{-q_4}) \qquad (6.8.16)$$

in the region (6.8.11) for all $j = 2, 3, \ldots, 2p$.

For the region (6.8.12), we set $j = 1$ in (6.8.9). Then we can see that the integral in (6.8.9) does not exceed the integral

$$\int \exp(-c_3 n^{2\varepsilon} f(y_2, \ldots, y_{2p}))\, \bar{P}_1^{k_2}(\sqrt{k_2}\, dy_2) \ldots \bar{P}_1^{k_{2p}}(\sqrt{k_{2p}}\, dy_{2p}).$$

By transforming this expression according to the scheme presented above, we find that it does not exceed the value (6.8.16). By virtue of the arbitrariness of ε, q, and q_j, from these estimates follows the statement of Lemma 6.8.1.

Lemma 6.8.2. *Under condition $E\,\|X_1\|^2 < \infty$, the following inequality*

$$|\varphi_n(t) - \varphi(t)| \le c\,(\alpha_n + \gamma_n)\,|t|\,(1 + |t|)^{-q}$$

holds in the region $|t| \le n^{\varepsilon_1}$.

Proof. According to formula (6.1.5), we get

$$\varphi_n(t) - \varphi(t) =$$

$$= \sum_{m=0}^{n-1} \int \exp(it\,\|x + z + y\|^2)\, F^m(\sqrt{n}\, dx)\, H(\sqrt{n}\, dy)\, N^{n-m-1}(\sqrt{n}\, dz), \qquad (6.8.17)$$

where $H(\cdot) = F(\cdot) - N(\cdot)$.

When examining the expression on the right-hand side of (6.8.17), we used formula (6.6.5). If $x_1, x_2 \in X$, then

$$\|x_1 + x_2\|^{2p} = \|x_1\|^{2p} + f(x_1, x_2),$$

with some multilinear expression $f(x_1, x_2)$. We choose $x_1 = x + z$ and $x_2 = y$. Then, by using (6.6.5) we represent (6.8.17) as follows

$$\varphi_n(t) - \varphi(t) = \sum_{m=0}^{n-1} \Big(\sum_{s=1}^{r-1} (s!)^{-1}\, (it)^s\, \Pi_{sm}(t) + ((r-1)!)^{-1}\, (it)^r\, \Pi_{rm}(t) \Big), \qquad (6.8.18)$$

where

$$\Pi_{sm}(t) = \int e^{it\|x+z\|^{2p}} f^s F^m(\sqrt{n}\,dx)\, H(\sqrt{n}\,dy)\, N^{n-m-1}(\sqrt{n}\,dz),$$

if $s = 1, \dots, r-1$, and

$$\Pi_{rm}(t) = \int_0^1 du\,(1-u)^{r-1} \times$$
$$\times \int e^{it\|x+z\|^{2p}+ituf} f^r F^m(\sqrt{n}\,dx)\, H(\sqrt{n}\,dy)\, N^{n-m-1}(\sqrt{n}\,dz).$$

When deriving (6.8.18), we also take into account that $\int H(dy) = 0$. We set $r = 4$ in (6.8.18) and then use the argument presented in Borovskikh (1983e) to estimate $\Pi_{sm}(t)$ in the region $|t| \le n^{\varepsilon_1}$. After a calculation, we arrive at the inequality

$$|\varphi_n(t) - \varphi(t)| \le c_4\,(\alpha_n + \gamma_n)\,|t|\,(|\Psi(c_5 t)|^{2^{-2p}} + n^{-q_5})$$

if $|t| \le n^{\varepsilon_1}$. By virtue of the arbitrariness of q and q_j, these inequalities imply the proof of Lemma 6.8.2.

Theorem 6.8.1 now follows from (6.8.2) and Lemmas 6.8.1, 6.8.2.

Corollary 6.8.1. *Assume that the distribution P is such that*

$$\int_{\|y\|>x} \|y\|^2\, P(dy) = \frac{c}{x \ln x}$$

as $x \to \infty$. In this case, $E\,\|X_1\|^3 = \infty$, and according to Theorem 6.8.1, we get

$$\sup_x \Big| P_n(\|y\| < x) - N(\|y\| < x) \Big| = O((\ln n\sqrt{n})^{-1})$$

as $n \to \infty$.

Corollary 6.8.2. *Let $E\,\|X_1\|^3 < \infty$. Since the inequality*

$$\gamma_n \le n^{-1/2-\alpha}\, E\,\|X_1\|^3 + n^{-1/2} \int_{\|x\|>n^{1/2-\alpha}} \|y\|^3\, P(dy)$$

holds for every $0 < \alpha < 1/2$, under these conditions we always have

$$\sup_x \Big| P_n(\|y\| < x) - N(\|y\| < x) \Big| = o(n^{-1/2}).$$

For $p = 1$, this estimate was obtained by Zalesskii (1982).

Corollary 6.8.3. *Let $E\,\|X_1\|^4 < \infty$. Then*

$$\sup_x \Big| P_n(\|y\| < x) - N(\|y\| < x) \Big| = O(n^{-1})$$

for $p \ge 2$.

A similar statement is presented in Götze (1984).

Nonuniform estimate.

Theorem 6.8.2. *If condition $E\|X_1\|^2 < \infty$ is satisfied, then for all $x > 0$, the following inequality holds*

$$\begin{aligned}&\left|P_n(\|y\| < x) - N(\|y\| < x)\right| \\ &\quad \le (1+x)^{-2} \int_{\bar{B}_x} \|y\|^2 \, P(dy) + c_1 \, (1+x)^{-r} \, (\alpha_n + \gamma_n + n^{-p+\varepsilon}) \\ &\qquad + c_2 \, (1+x)^{-r_1} \, n^{-(r_1-2)/2} \int_{B_x \setminus B} \|y\|^{r_1} \, P(dy),\end{aligned}$$

where r is an arbitrary fixed number, r_1 is an arbitrary real number such that $r_1 > 0$,

$$\begin{aligned}\bar{B}_x &= \{y \in X \colon \|y\| \le \sqrt{n}\,(1+x)\}, \\ B_x \setminus B &= \{y \in X \colon \sqrt{n} < \|y\| \le \sqrt{n}\,(1+x)\}, \qquad x \in R^+.\end{aligned}$$

Proof. For $x \in R^+$ and $j = 1, \ldots, n$, we set $X_j'' = X_j \, I(\|X_j\| \le \sqrt{n}\,(1+x))$. Denote $S_n'' = n^{-1/2}\,(X_1'' + \cdots + X_n'')$. Then, for all $x > 0$, we have

$$\left|P_n(\|y\| < x) - N(\|y\| < x)\right| \le \sum_{j=1}^{3} \Delta_j(n, x), \tag{6.8.19}$$

where

$$\begin{aligned}\Delta_1(n,x) &= \left|P(\|S_n'\| > x) - P(\|S_n''\| > x)\right|, \\ \Delta_2(n,x) &= \left|P(\|S_n''\| > x) - P(\|S_n\| > x)\right|, \\ \Delta_3(n,x) &= \left|P(\|S_n'\|^{2p} < x^{2p}) - P(\|\tau\|^{2p} < x^{2p})\right|.\end{aligned}$$

Estimate of $\Delta_1(n, x)$. For $k = 1, \ldots, n$, we denote

$$\theta_k' = n^{-1/2} \sum_{j=1}^{k-1} X_j'' + n^{-1/2} \sum_{j=k+1}^{n} X_j'.$$

According to formula (6.1.4), we have

$$\begin{aligned}&P(\|S_n'\| > x) - P(\|S_n''\| > x) = \\ &= \sum_{k=1}^{n} \left(P\left(\left\| \theta_k' + n^{-1/2} X_k' \right\| > x \right) - P\left(\left\| \theta_k' + n^{-1/2} X_k'' \right\| > x \right) \right).\end{aligned}$$

Clearly,

$$P\left(\left\|\theta'_k + n^{-1/2} X'_k\right\| > x\right) =$$
$$= \int_B P\left(\left\|\theta'_k + n^{-1/2} y\right\| > x\right) P(dy) + P(\|\theta'_k\| > x)\, P(\|X_1\| > \sqrt{n}).$$

With the help of this formulae, we write the representation

$$P(\|S'_n\| > x) - P(\|S''_n\| > x) =$$
$$= -\sum_{k=1}^{n} \int_{B_x \setminus B} P\left(\left\|\theta'_k + n^{-1/2} y\right\| > x\right) P(dy) + P(B_x \setminus B) \sum_{k=1}^{n} P(\|\theta'_k\| > x). \tag{6.8.20}$$

Further, for $r = 1, \ldots , k$ and $k = 1, \ldots , n$, we denote

$$\eta'_{n,k} = n^{-1/2}\Big(\sum_{j=2}^{k-1} X'_j + \sum_{j=k+1}^{n} X'_j\Big),$$
$$\theta'_{r,k} = n^{-1/2}\Big(\sum_{j=1}^{r-1} X''_j + \sum_{j=r+1}^{k-1} X'_j + \sum_{j=k+1}^{n} X'_j\Big),$$

and then, by analogy with (6.8.20), we find

$$P(\|\theta'_k\| > x) = P(\|\eta'_{n,k}\| > x) +$$
$$+\sum_{r=1}^{k} \int_{B_x \setminus B} P\left(\left\|\theta'_{r,k} + n^{-1/2} y\right\| > x\right) P(dy) - P(B_x \setminus B) \sum_{r=1}^{k} P(\|\theta'_{r,k}\| > x).$$

Inserting this formula into (6.8.20), we get

$$P(\|S'_n\| > x) - P(\|S''_n\| > x) = \sum_{j=1}^{5} v_j(n, x), \tag{6.8.21}$$

where

$$v_1(n, x) = -\sum_{k=1}^{n} \int_{B_x \setminus B} P\left(\left\|\eta'_{n,k} + n^{-1/2} y\right\| > x\right) P(dy),$$
$$v_2(n, x) = -\sum_{k=1}^{n}\sum_{r=1}^{k} \int_{B_x \setminus B} P(dy_1) \int_{B_x \setminus B} P(dy_2)\, P\left(\left\|\theta_{r,k} + n^{-1/2} y_1 + n^{-1/2} y_2\right\| > x\right),$$
$$v_3(n, x) = 2\, P(B_x \setminus B) \sum_{k=1}^{n}\sum_{r=1}^{k} \int_{B_x \setminus B} P\left(\left\|\theta'_{r,k} + n^{-1/2} y\right\| > x\right) P(dy),$$
$$v_4(n, x) = P(B_x \setminus B) \sum_{k=1}^{n} P(\|\eta'_{n,k}\| > x),$$
$$v_5(n, x) = P^2(B_x \setminus B) \sum_{k=1}^{n}\sum_{r=1}^{k} P(\|\theta'_{r,k}\| > x).$$

Consider each term on the right-hand side of (6.8.21). Since l_{2p} is a Banach space of type 2, it follows from the exponential inequality presented in Yurinskii (1976) that if $E X_1 = 0$ and $E\|X_1\|^2 < \infty$, then, for all $k = 1, \ldots, n$ and any $m \geq 2$, we have

$$E\left\|n^{-1/2}(X_1' + \cdots + X_k')\right\|^m \leq c(m), \tag{6.8.22}$$

where the positive constant $c(m)$ does not depend on n. Taking into account (6.8.22) and Chebyshev's inequality, we get

$$P\left(\left\|\eta_{n,k}' + n^{-1/2}y\right\| > x\right) \leq c(1+x)^{-r_1}(1 + n^{-r_1/2}\|y\|^{r_1})$$

for all $x > 0$ and any $r_1 > 0$. Clearly,

$$P(B_x \setminus B) \leq n^{-r_1/2}\int_{B_x\setminus B}\|y\|^{r_1}P(dy).$$

For all $x > 0$ and $r_1 > 0$, these inequalities yield

$$|v_1(n,x)| \leq c_1(1+x)^{-r_1}n^{-(r_1-2)/2}\int_{B_x\setminus B}\|y\|^{r_1}P(dy). \tag{6.8.23}$$

When estimating $v_2(n,x)$, just as in (6.8.23), we have

$$P\left(\left\|\theta_{r,k}' + n^{-1/2}y_1 + n^{-1/2}y_2\right\| > x\right) \leq$$
$$\leq c(1+x)^{-r_2}\left(E\|\theta_{r,k}'\|^{r_2} + n^{-r_2/2}(\|y_1\|^{r_2} + \|y_2\|^{r_2})\right),$$

for any $r_2 > 0$. First, we estimate the expectation on the right-hand side by using (6.8.22). We have

$$E\|\theta_{r,k}'\|^{r_2} \leq c_1 E\left\|n^{-1/2}\sum_{j=1}^{r-1}X_j I((B_x\setminus B)_j)\right\|^{r_2} + c_2,$$

where $(B_x \setminus B)_j = \left\{X_j \in X\colon \sqrt{n} < \|X_j\| \leq \sqrt{n}(1+x)\right\}$, $x \in R^+$. Then, after performing the calculation with the help of Acosta's (1981) inequality, we can write

$$E\left\|n^{-1/2}\sum_{j=1}^{r-1}X_j I((B_x\setminus B)_j)\right\|^{r_2} \leq c_3\int_{B_x\setminus B}\|y\|^{r_2}P(dy) + c_1.$$

This yields the inequality

$$|v_2(n,x)| \leq c_5(1+x)^{-r_2}n^{-(r_2+r_3-4)/2} \times$$
$$\times \int_{B_x\setminus B}\|y\|^{r_2}P(dy)\int_{B_x\setminus B}\|y\|^{r_3}P(dy), \tag{6.8.24}$$

which holds for all $r_2 > 0$ and $r_3 > 0$.

By a similar argument, for $x > 0$, we can show that

$$|v_3(n,x)| \leq c_6(1+x)^{-r_4}n^{-(r_4+r_5-4)/2}$$
$$\times \int_{B_x\setminus B}\|y\|^{r_4}P(dy)\int_{B_x\setminus B}\|y\|^{r_5}P(dy), \tag{6.8.25}$$

$$|v_4(n,x)| \le c_7\,(1+x)^{-r_6}\, n^{-(r_7-2)/2} \int_{B_x\setminus B} \|y\|^{r_7}\, P(dy), \tag{6.8.26}$$

$$|v_5(n,x)| \le c_8\,(1+x)^{-r_8}\, n^{-(r_8+r_9-4)/2} \times \int_{B_x\setminus B} \|y\|^{r_8}\, P(dy) \int_{B_x\setminus B} \|y\|^{r_9}\, P(dy), \tag{6.8.27}$$

where r_j are arbitrary numbers such that $r_i > 0$.

Further, we combine the inequalities (6.8.23) and (6.8.26) by setting $r_1 = r_6 = r_7$, and then the inequalities (6.8.24), (6.8.25), and (6.8.27) by setting $r_2 = r_4 = r_8 = r_1$ and $r_3 = r_5 = r_9 = 2$. Then, according to (6.8.21), we obtain

$$\Delta_1(n,x) \le c\,(1+x)^{-r_1}\, n^{-(r_1-2)/2} \int_{B_x\setminus B} \|y\|^{r_1}\, P(dy) \tag{6.8.28}$$

for all $x > 0$.

Estimate of $\Delta_2(n,x)$. For $k = 1, \dots, n$, we denote

$$\theta_k = n^{-1/2}\Big(\sum_{j=1}^{k-1} X_j'' + \sum_{j=k+1}^{n} X_j'\Big).$$

Further, we write the representation

$$P(\|S_n\| > x) - P(\|S_n''\| > x) =$$
$$= \sum_{k=1}^{n} \int_{\bar{B}_x} \Big(P\Big(\Big\|\theta_k + n^{-1/2} y\Big\| > x\Big) - P(\|\theta_k\| > x)\Big)\, P(dy).$$

The modulus of the integrand on the right-hand side of this equality does not exceed 1. Therefore, under condition $E\,\|X_1\|^2 < \infty$, we have, for all $x > 0$,

$$\Delta_2(n,x) \le (1+x)^{-2} \int_{\bar{B}_x} \|y\|^2\, P(dy). \tag{6.8.29}$$

Estimate of $\Delta_3(n,x)$. According to inequality (6.1.3), we get

$$\Delta_3(n,x) \le c\,(1+x^{2p})^{-r}\,(J_1 + J_2 + T^{-1}), \tag{6.8.30}$$

where r is an arbitrary integer such that $r \ge 2$, and

$$J_1 = \int_{-T}^{T} |\varphi_n(t) - \varphi(t)|\, |t|^{-1}\, dt, \quad J_2 = \int_{-T}^{T} \left|\frac{d^r}{dt^r}\Big(\frac{\varphi_n(t) - \varphi(t)}{t}\Big)\right| dt,$$
$$T = n^{p-\varepsilon}.$$

The integral J_1 was estimated in the proof of Theorem 6.8.1. We consider J_2. It suffices to establish the estimates for the derivatives $\varphi_n(t)$ and $\varphi(t)$.

Lemma 6.8.3. *Under the condition $E\|X_1\|^2 < \infty$, for all $r = 1, 2, \dots,$ the following inequality*

$$\left|\frac{d^r}{dt^r}\varphi_n(t)\right| \le c(1+|t|)^{-q}$$

holds in the region $|t| \le n^{p-\varepsilon}$.

Proof. By definition, we have

$$\frac{d^r}{dt^r}\varphi_n(t) = i^r E\left((\Phi(S_n', \dots, S_n'))^r e^{it\Phi(S_n', \dots, S_n')}\right).$$

By using the same argument, we derive from this formula an inequality similar to (6.8.3)

$$\left|\frac{d^r}{dt^r}\varphi_n(t)\right|^{2^{2p}} \le c_1 E\exp\left\{itn^{-p}(2p)!\,\Phi(\bar{y}_1, \dots, \bar{y}_{2p})\right\},$$

where c_1 is some positive constant; here, we have also used (6.8.22) and Hölder's inequality.

Lemma 6.8.4. *For all $r = 1, 2, \dots$ and $t \in R^+$, the following estimate holds*

$$\left|\frac{d^r}{dt^r}\varphi(t)\right| \le c(1+|t|)^{-q},$$

where the positive arbitrary number q is large.

Proof. By virtue of the property of the infinite divisibility of τ, we get

$$\varphi(t) = E\exp(it\,\|\eta_{2p}\|^{2p}),$$

where

$$\eta_{2p} = (2p)^{-1/2}(\tau_1 + \dots + \tau_{2p})$$

and τ_j are independent copies of τ. Thus,

$$\frac{d^r}{dt^r}\varphi(t) = i^r E\left((\Phi(\eta_{2p}, \dots, \eta_{2p}))^r e^{it\Phi(\eta_{2p}, \dots, \eta_{2p})}\right).$$

We transform the right-hand side of this equality with the help of the symmetrization inequality and obtain

$$\left|\frac{d^r}{dt^r}\varphi(t)\right|^{2^{2p}} \le c_2 E\exp\left(it\,(2p)!\,(2p)^{-p}\,\Phi(\tau_1, \dots, \tau_{2p})\right).$$

From this inequality and (6.8.5) follows Lemma 6.8.4.

Further, by using representation (6.8.18), we derive the following inequality in the region $|t| \le n^{\varepsilon_1}$:

$$\left|\frac{d^r}{dt^r}\left(\frac{\varphi_n(t)-\varphi(t)}{t}\right)\right| \le c_3(\alpha_n + \gamma_n)\left(|\Psi(c_4 t)|^{2^{-2p}} + n^{-q_s}\right). \tag{6.8.31}$$

For all $x > 0$, the inequalities (6.8.30), (6.8.31) and Lemmas 6.8.3, 6.8.4 yield

$$\Delta_3(n,x) \le c_5\,(1+x)^{-r}\,(\alpha_n + \gamma_n + n^{-p+\epsilon}). \qquad (6.8.32)$$

Combining (6.8.19), (6.8.28), (6.8.29), and (6.8.32), we obtain Theorem 6.8.2.

Corollary 6.8.4. *Let $E\,\|X_1\|^3 < \infty$. Then, according to Theorem 6.8.2 with $r_1 = 3$, we have*

$$\left|P_n(\|y\| < x) - N(\|y\| < x)\right| \le (1+x)^{-3}\,n^{-1/2}\int_{\bar{B}_x}\|y\|^3\,P(dy) +$$
$$+c_1\,(1+x)^{-r}\,(\alpha_n+\gamma_n+n^{-p+\epsilon}) + c_2\,(1+x)^{-3}\,n^{-1/2}\int_{B_x\setminus B}\|y\|^3\,P(dy).$$

Thus, the presence of the integrals on the right-hand side of this relation makes this estimate better than the corresponding assertion in the one-dimensional case concerning the decrease with respect to the parameter n.

Corollary 6.8.5. *Assume that $p \ge 2$ and $E\,\|X_1\|^4 < \infty$. Then, according to Theorem 6.8.2 with $r_1 = 4$, we find*

$$\left|P_n(\|y\| < x) - N(\|y\| < x)\right| \le cn^{-1}\left((1+x)^{-r} + (1+x)^{-2}\int_B \|y\|^4\,P(dy)\right)$$

with some constant $c > 0$.

6.9 Rate of Poisson approximation

In the same way as in Section 4.4 we start from a sample $X_j^{(n)}$, $1 \le j \le n$, in the series scheme with

$$P(X_j^{(n)} = 1) = 1 - P(X_j^{(n)} = 0) = \lambda/n, \qquad \lambda > 0.$$

Let

$$\nu_n := \sum_{j=1}^{n} X_j^{(n)}$$

and ν be a Poisson distributed random variable with parameter λ:

$$P(\nu = k) = e^{-\lambda}\,\lambda^k/k!, \qquad k \ge 0.$$

As was mentioned in Section 4.4

$$\sum_{1\le i_1\ne\cdots\ne i_m\le n} \Phi(X_{i_1}^{(n)},\dots,X_{i_m}^{(n)}) = \nu_n^{[m]}$$

with kernel $\Phi(x_1,\dots,x_m) = x_1\cdots x_m$, and

$$\nu_n^{[m]} = \nu_n\,(\nu_n - 1)\cdots(\nu_n - m + 1).$$

In this Poisson approximation scheme under fixed m and for all $k \geq 0$

$$\left|P(\nu_n^{[m]} = k) - P(\nu^{[m]} = k)\right| \to 0$$

as $n \to \infty$. Consider the problem of the estimate of the rate of convergence to zero in this relation, relative to n.

Theorem 6.9.1. *Under the Poisson approximation condition, the following estimate is valid for all $k \geq 0$, $1 \leq m \leq n$, and $0 < \lambda \leq n$*

$$\left|P(\nu_n^{[m]} = k) - P(\nu^{[m]} = k)\right| \leq \lambda\,(1 - e^{-\lambda})/n. \tag{6.9.1}$$

Proof. Let $I(B)$ denote the indicator function of the set $B \in \mathfrak{X}$. Then

$$\begin{aligned} P(\nu_n^{[m]} = k) &= E\, I(\nu_n^{[m]} = k) \\ &= \sum_{r=0}^{\infty} I(r^{[m]} = k)\, P(\nu_n = r), \\ P(\nu^{[m]} = k) &= E\, I(\nu^{[m]} = k) \\ &= \sum_{r=0}^{\infty} I(r^{[m]} = k)\, P(\nu = r). \end{aligned}$$

From here we obtain the following estimate

$$\left|P(\nu_n^{[m]} = k) - P(\nu^{[m]} = k)\right| \leq$$
$$\leq \sum_{r=0}^{\infty} I(r^{[m]} = k)\left|P(\nu_n = r) - P(\nu = r)\right|.$$

Following Barbour and Hall (1984) we will prove that

$$\left|P(\nu_n = r) - P(\nu = r)\right| \leq \lambda\,(1 - e^{-\lambda})\,n^{-1} \tag{6.9.2}$$

uniformly on $r \geq 0$ and $0 < \lambda \leq n$. Then by using the obvious inequality

$$\sum_{r=0}^{\infty} I(r^{[m]} = k) \leq 1$$

combined with (6.9.2) we get (6.9.1). So to get (6.9.2) we write, as in Barbour and Hall (1984), applying the formula (6.1.9) with $w = \nu_n$,

$$\begin{aligned} &E\left(\lambda\, g_{A,\lambda}(\nu_n + 1) - \nu_n\, g_{A,\lambda}(\nu_n)\right) \\ &= \sum_{j=1}^{n}\left\{\frac{\lambda}{n}\, E\, g_{A,\lambda}(\nu_n + 1) - E(\nu_{nj}\, g_{A,\lambda}(\nu_n))\right\} \\ &= \frac{\lambda}{n}\sum_{j=1}^{n} E\left\{g_{A,\lambda}(\nu_n + 1) - g_{A,\lambda}(\nu_{nj} + 1)\right\} = \end{aligned}$$

$$= \frac{\lambda^2}{n^2} \sum_{j=1}^{n} E\{g_{A,\lambda}(\nu_{nj}+2) - g_{A,\lambda}(\nu_{nj}+1)\},$$

where $\nu_{nj} = \nu_n - X_j^{(n)}$. Here, by inequality (6.1.10) with probability 1

$$\left|g_{A,\lambda}(\nu_{nj}+2) - g_{A,\lambda}(\nu_{nj}+1)\right| \le \lambda^{-1}(1-e^{-\lambda}).$$

This proves (6.9.2) and, therefore, Theorem 6.9.1.

Estimate (6.9.1) can be made more precise as follows. Denote

$$\Delta(k,m) := P(\nu_n^{[m]} = k) - P(\nu^{[m]} = k) +$$
$$+\frac{1}{2} E\left\{I(\nu^{[m]} = k)\,(\nu^2 - (2\lambda+1)\,\nu + \lambda^2)\right\} n^{-1}.$$

Theorem 6.9.2. *For all $0 < \lambda \le n$, $1 \le m \le n$, and uniformly in k, the following estimate*

$$|\Delta(k,m)| \le 4\lambda^2\,(1-e^{-\lambda})\,n^{-2}$$

takes place.

Proof. First we show that

$$\Delta(k,m) = \sum_{r=0}^{\infty} I(r^{[m]} = k)\,\Delta(r,1).$$

For this purpose we write

$$E\left\{I(\nu^{[m]} = k)\,(\nu^2 - (2\lambda+1)\,\nu + \lambda^2)\right\} =$$
$$= \sum_{r=0}^{\infty} I(r^{[m]} = k)\,(r^2 - (2\lambda+1)\,r + \lambda^2)\,P(\nu = r).$$

Here

$$(r^2 - (2\lambda+1)\,r + \lambda^2)\,P(\nu = r) = E\left\{I(\nu = r)\,(\nu^2 - (2\lambda+1)\,\nu + \lambda^2)\right\},$$

with the substitution of these relations into the definition of the variable $\Delta(k,m)$ we obtain the desired formula. Then, we get the inequality

$$|\Delta(k,m)| \le \sum_{r=0}^{\infty} I(r^{[m]} = k)\,|\Delta(r,1)|.$$

By Theorem 3 in Barbour and Hall (1984)

$$|\Delta(r,1)| \le 4\lambda^2\,(1-e^{-\lambda})\,n^{-2}$$

uniformly for all $r = 0, 1, 2, \ldots$ and for all $0 < \lambda \le n$.

Combining these estimates we obtain the proof of Theorem 6.9.2.

Chapter 7

Asymptotic expansions

Assume that the random variables $X_1, \dots, X_n$ with values in the measurable space $(X, \mathfrak{X})$ are defined on the probability space $(\Omega, \mathcal{F}, P)$, and $\eta_n = \eta(X_1, \dots, X_n)$, where η is a mapping $\eta\colon X^n \to R$, and the weak convergence

$$\eta_n \Longrightarrow \eta_\infty$$

takes place.

Among a number of problems connected with this convergence, there exists the problem of the *asymptotic expansion* of a probability $P(\eta_n < x)$, $x \in R$, in a series in powers of $n^{-1/2}$. This problem consists of clarifying the conditions and proving the existence of the functions $F_j(x)$, $j = 1, \dots, s-1$, for which

$$\sup_x \left| P(\eta_n < x) - P(\eta_\infty < x) - \sum_{j=1}^{s-1} n^{-j/2} F_j(x) \right| = o(n^{-(s-1)/2}),$$

as $n \to \infty$, where $s \geq 1$ is some fixed integer.

We set

$$\eta_n = n^{r/2} (U_n - \theta),$$

where r is the rank of kernel Φ. If $m = r = 1$, then η_n is a sum of independent random variables. For this case, this problem is well investigated. However, if $m \geq 2$, the problem of the validity of the asymptotic expansion requires additional investigation. In particular, if $m = r = 2$, $X = H$ is a separable real Hilbert space, and $\Phi(x, y) = (x, y)$, x, $y \in H$, then it coincides with the problem concerning the asymptotic expansion on balls in the central limit theorem in a Hilbert space. Clearly, the coefficients $F_j(x)$ and the remainder $o(n^{-(s-1)/2})$ depend on m, r, on the properties of kernel Φ, on the space X, and on the original distribution P of random variables $X_1, \dots, X_n$.

The problem of the existence of the coefficients of the expansion of $F_j(x)$ can be solved most simply in the case where $r = 1$.

7.1 Expansions for non-degenerate *UR*-statistics

Consider a one-sample *UR*-statistic

$$U_n = \binom{n}{2}^{-1} \sum_{1\le i<j\le n} \Phi(X_i, X_j)$$

with independent identically distributed random variables $X_1, \ldots, X_n$ and a symmetric kernel Φ such that $E\,\Phi(X_1, X_2) = 0$, and the variance σ_1^2 of the random variable $g_1(X_1) = E\,(\Phi(X_1, X_2) \mid X_1)$ is positive. According to the Hoeffding formula, variance $\sigma^2 = E\,U_n^2$ of this U-statistic is equal to

$$\sigma^2 = \frac{4}{n}\sigma_1^2 + \frac{2}{n\,(n-1)}\,E\,g_2^2(X_1, X_2),$$

where $g_2(x_1, x_2) = \Phi(x_1, x_2) - g_1(x_1) - g_1(x_2)$.

Let us rewrite the U-statistic into the form

$$U_n = \frac{2}{n}\sum_{j=1}^{n} g_1(X_j) + \binom{n}{2}^{-1} \sum_{1\le i<j\le n} g_2(X_i, X_j).$$

Coefficients of expansion for $m = 2$. For $x \in R$, we denote

$$G_n(x) = F(x) - f(x)\left(\frac{k_3}{6}\,(x^2-1)\,n^{-1/2} + \frac{k_4}{24}\,(x^3 - 3x)\,n^{-1} + \right.$$
$$\left. + \frac{k_3^2}{72}\,(x^5 - 10x^3 + 15x)\,n^{-1}\right),$$

where

$$k_3 = \sigma_1^{-3}\left(E\,g_1^3(X_1) + 3\,E(g_1(X_1)\,g_1(X_2)\,g_2(X_1, X_2))\right),$$
$$k_4 = \sigma_1^{-4}\left(E\,g_1^4(X_1) - 3\sigma_1^4 + 12\,E(g_1^2(X_1)\,g_1(X_2)\,g_2(X_1, X_2))\right.$$
$$\left. + 12\,E(g_1(X_2)\,g_1(X_3)\,g_2(X_1, X_2)\,g_2(X_1, X_3))\right),$$

$F(x)$ is the distribution function of the standard normal law, and $f(x) = F'(x)$ is the density of this distribution. For $t \in R$, we assume that

$$\varphi(t) = E\exp(it\,g_1(X_1)),$$
$$\varphi_n(t) = E\exp(it\sigma^{-1}\,U_n),$$
$$g_n(t) = \int_{-\infty}^{\infty} \exp(itx))\,dG_n(t) =$$
$$= \exp(-t^2/2)\left(1 + \frac{k_3}{6}\,(it)^3 n^{-1/2} + \frac{k_4}{24}\,(it)^4 n^{-1} + \frac{k_3^2}{72}\,(it)^6 n^{-1}\right).$$

Expansion with the Cramer condition for $m = 2$. Let the Cramer condition be imposed on the function $\varphi(t)$.

Theorem 7.1.1. *Suppose that*

$$E\, g_2^4(X_1, X_2) < \infty, \quad E\, |g_1^4(X_1)\, g_2(X_1, X_2)| < \infty,$$
$$E\, \left|g_1^3(X_1)\, g_2(X_1, X_2)\, g_2(X_1, X_3)\right| < \infty,$$

and the following relation is valid in the region $t \in [n^{3/4}/\ln n, n \ln n]$

$$E\left|E\left(\exp\left(it\,\alpha_n \sum_{j=k+1}^{n} \Phi(X_1, X_j)\right) \mid X_{k+1}, \dots, X_n\right)\right|^k = o((n \ln n)^{-1}), \qquad (7.1.1)$$

for $k = [n^\alpha]$, $0 < \alpha < 1/8$ *and* $\alpha_n = \sigma^{-1}\binom{n}{2}^{-1}$. *Then under the Cramer condition*

$$\lim_{|t|\to\infty} \sup |\varphi(t)| < 1 \qquad (7.1.2)$$

the following expansion holds

$$\sup_x |P(\sigma^{-1} U_n < x) - G_n(x)| = o(n^{-1}).$$

Proof. According to the Esséen inequality, we have

$$\sup_x |P(\sigma^{-1} U_n < x) - G_n(x)| \le \frac{2}{\pi} \int_0^{n \ln n} |\varphi_n(t) - g_n(t)|\, |t|^{-1}\, dt + o(n^{-1}).$$

First, we estimate the integral on the right-hand side as follows

$$\int_0^{n \ln n} |\varphi_n(t) - g_n(t)|\, t^{-1}\, dt \le \sum_{s=1}^{4} J_s,$$

where

$$J_1 = \int_0^{n^{1/4} \ln n} |\varphi_n(t) - g_n(t)|\, t^{-1}\, dt, \quad J_2 = \int_{n^{1/4}/\ln n}^{n^{3/4}/\ln n} |\varphi_n(t)|\, t^{-1}\, dt,$$
$$J_3 = \int_{n^{3/4}/\ln n}^{n \ln n} |\varphi_n(t)|\, t^{-1}\, dt, \quad J_4 = \int_{n^{1/4}/\ln n}^{n^{1/4} \ln n} |g_n(t)|\, t^{-1}\, dt.$$

Clearly, $J_4 = o(n^{-1})$ as $n \to \infty$. Therefore, it suffices to examine the values J_s, $s = 1, 2, 3$.

Estimate of J_1. For $r = 1, \dots, n$, we assume that

$$S_r = \sum_{i=1}^{r} g_1(X_i)$$

and denote

$$v_{r,s} = \sum_{i=1}^{r} \sum_{j=i+1}^{s} g_2(X_i, X_j)$$

for $1 \le r < s \le n$. Note that $\varphi_n(t) = E\left[\exp(it\alpha_n (n-1) S_n) \exp(it\alpha_n v_{n-1,n})\right]$. Here, we replace the second exponential factor in the preceding expectation by its truncated Taylor's series taken up to the term of power t^2 and denote the result by $\varphi_{n,1}(t)$, i.e.

$$\varphi_{n,1}(t) = E\left(\exp(it\alpha_n (n-1) S_n)\left(1 + it\alpha_n v_{n-1,n} + \frac{1}{2}(it)^2 \alpha_n^2 v_{n-1,n}^2\right)\right).$$

By virtue of the independence of the variables, we can represent $\varphi_{n,1}(t)$ in the form

$$\begin{aligned}
\varphi_{n,1}(t) &= (\varphi(\alpha_n (n-1) t))^n \\
&+ it (\varphi(\alpha_n (n-1) t))^{n-2} \alpha_n \frac{n(n-1)}{2} \\
&\times E\left((\exp(it\alpha_n (n-1)(g_1(X_1) + g_1(X_2)))) g_2(X_1, X_2)\right) \\
&+ \frac{1}{2}(it)^2 (\varphi(\alpha_n (n-1) t))^{n-2} \alpha_n^2 \frac{n(n-1)}{2} \\
&\times E\left((\exp(it\alpha_n (n-1)(g_1(X_1) + g_1(X_2)))) g_2^2(X_1, X_2)\right) \\
&+ \frac{1}{2}(it)^2 (\varphi(\alpha_n (n-1) t))^{n-3} \alpha_n^2 n (n-1)(n-2) \\
&\times E\left((\exp(it\alpha_n (n-1)(g_1(X_1) + g_1(X_2) + g_1(X_3)))) g_2(X_1, X_2) g_2(X_1, X_3)\right) \\
&+ \frac{1}{2}(it)^2 (\varphi(\alpha_n (n-1) t))^{n-4} \alpha_n^2 \frac{n}{4}(n-1)(n-2)(n-3) \\
&\times \left[E\left((\exp(it\alpha_n (n-1)(g_1(X_1) + g_1(X_2)))) g_2(X_1, X_2)\right)\right]^2.
\end{aligned}$$

For convenience, we rewrite this formula once again using the obvious notation

$$\varphi_{n,1}(t) = a_0^* + it\alpha_n \frac{n(n-1)}{2} a_2^* b_1^* +$$
$$+ \frac{(it)^2}{2} \alpha_n^2 n(n-1)\left(\frac{1}{2} a_2^* b_2^* + (n-2) a_3^* b_3^* + \frac{1}{4}(n-2)(n-3) a_4^* b_4^*\right).$$

By using Taylor's formula, we approximate the expectations b_i^* by the variables b_i, $1 \le i \le 4$, defined as follows

$$\begin{aligned}
b_1 &= (it)^2 \sigma_1^{-2} E\left(g_1(X_1) g_1(X_2) g_2(X_1, X_2)\right) n^{-1} \\
&\quad + (it)^2 \sigma_1^{-3} E\left(g_1^2(X_1) g_1(X_2) g_2(X_1, X_2)\right) n^{-3/2}, \\
b_2 &= E\, g_2^2(X_1, X_2), \\
b_3 &= (it)^2 \sigma_1^{-2} E\left(g_1(X_2) g_1(X_3) g_2(X_1, X_2) g_2(X_1, X_3)\right) n^{-1}, \\
b_4 &= (it)^4 \sigma_1^{-4} \left(E\left(g_1(X_1) g_1(X_2) g_2(X_1, X_2)\right)\right)^2 n^{-2}.
\end{aligned}$$

Further, the approximation for the values a_k^*, $k = 0, 2, 3, 4$, can be obtained by the standard expansion of the characteristic function

$$E\left\{\exp\left[it\alpha_n (n-1)\,\sigma_1\,(n-k)^{1/2}\left(\sigma_1^{-1}\,(n-k)^{-1/2}\sum_{i=1}^{n-k} g_1(X_i)\right)\right]\right\},$$

where $\sigma_1^{-1}\,(n-k)^{-1/2}\sum_{i=1}^{n-k} g_1(X_i)$ is a normed sum of independent equally distributed random variables.

Denote

$$g_{n,1}(t) = a_0 + it\alpha_n \frac{n(n-1)}{2}\, a_2 b_1 +$$
$$+\frac{(it)^2}{2}\,\alpha_n^2 n\,(n-1)\left(\frac{1}{2}\,a_2 b_2 + (n-2)\,a_3 b_3 + \frac{1}{4}\,(n-2)(n-3)\,a_4 b_4\right),$$

where

$$a_k = e^{-t^2/2}\left[1 - \frac{(it)^2}{2n}\left(\frac{1}{2}\,\sigma_1^{-2}\,E\,g_2^2(X_1,X_2) + k\right) + \frac{(it)^3}{6\sqrt{n}}\,\sigma_1^{-3}\,E\,g_1^3(X_1)\right.$$
$$\left. + \frac{(it)^4}{24n}\,\sigma_1^{-4}\,(E\,g_1(X_1) - 3\sigma_1^4) + \frac{(it)^6}{72n}\,\sigma_1^{-6}\,(E\,g_1^3(X_1))^2\right].$$

The function $g_{n,1}(t)$ is the approximation of $\varphi_{n,1}(t)$.

Note that

$$J_1 \le \int_0^{n^{1/4}/\ln n} |g_{n,1}(t) - g_n(t)|\,t^{-1}\,dt$$
$$+ \int_0^{n^{1/4}/\ln n} |\varphi_n(t) - \varphi_{n,1}(t)|\,t^{-1}\,dt$$
$$+ \int_0^{n^{1/4}/\ln n} |\varphi_{n,1}(t) - g_{n,1}(t)|\,t^{-1}\,dt.$$

Here, it is clear that the first integral on the right-hand side can be estimated as $o(n^{-1})$. To estimate the second integral, we first obtain

$$|\varphi_n(t) - \varphi_{n,1}(t)| \le |t|^3\,\alpha_n^3\,E\,|v_{n-1,n}|^3,$$

where $\alpha_n^3 = O(n^{-9/2})$ and $E\,|v_{n-1,n}|^3 = O(n^3)$. Therefore,

$$\int_0^{n^{1/6}/\ln n} |\varphi_n(t) - \varphi_{n,1}(t)|\,t^{-1}\,dt = o(n^{-1}).$$

For $t \in [n^{1/6}/\ln n, n^{1/4}/\ln n]$, according to the definition of $\varphi_n(t)$ and $\varphi_{n,1}(t)$, we write

$$|\varphi_n(t) - \varphi_{n,1}(t)| \le |t|^3\,\alpha_n^3\left|E\,\exp(it\alpha_n\,(n-1)\,S_n)\,v_{n-1,n}^3\right| + t^4\,\alpha_n^4\,E\,v_{n-1,n}^4.$$

The expectation factor next to $|t|^3\,\alpha_n^3$ does not exceed

$$\sum\nolimits_1 \left|E\,[\exp(it\alpha_n\,(n-1)\,S_n)]\,g_2(X_{i_1},X_{j_1})\,g_2(X_{i_2},X_{j_2})\,g_2(X_{i_3},X_{j_3})\right| =$$

$$
\begin{aligned}
&= \sum\nolimits_1 \left| E\left[\exp\Big(it\,\alpha_n\,(n-1)\sum_{j\in J'} g_1(X_j)\Big)\right]\right| \\
&\quad \times \left| E\left\{\left[\exp\Big(it\,\alpha_n\,(n-1)\sum_{j\in J} g_1(X_j)\Big)\right]\right.\right. \\
&\quad \left.\left. \times\, g_2(X_{i_1}, X_{j_1})\, g_2(X_{i_2}, X_{j_2})\, g_2(X_{i_3}, X_{j_3})\right\}\right| \\
&\le |\varphi(t\,\alpha_n\,(n-1))|^{n-6} \\
&\quad \times \sum\nolimits_1 \left| E\left\{\left[\exp\Big(it\,\alpha_n\,(n-1)\sum_{j\in J} g_1(X_j)\Big)\right]\right.\right. \\
&\quad \left.\left. \times\, g_2(X_{i_1}, X_{j_1})\, g_2(X_{i_2}, X_{j_2})\, g_2(X_{i_3}, X_{j_3})\right\}\right|,
\end{aligned}
$$

where J is a set of different indices among i_1, j_1, i_2, j_2, i_3, and j_3, and $J' = \{1, 2, \dots, n\} \setminus J$. Let s be a number of elements in J. Then we have $2 \le s \le 6$, and for every s, the number of terms in the series $\sum_1$ has order $O(n^s)$. Further, there always exists $\varepsilon > 0$ such that for $|t| < \varepsilon\sigma_1^{-1}$, we get

$$|\varphi(t)| \le \exp\Big(-\frac{1}{3}\,t^2\,\sigma_1^2\Big).$$

Since $2/\sigma < n^{1/2}/\sigma_1$, we have $t\alpha_n\,(n-1) < \varepsilon/\sigma_1$ if $t < \varepsilon n^{1/2}$. Consequently, for $t < \varepsilon n^{1/2}$ and large n, the following inequality holds

$$|\varphi(t\,\alpha_n\,(n-1))|^{n-6} \le \exp\Big(-\frac{1}{12}\,t^2\Big).$$

From this remark, we see that the terms with $s = 2$, 3 can be replaced by an absolute constant, because $\alpha_n^3 = O(n^{-9/2})$, and the number of all these terms has order $O(n^3)$. In order to estimate the terms for $s = 4$, 5, 6, we employ the property of degeneracy of function $g_2(x_1, x_2)$: for every measurable function $f(x_{s_1}, \dots, x_{s_r})$ such that $E\,|fg_2| < \infty$, the following equality holds

$$E\,\Big(f(X_{s_1}, \dots, X_{s_r})\, g_2(X_i, X_j)\Big) = 0,$$

if at least one of the indices i or j does not belong to $\{s_1, \dots, s_r\}$. This relation also holds if both indices i and j differ from s_j, $1 \le l \le r$, by virtue of the equality $E\,g_2(X_i, X_j) = 0$ and the independence of the variables. On the other hand, if $i \in \{s_1, \dots, s_r\}$ and $j \notin \{s_1, \dots, s_r\}$, then $E(g_2(X_i, X_j) \mid X_{s_1}, \dots, X_{s_r}) = E(g_2(X_i, X_j) \mid X_i) = 0$ by the definition of $g_2(x_1, x_2)$.

We consider the case of $s = 4$ in more detail and show the analogy for $s = 5$, 6. If J contains exactly four different indices, we have the expressions of the form $g_2(X_1, X_2)\, g_2(X_1, X_3)\, g_2(X_1, X_4)$ and $g_2^2(X_1, X_2)\, g_2(X_3, X_4)$. Note that the expression of the first form is not factorized when we take the expectation of it. Consequently, the expressions

$$g_2(X_1, X_2)\, g_2(X_2, X_3)\, g_2(X_3, X_4) \quad \text{and} \quad g_2(X_1, X_2)\, g_2(X_1, X_3)\, g_2(X_3, X_4)$$

are also of this type. For definiteness, we consider the expression

$$g_2(X_1, X_2)\, g_2(X_1, X_3)\, g_2(X_1, X_4)$$

in detail. Let us estimate each term of the sum which contains the expression of this form by using the property of the degeneracy of $g_2(x_1, x_2)$. We have

$$\Big|E\Big\{\Big[\exp\Big(it\,\alpha_n\,(n-1)\sum_{s=1}^{4} g_1(X_s)\Big)\Big]\, g_2(X_1, X_2)\, g_2(X_1, X_3)\, g_2(X_1, X_4)\Big\}\Big| =$$

$$= \Big|E\Big\{\Big[\exp\Big(it\,\alpha_n\,(n-1)\sum_{s=1}^{4} g_1(X_s)\Big) - 1\Big]\, g_2(X_1, X_2)\, g_2(X_1, X_3)\, g_2(X_1, X_4)\Big\}\Big| \le$$

$$\le ct\,\alpha_n\,(n-1).$$

We recall that each term has the factor $t^{-1}\,|t|^3\,\alpha_n^3\,|\varphi(t\alpha_n\,(n-1))|^{n-6}$. Then, integrating over the region $t \in [n^{1/6}/\ln n, n^{1/4}/\ln n]$, we find that each term is bounded by

$$c\,\alpha_n^4\,(n-1)\int_{n^{1/6}/\ln n}^{n^{1/4}/\ln n} t^3 \exp\Big(-\frac{1}{2}t^2\Big)\, dt.$$

Since the number of such terms is at most $O(n^4)$, we conclude that the sum of all the terms of this type has order $o(n^{-1})$.

Consider the terms of the second type for $s = 4$. Taking into account the property of the degeneracy of $g_2(x_1, x_2)$ and the independence of the variables, we get

$$\Big|E\Big\{\Big[\exp\Big(it\,\alpha_n\,(n-1)\sum_{s=1}^{4} g_1(X_s)\Big)\Big]\, g_2^2(X_1, X_2)\, g_2(X_3, X_4)\Big\}\Big|$$

$$= \Big|E\Big\{\Big[\exp\Big(it\,\alpha_n\,(n-1)\sum_{s=1}^{2} g_1(X_s)\Big)\Big]\, g_2^2(X_1, X_2)\Big\}\Big|$$

$$\times \Big|E\Big\{\Big[\exp\Big(it\,\alpha_n\,(n-1)\sum_{s=3}^{4} g_1(X_s)\Big) - 1 - it\,\alpha_n\,(n-1)\sum_{s=3}^{4} g_1(X_s)\Big]$$

$$\times\, g_2(X_3, X_4)\Big\}\Big|$$

$$\le c t^2 \alpha_n^2\,(n-1)^2.$$

Hence, after performing the same operations as before, we find that each term is bounded by the value

$$c\,\alpha_n^5\,(n-1)^2\int_0^\infty t^4 \exp\Big(-\frac{1}{2}t^2\Big)\, dt = O(n^{-11/12}).$$

This implies that the sum of all the terms of the second type has order $O(n^{-3/2})$. For $s = 5$, 6, the argument is the same. For example, for $s = 5$, we obtain

$$\Big|E\Big\{\Big[\exp\Big(it\,\alpha_n\,(n-1)\sum_{r=1}^{5} g_1(X_r)\Big)\Big]\, g_2(X_1, X_2)\, g_2(X_4, X_5)\Big\}\Big| =$$

$$= \Big| E\Big\{ \Big[\exp\Big(it\,\alpha_n\,(n-1) \sum_{r=1}^{3} g_1(X_r) \Big) - 1 - it\,\alpha_n (n-1) \sum_{r=1}^{3} g_1(X_r) \Big]$$
$$\times\, g_2(X_1, X_2)\, g_2(X_1, X_3) \Big\}$$
$$\times\, E\Big\{ \Big[\exp\Big(it\,\alpha_n\,(n-1) \sum_{r=1}^{5} g_1(X_r) \Big) - 1 - it\,\alpha_n (n-1) \sum_{r=4}^{5} g_1(X_r) \Big]$$
$$\times\, g_2(X_4, X_5) \Big\} \Big|$$
$$\leq c\,t^4 \alpha_n^4\,(n-1)^4.$$

Thus,

$$\int_0^{n^{1/4}/\ln n} |\varphi_n(t) - \varphi_{n,1}(t)|\, t^{-1}\, dt = o(n^{-1}).$$

For simplicity, we write φ instead of $\varphi(\alpha_n\,(n-1)\,t)$. Then,

$$|\varphi_{n,1}(t) - g_{n,1}(t)| \leq |\varphi^n - a_0|$$
$$+ |t|\,\alpha_n\,\frac{n(n-1)}{2} \Big[|\varphi^{n-2}(b_1^* - b_1)| + |b_1\,(\varphi^{n-2} - a_2)| \Big]$$
$$+ \frac{t^2}{2}\,\alpha_n^2\,\frac{n(n-1)}{2} \Big[|\varphi^{n-2}(b_2^* - b_2)| + |b_2\,(\varphi^{n-2} - a_2)| \Big]$$
$$+ \frac{t^2}{2}\,\alpha_n^2\,(n-1)\,(n-2) \Big[|\varphi^{n-3}(b_3^* - b_3)| + |b_3\,(\varphi^{n-3} - a_3)| \Big]$$
$$+ \frac{t^2}{2}\,\alpha_n^2\,\frac{n(n-1)(n-2)(n-3)}{4} \Big[|\varphi^{n-4}(b_4^* - b_4)| + |b_4\,(\varphi^{n-4} - a_4)| \Big],$$

where, firstly, for all t, we have

$$\Big| E\Big\{ [\exp(it\,\alpha_n\,(n-1)\,(g_1(X_1) + g_1(X_2)))]\, g_2(X_1, X_2) \Big\} - b_1 \Big|$$
$$\leq c\,(n^{-2}\,(t^2 + t^4) + n^{-5/2}\,|t|^3),$$
$$\Big| E\Big\{ [\exp(it\,\alpha_n\,(n-1)\,(g_1(X_1) + g_1(X_2)))]\, g_2^2(X_1, X_2) \Big\} - b_2 \Big|$$
$$\leq c\,n^{-1/2}\,|t|,$$
$$\Big| E\Big\{ [\exp(it\,\alpha_n\,(n-1)\,(g_1(X_1) + g_1(X_2) + g_1(X_3)))]\, g_2(X_1, X_2)\, g_2(X_1, X_3) \Big\} - b_3 \Big|$$
$$\leq c\,(n^{-3/2}\,|t|^3 + n^{-2}\,t^2),$$
$$\Big| \Big(E\Big\{ [\exp(it\,\alpha_n\,(n-1)\,(g_1(X_1) + g_1(X_2)))]\, g_2(X_1, X_2) \Big\} \Big)^2 - b_4 \Big|$$
$$\leq c\,(n^{-5/2}\,|t|^5 + n^{-3}\,t^4),$$

and, secondly, in the region $0 \leq t \leq \varepsilon\,n^{1/2}$, for all $n > k$, $k = 0, 1, 2, \ldots$, we have

$$|(\varphi(\alpha_n\,(n-1)\,t))^{n-k} - a_n| \leq c\,\delta_n n^{-1} t\,p(t)\,e^{-t^2/4},$$

where $\delta_n \to 0$ as $n \to \infty$, and $p(t)$ is some polynomial in t. Consequently,

$$\int_0^{n^{1/4}/\ln n} |\varphi_{n,1}(t) - g_{n,1}(t)|\, t^{-1}\, dt = o(n^{-1}).$$

This finally yields $J_1 = o(n^{-1})$ as $n \to \infty$.

Estimate of J_2. We consider the problem of the upper bound for $|\varphi_n(t)|$.

Lemma 7.1.1. *Under the condition of the theorem, there exist positive constants c_1 and c_2 such that the following inequality holds*

$$|\varphi_n(t)| \le |\varphi(t\,\alpha_n\,(n-1))|^{k-6}\Big(1 + c_1 \sum_{s=1}^{3} |t|^s\,\alpha_n^s\,(kn)^s\Big) + c_2\,|t|^4\,\alpha_n^4\,(kn)^2$$

for all t and all integers n and k, $6 < k < n$.

Proof. By definition, we get

$$\varphi_n(t) = E\Big[\exp(it\,\alpha_n\,(n-1)\,S_k)\exp(it\,\alpha_n\,(n-1)\,(S_n - S_k)) \times \\ \times \exp(it\,\alpha_n\,(v_{n-1,n} - v_{k,n}))\exp(it\alpha_n\,v_{k,n})\Big].$$

By using the expansion of $\exp(it\,\alpha_n\,v_{k,n})$, we obtain from this the following inequality

$$|\varphi_n(t)| \le \sum_{s=0}^{3} |t|^s\,\alpha_n^s\,\Big|E\Big[\exp(it\,\alpha_n\,(n-1)\,S_k) \times \\ \times \exp(it\,\alpha_n\,(n-1)(S_n - S_m))\exp(it\,\alpha_n\,(v_{n-1,n} - v_{k,n}))\,v_{k,n}^s\Big]\Big| + |t|^4\,\alpha_n^4\,v_{k,n}^4.$$

In view of the independence of the variables, we can estimate the term with $s = 0$ as follows

$$|E\,[\exp(it\,\alpha_n\,(n-1)\,S_k)]| \le |\varphi(t\,\alpha_n\,(n-1))|^{k-6}.$$

For $s = 1, 2, 3$, the factor next to $|t|^s\,\alpha^s$ is bounded from above by the sum

$$\sum\nolimits_2 \Big|E\Big[\exp(it\,\alpha_n\,(n-1)\,S_k)\exp(it\,\alpha_n\,(n-1)\,(S_n - S_k)) \times \\ \times \exp(it\,\alpha_n\,(v_{n-1,n} - v_{k,n}))\,g_2(X_{i_1}, X_{j_1})\dots g_2(X_{i_s}, X_{j_s})\Big]\Big|,$$

where the number of terms in $\sum_2$ is less then $(kn)^s$.

For an arbitrary term in $\sum_2$, we denote by $\mathfrak{D}$ the set of different indices among $i_1, j_1, \dots, i_s, j_s$. If $\mathfrak{D}_1 = \{1, \dots, k\} \cap \mathfrak{D}$ and $\mathfrak{D}_2 = \{1, \dots, k\} \setminus \mathfrak{D}$, then it is easy to see that $\mathfrak{D}_1$ contains at least one and at most $2s$ elements, and hence, the number of elements in $\mathfrak{D}_2$ lies between $k-1$ and $k-2s$. Therefore,

$$\Big|E\Big[\exp\Big(it\,\alpha_n\,(n-1)\sum_{r\in\mathfrak{D}_2} g_1(X_r)\Big) \\ \times \exp\Big(it\,\alpha_n\,(n-1)\Big(\sum_{r\in\mathfrak{D}_1} g_1(X_r) + (S_n - S_k)\Big)\Big) \times$$

$$\times \exp(it\,\alpha_n\,(v_{n-1,n} - v_{k,n}))\, g_2(X_{i_1}, X_{j_1}) \dots g_2(X_{i_s}, X_{j_s})\Big]\Big|$$
$$\leq \Big| E\Big[\exp\Big(it\,\alpha_n\,(n-1) \sum_{r\in\mathfrak{D}_2} g_1(X_r)\Big)\Big]\Big|\, E\,\big|g_2(X_{i_1}, X_{j_1}) \dots g_2(X_{i_s}, X_{j_s})\big|$$
$$\leq c\,|\varphi(t\alpha_n\,(n-1))|^{k-6}.$$

Together with the inequality

$$E\,v_{k,n}^4 \leq c\,(kn)^2,$$

this proves Lemma 7.1.1.

Further, choosing $k = [n^{1/2+\delta}]$, $0 < \delta < 1/4$, for $t \in [n^{1/4}/\ln n, n^{3/8}]$, we find, according to Lemma 7.2.1, that

$$\int_{n^{1/4}/\ln n}^{n^{3/8}} |\varphi_n(t)|\, t^{-1}\, dt \leq \int_{n^{1/4}/\ln n}^{n^{3/8}} t^{-1} \exp\Big(-\frac{4\sigma_1^2}{3\,n^2\sigma^2}\,(k-6)\,t^2\Big)\, dt$$
$$+ c_1 \sum_{s=1}^{3} \alpha_n^s\,(kn)^s \int_{n^{1/4}/\ln n}^{n^{3/8}} t^{s-1} \exp\Big(-\frac{4\sigma_1^2}{3\,n^2\sigma^2}\,(k-6)\,t^2\Big)\, dt$$
$$+ c_2\,\alpha_n^4\,(kn)^2 \int_{n^{1/4}/\ln n}^{n^{3/8}} t^3\, dt = o(n^{-1}).$$

For $t \in [n^{3/8}, \varepsilon n^{1/2}]$, we choose $k = [n^\delta]$, $1/4 < \delta < 1/2$ and, in a similar way, obtain the estimate $o(n^{-1})$ by using Lemma 7.1.1.

Note that by virtue of the Cramer condition, there exists a constant $a > 0$ such that $|\varphi(t\alpha_n\,(n-1))| \leq e^{-a}$ if $t \geq \varepsilon\, n^{1/2}$. Taking this into account, we use Lemma 7.1.1 with $k = (2/a)\ln n$. Then we get

$$\int_{\varepsilon n^{1/2}}^{n^{3/4}/\ln n} |\varphi_n(t)|\, t^{-1}\, dt = o(n^{-1}).$$

Thus, $J_2 = o(n^{-1})$ as $n \to \infty$.

Estimate of J_3. The following inequality plays an important role.

Lemma 7.1.2. *Under the conditions of the theorem, there exist positive constants c_3 and c_4 such that the following inequality holds*

$$|\varphi_n(t)| \leq E\Big| E\Big[\exp\Big(it\,\alpha_n \sum_{j=k+1}^{n} \Phi(X_1, X_j)\Big) \;\Big|\; X_{k+1}, \dots, X_n\Big]\Big|^{k-6} \times$$
$$\times \Big(1 + c_3 \sum_{s=1}^{3} |t|^s\, \alpha_n^s m^{2s}\Big) + c_4\,|t|^4\,\alpha_n^4 k^8,$$

for all t and all integers n and k, $6 < k < n$.

Proof. For $1 \le r < s \le n$, we set

$$\tilde{v}_{r,s} = \sum_{i=1}^{r} \sum_{j=i+1}^{s} \Phi(X_i, X_j).$$

Then

$$\varphi_n(t) = E\Big[E\left[\exp(it\,\alpha_n\,(\tilde{v}_{k,n} - \tilde{v}_{k-1,n}))\exp(it\,\alpha_n\,v_{k-1,k}) \times \right.$$
$$\left. \times \exp(it\,\alpha_n\,(\tilde{v}_{n-1,n} - \tilde{v}_{k,n})) \mid X_{k+1}, \dots, X_n\right]\Big].$$

Now expanding $\exp(it\,\alpha_n\,\tilde{v}_{k-1,k})$, we obtain the following inequality:

$$|\varphi_n(t)| \le E\left|E\left[\exp(it\,\alpha_n\,(\tilde{v}_{k,n} - \tilde{v}_{k-1,n}))\exp(it\,\alpha_n\,\tilde{v}_{k-1,k}) \mid X_{k+1}, \dots, X_n\right]\right|$$
$$\le \sum_{s=0}^{3} |t|^s\,\alpha_n^s\,E\left|E\left[\exp(it\,\alpha_n\,(\tilde{v}_{k,n} - \tilde{v}_{k-1,n}))\,\tilde{v}_{k-1,k}^s \mid X_{k+1}, \dots, X_n\right]\right|$$
$$+ |t|^4\,\alpha_n^4\,E\,\tilde{v}_{k-1,k}^4.$$

The rest of the argument is exactly the same as in the proof of Lemma 7.1.1.

We use Lemma 7.1.2 for $k = [n^\alpha]$, $0 < \alpha < 1/8$, and $t \in [n^{3/4}/\ln n, n \ln n]$. As a result, under the condition of the theorem, we get

$$\int_{n^{3/4}/\ln n}^{n \ln n} |\varphi_n(t)|\, t^{-1}\, dt = o(n^{-1}).$$

Consequently, $J_3 = o(n^{-1})$ as $n \to \infty$. This completes the proof of Theorem 7.1.1.

Now we consider UR-statistics of degree m, i.e.

$$U_n = \binom{n}{m}^{-1} \sum_{1 \le i_1 < \dots < i_m \le n} \Phi(X_{i_1}, \dots, X_{i_m})$$

with independent identically distributed random variables $X_1, \dots, X_n$, and symmetric kernel $\Phi\colon X^m \to R$ such that $E\,\Phi(X_1, \dots, X_m) = 0$. Theorem 7.1.1 is extended to a general degree $m \ge 2$.

Coefficients of expansion for $m \ge 2$. Let us define

$$k_3 = \sigma_1^{-3}\big\{E\,g_1^3(X_1) + 3(m-1)\,E\,(g_1(X_1)\,g_1(X_2)\,g_2(X_1, X_2))\big\},$$
$$k_4 = \sigma_1^{-4}\Big\{E\,g_1^4(X_1) - 3\sigma_1^4 + 12(m-1)\,E\,(g_1^2(X_1)\,g_1(X_2)\,g_2(X_1, X_2))$$
$$+ 12(m-1)^2\,E(g_1(X_2)\,g_1(X_3)\,g_2(X_1, X_2)\,g_2(X_1, X_3))$$
$$+ 4(m-1)(m-2)\,E\,(g_1(X_1)\,g_1(X_2)\,g_1(X_3)\,g_3(X_1, X_2, X_3))\Big\},$$

$$G_n(x) = F(x) - f(x)\Big(\frac{k_3}{6}(x^2-1)\,n^{-1/2} + \frac{k_4}{24}(x^3-3x)\,n^{-1} + \frac{k_3^2}{72}(x^5-10x^3+15x)\,n^{-1}\Big),$$

where $f(x)$ is the probability density function of the standard normal distribution $F(x)$.

Expansions with the Cramer condition for $m \geq 2$.

Theorem 7.1.2. *Suppose that non-degenerate kernel Φ satisfies the following conditions:*

$$E\,\Phi^4(X_1,\dots,X_m) < \infty,$$

$$\lim_{|t|\to\infty} \sup |\varphi(t)| < 1,$$

and there exist positive constants $\gamma < 1$ and $\alpha < 1/8$ such that for $k = [n^\alpha]$,

$$P\Big\{\Big|E\Big(\exp\Big(it\sigma^{-1}\frac{m(m-1)}{n(n-1)}\sum_{j=k+1}^{n} g(X_1,X_j)\Big)\;\Big|\; X_{k+1},\dots,X_n\Big)\Big| \leq \gamma\Big\} \leq$$

$$\leq 1 - o\Big(\frac{1}{n\ln n}\Big)$$

uniformly for all $t \in [n^{3/4}/\ln n,\ n\ln n]$, where

$$g(x,y) = g_2(x,y) - \frac{m-2}{m-1}(g_1(x)+g_1(y)).$$

Then

$$\sup_x \Big|P(\sigma^{-1}U_n < x) - G_n(x)\Big| = o(n^{-1})$$

is valid as $n \to \infty$.

For the expansion with order $n^{-1/2}$ the smoothness condition of the function $g(x,y)$ is not necessary. In this case we have the following statement.

Theorem 7.1.3. *Suppose that*

$$E\,|\Phi(X_1,\dots,X_m)|^3 < \infty$$

and

$$\lim_{|t|\to\infty} \sup |\varphi(t)| < 1.$$

Then the next relation holds when $n \to \infty$:

$$\sup_x \Big|P(\sigma^{-1}U_n < x) - F(x) + f(x)\frac{k_3}{6}(x^2-1)\,n^{-1/2}\Big| = o(n^{-1/2}).$$

Expansion with a condition on the eigenvalues. It was established in the proof of Theorem 7.1.1 that the behavior of $|\varphi_n(t)|$ in the region $|t| \le n^{3/4}(\ln n)^{-1}$ is determined by the properties of the characteristic function $\varphi(t)$ that satisfies (7.1.2). However, in the region $n^{3/4}(\ln n)^{-1} \le t \le n \ln n$, condition (7.1.1) is satisfied and this allows us to estimate $|\varphi_n(t)|$ in view of Lemma 7.1.2. Let us present an analogue of Theorem 7.1.1 in which condition (7.1.1) is replaced by the condition on the properties of eigenvalues λ_1, λ_2, ... of an integral operator generated by the kernel $g_2(x, y)$, i.e.

$$\int_X g_2(x,y)\, e_j(x)\, P(dx) = \lambda_j\, e_j(y), \tag{7.1.3}$$

where e_1, e_2, ... are the eigenfunctions, and $|\lambda_1| \ge |\lambda_2| \ge \cdots$.

In what follows $\delta > 0$ and $\delta_i > 0$ are fixed real numbers, $i = 1, 2, \ldots$.

Theorem 7.1.4. *Suppose that there exists some number $p \ge 2 + \delta > 2$ and an integer k such that $(p-2)(k-4) \ge 8 + \delta_1 > 8$. Assume also that kernel $\Phi \colon X^2 \to R$ satisfies the conditions*

$$E\, g_1^4 < \infty, \quad E\,|g_2|^p < \infty, \quad |\lambda_k| \ge \delta_2 > 0$$

and condition (7.1.2) *is satisfied. Then the following expansion holds*

$$\sup_x |P(\sigma^{-1} U_n < x) - G_n(x)| = o(n^{-1}).$$

Proof. We set $t_n = n^{(p-1)/p}(\ln n)^{-1}$, $v_n = n \ln n$. Just as in Bickel *et al.* (1986), by using the argument presented in the proof of Lemma 7.1.1, we obtain the estimate

$$\int_{-t_n}^{t_n} |\varphi_n(t) - g_n(t)|\, |t|^{-1}\, dt = o(n^{-1}).$$

In view of the Esséen inequality, it suffices to establish that

$$\int_{t_n \le |t| \le v_n} |\varphi_n(t)|\, |t|^{-1}\, dt = o(n^{-1}). \tag{7.1.4}$$

For this purpose, we employ the symmetrization inequality (2.3.2) in Borovskikh and Korolyuk (1994a) (see also Bickel *et al.* (1986)). Let the random variables Z_1, ..., Z_n be such that the sequence $X_1, \ldots, X_n, Z_1, \ldots, Z_n$ consists only of independent random variables with the same distribution P. We denote $\xi_i = (X_i, Z_i)$, $i = 1$, ..., n, and assume that the integer m is such that $1 \le m \le (n-1)/4$. Then, by using the argument used when deriving (2.3.2) in Borovskikh and Korolyuk (1994a), we get

$$|\varphi_n(t)|^2 \le E \exp\Big\{ it\, \alpha_n \sum_{j=1}^{2m} \sum_{l=4m+1}^{n} h(\xi_j, \xi_l) \Big\}, \tag{7.1.5}$$

where

$$h(\xi_j,\xi_l) = g_2(X_j,X_l) - g_2(Z_j,X_l) - g_2(X_j,Z_l) + g_2(Z_j,Z_l).$$

To estimate the expression on the right-hand side of (7.1.5), it is helpful to truncate the random variables $X_1, \ldots, X_n, Z_1, \ldots, Z_n$ (see Borovskikh and Korolyuk (1984c) or Bickel *et al.* (1986)). We follow the paper by Bickel *et al.* (1986). Consider a measurable set $B \subset \mathfrak{X}$ such that $\alpha = P(\xi_1 \in B \times B) = P^2(B)$. Since for any $\alpha \in [0,1]$, $x \in [0,1]$, and $\rho \in [0,1]$, we always have $\alpha x + (1-\alpha) \leq \max\{x^\rho, (\rho/\alpha)^{\rho/(1-\rho)}\}$, the following inequalities hold due to (7.1.5)

$$\begin{aligned}
|\varphi_n(t)|^2 &\leq E\Big[E\Big(\exp\Big\{it\,\alpha_n \sum_{l=4m+1}^{n} h(\xi_1,\xi_l)\Big\} \mid \xi_{4m+1},\ldots,\xi_n\Big)\Big]^{2m} \\
&\leq E\Big[\alpha E\Big(\exp\Big\{it\,\alpha_n \sum_{l=4m+1}^{n} h(\xi_1,\xi_l)\Big\} \mid \xi_{4m+1},\ldots,\xi_n;\, \xi_1 \in B\times B\Big) \\
&\quad + (1-\alpha)\Big]^{2m} \\
&\leq E\Big[E\Big(\exp\Big\{it\,\alpha_n \sum_{l=4m+1}^{n} h(\xi_1,\xi_l)\Big\} \mid \xi_{4m+1},\ldots,\xi_n;\, \xi_1 \in B\times B\Big)\Big]^{2\rho m} \\
&\quad + \Big(\frac{\rho}{\alpha}\Big)^{2\rho m/(1-\rho)},
\end{aligned}$$

for each $\rho \in [0,1]$.

We define $\tilde{\xi}_j = (\tilde{X}_j, \tilde{Z}_j)$, $j = 1, \ldots, m$, where $\tilde{X}_1, \ldots, \tilde{X}_m, \tilde{Z}_1, \ldots, \tilde{Z}_m$ are independent random variables which have the same distribution

$$P(\tilde{X}_j \in A) = P(\tilde{Z}_j \in A) = P(A|B)$$

and are independent of $\xi_{4m+1}, \ldots, \xi_n$. Then for $\rho = 1/2$, the last inequality can be rewritten as

$$\begin{aligned}
|\varphi_n(t)|^2 &\leq E\exp\Big\{it\,\alpha_n \sum_{j=1}^{m}\sum_{l=4m+1}^{n} h(\tilde{\xi}_j,\xi_l)\Big\} + (2\alpha)^{-2m} \\
&= E\Big[E\left(\exp\{it\alpha_n h_m\} \mid \tilde{\xi}_1,\ldots,\tilde{\xi}_m\right]^{n-4m} + (2\alpha)^{-2m},
\end{aligned} \tag{7.1.6}$$

where

$$h_m = \sum_{j=1}^{m} h(\tilde{\xi}_j,\xi_n) =$$

$$= \sum_{j=1}^{m}\Big[g_2(\tilde{X}_j,X_n) - g_2(\tilde{Z}_j,X_n) - g_2(\tilde{X}_j,Z_n) + g_2(\tilde{Z}_j,Z_n)\Big].$$

We define the set $B \in \mathfrak{X}$ as follows

$$B = \Big\{x\colon \int_X |g_2(x,y)|^p\, P(dy) \leq \delta_3\,\tau\Big\}, \tag{7.1.7}$$

where $\tau > 0$ is a large but fixed number which will be defined below. Let us estimate the conditional expectation in (7.1.6). Since $|\exp(ix) - 1 - ix + x^2/2| \le x^2/5 + |x|^p$ for $p > 2$, we have

$$0 \le E\left(\exp\{it\alpha_n h_m\} \mid \tilde{\xi}_1, \dots, \tilde{\xi}_m\right) \le$$
$$\le 1 - \frac{1}{3} t^2 \alpha_n^2 E\left(h_m^2 \mid \tilde{\xi}_1, \dots, \tilde{\xi}_m\right) + |t|^p \alpha_n^p E\left(|h_m|^p \mid \tilde{\xi}_1, \dots, \tilde{\xi}_m\right). \qquad (7.1.8)$$

By virtue of (7.1.7) and the property of degeneracy of $g_2(x, y)$, we get

$$E\left(h(\tilde{\xi}_1, \xi_n)|\tilde{\xi}_1\right) = E\left(h(\tilde{\xi}_1, \xi_n)|\xi_n\right) = 0 \quad a.s.,$$
$$E\left(|h(\tilde{\xi}_1, \xi_n)|^p \, |\tilde{\xi}_1\right) \le 4^p \, \delta_{3\tau}.$$

Under the conditions of the theorem, we can assume that $2 + \delta \le p \le 10 + \delta_1$ and $5 \le k \le 5 + (8 + \delta_1)/\delta$. In addition, for all real $p \ge 2$ and integer $q \ge 1$, by virtue of Corollary 2.7.1 we have

$$E\left\{\int_X \left|\sum_{i=1}^{m} g_2(X_i, y)\right|^p P(dy)\right\}^q \le a(p,q)\, m^{qp/2}\, E\left\{\int_X |g_2(X_1, y)|^p \, P(dy)\right\}^q,$$

where $a(p, q) > 0$ depends only on p and q. Therefore, for every integer $q \ge 1$, we obtain

$$E\left[E\left(|h_m|^p \mid \tilde{\xi}_1, \dots, \tilde{\xi}_m\right)\right]^q = O(m^{pq/2}).$$

Choosing $q = 10\, k/\delta_1$ and using Chebyshev's inequality, we find that

$$P\Big(E(|h_m|^p \mid \tilde{\xi}_1, \dots, \tilde{\xi}_m) \ge m^{p/2} n^{\delta_1/(4k)}\Big) = O(n^{-5/2}). \qquad (7.1.9)$$

Under the conditions of the theorem, the function $g_2(x, y)$ can be written as follows

$$g_2(x, y) = \sum_{r=1}^{k} \lambda_r \, e_r(x)\, e_r(y) + Q(x, y),$$

where $Q \colon X^2 \to R$ is a symmetric function such that

$$\int_X Q(x, y)\, e_r(y)\, P(dy) = 0$$

for all $r = 1, \dots, k$. Taking into account these relations, we obtain

$$E(h_m^2 \mid \tilde{\xi}_1, \dots, \tilde{\xi}_m) = 2 \sum_{r=1}^{k} \lambda_r^2 \left(\sum_{j=1}^{m} \{e_r(\tilde{X}_j) - e_r(\tilde{Z}_j)\}\right)^2$$
$$+ 2k^2 \int_X \left[\sum_{j=1}^{m} \{Q(\tilde{X}_j, y) - Q(\tilde{Z}_j, y)\}\right]^2 P(dy)$$
$$\ge 2\delta_2^2 \sum_{r=1}^{k} \left(\sum_{j=1}^{m} \{e_r(\tilde{X}_j) - e_r(\tilde{Z}_j)\}\right)^2. \qquad (7.1.10)$$

Consider a covariance matrix $\Sigma = (\sigma_{r,r'})$ of a random vector $(e_1(\tilde{X}_1) - e_1(\tilde{Z}_1), \ldots, e_k(\tilde{X}_1) - e_k(\tilde{Z}_1))$ with a general element $\sigma_{r,r'} = E(e_r(\tilde{X}_1) - e_r(\tilde{Z}_1))(e_{r'}(\tilde{X}_1) - e_{r'}(\tilde{Z}_1))$, r, $r' = 1, \ldots, k$. By virtue of (7.1.3), we have

$$E\,|e_r(X_1)|^p \le \lambda_r^{-p}\, E\left|\int_X g_2(X_1,y)\, e_r(y)\, P(dy)\right|^p \le$$
$$\le \delta_2^{-p}\, E\left(\int_X g_2^2(X_1,y)\, P(dy)\right)^{p/2} \le \delta_2^{-p}\, E\,|g_2|^p. \tag{7.1.11}$$

Together with Hölder's inequality, this implies that for $r \ne r'$

$$|\sigma_{r,r'}| \le \alpha^{-1} \left|E(e_r(X_1) - e_r(Z_1))\,(e_{r'}(X_1) - e_{r'}(Z_1))\, I((X_1, Z_1) \in \overline{B \times B})\right| \le$$
$$\le 4\delta_2^{-2}\,(E\,|g_2|^p)^{2/p}\,\alpha^{-1}(1-\alpha)^{(p-2)/p}.$$

Similarly, we find that

$$-4\delta_2^{-2}\,(E\,|g_2|^p)^{2/p}\,\alpha^{-1}(1-\alpha)^{(p-2)/p} \le \sigma_{r,r} - 2 \le 2\,\alpha^{-1}(1-\alpha)$$

for $r = r'$.

We assume that

$$1 - \alpha^{1/2} = P(\bar{B}) \le \tau^{-1}$$

in definition (7.1.7). Taking these inequalities into account, we can choose τ to be sufficiently large so that $2\alpha \ge e^{1/2}$ and that the inequalities $|\sigma_{r,r'} - 2\,\delta_{r,r'}| \le k^{-1}$ hold for all r, $r' = 1, \ldots, k$; here $\delta_{r,r'} = 0$ if $r \ne r'$, and $\delta_{r,r} = 1$. Consequently, if ρ_k denotes the least eigenvalue of the covariance matrix Σ, then

$$\rho_k \ge 2 - \left\{\sum_{r=1}^{k}\sum_{r'=1}^{k}(\sigma_{r,r'} - 2\,\delta_{r,r'})^2\right\}^{1/2} \ge 1. \tag{7.1.12}$$

Note also that (7.1.11) and the choice of α involve the boundedness of the expectations $E\,|e_r(\tilde{X}_1) - e_r(\tilde{Z}_1)|^{2+\delta}$, $r = 1, \ldots, k$.

Further, we employ the following lemma (see Bickel *et al.* (1986)).

Lemma 7.1.3. *Let $X_1, \ldots, X_n$ be independent k-dimensional random vectors with the same distribution P and with a positive definite covariance matrix Σ whose least eigenvalue is equal to ρ_k. Suppose that $S_n = n^{-1/2}(X_1 + \cdots + X_n)$. Then there exists a positive number $b = b(k, P)$ which depends only on k and P and is such that for every $\varepsilon > 0$ and all $n = 1, 2, \ldots$ we have*

$$P(\|S_n\| \le \varepsilon) \le b\,(\varepsilon^k + n^{-k/2}).$$

Here, b is a uniform constant with respect to any class of distribution if k is bounded, ρ_k is separated from zero, and $E\,\|X_1\|^{2+\delta}$ is bounded for fixed $\delta > 0$.

The proof can be found in Bickel et al. (1986).

In view of Lemma 7.1.3 and inequalities (7.1.10) and (7.1.12), we conclude that

$$P\left(E(h_m^2|\tilde{\xi}_1, \dots, \tilde{\xi}_m) \leq mn^{-4/k} (\ln n)^{-6/k}\right) = O(n^{-2} (\ln n)^{-3} + m^{-k/2}). \quad (7.1.13)$$

In region $t_n \leq |t| \leq v_n$, we choose m to be a function of t of the form

$$m = m(t) = [(\alpha_n t)^{-2} n^{4/k-1} (\ln n)^{2+6/k}],$$

where $[x]$ denotes the integral part of the number x. It can be verified that for indicated t, we always have $1 \leq m \leq (n-1)/4$. For such a choice of m, it follows from (7.1.6), (7.1.8), (7.1.9), and (7.1.13) that

$$|\varphi_n(t)| = O(n^{-1} (\ln n)^{-3/2})$$

as $n \to \infty$ uniformly in t in the region $t_n \leq |t| \leq v_n$. This proves (7.1.4) and the statement of Theorem 7.1.4.

7.2 General method of expansions

Let h_n, $n = 1, 2, \dots$, be a sequence of some probabilities; here, n is the number of observations on the basis of which h_n is defined. Assume that for each observation we can introduce weights ε_j, $j = 1, \dots, n$. More exactly, we assume that there exists a sequence of functions $h_n(\varepsilon_1, \dots, \varepsilon_n)$ of n variables such that $|\varepsilon_j| \leq n^{-1/2}$, and that for every $n \geq 1$ the following conditions hold

$$\begin{gathered} h_n = h_n(n^{-1/2}, \dots, n^{-1/2}) + O(n^{-(s-2)/2}), \quad s \geq 3, \\ h_{n+1}(\varepsilon_1, \dots, \varepsilon_{j-1}, 0, \varepsilon_{j+1}, \dots, \varepsilon_n) = \\ = h_n(\varepsilon_1, \dots, \varepsilon_{j-1}, \varepsilon_{j+1}, \dots, \varepsilon_n), \quad j = 1, \dots, n, \\ \frac{\partial}{\partial \varepsilon_j} h_n(\varepsilon_1, \dots, \varepsilon_{j-1}, \varepsilon_j, \varepsilon_{j+1}, \dots, \varepsilon_n)\Big|_{\varepsilon_j = 0} = 0, \quad j = 1, \dots, n. \end{gathered} \quad (7.2.1)$$

The functions

$$h_n(\varepsilon_1, \dots, \varepsilon_n)$$

are symmetric with respect to all arguments $\varepsilon_1, \dots, \varepsilon_n$.

Example 7.2.1. Let $X_1, \dots, X_n$ be a sequence of independent identically distributed random variables. We set

$$h_n = E f(n^{-1/2} (X_1 + \dots + X_n))$$

for some function f.

Then

$$h_n(\varepsilon_1,\ldots,\varepsilon_n) = E\,f(\varepsilon_1 X_1 + \cdots + \varepsilon_n X_n).$$

If $f(x)$ is differentiable, then this sequence $h_n(\varepsilon_1,\ldots,\varepsilon_n)$ satisfies all the conditions in (7.2.1).

The following theorem guarantees the existence of the limit $\lim_n h_n$ for symmetric functions h_n, $n \ge 1$, and contains an estimate of the convergence rate. Let $E_{q,m}$ denote a set of weighted vectors $(\varepsilon_1,\ldots,\varepsilon_m)$, where all the weights except q ones are equal to $m^{-1/2}$, and the other q weights are bounded in an absolute value by $m^{-1/2}$. Assume that $\mathfrak{D}^\alpha$ denotes partial derivatives

$$\frac{\partial^{\alpha_1}}{\partial \varepsilon_1^{\alpha_1}} \cdots \frac{\partial^{\alpha_n}}{\partial \varepsilon_m^{\alpha_n}},$$

where $\alpha = (\alpha_1,\ldots,\alpha_n)$ is a nonnegative integer-valued vector, and $|\alpha| = \alpha_1 + \cdots + \alpha_n$.

Theorem 7.2.1. *Suppose that the sequence $h_n(\cdot)$, $n \ge 1$, satisfies conditions* (7.2.1) *and the condition*

$$d_q(h,n) = \sup\Big\{|\mathfrak{D}^\alpha h_m(\varepsilon_1,\ldots,\varepsilon_m)| \colon |\alpha| = q, (\varepsilon_1,\ldots,\varepsilon_m) \in E_{q,m},\ m \ge n\Big\} < \infty$$

for $q = 3$. Then the limit

$$\lim_{m\to\infty} h_m(m^{-1/2},\ldots,m^{-1/2}) = h_\infty$$

exists, and

$$|h_n(n^{-1/2},\ldots,n^{-1/2}) - h_\infty| \le c\, d_3(h,n)\, n^{-1/2},$$

where c is an absolute constant.

Proof. Let $\bar\varepsilon_m = (\varepsilon_1,\ldots,\varepsilon_m)$, $\varepsilon_1 = \cdots = \varepsilon_m = m^{-1/2}$ and $h_{m+r}(\bar\varepsilon_m) = h_{m+r}(m^{-1/2},\ldots,m^{-1/2},0,\ldots,0)$. According to Taylor's formula, we get

$$h_{m+1}(\bar\varepsilon_m) - h_{m+1}(\bar\varepsilon_{m+1}) = -\sum_{|\alpha|\le 2} \mathfrak{D}^\alpha h_{m+1}(\bar\varepsilon_m)\,(\bar\varepsilon_{m+1} - \bar\varepsilon_m)^\alpha\,(\alpha!)^{-1} + R_3(m),$$

where

$$|R_3(m)| \le c\, d_3(h,n)\, m^{-3/2}, \quad m \ge n \ge 1,$$

by virtue of the fact that $|m^{-1/2} - (m+1)^{-1/2}| \le c\, m^{-3/2}$.

If $j \ne k$, then, according to the third condition in (7.2.1), we find

$$\frac{\partial}{\partial \varepsilon_j}\frac{\partial}{\partial \varepsilon_k} h_{m+1}(\bar\varepsilon_m) = O(d_3(h,n)\, m^{-1/2}).$$

Since

$$\frac{\partial}{\partial \varepsilon_j} h_{m+1}(\bar{\varepsilon}_m) = \frac{\partial^2}{\partial \varepsilon_j^2} h_{m+1}(\bar{\varepsilon}_m)\Big|_{\varepsilon_j=0} m^{-1/2} + O(d_3(h,n)\, m^{-1}),$$

$$\frac{\partial^2}{\partial \varepsilon_j^2} h_{m+1}(\bar{\varepsilon}_m) = \frac{\partial^2}{\partial \varepsilon_j^2} h_{m+1}(\bar{\varepsilon}_m)\Big|_{\varepsilon_j=0} + O(d_3(h,n)\, m^{-1/2}),$$

we have

$$h_{m+1}(\bar{\varepsilon}_m) - h_{m+1}(\bar{\varepsilon}_{m+k}) = -\sum_{j=1}^{m} \frac{\partial^2}{\partial \varepsilon_j^2} h_{m+1}(\bar{\varepsilon}_m)\Big|_{\varepsilon_j=0} \times$$

$$\times \Big[m^{-1/2}(m^{-1/2} - (m+1)^{-1/2}) + \frac{1}{2}(m^{-1/2} - (m+1)^{-1/2})\Big] +$$

$$+\frac{1}{2}(m+1)^{-1} \frac{\partial^2}{\partial \varepsilon_{m+1}^2} h_{m+1}(\bar{\varepsilon}_m)\Big|_{\varepsilon_{m+1}=0} + O(d_3(h,n)\, m^{-3/2}).$$

Since $h_{m+1}(\cdot)$ is a function symmetric with respect to its variables, the sum over j on the right-hand side cancels the second derivative with respect to ε_{m+1}. Consequently, for any $r \geq m$, we obtain

$$h_m(\bar{\varepsilon}_m) - h_{m+r}(\bar{\varepsilon}_{m+r}) = O(d_3(h,n)) \sum_{p=m}^{m+r-1} p^{-3/2}.$$

Therefore, $h_n(\bar{\varepsilon}_n) - h_m(\bar{\varepsilon}_m)$, $m = n, n+1, \ldots$, is a Cauchy sequence which has the limit $h_\infty(0)$ with respect to m. Setting $m = n$ in the last relation and then proceeding to the limit as $r \to \infty$, we arrive at the proof of Theorem 7.2.1.

To formulate an asymptotic expansion for h_n, we introduce the following differential operator by using a formal power series. First, we introduce cumulant differential operator $\mathcal{K}_p(\mathfrak{D})$ by

$$\sum_{p=2}^{\infty}(p!)^{-1} z^p \mathcal{K}_p(\mathfrak{D}) = \ln\Big(1 + \sum_{p=2}^{\infty}(p!)^{-1} z^p \mathfrak{D}^p\Big),$$

where z is a formal variable. Here, $\mathfrak{D}^p$ denotes the p times differentiation with respect to one variable η, and $\mathfrak{D}^{p_1} \ldots \mathfrak{D}^{p_r} = \mathfrak{D}^{(p_1,\ldots,p_k)}$ denotes the differentiation with respect to r different variables $\eta_1, \ldots, \eta_r$. The first cumulant operators have the form

$$\mathcal{K}_2(\mathfrak{D}) = \mathfrak{D}^2,$$

$$\mathcal{K}_3(\mathfrak{D}) = \mathfrak{D}^3,$$

$$\mathcal{K}_4(\mathfrak{D}) = \mathfrak{D}^4 - 3\mathfrak{D}^2\mathfrak{D}^2.$$

We denote $\bar{\mathcal{K}} = (\mathcal{K}_3(\mathfrak{D}), \ldots, \mathcal{K}_{r+2}(\mathfrak{D}))$ and then define differential operators by the formal expansion

$$\sum_{r=0}^{\infty} z^r P_r(\bar{\mathcal{K}}) = \exp\Big(\sum_{r=3}^{\infty}(r!)^{-1} z^{r-2} \mathcal{K}_r(\mathfrak{D})\Big).$$

For $P_r(\bar{\mathcal{K}})$, it follows from this expansion that

$$P_r(\bar{\mathcal{K}}) = \sum_{m=1}^{r} (m!)^{-1} \Big\{ \sum\nolimits^{*} (j_1+2)!^{-1} \mathcal{K}_{j_1+2}(\mathfrak{D}) \times \times (j_2+2)!^{-1} \mathcal{K}_{j_2+2}(\mathfrak{D}) \dots (j_m+2)!^{-1} \mathcal{K}_{j_m+2}(\mathfrak{D}) \Big\}, \quad (7.2.2)$$

where $\sum^*$ denotes the summation over set of all m positive integers $(j_1, \dots, j_m)$ which satisfy the equality $j_1 + j_2 + \dots + j_m = r$; here, $j_i = 1, 2, \dots, r$, for all $i = 1, \dots, m$.

Sometimes, $P_r(\bar{\mathcal{K}})$ will be denoted by $P_r(\cdot)$ or by $P_r(\mathfrak{D}_\eta^p)$, $P_r(\mathfrak{D}_\eta)$, where $\mathfrak{D}_\eta^p$ and $\mathfrak{D}_\eta$ are partial derivatives of order $p = (p_1, \dots, p_r)$ with respect to the variables $\eta = (\eta_1, \dots, \eta_r)$. For example,

$$P_0(\mathfrak{D}_\eta) = 1, \quad P_1(\mathfrak{D}_\eta) = \frac{1}{6}\frac{\partial^3}{\partial\eta_1^3},$$

$$P_2(\mathfrak{D}_\eta) = \frac{1}{24}\left(\frac{\partial^4}{\partial\eta_1^4} - 3\frac{\partial^4}{\partial\eta_1^2\,\partial\eta_2^2}\right) + \frac{1}{72}\frac{\partial^6}{\partial\eta_1^3\,\partial\eta_2^3}. \quad (7.2.3)$$

Assume that

$$h_\infty(\varepsilon_1, \dots, \varepsilon_s) = \lim_{m\to\infty} h_{m+s}(\varepsilon_1, \dots, \varepsilon_s, m^{-1/2}, \dots, m^{-1/2})$$

and let $C_B^r(G)$ denote the set of r times differentiable functions on $G \subset R^m$, the derivatives of which are bounded by the constant $B > 0$. For $s \geq 3$, we set

$$q_s(h) = \sup\Big\{ \big|\mathfrak{D}^\alpha h_\infty(\eta_1, \dots, \eta_r)|_{\eta_1=\dots=\eta_r=0}\big|,\ 2r \leq 3(s-3),$$
$$\alpha = (\alpha_1, \dots, \alpha_r), \quad \alpha_j \geq 2, \quad j = 1, \dots, r, \quad \sum_{j=1}^{r} (\alpha_j - 2) \leq s-3 \Big\}.$$

Theorem 7.2.2. *Suppose that h_m, $m \geq 1$, satisfies (7.2.1) and the conditions*

$$h_m(\cdot) \in C_B^s(E_{s,m}) \quad \textit{for any } m \geq 1,$$
$$q_s(h) \leq B.$$

Then the following asymptotic expansion holds

$$\left| h_n(n^{-1/2}, \dots, n^{-1/2}) - \sum_{r=0}^{s-3} n^{-r/2} P_r(\mathfrak{D}_\eta)\, h_\infty(\eta_1, \dots, \eta_r)\Big|_{\bar{\eta}=0} \right| \leq c_s\, B n^{-(s-2)/2},$$

where one can take $B = q_s(h) + d_s(h, n)$.

For the proof, see Götze (1985).

Example 7.2.2. Theorem 7.2.2 can be applied to the functionals of empirical processes which correspond to independent identically distributed random elements

X_j, $j = 1, 2, \ldots$, which take values in a measurable space $(X, \mathfrak{X})$. Let P denote their distribution on $(X, \mathfrak{X})$. We define a generalized empirical process $P_{\bar{\varepsilon}}$ by

$$P_{\bar{\varepsilon}} = \sum_{j=1}^{n} (\delta_{X_j} - P)\, \varepsilon_j, \tag{7.2.4}$$

where $\delta_x(A) = 1$ if $x \in A$, and $\delta_x(A) = 0$, otherwise.

Consider a sequence

$$h_n(\varepsilon_1, \ldots, \varepsilon_n) = E\, f(P_{\bar{\varepsilon}}), \quad n \geq 1,$$

for some bounded functional f. Here, conditions (7.2.1) are satisfied under proper smoothness conditions imposed on f.

Example 7.2.3. (Central limit theorem in Banach spaces). Let $X_1, X_2, \ldots$, take values in a separable real Banach space X of type 2 with norm $\|\cdot\|$ and σ-algebra of Borel sets $\mathfrak{X}$. For a bounded function f on X which has $3\,(s-2)$ bounded Fréchet derivatives, we set

$$h_n(\varepsilon_1, \ldots, \varepsilon_n) = E\, f(\varepsilon_1 X_1 + \cdots + \varepsilon_n X_n),$$

where $s \geq 3$ and $E\,\|X_1\|^s < \infty$, $E\,X_1 = 0$. Then according to Theorem 7.2.2, we have $h_n = E\, f(S_n)$, where $S_n = n^{-1/2}\,(X_1 + \cdots + X_n)$ has an asymptotic expansion up to the terms of order $O(n^{-(s-2)/2})$. Let us formulate this assertion. Denote

$$\varphi(\bar{\eta}) = \int f(y + \eta_1 x_1 + \cdots + \eta_q x_q)\, N(dy)\, P^q(d\bar{x}),$$

$$\varphi_i(\bar{\varepsilon}; \bar{\eta}) = \int f(t^{(i)}_{\bar{\varepsilon},\bar{\eta}}(\bar{x}, \bar{y}))\, P^{n+q}(d\bar{x})\, N^{n+q}(d\bar{y}),$$

where N is a Gaussian distribution on $(X, \mathfrak{X})$ with the same covariance structure as P; $\bar{x} = (x_1, \ldots, x_q)$, $\bar{y} = (y_1, \ldots, y_q)$, and

$$t^{(i)}_{\bar{\varepsilon},\bar{\eta}}(\bar{x}, \bar{y}) = n^{-1/2}\,(x_1 + \cdots + x_i + y_{i+1} + \cdots + y_n) +$$
$$+\varepsilon_1 y_{n+1} + \cdots + \varepsilon_q y_{n+q} + \eta_1 x_{n+1} + \cdots + \eta_q x_{n+q}, \qquad i = 0, 1, \ldots, n.$$

Assume that the functions $\varphi_i(\bar{\varepsilon}; \bar{\eta})$ are differentiable so that the variable

$$R_n = \sup \left| \mathfrak{D}_{\varepsilon}^{(j,0,\ldots,0)}\, \mathfrak{D}_{\eta}^{(i_1,\ldots,i_q)}\, \varphi_i(\varepsilon, 0, \ldots, 0, \eta, 0, \ldots, 0) \right|$$

exists, where the supremum is taken over all $0 \leq \varepsilon, \eta \leq n^{-1/2}$ such that $\varepsilon\eta = 0$, $i = 0$, $1, \ldots, n$ and over all integers $0 \leq j, i_1, \ldots, i_q \leq s$ such that $s \leq j + i_1 + \cdots + i_q \leq 3\,(s-2)$. In addition, let

$$\mathfrak{D}_{\varepsilon}^{\alpha}\, \mathfrak{D}_{\eta}^{\beta}\, \varphi_i(0,0) = \mathfrak{D}_{\eta}^{\alpha+\beta}\, \varphi_i(0,0),$$

for $\alpha = (i_1, \dots, i_q)$, $0 \le i_k \le 2$, $\beta = (j_1, \dots, j_q)$, $i_k + j_k \le s$, $1 \le k \le q$, $0 \le i \le n$, and $|\alpha| + |\beta| \le 3(s-2)$. Under these conditions, according to Theorem 7.2.2, we have

$$\left| h_n - \sum_{j=0}^{s-3} n^{-j/2} P_j(\mathfrak{D}_\eta)\, \varphi(\eta) \right|_{\bar\eta=0} \le c(s)\, R_n\, n^{-(s-2)/2}. \tag{7.2.5}$$

If we set $f(x) = \exp(it\,\|x\|)$, $x \in X$, $t \in R$ then in (7.2.5) we obtain the asymptotic expansion of the characteristic function of the distribution of the random variable $\|S_n\|$. Here, the smoothness conditions on f are reduced to the requirement of the smoothness of the norm $\|\cdot\|$ in the Banach space X.

Example 7.2.4. (*UR*-statistics). Let $X_1, \dots, X_n, \dots$ be a sequence of independent identically distributed random elements in $(X, \mathfrak{X})$ with distribution P. Consider *UR*-statistic of degree $m = 2$ with degenerate kernel Φ and let $w_n = (n-1)\, U_n$, i.e.

$$w_n = n^{-1} \sum_{1 \le i \ne j \le n} \Phi(X_i, X_j).$$

The weighted version of w_n is

$$w_n(\bar\varepsilon) = \sum_{1 \le i \ne j \le n} \varepsilon_i \varepsilon_j\, \Phi(X_i, X_j).$$

With the help of Theorem 7.2.2, we define the asymptotic expansion for $h_n = E\exp(it\, w_n)$, i.e. for the characteristic function of the random variable w_n. For $s \ge 3$, we assume that $E\,|\Phi(X_1, X_2)|^s < \infty$.

By virtue of Theorem 4.2.1 on weak convergence, we have

$$w_{m+p}(\varepsilon_1, \dots, \varepsilon_p, m^{-1/2}, \dots, m^{-1/2}) \Longrightarrow w_\infty(\varepsilon_1, \dots, \varepsilon_p)$$

as $m \to \infty$, where

$$w_\infty(\varepsilon_1, \dots, \varepsilon_p) = w_p(\varepsilon_1, \dots, \varepsilon_p) + 2\sum_{j=1}^{p} \varepsilon_j \int_X \Phi(X_j, x)\, W(dx) + I_2^W(\Phi).$$

Thus, by Theorem 7.2.2, we have the asymptotic expansion

$$E\exp(it\,(n-1)\,U_n) =$$

$$= \sum_{j=0}^{s-3} n^{-j/2} P_j(\mathfrak{D}_\varepsilon) \left(E\exp(it\, w_\infty(\bar\varepsilon)) \right)\Big|_{\bar\varepsilon=0} + n^{-(s-2)/2} R_n,$$

where R_n depends on s and t. This dependence can be written in an explicit form.

7.3 Expansions with canonical kernels

Let $(X, \mathfrak{X})$ be a measurable space, and let $X_1, X_2, \dots$ be a sequence of independent random elements in X with the same distribution P. We denote by $\Phi(x_1, \dots, x_m)$, a symmetric real-valued function defined on X^m, and assume that P_n is an empirical measure which corresponds to the sampling $X_1, \ldots, X_n$.

We define a statistic

$$w_n = n^{m/2} \int \cdots \int \Phi(x_1, \ldots, x_m) (P_n - P)(dx_1) \ldots (P_n - P)(dx_m).$$

By definition of the measure P_n, this statistic can be written in the form

$$w_n = n^{-m/2} \sum_{(i_1, \ldots, i_m)=1}^{n} \int \cdots \int \Phi(x_1, \ldots, x_m) \times$$
$$\times (\delta_{X_{i_1}} - P)(dx_1) \ldots (\delta_{X_{i_m}} - P)(dx_m).$$

This formula shows that w_n is a linear normed von Mises' statistic with the canonical kernel

$$h(y_1, \ldots, y_m) = \int \cdots \int \Phi(x_1, \ldots, x_m) (\delta_{y_1} - P)(dx_1) \ldots (\delta_{y_1} - P)(dx_m).$$

Example 7.3.1. Let $X_1, \ldots, X_n$ be an independent random sampling of a population with uniform distribution P on $[0, 1]$, and let P_n be the ordinary distribution function which corresponds to this sampling. Assume that $\xi(t, x)$, $0 \leq x \leq 1$, $0 \leq t \leq 1$, is a real-valued function such that

$$\int_0^1 E |\xi(t, X_1)|^m < \infty.$$

Consider the statistic

$$w_n = n^{m/2} \int_0^1 \left[\int \xi(t, x) (P_n - P)(dx) \right]^m dt.$$

For integer $m \geq 2$, this statistic is a random variable of the type above. In particular, if $\xi(t, x) = I(t \geq x)$, then we have

$$h(y_1, \ldots, y_m) = \int (1 - \max(x_1, \ldots, x_m)) (\delta_{y_1} - P)(dx_1) \ldots (\delta_{y_m} - P)(dx_m).$$

If $m = 2$, then this kernel corresponds to the Cramer–von Mises ω^2-statistic and is equal to

$$h(y_1, y_2) = \frac{1}{2} (y_1^2 + y_2^2) - \max(y_1, y_2) + \frac{1}{3}.$$

The asymptotic expansion of the characteristic function of the random variable w_n is based on Theorem 7.2.2. The weighted version of w_n is

$$w_n(\bar{\varepsilon}) = \int \Phi(x_1, \ldots, x_m) P_{\bar{\varepsilon}}(dx_1) \ldots P_{\bar{\varepsilon}}(dx_m),$$

where $P_{\bar{\varepsilon}}(A)$, $A \in \mathfrak{X}$, is the generalized empirical process from (7.2.4). We define a weighted characteristic function

$$\varphi_n(t; \varepsilon_1, \dots, \varepsilon_n) = E \exp(it\, w_n(\bar{\varepsilon})).$$

Clearly,

$$\varphi_n(t) = E \exp(it\, w_n) = \varphi_n(t; n^{-1/2}, \dots, n^{-1/2}).$$

Let $w_{k+p}(\varepsilon_1, \dots, \varepsilon_p, k^{-1/2}, \dots, k^{-1/2})$ converge in distribution to $w_\infty(\varepsilon_1, \dots, \varepsilon_p)$ as $k \to \infty$.

In what follows we assume that function Φ satisfies the condition

$$\max\Big\{E\,\Phi^2(X_{i_1}, \dots, X_{i_m}) \colon 1 \le i_1 \le \cdots \le i_m \le m\Big\} < \infty.$$

Then by virtue of Theorem 4.2.6

$$w_\infty(\varepsilon_1, \dots, \varepsilon_p) = \int_X \cdots \int_X \Phi(x_1, \dots, x_m) \prod_{s=1}^{m} (P_{\bar{\varepsilon}} + Q)(dx_s) \tag{7.3.1}$$

where $\bar{\varepsilon} = (\varepsilon_1, \dots, \varepsilon_p)$.

In addition, we define

$$f_j(t) = P_j(\mathfrak{D}_\varepsilon)\, E \exp(it\, w_\infty(\varepsilon_1, \dots, \varepsilon_p))\Big|_{\bar{\varepsilon}=0}, \quad t \in R,$$
$$F_j(x) = P_j(\mathfrak{D}_\varepsilon)\, P(w_\infty(\varepsilon_1, \dots, \varepsilon_p) < x)\Big|_{\bar{\varepsilon}=0}, \quad x \in R,$$

and set

$$f_{n,s-3}(t) = \sum_{j=0}^{s-3} n^{-j/2} f_j(t),$$
$$F_{n,s-3}(x) = \sum_{j=0}^{s-3} n^{-j/2} F_j(x). \tag{7.3.2}$$

The moment condition. Let $s \ge 3$. Assume that there exist a nonnegative measurable function $T(x)$ and a constant A_m such that

$$E\,|T(X_1)|^s < \infty,$$
$$|\Phi(X_1, \dots, X_m)| \le A_m\, T(X_1) \dots T(X_m) \quad P\text{-a.s.} \tag{7.3.3}$$

Example 7.3.2. For $m = 2$, assume that $\{\lambda_j\}$ and $\{e_j\}$ are eigenvalues and eigenfunctions respectively of the integral operator generated by the function $\Phi(x, y)$. Let $e_j \in L^s(X, P)$. If $\sum_{j=1}^{\infty} |\lambda_j| < \infty$, then we can take

$$T(x) = \Big(\sum_{j=1}^{\infty} |\lambda_j|\, e_j^2(x)\Big)^{1/2}.$$

If X_j takes values in a Banach space X and Φ is of an m-linear form, then we set $T(x) = \|x\|$, where $\|\cdot\|$ is a norm in X.

A condition on limit characteristic functions. For $t \in R$ denote

$$\varphi^{\text{dec}}(t) = E\exp(it\, w_\infty^{\text{dec}}),$$

where

$$w_\infty^{\text{dec}} = \int_X \cdots \int_X \Phi(x_1, \dots, x_m) \prod_{s=1}^{m} W_s(dx_s).$$

Additionally $W_1(\cdot), \dots, W_m(\cdot)$ are independent copies of a Gaussian random measure $W(\cdot)$ with zero mean value and covariance $E\, W(A)\, W(B) = P(A \cap B)$.

Assume that there exists a large arbitrary number $\gamma > 0$ such that

$$|\varphi^{\text{dec}}(t)| = O(|t|^{-\gamma}), \tag{7.3.4}$$

as $|t| \to \infty$.

Let $\varphi(t) = E\exp(it\, w_\infty)$. According to the symmetrization inequality, we have

$$|\varphi(t)|^{2^m} \leq |\varphi^{\text{dec}}(m!\, t)|.$$

Therefore, condition (7.3.4) involves the estimate of the limit characteristic function $\varphi(t)$ and its derivatives.

A condition on prelimit characteristic functions. Let $\varphi_n(t) = E\exp(it\, w_n)$, $t \in R$. Assume that in the region $n^{m/2-\varepsilon} \leq |t| \leq n^{(s-2)/2}$, the following relation holds for $s \geq 3$ and $s \geq r+2$

$$|\varphi_n(t)| = O(n^{-(s-2)/2} (\ln n)^{-1}), \tag{7.3.5}$$

as $n \to \infty$.

To verify (7.3.5), we note the following. Let $Z_j = (X_j, X_j')$, $j = 1, \dots, n$, where X_j' is an independent copy of X_j, and let Δ_j be a difference operator applied to the j-th argument of the function $\Phi(X_1, \dots, X_m)$ and acting as follows $\Delta_j \Phi = \Phi(\dots, X_j, \dots,) - \Phi(\dots, X_j', \dots,)$. We define $\Phi(Z_1, \dots, Z_m) = \Delta_1 \dots \Delta_m \Phi$ and set

$$w_I = \sum_{j_1 \in I_1} \cdots \sum_{j_m \in I_m} \Phi(Z_{j_1}, \dots, Z_{j_m})\, n^{-m/2},$$

where $I = (I_1, \dots, I_m)$ is a decomposition of the set $\{1, 2, \dots, n\}$ such that $|I_1| \geq cn$, $|I_j| \geq \ln n$, $j = 2, \dots, m$; here, c is a constant, $0 < c < 1$ and $|I_j|$ denotes the number of elements in the set I_j, $j = 1, \dots, m$.

By the symmetrization inequality, we have

$$|\varphi_n(t)|^{2^m} \le E\exp(it\, m!\, w_I).$$

Hence, in order that (7.3.5) holds, it suffices to have the estimate

$$E\exp(it\, m!\, w_I) = O(n^{-(s-2)\,2^{m-1}}\,(\ln n)^{-2^m})$$

in the region $n^{m/2-\varepsilon} \le |t| \le n^{(s-2)/2}$, at least for one indicated decomposition I.

Theorem 7.3.1. *Under conditions* (7.3.3)-(7.3.4), *we have*

$$\sup_x \Big|P(w_n < x) - F_{n,s-3}(x)\Big| = O(n^{-m/2+\varepsilon}) + O(n^{-(s-2)/2})$$

for $s \ge 3$. *If* $s \ge m+2$, *then, under the additional condition* (7.3.5), *the following relation holds*

$$\sup_x \Big|P(w_n < x) - F_{n,s-3}(x)\Big| = O(n^{-(s-2)/2}).$$

Proof. According to the Esséen inequality, we get

$$\sup_x |P(w_n < x) - F_{n,s-3}(x)| \le \sum_{j=1}^{5} J_j$$

where

$$J_1 = c\int_{-T_1}^{T_1} \Big|E\exp(it\,w_n) - f_{n,s-3}(t)\Big|\,|t|^{-1}\,dt,$$

$$J_j = c\int_{T_{j-1}<|t|<T_j} \Big|E\exp(it\,w_n)\Big|\,|t|^{-1}\,dt, \quad j = 2,3,$$

$$J_4 = c\int_{|t|>T_1} |f_{n,s-3}(t)|\,|t|^{-1}\,dt,$$

$$J_5 = c\,T_3^{-1}\sup_x\Big|\frac{\partial}{\partial x}F_{n,s-3}(x)\Big|,$$

$$T_1 = n^{m/4-\varepsilon}, \quad T_2 = n^{m/2-\varepsilon}, \quad T_3 = n^{(s-2)/2}.$$

Estimate of J_j. The sequence of symmetric functions $\varphi_n(t;\varepsilon_1\dots,\varepsilon_n)$, $n \ge 1$, satisfies the conditions of Theorem 7.2.2, because

$$\frac{\partial}{\partial\varepsilon_1}\varphi_n(t;\varepsilon_1,\dots,\varepsilon_n)\Big|_{\varepsilon_1=0} = 0$$

by the definition of $w_n(\bar\varepsilon)$. Consequently, according to Theorem 7.2.2, we find

$$\Big|\varphi_n(t;n^{-1/2},\dots,n^{-1/2}) - f_{n,s-3}(t)\Big| \le c(s)\,n^{-(s-2)/2}\,(g_n(t) + g(t)),$$

where

$$g_n(t) = \sup\Big\{|\mathfrak{D}^\alpha \varphi_k(t;\varepsilon_1,\dots,\varepsilon_k)|: |\alpha| \le s,\ (\varepsilon_1,\dots,\varepsilon_k) \in E_{s,k},\ k \ge n\Big\},$$

$$g(t) = \sup\Big\{\Big|\mathfrak{D}^\alpha E\exp(it\,w_\infty(\bar\varepsilon))|_{\bar\varepsilon=0}\Big| : \alpha = (\alpha_1,\dots,\alpha_p),$$
$$0 \le 2p \le 3(s-3),\ \alpha_j \ge 2,\ \sum_j(\alpha_j - 2) \le s-3\Big\}.$$

In Götze (1984), Lemma 5.50 (i) and (ii), the functions $g_n(t)$ and $g(t)$ were estimated with the help of a symmetrization inequality, and, as a result, it was shown that $J_1 = O(n^{-(s-2)/2})$ as $n \to \infty$. Further, by using Götze (1984), Lemma 5.50 (v), we obtain $J_2 = O(n^{-(s-2)/2})$. Condition (7.3.5) guarantees that $J_3 = O(n^{-(s-2)/2})$ as $n \to \infty$. According to Götze (1984), Lemma 5.50 (i), we get $J_4 = O(n^{-(s-2)/2})$, and, by virtue of Götze (1984), Lemma 5.50 (ii), we obtain $J_5 = O(n^{-(s-2)/2})$ as $n \to \infty$. Combining all the estimates of J_j, $j = 1, \dots, 5$, we can prove Theorem 7.3.1.

Example 7.3.3. Assume that X is a separable real Banach space, $X_1, \dots, X_n$ are independent identically distributed X-valued random elements, and $S_n = n^{-1/2}(X_1 + \dots + X_n)$. Then Theorem 7.3.1 can be applied to m-linear kernel $\Phi(x_1,\dots,x_m)$ and von Mises' statistics of the form

$$w_n = \Phi(S_n,\dots,S_n)$$

provided that $E\,\|X_1\|^s < \infty$ with some $s \ge 3$ and condition (7.3.4) holds.

Example 7.3.4. Assume that $X = L^p(Y,\mu)$, $p \ge 2$, is an integer Banach space of real functions integrable with respect to a σ-finite measure μ and given on the set Y, and let $S_n = n^{-1/2}(X_1 + \dots + X_n) \in X$, where $X_1, \dots, X_n$ are independent identically distributed elements in X. We set

$$w_n(a) = \int_Y (S_n(t) + a(t))^p \, d\mu(t), \quad a \in X.$$

The variable $w_n(a)$ is a von Mises' statistic. Since L^p is a Banach space of type 2 for $p \ge 2$, in L^p there exists a Gaussian random element τ which has the same covariance structure as X_1. Therefore,

$$w_\infty(a) = \int_Y (\tau + a)^p \, d\mu,$$

and the first three coefficients in the asymptotic expansion of Theorem 7.3.1 with $m = p$ have the explicit form

$$F_0(x) = P(w_\infty(a) < x),$$
$$F_1(x) = \frac{1}{6}\frac{\partial^3}{\partial\varepsilon_1^3} P(w_\infty(a + \varepsilon_1 X_1) < x)\Big|_{\varepsilon_1=0},$$

$$F_2(x) = \left[\frac{1}{24}\left(\frac{\partial^4}{\partial\varepsilon_1^4} - 3\frac{\partial^2}{\partial\varepsilon_1^2}\frac{\partial^2}{\partial\varepsilon_2^2}\right) + \frac{1}{72}\frac{\partial^3}{\partial\varepsilon_1^3}\frac{\partial^3}{\partial\varepsilon_2^3}\right]$$
$$\times P\Big(w_\infty(a + \varepsilon_1 X_1 + \varepsilon_2 X_2) < x\Big)\Big|_{\varepsilon_1=\varepsilon_2=0},$$

where X_1 and X_2 do not depend on τ.

Example 7.3.5. Assume that $X = l_{2p}$, where p is a natural number. Let the condition $E\,\|X_1\|^{2s} < \infty$ be satisfied for some integer s, $1 \le s \le p$. Then there exist functions $N_j(x)$, which can be expanded in terms of $F_j(x)$ from (7.3.2) and are such that

$$\sup_x \left| P_n(\|y\|^{2p} < x) - N(\|y\|^{2p} < x) - \sum_{j=1}^{s-1} n^{-j}\, N_j(x)\right| = \rho_n n^{-(s-1)},$$

where $\rho_n \to 0$ as $n \to \infty$.

Example 7.3.6. Let $X_1, \dots, X_n$ be independent identically uniformly distributed random variables which take values in the k-dimensional unit cube $X = [0,1]^k$. We assume that the coordinates of the vector $X_j = (x_{j1}, \dots, x_{jk})$, $j = 1, \dots, n$, are independent. We define a k-dimensional empirical distribution function

$$H_n(x) = \frac{1}{n}\sum_{j=1}^{n}\prod_{r=1}^{k}\delta(x_r - x_{jr}), \quad \delta(u) = \begin{cases} 1, & u \ge 0 \\ 0, & u < 0, \end{cases}$$

and k-dimensional empirical process $y_n(x) = \sqrt{n}\,(H_n(x) - \bar{x})$, where $x = (x_1, \dots, x_k)$ and $\bar{x} = x_1 \dots x_k$. Consider the integral functional

$$w_n = \int_X y_n^{2p}(x)\, dx,$$

where $dx = dx_1 \dots dx_k$, and p is a natural number. By virtue of the invariance principle, the random variable w_n converges weakly to

$$w_\infty = \int_X \omega^{2p}(x)\, dx,$$

as $n \to \infty$; here, $\omega(x)$ is a k-dimensional Brownian bridge on X, i.e. the Gaussian process with zero mean value and correlation function

$$B(s,t) = \prod_{i=1}^{k}\min(s_i, t_i) - \prod_{i=1}^{k} s_i t_i,$$
$$s = (s_1, \dots, s_k), \quad t = (t_1, \dots, t_k).$$

For $q \ge 2p$, we denote

$$w_\infty(\varepsilon_1, \dots, \varepsilon_q) = \int_X (\omega(x) + \varepsilon_1\xi_1 + \dots + \varepsilon_q\xi_q)^{2p}\, dx,$$

$$\xi_j = \prod_{r=1}^{k} \delta(x_r - x_{jr}) - \prod_{r=1}^{k} x_r, \quad j = 1, \dots, q.$$

By using the differential operator $P_j(\mathfrak{D}_{\bar\varepsilon})$, we define the functions

$$F_j(x) = P_{2j}(\mathfrak{D}_{\bar\varepsilon}) \left(P(w_\infty(\varepsilon_1, \dots, \varepsilon_q) < x)\right)\Big|_{\bar\varepsilon=0},$$

where $j = 1, \dots, s$. Then for all $x > 0$ and $n \geq 2$, the following inequality holds

$$\left|P(w_n < x) - P(w_\infty < x) - \sum_{j=1}^{s-1} n^{-1} F_j(x)\right| \leq \gamma_1 n^{-s} e^{-\gamma_2 x^{1/p}},$$

where $s \geq 1$ is a fixed arbitrary integer, and $\gamma_i > 0$ are some constants (see Borovskikh (1983a) for $p = 1$).

7.4 Expansions with arbitrary kernels

Consider a von Mises' statistic

$$V_n = n^{-m} \sum_{i_1=1}^{n} \cdots \sum_{i_m=1}^{n} \Phi(X_{i_1}, \dots, X_{i_m})$$

with a symmetric kernel $\Phi\colon (x_1, \dots, x_m) \to \Phi(x_1, \dots, x_m)$ satisfying the condition $E\,\Phi(X_1, \dots, X_m) = 0$.

According to the Hoeffding formula, we have

$$V_n = \sum_{c=1}^{m} \binom{m}{c} V_{nc},$$

where

$$V_{nc} = n^{-c} \sum_{i_1=1}^{n} \cdots \sum_{i_c=1}^{n} g_c(X_{i_1}, \dots, X_{i_c}).$$

If r is the rank of kernel Φ, then

$$V_n = \sum_{c=r}^{m} \binom{m}{c} V_{nc}.$$

Denote $w_n = n^{r/2} V_n$. We write w_n in the form

$$w_n = \binom{m}{r} n^{-r/2} \sum_{i_1=1}^{n} \cdots \sum_{i_r=1}^{n} g_r(X_{i_1}, \dots, X_{i_r}) + \\ + \sum_{c=r+1}^{m} \binom{m}{c} n^{-c+r/2} \sum_{i_1=1}^{n} \cdots \sum_{i_c=1}^{n} g_c(X_{i_1}, \dots, X_{i_c}). \tag{7.4.1}$$

The weighted version of w_n is

$$w_n(\bar\varepsilon) = \binom{m}{r} \sum_{i_1=1}^{n} \cdots \sum_{i_r=1}^{n} \varepsilon_{i_1} \cdots \varepsilon_{i_r} g_r(X_{i_1}, \dots, X_{i_r}) +$$

$$+\sum_{c=r+1}^{m}\binom{m}{c}\sum_{i_1=1}^{n}\cdots\sum_{i_{c-r}=1}^{n}\varepsilon_{i_1}^2\cdots\varepsilon_{i_{c-r}}^2$$

$$\times\sum_{i_{c-r+1}=1}^{n}\cdots\sum_{i_c=1}^{n}\varepsilon_{i_{c-r+1}}\cdots\varepsilon_{i_c}\,g_r(X_{i_1},\ldots,X_{i_c}),$$

or in the integral form

$$w_n(\bar\varepsilon)=\sum_{c=r}^{m}\binom{m}{c}V_{nc}(\bar\varepsilon), \tag{7.4.2}$$

where

$$V_{nc}(\bar\varepsilon)=\int_X\cdots\int_X g_c(x_1,\ldots,x_c)\prod_{s=1}^{c-r}P_{\bar\varepsilon_n^2}(dx_s)\prod_{s=c-r+1}^{c}P_{\bar\varepsilon_n}(dx_s).$$

In addition $P_{\bar\varepsilon_n}(A)=\varepsilon_1\delta_{X_1}(A)+\cdots+\varepsilon_n\delta_{X_n}(A)$ is a weighted empirical measure and $\bar\varepsilon_n^2=(\varepsilon_1^2,\ldots,\varepsilon_n^2)$.

Under the condition

$$\max\Big\{E\,g_c^2(X_1,\ldots,X_c)\colon 1\le i_1\le\cdots\le i_c\le c,\ c=r,\ldots,m\Big\}<\infty$$

by virtue of Theorem 4.2.6,

$$V_{k+p,c}(\varepsilon_1,\ldots,\varepsilon_p,k^{-1/2},\ldots,k^{-1/2})\Longrightarrow V_{\infty,c}(\varepsilon_1,\ldots,\varepsilon_p),$$

where

$$V_{\infty,c}(\varepsilon_1,\ldots,\varepsilon_p)=\int_X\cdots\int_X g_c(x_1,\ldots,x_c)\prod_{s=1}^{c-r}P_{\bar\varepsilon_p^2}(dx_s)\prod_{s=c-r+1}^{c}(P_{\bar\varepsilon_p}+Q)(dx_s)$$

and hence

$$w_\infty(\varepsilon_1,\ldots,\varepsilon_p)=\sum_{c=r}^{m}\binom{m}{c}V_{\infty,c}(\varepsilon_1,\ldots,\varepsilon_p). \tag{7.4.3}$$

Now we represent $w_\infty(\varepsilon_1,\ldots,\varepsilon_p)$ in a coordinate form.

Let $\{e_j, j=1,2,\ldots\}$ be an orthonormal basis in $L^2(X,P)$, and let $\{\tau_j, j=1,2,\ldots\}$ be independent standard normal random variables which do not depend on $X_1,\ldots,X_n$.

Under the condition $E\,g_c^2<\infty$, we represent each function $g_c(x_1,\ldots,x_c)$, $c=r,\ldots,m$ as a Fourier series

$$g_c(x_1,\ldots,x_c)=\sum_{(i_1,\ldots,i_c)=1}^{\infty}g_{i_1\ldots i_c}\,e_{i_1}(x_1)\cdots e_{i_c}(x_c)$$

with Fourier coefficients

$$g_{i_1\ldots i_c}=E\Big(g_c(X_1,\ldots,X_c)\,e_{i_1}(X_1)\cdots e_{i_c}(X_c)\Big).$$

Then we get

$$V_n(\bar\varepsilon)=\sum_{i_1=1}^{n}\cdots\sum_{i_r=1}^{n}\varepsilon_{i_1}\cdots\varepsilon_{i_r}\,g_r(X_{i_1},\ldots,X_{i_r})=$$

$$= \sum_{(i_1,\dots,i_r)=1}^{\infty} g_{i_1\dots i_r} \prod_{s=1}^{r} \Big(\sum_{j_s=1}^{n} \varepsilon_{j_s} e_{i_s}(X_{j_s}) \Big).$$

We set

$$S_{n,i_s}(\bar{\varepsilon}) = \sum_{j_s=1}^{n} \varepsilon_{j_s} e_{i_s}(X_{j_s}).$$

By virtue of the one-dimensional central limit theorem, we have

$$S_{k+p,i_s}(\varepsilon_1, \dots, \varepsilon_p, k^{-1/2}, \dots, k^{-1/2}) \Longrightarrow \tau_{i_s} + \varepsilon_1 e_{i_s}(X_1) + \dots + \varepsilon_p e_{i_s}(X_p)$$

as $k \to \infty$. Consequently, if $k \to \infty$, then

$$V_{k+p}(\varepsilon_1, \dots, \varepsilon_p, k^{-1/2}, \dots, k^{-1/2}) \Longrightarrow V_{\infty,r}(\varepsilon_1, \dots, \varepsilon_p),$$

where

$$V_{\infty,r}(\varepsilon_1, \dots, \varepsilon_p) = \sum_{(i_1,\dots,i_r)=1}^{\infty} g_{i_1\dots i_r} \delta_{i_1} \cdots \delta_{i_r},$$

$$\delta_i = \tau_i + \varepsilon_1 e_i(X_1) + \dots + \varepsilon_p e_i(X_p).$$

Similarly, for $c = r+1, \dots, m$, we obtain with the help of a Fourier series

$$\begin{aligned} V_{n,c}(\bar{\varepsilon}) &= \sum_{i_1=1}^{n} \cdots \sum_{i_{c-r}=1}^{n} \varepsilon_{i_1}^2 \cdots \varepsilon_{i_{c-r}}^2 \\ &\quad \times \sum_{i_{c-r+1}=1}^{n} \cdots \sum_{i_c=1}^{n} \varepsilon_{i_{c-r+1}} \cdots \varepsilon_{i_c} g_c(X_{i_1}, \dots, X_{i_c}) \\ &= \sum_{(i_1,\dots,i_c)=1}^{\infty} g_{i_1\dots i_c} \sum_{s=1}^{c-r} \Big(\sum_{j_s=1}^{n} \varepsilon_{j_s}^2 e_{i_s}(X_{j_s}) \Big) \\ &\quad \times \prod_{s=c-r+1}^{c} \Big(\sum_{j_s=1}^{n} \varepsilon_{j_s} e_{i_s}(X_{j_s}) \Big). \end{aligned}$$

Thus,

$$V_{k+p,c}(\varepsilon_1, \dots, \varepsilon_p, k^{-1/2}, \dots, k^{-1/2}) \Longrightarrow V_{\infty,c}(\varepsilon_1, \dots, \varepsilon_p),$$

as $k \to \infty$; here

$$V_{\infty,c}(\varepsilon_1, \dots, \varepsilon_p) = \sum_{(i_1,\dots,i_c)=1}^{\infty} g_{i_1\dots i_c} \gamma_{i_1} \cdots \gamma_{i_{c-r}} \delta_{i_{c-r+1}} \cdots \delta_{i_c},$$

$$\gamma_i = \varepsilon_1^2 e_i(X_1) + \dots + \varepsilon_p^2 e_i(X_p).$$

Combining these relations, we find

$$w_\infty(\varepsilon_1, \dots, \varepsilon_p) = \sum_{c=r}^{m} \binom{m}{c} V_{\infty,c}(\varepsilon_1, \dots, \varepsilon_p),$$

i.e. formula (7.4.1).

In particular,

$$w_\infty(0) = \binom{m}{r} \sum_{(i_1,\dots,i_r)=1}^{\infty} g_{i_1\dots i_r} \tau_{i_1} \cdots \tau_{i_r}.$$

We assume that in formulas (7.3.2) the function $w_\infty(\varepsilon_1, \dots, \varepsilon_p)$ from (7.4.3) is applied.

Theorem 7.4.1. *If kernel $\Phi\colon X^m \to R$ satisfies conditions* (7.3.3) *and* (7.3.4), *then, for $s \geq 3$, the following asymptotic expansion holds*

$$\sup_x \left| P(w_n < x) - F_{n,s-3}(x) \right| = O(n^{-m+r/2+\varepsilon}) + O(n^{-(s-2)/2}).$$

If $s \geq 2m + 2 - r$, then under condition (7.3.5), *we have*

$$\sup_x \left| P(w_n < x) - F_{n,s-3}(x) \right| = O(n^{-(s-2)/2}).$$

Proof. Let us show that in this situation the conditions of Theorem 7.2.2 are satisfied for the weighted characteristic function $\varphi_n(t; \varepsilon_1, \dots, \varepsilon_n) = E \exp(it\, w_n(\bar\varepsilon))$, where $w_n(\bar\varepsilon)$ is defined by (7.4.2). Assume that $h_n = \varphi_n(t) = E \exp(it\, w_n)$, where w_n is given in (7.4.1), and $h_n(\varepsilon_1, \dots, \varepsilon_n) = \varphi_n(t; \varepsilon_1, \dots, \varepsilon_n)$. Clearly, $\varphi_n(t) = \varphi_n(t; n^{-1/2}, \dots, n^{-1/2})$, i.e. the first relation of (7.2.1) holds. The validity of the second relation of (7.2.1) follows from the representations of $V_n(\bar\varepsilon)$ and $V_{n,c}(\bar\varepsilon)$. The third relation (7.2.1) holds due to the property of complete degeneracy of the functions $g_c(x_1, \dots, x_c)$, $c = r, \dots, m$, i.e. due to the fact that $E\, g_c(x_1, \dots, x_{c-1}, X_c) = 0$. The symmetry condition $w_n(\bar\varepsilon)$ follows from the independence of random elements $X_1, \dots, X_n$ and from the fact that they are equally distributed. The remaining part of the proof is the same as in Theorem 7.3.1; one should use the Esséen inequality and then estimate the integrals J_j by employing the symmetrization inequality and the results presented in Götze (1984), Lemma 5.50 (i)–(iv).

Chapter 8

Large deviations

8.1 Exponential inequalities

***UR*-statistics.** At first we consider U-statistics with real-valued kernel Φ of degree m. In *large deviations* theory the following lemma is commonly used and will be needed here in the following.

Lemma 8.1.1. *Let X be a random variable with zero mean and variance $\sigma^2 = EX^2 > 0$. Suppose that for $L > 0$,*

$$EX^r \le \frac{r!}{2} L^{r-2}\sigma^2, \quad r = 2, 3, \ldots .$$

Then for $t > 0$ with $tL < 1$,

$$Ee^{tX} \le e^{\sigma^2 t^2/(2(1-tL))} .$$

Proof. Since

$$e^{tX} = 1 + tX + \sum_{r=2}^{\infty} \frac{t^r}{r!} X^r ,$$

then we have

$$\begin{aligned} Ee^{tX} &= 1 + \sum_{r=2}^{\infty} \frac{t^r}{r!} EX^r \\ &\le 1 + \frac{1}{2} \sum_{r=2}^{\infty} t^r L^{r-2} \sigma^2 \\ &= 1 + \frac{t^2\sigma^2}{2} \sum\nolimits_{r=2}^{\infty} (tL)^{r-2} \\ &= 1 + \frac{t^2\sigma^2}{2(1-tL)} \\ &\le e^{t^2\sigma^2/(2(1-tL))} . \end{aligned}$$

Now we present some probability inequalities with *exponential* bounds for non-degenerate *UR*-statistics.

Theorem 8.1.1. *Let $\Phi : X^m \to R$ be a symmetric kernel with rank $r = 1$ and $\sigma^2 = E(\Phi - \theta)^2$. Suppose that there exists $L > 0$ such that*

$$E(\Phi - \theta)^r \leq \frac{r!}{2} L^{r-2} \sigma^2, \quad r = 2, 3, \dots .$$

Then for $x > 0$

$$P(U_n - \theta > x) \leq \exp(-\frac{k}{2L^2}(\sqrt{2xL + \sigma^2} - \sigma)^2),$$

where $k = [n/m]$.

Proof. For any $t > 0$, by Chebyshev's inequality and according to the Hoeffding representation (1.1.17) we may write

$$\begin{aligned} P(U_n - \theta > x) &= P(e^{t(U_n - \theta)} > e^{tx}) \\ &\leq e^{-tx} E e^{t(U_n - \theta)} \\ &= e^{-tx} E \exp(t E(\varphi_n | \mathfrak{B}_n)) \\ &\leq e^{-tx} E \exp(t \varphi_n(X_1, \dots, X_n)), \end{aligned}$$

where the second inequality is a result of using the conditional Jensen's inequality. Now we have by independence and Lemma 8.1.1

$$\begin{aligned} E \exp(t\varphi_n(X_1, \dots, X_n)) &= E \exp\Big(\frac{t}{k} \sum\nolimits_{i=1}^{k} (\Phi(X_{im-m+1}, \dots, X_{im}) - \theta)\Big) \\ &= (E \exp(\frac{t}{k}(\Phi(X_1, \dots, X_m) - \theta)))^k \\ &= e^{\sigma^2 t^2/(2(k - tL))}, \quad tL < k. \end{aligned}$$

Hence, for $t > 0$ with $tL < k$

$$P(U_n - \theta > x) \leq e^{-f(t)}$$

where $f(t) = tx - \sigma^2 t^2/(2(k - tL))$. This $f(t)$ is maximized at

$$t^* = \frac{k}{L}\Big(1 - \frac{\sigma}{\sqrt{2xL + \sigma^2}}\Big)$$

and its maximum is equal to

$$f(t^*) = \frac{k}{2L^2}\Big(\sqrt{2xL + \sigma^2} - \sigma\Big)^2.$$

Thus

$$P(U_n - \theta > x) \le e^{-f(t^*)}.$$

This is the desired result.

Now we present the same probability inequalities with exponential bounds for degenerate UR-statistics.

Theorem 8.1.2. *Let $\Phi : X^m \to R$ be a symmetric kernel of rank $r = m$ and $\|\Phi\|_\infty \le \gamma$.*

Then for any $x > 0$ there are constants c_i dependent only on m such that

$$P(|n^{m/2}U_n| > x) \le c_1 \exp(-c_2(x/\gamma)^{2/m}).$$

Proof. The following lemma is needed.

Lemma 8.1.2. *Let $\{\varepsilon_i\}$ be a sequence of independent Bernoulli random variables independent of $\{X_i\}$. Denote*

$$X = \sum_{1\le i_1<\ldots<i_m\le n} \varepsilon_{i_1}\cdots\varepsilon_{i_m}\Phi(X_{i_1},\ldots,X_{i_m})$$

and

$$s^2 = \sum_{1\le i_1<\ldots<i_m\le n} \Phi^2(X_{i_1},\ldots,X_{i_m}).$$

Then for all $t > 0$ and $0 < \alpha < 2/m$ there exist constants $c_1(\alpha,m)$, $c_2(\alpha,m) \in (0,\infty)$ such that

$$E_\varepsilon e^{t|X|^\alpha} \le c_1 \exp(c_2(s^\alpha t)^{1/(1-\alpha m/2)}).$$

Proof. It is well known that

$$|ab| \le p^{-1}|a|^p + q^{-1}|b|^q$$

with $p^{-1} + q^{-1} = 1$, $1 < p < \infty$, $a, b \in R$. By this inequality we have

$$t|X|^\alpha \le \frac{\alpha m}{2}|cX/s|^{2/m} + (1 - \frac{\alpha m}{2})(s^\alpha t/c^\alpha)^{1/(1-\alpha m/2)}$$

for any $c > 0$.

Hence,

$$E_\varepsilon e^{t|X|^\alpha} \le \exp\{(1-\frac{\alpha m}{2})(s^\alpha t/c^\alpha)^{1/(1-\alpha m/2)}\} E_\varepsilon \exp\{\frac{\alpha m}{2}|cX/s|^{2/m}\}.$$

Taking c so that $\lambda := \alpha m c^{2/m} < m/(2e)$ and developing $\exp\{\lambda|X/s|^{2/m}\}$ we can write

$$E_\varepsilon \exp\{\lambda|X/s|^{2/m}\} = \sum_{p=0}^\infty \frac{\lambda^p}{p!} s^{-\frac{2p}{m}} E_\varepsilon|X|^{\frac{2p}{m}} =$$

$$= 1 + \sum_{p=1}^{m} \frac{\lambda^p}{p!} s^{-\frac{2p}{m}} E_\varepsilon |X|^{\frac{2p}{m}} + \sum_{p=m+1}^{\infty} \frac{\lambda^p}{p!} s^{-\frac{2p}{m}} E_\varepsilon |X|^{\frac{2p}{m}} \leq$$

[by using Hölder's inequality for $p \leq m$ and Borell's inequality (2.6.5) for $p > m$]

$$\leq 1 + \sum_{p=1}^{m} \frac{\lambda^p}{p!} + \sum_{p=m+1}^{\infty} \frac{\lambda^p}{p!} \left(\frac{2p}{m} - 1\right)^p =: c_1(\alpha, m) < \infty.$$

This proves Lemma 8.1.2.

Further, for any $t > 0$ by Chebyshev's inequality

$$P(|n^{m/2} U_n| > x) \leq$$

$$\leq e^{-tx^{2/(m+1)}} E \exp(t |n^{m/2} U_n|^{2/(m+1)}) \leq$$

$$\text{[since } \binom{n}{m}^{-1} \leq n^{-m} m^m m! \text{]}$$

$$\leq e^{-tx^{2/(m+1)}} E \exp\{(m^m m!)^{2/(m+1)} t |n^{-m/2} \sum_{1 \leq i_1 < \ldots < i_m \leq n} \Phi(X_{i_1}, \ldots, X_{i_m})|^{2/(m+1)}\}.$$

Consider the convex function

$$\Psi(x) = \exp\left(\max\left\{\frac{m-1}{2}, x^{2/(m+1)}\right\}\right), \quad x \geq 0.$$

This function satisfies the inequalities

$$e^{-\frac{m-1}{2}} \Psi(x) \leq e^{x^{2/(m+1)}} \leq \Psi(x)$$

for all $x > 0$.

Now we apply decoupling Theorem 2.5.4 and symmetrization Lemma 2.4.5 with this function $\Psi(x)$. Then by Lemma 8.1.2 with $\alpha = 2/(m+1)$ we get

$$E \exp\{(m^m m!)^{2/(m+1)} t |n^{-m/2} \sum\nolimits_{1 \leq i_1 < \ldots < i_m \leq n} \Phi(X_{i_1}, \ldots, X_{i_m})|^{2/(m+1)}\} \leq$$

$$\leq c_1 E \exp\{c_2 t^{m+1} n^{-m} \sum\nolimits_{1 \leq i_1 < \ldots < i_m \leq n} \Phi^2(X_{i_1}, \ldots, X_{i_m})\}.$$

Hence,

$$P(|n^{m/2} U_n| > x) \leq$$

$$\leq c_1 E \exp\{-tx^{2/(m+1)} + c_2 t^{m+1} n^{-m} \sum\nolimits_{1 \leq i_1 < \ldots < i_m \leq n} \Phi^2(X_{i_1}, \ldots, X_{i_m})\}. \qquad (8.1.1)$$

Since $\|\Phi\|_\infty \leq \gamma$ then from (8.1.1) we have

$$P(|n^{m/2} U_n| > x) \leq c_1 \exp(-tx^{2/(m+1)} + c_2 \gamma^2 t^{m+1}).$$

The function $f(t) = -tx^{2/(m+1)} + c_2 \gamma^2 t^{m+1}$ is minimized at

$$t^* = (x^{2/(m+1)} / (c_2 (m+1) \gamma^2))^{1/m}$$

and

$$f(t^*) = -\frac{m}{m+1}(c_2(m+1))^{-1/m}(x/\gamma)^{2/m}.$$

This completes the proof of Theorem 8.1.2.

UH-statistics.

Theorem 8.1.3. *Assume that kernel $\Phi : X^m \to H$ has rank $r = m$ and satisfies the condition*

$$E\exp\{t_0\|\Phi(X_1,\dots,X_m)\|^2\} < \infty$$

for some $t_0 > 0$.

Then for any $x > 0$ there are constants c_i dependent only on m such that

$$P(\|n^{m/2}U_n\| > x) \le c_1 \inf_{t>0}\{e^{-tx^{2/(m+1)}} E\exp(c_2 t^{m+1}\|\Phi\|^2)\}.$$

Proof. For any $t > 0$ by Chebyshev's inequality

$$P(\|n^{m/2}U_n\| > x) \le e^{-tx^{2/(m+1)}} E\exp(t\|n^{m/2}U_n\|^{2/(m+1)}) \le$$

[because of inequality (8.1.1)]

$$\le c_1 E\exp\{-tx^{2/(m+1)} + c_2 t^{m+1} n^{-m} \textstyle\sum_{1\le i_1<\dots<i_m\le n} \|\Phi(X_{i_1},\dots,X_{i_m})\|^2\} \le$$

[by formula (1.1.15)]

$$\le c_1 E\exp\{-tx^{2/(m+1)} + c_2 t^{m+1} E(\|\Phi\|^2|\mathfrak{B}_n)\} \le$$

[by the conditional Jensen's inequality]

$$\le c_1 e^{-tx^{2/(m+1)}} E\exp(c_2 t^{m+1}\|\Phi\|^2).$$

From here we obtain the result.

Corollary 8.1.1. *If $\|\Phi\| \le \gamma$, then*

$$P(\|n^{m/2}U_n\| > x) \le c_1\exp(-c_2(x/\gamma)^{2/m})$$

for any $x > 0$.

UB-statistics.

Theorem 8.1.4. *Suppose that symmetric kernel $\Phi : X^m \to B$ has rank $r = 1$ and $E\Phi(X_1,\dots,X_m) = 0$. Assume also that the following condition is satisfied*

$$E\exp(t_0\|\Phi(X_1,\dots,X_m)\|) < \infty$$

for some $t_0 > 0$.

Then for any $x > 0$

$$P(\|U_n\| > x) \le$$

$$\le \inf_{t>0}\left\{\exp(-tx + \frac{t}{k}E\|S_k(\Phi)\|)(\frac{1}{2}E\exp(\frac{2t}{k}\|\Phi\|) + \frac{1}{2}E\exp(-\frac{2t}{k}\|\Phi\|))^k\right\},$$

where $k = [n/m]$,

$$S_k(\Phi) = \sum_{i=1}^{k}\Phi(X_{im-m+1},\dots,X_{im}).$$

Proof. According to the Hoeffding formula (1.1.17) and the conditional Jensen's inequality

$$P(\|U_n\| > x) \le e^{-tx}E\exp(\frac{t}{k}\|S_k(\Phi)\|) \tag{8.1.2}$$

for all $t > 0$.

By majorization Theorem 2.1.3

$$E\exp(\frac{t}{k}\|S_k(\Phi)\|) \le$$

$$\le \exp(\frac{t}{k}E\|S_k(\Phi)\|)E\exp(\frac{2t}{k}\sum_{i=1}^{k}\varepsilon_i\|\Phi(X_{im-m+1},\dots,X_{im})\|) =$$

$$= \exp(\frac{t}{k}E\|S_k(\Phi)\|)(\frac{1}{2}E\exp(\frac{2t}{k}\|\Phi\|) + \frac{1}{2}E\exp(-\frac{2t}{k}\|\Phi\|))^k.$$

This proves Theorem 8.1.4.

Corollary 8.1.2. *If* $E\exp(t_0\|\Phi(X_1,\dots,X_m)\|^2) < \infty$ *for some* $t_0 > 0$, *then for any* $x > 0$

$$P(\|U_n\| > x) \le$$

$$\le \inf_{t>0}\{\exp(-tx + \frac{t}{k}E\|S_k(\Phi)\|)(E\exp(\frac{2t^2}{k^2}\|\Phi\|^2))^k\}.$$

Proof. It is sufficient to note that

$$\frac{1}{2}(\exp(\frac{2t}{k}\|\Phi\|) + \exp(-\frac{2t}{k}\|\Phi\|)) \le \exp(\frac{2t^2}{k^2}\|\Phi\|^2).$$

Corollary 8.1.3. *Let* $\sigma^2 = E\|\Phi(X_1,\dots,X_m)\|^2$ *and*

$$E\|\Phi\|^r \le \frac{r!}{2}L^{r-2}\sigma^2, \quad r = 2,3,\dots$$

and $E\|S_k(\Phi)\| \le \beta$.

Then for any $x > \beta/(\sigma\sqrt{k})$

$$P(\sqrt{k}\|U_n\| > \sigma x) \le \exp\left\{-\frac{x^2}{8}\frac{(1-\frac{\beta}{x\sigma\sqrt{k}})^2}{\left(1+\frac{(x-\beta/(\sigma\sqrt{k}))L}{2\sigma\sqrt{k}}\right)}\right\}.$$

Proof. This estimate follows from (8.1.2) and Yurinskii's Theorem (1976).

Corollary 8.1.4. *If $\|\Phi\| \le \gamma$ and $E\|S_k(\Phi)\| \le \beta$, then for any $x > \beta/\sqrt{k}$,*

$$P(\sqrt{k}\|U_n\| > x) \le \exp\{-\frac{1}{8\gamma^2}(x-\frac{\beta}{\sqrt{k}})^2\}.$$

Proof. From Corollary 8.1.2 we have

$$P(\sqrt{k}\|U_n\| > x) \le \inf_{t>0}\{\exp(-t\frac{x}{\sqrt{k}} + \frac{t\beta}{k} + \frac{2t^2\gamma^2}{k})\}.$$

The result now follows by minimizing the exponent with respect to t.

Theorem 8.1.5. *Suppose that kernel $\Phi : X^m \to B$ has rank $r = m$. Then for any $x > 0$ there are constants c_i such that*

$$P(\|n^{m/2}U_n\| > x) \le c_1\inf_{t>0}\{e^{-tx^{2/(m+1)}}E\exp(c_2t^{m+1}n^{-m}s^2)\},$$

where

$$s^2 = E_\varepsilon\|\sum\nolimits_{1\le i_1<\ldots<i_m\le n}\varepsilon_{i_1}\cdots\varepsilon_{i_m}\Phi(X_{i_1},\ldots,X_{i_m})\|^2.$$

Proof. It is sufficient to apply the method of the proof of Theorem 8.1.2.

Consider now some cases when s^2 admits some useful estimates.

Corollary 8.1.5. *Let B be of type 2. Then for any $x > 0$*

$$P(\|n^{m/2}U_n\| > x) \le c_1\inf_{t>0}\{e^{-tx^{2/(m+1)}}E\exp(c_2t^{m+1}\|\Phi\|^2)\}.$$

Proof. By Corollary 2.5.2

$$s^2 \le m^{2m}E_\varepsilon\|\sum\nolimits_{1\le i_1<\ldots<i_m\le n}\varepsilon^1_{i_1}\cdots\varepsilon^m_{i_m}\Phi(X_{i_1},\ldots,X_{i_m})\|^2 \le$$

[because of inequality (2.7.2)]

$$\le m^{2m}c^m(B)\sum\nolimits_{1\le i_1<\ldots<i_m\le n}\|\Phi(X_{i_1},\ldots,X_{i_m})\|^2 =$$

[by formula (1.1.15)]

$$= m^{2m} c^m(B) \binom{n}{m} E(\|\Phi\|^2 | \mathfrak{B}_n).$$

From Theorem 8.1.5 with the help of the conditional Jensen's inequality we obtain Corollary 8.1.5.

Corollary 8.1.6. *Let B be of type 2 and $\|\Phi\| \le \gamma$. Then for any $x > 0$*

$$P(\|n^{m/2} U_n\| > x) \le c_1 \exp(-c_2 (x/\gamma)^{2/m}),$$

where the constants $c_i > 0$ depend only on m and $c(B)$ in (2.7.2).

Proof. This inequality follows from Corollary 8.1.5.

Theorem 8.1.6. *Let kernel $\Phi : X^m \to B$ have rank r, $1 \le r \le m$. Then for any $x > 0$ there are constants c_i dependent only on r and m such that*

$$P(\|n^{r/2} U_n\| > x) \le$$

$$\le c_1 \inf_{t>0} \{ e^{-tx^{2/(r+1)}} E \exp(c_2 t^{r+1} n^r \textstyle\sum_{c=r}^m n^{-2c} s_c^2) \},$$

where

$$s_c^2 = E_\varepsilon \|\textstyle\sum_{1 \le i_1 < \ldots < i_c \le n} \varepsilon_{i_1} \cdots \varepsilon_{i_c} g_c(X_{i_1}^c, \ldots, X_{i_c}^c)\|^2 .$$

In addition, $\{X_1^c, \ldots, X_n^c\}_{c=r}^m$ are independent copies of a sequence $X_1, \ldots, X_n$.

Proof. For any $t > 0$ by Chebyshev's inequality we have

$$P(\|n^{r/2} U_n\| > x) \le e^{-tx^{2/(r+1)}} E \exp(t \|n^{r/2} U_n\|^{2/(r+1)}).$$

Because of the Hoeffding representation (1.1.12) and inequality (2.2.3),

$$E \exp\{t \|n^{r/2} U_n\|^{2/(r+1)}\} \le$$

$$\le E \prod_{c=r}^m \exp\left\{ t \binom{m}{c}^{\frac{2}{r+1}} \binom{n}{c}^{-\frac{2}{r+1}} n^{\frac{r}{r+1}} \|S_n(g_c)\|^{\frac{2}{r+1}} \right\} \le$$

[here $S_n(g_c) = \sum_{1 \le i_1 < \ldots < i_c \le n} g_c(X_{i_1}, \ldots, X_{i_c})$ and then by Hölder's inequality with $q_c = m - c + 1$]

$$\le \prod_{c=r}^m \left(E \exp\left\{ t q_c \binom{m}{c}^{\frac{2}{r+1}} \binom{n}{c}^{-\frac{2}{r+1}} n^{\frac{r}{r+1}} \|S_n(g_c)\|^{\frac{2}{r+1}} \right\} \right)^{1/q_c} \le$$

[using Lemma 8.1.2 and then the inequalities $\exp(x) \geq 1$ for $x \geq 0$ and $q_c \geq 1$ for $c = r, \dots, m$]

$$\leq c_1\, E \exp(c_2 t^{r+1} n^r \sum\nolimits_{c=r}^{m} n^{-2c} s_c^2).$$

This proves Theorem 8.1.6.

Corollary 8.1.7. *Let B be of type 2. Suppose that kernel $\Phi : X^m \to B$ has rank r, $1 \leq r \leq m$, and the canonical functions g_c satisfy the conditions $\|g_c\| \leq \gamma_c$, $c = r, \dots, m$.*

Then for any $x > 0$

$$P(\|n^{r/2} U_n\| > x) \leq c_1 \exp(-c_2 (x/\gamma)^{2/r}),$$

where c_i are some constants and

$$\gamma^2 = \sum\nolimits_{c=r}^{m} n^{-c+r} \gamma_c^2.$$

Proof. See the proof of Corollary 8.1.5.

8.2 Moderate deviations

Consider UR-statistics with non-degenerate kernel $\Phi : X^m \to R$. Denote

$$q_n(\gamma) = P\Big(\frac{\sqrt{n}}{m\sigma_1}(U_n - \theta) > \gamma (\ln n)^{1/2}\Big)$$

for some constant $\gamma > 0$. The probability $q_n(\gamma)$ of a deviation of order $\sqrt{\ln n}$ is called moderate. By the central limit theorem, $q_n(\gamma) \to 0$ as $n \to \infty$, and by Chebyshev's inequality, $q_n(\gamma) = O((\ln n)^{-1})$. This estimate can be made more precise if we impose additional conditions on the canonical functions g_c, $c = 1, \dots, m$.

Theorem 8.2.1. *Suppose that non-degenerate kernel $\Phi : X^m \to R$ satisfies the following conditions*

$$E|g_1|^p < \infty, \quad p > 2 + \gamma^2$$

and

$$E|g_c|^{\gamma_c + \gamma^2} < \infty, \quad c = 2, \dots, m$$

with $\gamma_c = 2c/(2c-1)$ and some constant $\gamma > 0$.

Then

$$P\Big(\frac{\sqrt{n}}{m\sigma_1}(U_n - \theta) > x\Big) = (1 - F(x))(1 + o((\ln n)^{-1}))$$

uniformly in x in the region $-a \leq x \leq \gamma\sqrt{\ln n}$, where $a \geq 0$.

Proof. According to the Hoeffding representation (1.1.12), we have

$$\frac{\sqrt{n}}{m\sigma_1}(U_n - \theta) = \frac{1}{\sigma_1\sqrt{n}}\sum\nolimits_{i=1}^{n} g_1(X_i) + R_n,$$

where

$$R_n = \frac{1}{m\sigma_1}\sum\nolimits_{c=2}^{m} m^{[c]}\prod_{j=1}^{c-1}(1-n^{-1}j)^{-1}n^{-c/\gamma_c}S_n(g_c)\,.$$

Taking into account the preceding equality, we get

$$P\Big(\frac{1}{\sigma_1\sqrt{n}}\sum\nolimits_{i=1}^{n} g_1(X_i) > x+\varepsilon\Big) - P(|R_n| > \varepsilon) \le$$

$$\le P\Big(\frac{\sqrt{n}}{m\sigma_1}(U_n-\theta) > x\Big) \le$$

$$\le P\Big(\frac{1}{\sigma_1\sqrt{n}}\sum\nolimits_{i=1}^{n} g_1(X_i) > x-\varepsilon\Big) + P(|R_n| > \varepsilon) \qquad (8.2.1)$$

for any $\varepsilon > 0$. Choose $\varepsilon = (\ln n)^{-2}$ in (8.2.1). It is obvious that

$$P(|R_n| > (\ln n)^{-2}) \le \sum\nolimits_{c=2}^{m} P\Big(\sigma_1^{-1} m^{[c]} \prod_{j=1}^{c-1}\big(1-n^{-1}j\big)^{-1}|n^{-c/\gamma_c}S_n(g_c)| > (\ln n)^{-2}\Big)\,.$$

Here, by Corollary 2.7.1 for all $c = 2,\dots,m$ we have

$$P\Big(|n^{-c/\gamma_c}S_n(g_c)| > (\ln n)^{-2}\Big) \le \begin{cases} c_1 E|g_c|^{\gamma_c}(\ln n)^{2(\gamma_c+\gamma^2)}n^{-\frac{(2c-1)}{2}\gamma^2}, & \gamma_c+\gamma^2 \le 2, \\ c_2 E|g_c|^{\gamma_c}(\ln n)^{2(\gamma_c+\gamma^2)}n^{-\frac{c-1}{2}(\gamma_c+\gamma^2)}, & \gamma_c+\gamma^2 \ge 2\,. \end{cases}$$

Further, by the property of normal distribution function $F(x)$, we have

$$P(|R_n| > (\ln n)^{-2}) = (1-F(x))o((\ln n)^{-1}),$$

$$1 - F(x \pm (\ln n)^{-2}) = (1-F(x))(1+o((\ln n)^{-1}))$$

uniformly in x in the region $-a \le x \le \gamma\sqrt{\ln n}$.

Hence, it is sufficient to show that

$$P\Big(\frac{1}{\sigma_1\sqrt{n}}\sum\nolimits_{i=1}^{n} g_1(X_i) > x\Big) = (1-F(x))(1+o((\ln n)^{-1}))\,.$$

Since $E|g_1|^p < \infty$ for $p > 2+\gamma^2$, by virtue of the generalized Essén inequality, we get

$$\sup_x\Big|P\Big(\frac{1}{\sigma_1\sqrt{n}}\sum\nolimits_{i=1}^{n} g_1(X_i) > x\Big) - (1-F(x))\Big| = O(n^{-\delta/2}),$$

where $0 < \delta < 1$. Thus, for $0 < \gamma_0 < \sqrt{\delta}$,

$$P\Big(\frac{1}{\sigma_1\sqrt{n}}\sum_{i=1}^{n} g_1(X_i) > x\Big) = (1 - F(x))(1 + o((\ln n)^{-1}))$$

uniformly in x in the region $-a \leq x \leq \gamma_0\sqrt{\ln n}$. By Vandemaele and Veraverbeke (1982),

$$P\Big(\frac{1}{\sigma_1\sqrt{n}}\sum_{i=1}^{n} g_1(X_i) > x\Big) = (1 - F(x))(1 + o(x^{-2}))$$

in the region $b \leq x \leq \gamma\sqrt{\ln n}$, where $b > 0$. This implies that

$$P\Big(\frac{1}{\sigma_1\sqrt{n}}\sum_{i=1}^{n} g_1(X_i) > x\Big) = (1 - F(x))(1 + o((\ln n)^{-1}))$$

for $\gamma_0\sqrt{\ln n} \leq x \leq \gamma\sqrt{\ln n}$. This completes the proof of Theorem 8.2.1.

Corollary 8.2.1. *Under the conditions of Theorem* 8.2.1

$$q_n(\gamma) = (2\pi\gamma^2 \ln n)^{-1/2} n^{-\gamma^2/2}(1 + O((\ln n)^{-1})) \quad \text{as} \quad n \to \infty.$$

Proof. It suffices to note that

$$1 - F(x) = (x\sqrt{2\pi})^{-1}\exp(-x^2/2)(1 + O(x^{-2})) \quad \text{as} \quad x \to \infty.$$

8.3 Power zones of normal convergence

Consider UR-statistics with kernel $\Phi : X^2 \to R$.

Theorem 8.3.1. *Suppose that a non-degenerate kernel Φ satisfies the conditions*

$$E|\Phi(X_1, X_2)|^p \leq c^p p^{\gamma p}, \quad p = 1, 2, \ldots$$

with some constants $c > 0$ and $\gamma \geq 0$. Then

$$P\Big(\frac{\sqrt{n}}{2\sigma_1}(U_n - \theta) > x\Big) = (1 - F(x))(1 + o(1))$$

uniformly in x in the region $-a \leq x \leq o(n^{\alpha})$, where $a \geq 0$ and $\alpha = (2(3 + 2\gamma))^{-1}$.

Proof. We use (8.2.1) for $m = 2$. Let $\varepsilon = n^{-\alpha}$ in (8.2.1). By Corollary 2.7.1 under the conditions of this theorem we have

$$E|R_n|^p \leq c_1^p n^{-p/2} p^{(1+\gamma)p}$$

for all $p \geq 2$. According to Chebyshev's inequality for $n \geq 2$ and $p \geq 2$, we get

$$P(|R_n| > \varepsilon) \le c_2^p n^{\alpha p} E|R_n|^p.$$

By choosing here $p = cn^{(1-2\alpha)/(2+2\gamma)}$ with some constant $c > 0$, we find

$$P(|R_n| > \varepsilon) = O(e^{-cn^{(1-2\alpha)/(2+2\gamma)}}).$$

Because of the properties of function $F(x)$ this relation can be represented as

$$P(|R_n| > \varepsilon) = (1 - F(x))\, o(1)$$

uniformly in x in the region $-a \le x \le o(n^\alpha)$. Further, under the conditions of the theorem

$$P\Big(\frac{1}{\sigma_1\sqrt{n}}\sum_{i=1}^n g_1(X_1) > x\Big) = (1 - F(x))(1 + o(1))$$

for x from the same region. This completes the proof of Theorem 8.3.1.

If $x = O(\sqrt{n})$, then the method of the proof presented above allows us to find the asymptotics of the considered probability if we demand the stronger Cramer condition

$$E\exp(t_0|\Phi(X_1, X_2)|) < \infty$$

for some $t_0 > 0$. Also we assume that there exists the finite limit

$$\lim_{n\to\infty} n^{-1} P\Big(\sum_{i=1}^n g_1(X_i) > t\sigma_1 n\Big) = \rho(t)$$

for fixed $t > 0$. The properties of function $\rho(t)$ have been well investigated in the theory of large deviations of sums of identically distributed random variables. For example, under the Cramer condition

$$\rho(t) = -\frac{t^2}{2\sigma_1^2}(1 + o(1)), \quad \text{as } t \to 0.$$

Theorem 8.3.2. *Suppose that non-degenerate kernel Φ satisfies the Cramer condition and there exists the limit function $\rho(t)$. Then for $t > 0$ the following relation holds*

$$\lim_{n\to\infty} n^{-1} P(U_n - \theta > 2t\sigma_1) = \rho(t)\,.$$

Proof. It suffices to use again the method of the proof of Theorem 8.2.1.

8.4 Probabilities of large deviations for *UH*-statistics

Now we consider one-sample *UH*-statistics of degree 2 with non-degenerate kernel Φ. If $E\|g_1\|^2 < \infty$ and $E\|g_2\|^{4/3} < \infty$, then, by virtue of the central limit theorem for *UH*-statistics, we have

$$P(\|2^{-1}\sqrt{n}(U_n - \theta)\| > x) \,/\, P(\|\tau\| > x) - 1 \to 0,$$

as $n \to \infty$, for a fixed arbitrary $x > 0$. The left-hand side of this relation is treated as the relative error of the Gaussian approximation of *UH*-statistics. In connection with this relation, the problem of estimating the following variable appears,

$$\Delta_n(H;x) := \left|P\Big(\|2^{-1}\sqrt{n}(U_n - \theta)\| > x\Big) \,/\, P(\|\tau\| > x) - 1\right|,$$

as $n \to \infty$ in the region $x = O(n^\beta)$ with some $0 < \beta \le 1$.

In what follows, by virtue of the Hoeffding decomposition (1.1.12), we have

$$U_n - \theta = \frac{2}{n}\sum\nolimits_{j=1}^{n} g_1(X_j) + \binom{n}{2}^{-1} \sum\nolimits_{1\le i<j\le n} g_2(X_i, X_j),$$

$$g_1(x) = E(\Phi(X_1, X_2) - \theta | X_1 = x), \quad g_2(x,y) = \Phi(x,y) - \theta - g_1(x) - g_1(y).$$

Theorem 8.4.1. *Assume that* $E\|g_1(X_1)\|^2 \ne 0$, *the Cramer condition*

$$E\exp(t\|g_1(X_1)\|^{1/2}) < \infty$$

is satisfied for some $t > 0$, *and*

$$E\|g_2(X_1, X_2)\|^k \le c^k k^{\gamma k},$$

for all integers $k = 1, 2, \ldots$, *with some constants* $c > 0$ *and* $\gamma \ge 0$ *independent of* k. *Then the estimate*

$$\Delta_n(H;x) \le A(1 + x)n^{-1/2(3+2\gamma)} \tag{8.4.1}$$

holds in the region $0 \le x \le \delta n^{1/2(3+2\gamma)}$, *where constant* $A > 0$ *does not depend on* n *and* r, *and* $\delta^2 \le \min\{2, (4ce^2)^{-1}\}^{-1/(1+\gamma)}$.

Proof. We split the proof of (8.4.1) into several lemmas. The key point of our presentation is to derive the estimates for the expectations of H-valued random variables with martingale properties.

First, note that, for (8.4.1) to be valid, it is sufficient to establish the inequality

$$|P(\|2^{-1}\sqrt{n}(U_n - \theta)\| > x) - P(\|\tau\| > x)| \le$$

$$\leq A(1+x)n^{-1/2(3+2\gamma)}P(\|\tau\| > x). \tag{8.4.2}$$

In the region $0 \leq x \leq 1$, by Theorem 6.6.2 and the inequalities

$$P(\|\tau\| \geq x) \geq P(\|\tau\| \geq 1) \neq 0,$$

one can write

$$|P(\|2^{-1}\sqrt{n}(U_n - \theta)\| > x) - P(\|\tau\| > x)| \leq c(V)n^{-1/2} =$$

$$= An^{-1/2}P(\|\tau\| \geq 1) \leq A(1+x)n^{-1/2(3+2\gamma)}P(\|\tau\| > x),$$

where $A = c(V)/P(\|\tau\| \geq 1)$.

Thus, when proving (8.4.2), it is sufficient to consider the region $1 \leq x \leq \delta n^{1/2(3+2\gamma)}$.

Lemma 8.4.1. *Let $x \geq 1$ and $\epsilon > 0$. Then, for any H-valued random elements X and Y, the inequalities*

$$P(\|X\| > x+\epsilon) - P(\|Y\| > \epsilon) \leq P(\|X+Y\| > x)$$

$$\leq P(\|X\| > x-\epsilon) + P(\|Y\| > \epsilon) \tag{8.4.3}$$

hold.

Proof. For the inequality on the right in (8.4.3), we can write

$$\begin{aligned} P(\|X+Y\| > x) &\leq P(\|X\| + \|Y\| > x) = \\ &= EI(\|X\| + \|Y\| > x)I(\|Y\| \leq \epsilon) + EI(\|X\| + \|Y\| > x)I(\|Y\| > \epsilon) \\ &\leq EI(\|X\| + \epsilon > x) + EI(\|Y\| > \epsilon) \\ &= P(\|X\| > x - \epsilon) + P(\|Y\| > \epsilon). \end{aligned}$$

For the first inequality in (8.4.3) we have

$$\begin{aligned} P(\|X+Y\| > x) &= P(\|X+Y\|^2 > x^2) \\ &= P(\|X\|^2 + \|Y\|^2 + 2(X,Y) > x^2) \\ &= P((\|X\| - \|Y\|)^2 + 2(\|X\|\|Y\| + (X,Y)) > x^2). \end{aligned}$$

Since $\|X\|\|Y\| + (X,Y) \geq 0$, we get

$$\begin{aligned} P(\|X+Y\| > x) &\geq P(|\|X\| - \|Y\|| > x) \\ &= P(\{\|X\| - \|Y\| < -x\} \cup \{\|X\| - \|Y\| > x\}) \geq \end{aligned}$$

$$
\begin{aligned}
&\geq P(\|X\| > x + \|Y\|) = EI(\|X\| > x + \|Y\|) \\
&= EI(\|X\| > x + \|Y\|)I(\|Y\| \leq \epsilon) \\
&\quad + EI(\|X\| > x + \|Y\|)I(\|Y\| > \epsilon) \\
&\geq EI(\|X\| > x + \|Y\|)I(\|Y\| \leq \epsilon) \\
&\geq P(\|X\| > x + \epsilon) - P(\|Y\| > \epsilon),
\end{aligned}
$$

and this proves (8.4.3).

In what follows, we shall apply Lemma 8.4.1 with $\epsilon = n^{-1/2(3+2\gamma)}$ and

$$X = \frac{1}{\sqrt{n}} \sum\nolimits_{j=1}^{n} g_1(X_j),$$

$$Y = 2^{-1}\sqrt{n}U_{n2} = \frac{1}{(n-1)\sqrt{n}} \sum\nolimits_{1 \leq i < j \leq n} g_2(X_i, X_j).$$

Lemma 8.4.2. *Assume that an H-valued stochastic sequence $\xi_1, \dots, \xi_n$ is a martingale difference, i.e.*

$$E(\xi_j | \xi_1, \dots, \xi_{j-1}) = 0, \quad 1 \leq j \leq n.$$

Then, for all integers k such that $1 \leq k \leq n$, the inequality

$$E\|S_n\|^{2k} \leq (8ek)^k n^k m_{n,k} \tag{8.4.4}$$

holds where

$$S_n = \xi_1 + \dots + \xi_n \quad \textit{and} \quad m_{n,k} = \max\nolimits_{1 \leq j \leq n} E\|\xi_j\|^{2k}.$$

Proof. Let us apply the method used in Bickel (1974). We proceed by induction in n for every fixed k such that $k \leq n$. First, we assume that $n = k$. Then

$$
\begin{aligned}
E\|S_k\|^{2k} &= E(S_k, S_k)^k \\
&= E(\xi_1 + \dots + \xi_k, \xi_1 + \dots + \xi_k)^k \\
&= E\left(\sum\nolimits_{i=1}^{k} \sum\nolimits_{j=1}^{k} (\xi_i, \xi_j)\right)^k \leq E\left(\sum\nolimits_{j=1}^{k} \|\xi_j\|\right)^{2k} \\
&\leq k^{2k-1} \sum\nolimits_{j=1}^{k} E\|\xi_k\|^{2k} \leq k^{2k} m_{k,k}.
\end{aligned}
$$

Hence, (8.4.4) holds for $n = k$. Further, let us show that (8.4.4) is true for $n = l \geq k$, as well. For this purpose, we write

$$\begin{aligned} E\|S_{l+1}\|^{2k} &= E\|S_l + \xi_{l+1}\|^{2k} \\ &= E(S_l + \xi_{l+1}, S_l + \xi_{l+1})^k \\ &= E\left(\|S_l\|^2 + \|\xi_{l+1}\|^2 + 2(S_l, \xi_{l+1})\right)^k \\ &= \sum_{j=0}^{k} \binom{k}{j} E(\|S_l\|^{2(k-j)}(2(S_l, \xi_{l+1}) + \|\xi_{l+1}\|^2)^j) \\ &= E\|S_l\|^{2k} + \binom{k}{1} E(\|S_l\|^{2(k-1)}\|\xi_{l+1}\|^2) \\ &\quad + \sum_{j=2}^{k} \binom{k}{j} E(\|S_l\|^{2(k-j)}(2(S_l, \xi_{l+1}) + \|\xi_{l+1}\|^2)^j). \end{aligned}$$

Here, when writing the last equality, we have used the fact that

$$E(\|S_l\|^{2(k-1)}(S_l, \xi_{l+1})) = 0,$$

by virtue of the martingale property. In addition, we have

$$|2(S_l, \xi_{l+1}) + \|\xi_{l+1}\|^2|^j \leq 2^{j-1}(2^j|(S_l, \xi_{l+1})|^j + \|\xi_{l+1}\|^{2j}) \leq$$

$$\leq 2^{2j-1}\|S_l\|^j\|\xi_{l+1}\|^j + 2^{j-1}\|\xi_{l+1}\|^{2j}.$$

Thus,

$$E\|S_{l+1}\|^{2k} \leq E\|S_l\|^{2k} + \sum_{j=2}^{k} \binom{k}{j} 2^{2j-1} E(\|S_l\|^{2k-j}\|\xi_{l+1}\|^j)$$

$$+ \sum_{j=1}^{k} \binom{k}{j} 2^{j-1} E(\|S_l\|^{2(k-j)}\|\xi_{l+1}\|^{2j}).$$

By the Hölder inequality, we obtain

$$\begin{aligned} E(\|S_l\|^{2k-j}\|\xi_{l+1}\|^j) &\leq (E\|S_l\|^{2k})^{1-j/2k}(E\|\xi_{l+1}\|^2 k)^{j/2k} \\ &\leq (d_k l^k m_{l,k})^{1-j/2k}(m_{l+1,k})^{j/2k} \\ &= (d_k l^k m_{l+1,k})(d_k^{1/2k} l^{1/2})^{-j}, \end{aligned}$$

$$\begin{aligned} E(\|S_l\|^{2(k-j)}\|\xi_{l+1}\|^{2j}) &\leq (E\|S_l\|^{2k})^{1-j/k}(E\|\xi_{l+1}\|^{2k})^{j/k} \\ &\leq (d_k l^k m_{l,k})^{1-j/k}(m_{l+1,k})^{j/k} \\ &\leq (d_k l^k m_{l+1,k})^{1-j/k}(m_{l+1,k})^{j/k} = \end{aligned}$$

$$= d_k l^k m_{l+1,k}(d_k^{1/k} l)^{-j},$$

where $d_k = (8ek)^k$.

If we take into account these estimates and employ the induction hypothesis that (8.4.4) holds for $n = l \geq k$, then we get

$$E\|S_{l+1}\|^{2k} \leq d_k l^k m_{l+1,k}(1 + a_k + b_k),$$
$$a_k = \sum\nolimits_{j=2}^{k} \binom{k}{j} 2^{2j-1}(d_k^{1/2k} l^{1/2})^{-j},$$
$$b_k = \sum\nolimits_{j=1}^{k} \binom{k}{j} 2^{j-1}(d_k^{1/k} l)^{-j}.$$

Since

$$\binom{k}{j+2} \leq \frac{k(k-1)}{2}\binom{k-2}{j},$$

for all $0 \leq j \leq k-2$, we have the following estimates

$$\begin{aligned} a_k &= 2^3(d_k^{1/k} l)^{-1}\sum\nolimits_{j=0}^{k-2}\binom{k}{j+2} 2^{2j}(d_k^{1/2k} l^{1/2})^{-j} \\ &\leq 4k(k-1)(d_k^{1/k} l)^{-1}\sum\nolimits_{j=0}^{k-2}\binom{k-2}{j}(4^{-1} d_k^{1/2k} l^{1/2})^{-j} \\ &= 4k(k-1)(d_k^{1/k} l)^{-1}(1 + 4(d_k^{1/2k} l^{1/2})^{-1})^{k-2} \\ &= \frac{k-1}{2el}(1 + 2(2ekl)^{-1/2})^{k-2} \leq \frac{k-1}{2el}\left(1 + \frac{\sqrt{2}}{k\sqrt{e}}\right)^k. \end{aligned}$$

Owing to the fact that

$$\binom{k}{j+1} \leq k\binom{k-1}{j},$$

for all $0 \leq j \leq k-1$, we have

$$\begin{aligned} b_k &= (d_k^{1/k} l)^{-1}\sum\nolimits_{j=0}^{k-1}\binom{k}{j+2} 2^j (d_k^{1/k} l)^{-j} \\ &\leq k(d_k^{1/k} l)^{-1}\sum\nolimits_{j=0}^{k-1}\binom{k-1}{j}(2^{-1} d_k^{1/k} l)^{-j} \\ &= k(d_k^{1/k} l)^{-1}(1 + 2(d_k^{1/k} l)^{-1})^{k-1} \\ &= \frac{1}{8el}\left(1 + \frac{1}{4ekl}\right)^{k-1} \leq \frac{1}{8el}\left(1 + \frac{\sqrt{2}}{k\sqrt{e}}\right)^k. \end{aligned}$$

Therefore, we obtain

$$E\|S_{l+1}\|^{2k} \le d_k l^k m_{l+1,k} \left\{1 + \frac{k}{2el}\left(1 + \frac{\sqrt{2}}{k\sqrt{e}}\right)^k - \frac{3}{8el}\left(1 + \frac{\sqrt{2}}{k\sqrt{e}}\right)^k\right\} \le$$

$$\le d_k l^k m_{l+1,k}\left(1 + \frac{k}{el}\left(1 + \frac{\sqrt{2}}{k\sqrt{e}}\right)^k\right).$$

Here on the right-hand side, we have

$$\left(1 + \frac{1}{k}\left(\frac{2}{e}\right)^{1/2}\right)^k \le \exp\left(\left(\frac{2}{e}\right)^{1/2}\right) \le e$$

and, for all $l \ge k$,

$$1 + \frac{k}{el}\left(1 + \frac{\sqrt{2}}{k\sqrt{e}}\right)^k \le 1 + \frac{k}{l} \le \left(1 + \frac{1}{l}\right)^k.$$

Hence, statement (8.4.4) is true for all $n = l \ge k$.

Lemma 8.4.3. *The inequality*

$$E\|\textstyle\sum_{1\le i<j\le n} g_2(X_i, X_j)\|^{2k} \le (8ek)^{2k}(n-1)^{2k} E\|g_2(X_1, X_2)\|^{2k}$$

holds for all integers $k = 1, 2, \dots$.

Proof. First, we write the formula

$$\textstyle\sum_{1\le i<j\le n} g_2(X_i, X_j) = \sum_{i=2}^{n} \xi_i,$$

where

$$\xi_i = \textstyle\sum_{j=1}^{i-1} g_2(X_i, X_j).$$

Since function $g_2(x, y)$ possesses the property of degeneracy, the stochastic sequence $\xi_1, \xi_2, \dots \xi_n$ is a martingale difference. Therefore, taking into account inequality (8.4.4), we can write

$$E\|\textstyle\sum_{1\le i<j\le n} g_2(X_i, X_j)\|^{2k} \le (8ek)^k (n-1)^k \max_{2\le i\le n} E\|\xi_i\|^{2k}.$$

Further, for fixed i, stochastic sequence $g_2(X_i, X_1), g_2(X_i, X_2), \dots, g_2(X_i, X_{i-1})$ also possesses the martingale property. Thus, by using inequality (8.4.4), we obtain

$$E\|\xi_i\|^{2k} \le (8ek)^k (i-1)^k E\|g_2(X_1, X_2)\|^{2k}.$$

By deriving this inequality, we have taken into account the fact that $X_1, \dots, X_n$ are independent identically distributed random variables. As a result we obtain Lemma 8.4.3.

Lemma 8.4.4. *Assume that*

$$E\|g_2(X_1, X_2)\|^k \leq c^k k^{\gamma k}$$

for all $k = 1, 2, \dots$ *and some constants* $c > 0$ *and* $\gamma \geq 0$, *which are independent of* k.

Then the inequality

$$E\|\textstyle\sum_{1 \leq i < j \leq n} g_2(X_i, X_j)\|^p \leq e^{2(1+\gamma)}(4ec)^p p^{(1+\gamma)p}(n-1)^p$$

holds for all $n = 1, 2, \dots$ and all real $p \in [2, 2n]$.

Proof. For given $p \in [2, 2n]$ we choose an integer r, $r = 1, \dots, n$, such that $2(r-1) < p \leq 2r$. According to the Hölder inequality, we have

$$E\|\textstyle\sum_{1 \leq i < j \leq n} g_2(X_i, X_j)\|^p \leq \left(E\|\sum_{1 \leq i < j \leq n} g_2(X_i, X_j)\|^{2r}\right)^{p/2r}.$$

Since $1 \leq r \leq n$, by virtue of Lemma 8.4.3 and the condition of this lemma, we can write

$$E\|\textstyle\sum_{1 \leq i < j \leq n} g_2(X_i, X_j)\|^{2r} \leq (8er)^{2r}(n-1)^{2r} E\|g_2(X_1, X_2)\|^{2r} \leq$$
$$\leq (8er)^{2r}(n-1)^{2r} c^{2r}(2r)^{\gamma 2r}.$$

Thus,

$$E\|\textstyle\sum_{1 \leq i < j \leq n} g_2(X_i, X_j)\|^p \leq (8er)^p (n-1)^p c^p (2r)^{\gamma p} =$$
$$= (4ec)^p (n-1)^p (2r)^{(1+\gamma)p}.$$

By the choice of r, $p/2 \leq r < p/2 + 1$ for all real $p \in [2, 2n]$. Therefore, we have

$$(2r)^{(1+\gamma)p} \leq (p+2)^{(1+\gamma)p} = p^{(1+\gamma)p}\left(1 + \frac{2}{p}\right)^{p(1+\gamma)} \leq p^{(1+\gamma)p} e^{2(1+\gamma)},$$

and this proves Lemma 8.4.4.

Lemma 8.4.5. *The estimates*

$$c_1 x^{\nu-2} \exp\left(-x^2/2\right) \leq P(\|\tau\| > x) \leq c_2 x^{\nu-2} \exp\left(-x^2/2\right) \tag{8.4.5}$$

hold for $1 \leq x \leq An^{1/6}$, *where* ν *is the multiplicity of the first eigenvalue of the covariance operator* V.

Moreover, if $\epsilon \leq x$ *for given* $\epsilon > 0$, *then*

$$P(x - \epsilon \leq \|\tau\| \leq x + \epsilon) \leq c_3 \epsilon (1 + x) P(\|\tau\| > x). \tag{8.4.6}$$

The proof of this lemma can be found in Rackauskas (1988), Zalesskii (1989). Relation (8.4.6) can be derived from (8.4.5).

Lemma 8.4.6. *Assume that*

$$E\|g_2(X_1,X_2)\|^k \le c^k k^{\gamma k}$$

for all $k = 1,2,\dots$ *and some constants* $c > 0$ *and* $\gamma \ge 0$*, which do not depend on* k.

Then the relation

$$P(\|2^{-1}\sqrt{n}U_{n2}\| > n^{-1/2(3+2\gamma)}) = O(n^{-1})P(\|\tau\| > x),$$

holds in the region $1 \le x \le \delta n^{1/2(3+2\gamma)}$, $\delta > 0$, as $n \to \infty$.

Proof. For all $n = 1,2,\dots$ and all real $p \in [2,2n]$, we can write

$$P\Big(\|2^{-1}\sqrt{n}U_{n2}\| > n^{-1/2(3+2\gamma)}\Big) = P\Big(\|nU_{n2}\| > 2n^{(1+\gamma)/(3+2\gamma)}\Big) \le$$

$$\le 2^{-p}n^{-p(1+\gamma)/(3+2\gamma)}E\|nU_{n2}\|^p .$$

By Lemma 8.4.4, we have

$$E\|nU_{n2}\|^p = \Big(\frac{2}{n-1}\Big)^p E\|\textstyle\sum_{1\le i<j\le n} g_2(X_i,X_j)\|^p \le e^{2(1+\gamma)}(8ec)^p p^{(1+\gamma)p} .$$

Therefore,

$$P\Big(\|2^{-1}\sqrt{n}U_{n2}\| > n^{-1/2(3+2\gamma)}\Big) \le e^{2(1+\gamma)}((4ec)p^{1+\gamma}n^{-(1+\gamma)/(3+2\gamma)})^p. \tag{8.4.7}$$

Consider the expression on the right-hand side of (8.4.7) for the following choice of p: $p = \alpha n^{1/(3+2\gamma)}$, where $\alpha = \min\{2,(4ce^2)^{-1/(1+\gamma)}\}$. Clearly, for sufficiently large n, we have $p \in [2,\dots,2n]$. For this choice of p, we get

$$(4ec)p^{1+\gamma}n^{-(1+\gamma)/(3+2\gamma)} \le e^{-1} .$$

Then it follows from (8.4.7) that

$$P(\|2^{-1}\sqrt{n}U_{n2}\| > n^{-1/2(3+2\gamma)}) \le e^{2(1+\gamma)}e^{-p} = e^{2(1+\gamma)}e^{-\alpha n^{1/(3+2\gamma)}}.$$

Under the condition of this theorem, $\delta^2 \le \alpha$. Therefore, in the region $1 \le x \le \delta n^{1/2(3+2\gamma)}$, the inequality

$$e^{-\frac{\alpha}{2}n^{1/(3+2\gamma)}} \le e^{-\frac{x^2}{2}}$$

takes place. Hence, by virtue of Lemma 8.4.5, we have

$$e^{-\frac{\alpha}{2}n^{1/(3+2\gamma)}} \le \delta n^{1/2(3+2\gamma)}x^{\nu-2}e^{-\frac{x^2}{2}} \le\le \delta c_1^{-1}n^{1/2(3+2\gamma)}P(\|\tau\| > x).$$

This yields the following estimate

$$P(\|2^{-1}\sqrt{n}U_{n2}\| > n^{-1/2(3+2\gamma)}) \leq$$

$$\leq e^{2(1+\gamma)}\delta c_1^{-1} n^{1/2(3+2\gamma)} e^{-\frac{a}{2}n^{1/(3+2\gamma)}} P(\|\tau\| > x),$$

in the region $1 \leq x \leq \delta n^{1/2(3+2\gamma)}$, which implies Lemma 8.4.6.

Lemma 8.4.7. *Assume that*

$$E\exp(t\|g_1(X_1)\|^{1/2}) < \infty$$

for some $t > 0$.

Then the estimate

$$\Big|P\Big(\|\frac{1}{\sqrt{n}}\sum_{j=1}^{n} g_1(X_j)\| > x \pm n^{-1/2(3+2\gamma)}\Big) - P(\|\tau\| > x)\Big| \leq$$

$$\leq c(1+x)n^{-1/2(3+2\gamma)} P(\|\tau\| > x)$$

holds in the region $1 \leq x \leq \delta n^{1/6}$.

Proof. First, we note that

$$P\Big(\|\frac{1}{\sqrt{n}}\sum_{j=1}^{n} g_1(X_j)\| > x + n^{-1/2(3+2\gamma)}\Big) - P(\|\tau\| > x) =$$

$$= P\Big(\|\frac{1}{\sqrt{n}}\sum_{j=1}^{n} g_1(X_j)\| > x + n^{-1/2(3+2\gamma)}\Big) -$$

$$-P(\|\tau\| > x + n^{-1/2(3+2\gamma)}) - P(x \leq \|\tau\| \leq x + n^{-1/2(3+2\gamma)}),$$

and

$$P\Big(\|\frac{1}{\sqrt{n}}\sum_{j=1}^{n} g_1(X_j)\| > x - n^{-1/2(3+2\gamma)}\Big) - P(\|\tau\| > x) =$$

$$= P\Big(\|\frac{1}{\sqrt{n}}\sum_{j=1}^{n} g_1(X_j)\| > x - n^{-1/2(3+2\gamma)}\Big) -$$

$$-P(\|\tau\| > x + n^{-1/2(3+2\gamma)}) + P(x - n^{-1/2(3+2\gamma)} \leq \|\tau\| \leq x).$$

By the theorem on large deviations in a Hilbert space (see Rackauskas (1988), Zalesskii (1989)), we have

$$\Big|P\Big(\|\frac{1}{\sqrt{n}}\sum_{j=1}^{n} g_1(X_j)\| > x \pm n^{-1/2(3+2\gamma)}\Big) - P(\|\tau\| > x \pm n^{-1/2(3+2\gamma)})\Big| \leq$$

$$\leq c\frac{(1+x)^3}{\sqrt{n}} P(\|\tau\| > x - n^{-1/2(3+2\gamma)}) =$$

$$= c\frac{(1+x)^3}{\sqrt{n}}[P(\|\tau\| > x) + P(x - n^{-1/2(3+2\gamma)} \leq \|\tau\| \leq x)]$$

in the region $1 \leq x \leq \delta n^{1/6}$.

According to inequality (8.4.6) with $\epsilon = n^{-1/2(3+2\gamma)}$, we obtain

$$P(x - n^{-1/2(3+2\gamma)} \le \|\tau\| \le x) \le c_3(1+x)n^{-1/2(3+2\gamma)}P(\|\tau\| > x)$$

and

$$P(x \le \|\tau\| \le x + n^{-1/2(3+2\gamma)}) \le c_3(1+x)n^{-1/2(3+2\gamma)}P(\|\tau\| > x)\,.$$

From here Lemma 8.4.7 follows.

Finally, Lemmas (8.4.1), (8.4.6) and (8.4.7) lead to inequality (8.4.2), and this proves Theorem 8.4.1.

Chapter 9

Law of iterated logarithm

The sequence $X_1, X_2, \ldots,$ of independent identically distributed B-valued random variables such that $EX_1 = 0$, $E\|X_1\|^2 < \infty$ satisfies the *bounded* law of iterated logarithm (LIL) if

$$\lim_{n\to\infty} \sup \|S_n\|/a_n < \infty \quad a.s.$$

where $S_n = X_1 + \cdots + X_n$, $a_n = (2n \ln\ln n)^{1/2}$. In addition, it is assumed that $\ln\ln x = 1$ if $x \leq \exp(e)$.

The sequence $\{S_n\}$ (or $\{X_n\}$) satisfies the *compact* LIL if there exists a nonrandom compact set K in B such that

$$\lim_{n\to\infty} \sup \|S_n/a_n - K\| = 0 \quad a.s.$$

and

$$C(\{S_n/a_n\}) = K \quad a.s.$$

where $\|x - K\| = \inf\{\|x-y\| : y \in K\}$ and $C(\{x_n\})$ denotes all the limiting points of the sequence $\{x_n\}$, and is called *a limiting set* (see Acosta and Kuelbs (1983), Bingham (1986), Ledoux and Talagrand (1991)). If $\{X_n\}$ satisfies the compact LIL, then this sequence also satisfies the bounded LIL. For infinite-dimensional spaces, the converse statement does not hold (see Acosta and Kuelbs (1983)). The set K is uniquely determined by a covariance function

$$T(x^*, y^*) = E(x^*(X_1)y^*(X_1)), \quad x^*, y^* \in B^*$$

and coincides with the unit ball in the reproducing kernel Hilbert space determined by T (see Acosta and Kuelbs (1983), Bingham (1986)). For example, if $B = H$ is a separable real Hilbert space, then $K = V^{1/2}(D)$, where $D = \{x \in H : \|x\| \leq 1\}$ and V is a covariance operator corresponding to a random element $X_1 \in H$. In this case, $T(f,g) = (Vf,g) = (V^{1/2}f, V^{1/2}g)$, $f,g \in H$.

If $B = R$ and $EX_1^2 = \sigma^2 > 0$, then $K = [-\sigma, \sigma]$. In this situation, the compact LIL represents the classic Hartman–Wintner–Strassen LIL (see Acosta and Kuelbs

(1983), Bingham (1986)). In particular,

$$\lim_{n\to\infty} \sup \pm \sum_{j=1}^{n} X_j / a_n = \sigma \quad a.s.$$

This is the classic Hartman–Wintner LIL. From here, the two-sided version of the LIL immediately follows,

$$\lim_{n\to\infty} \sup |\sum_{j=1}^{n} X_j| / a_n = \sigma \quad a.s.$$

Moreover, it is known that the conditions $EX_1 = 0$ and $\sigma^2 < \infty$ are also necessary for this version of the LIL to hold.

For *UB*-statistics, all kinds of variants of the LIL depend on rank r of kernel Φ and are formulated via the canonical functions g_c, $c = r, \ldots, m$.

9.1 *UR*-statistics

Consider *UR*-statistics with non-degenerate kernel $\Phi : X^m \to R$.

Theorem 9.1.1. *Suppose that a non-degenerate kernel Φ satisfies the following conditions*

$$Eg_1^2 < \infty$$

and

$$E(|g_c|^{\gamma_c}(\ln\ln|g_c|)^{-1/2}) < \infty, \quad c = 2, \ldots, m$$

where $\gamma_c = 2c/(2c-1)$.

Then

$$\lim_{n\to\infty} \sup n(U_n - \theta)/a_n = m\sigma_1$$

with probability 1.

Proof. By using the Hoeffding representation (1.1.9), we can write

$$n(U_n - \theta)/a_n = m(g_1(X_1) + \cdots + g_1(X_n))/a_n + R_n \tag{9.1.1}$$

where

$$R_n = \sum_{c=2}^{m} \binom{m}{c} c! \prod_{k=1}^{c-1} (1 - n^{-1}k)^{-1} \frac{n^{-c/\gamma_c}}{\sqrt{2\ln\ln n}} S_{nc}.$$

According to LIL for the sums of independent identically random variables we have

$$\lim_{n\to\infty} \sup(g_1(X_1) + \cdots + g_1(X_n))/a_n = \sigma_1$$

with probability 1. To estimate R_n, we use the fact that

$$(\ln\ln n)^{-1/2} n^{-c/\gamma_c} S_{nc} \to 0 \tag{9.1.2}$$

as $n \to \infty$, with probability 1. This statement can be established just as Corollary 3.4.2 for $p = 2$ and $q = \gamma_c$, $c = 2, \ldots, m$. By combining (9.1.1)–(9.1.2), we get Theorem 9.1.1.

9.2 *UH*-statistics

Now consider *UH*-statistics with non-degenerate kernel $\Phi : X^m \to H$. The following theorem is the exact analogue of the Hartman–Wintner LIL in a Hilbert space.

Theorem 9.2.1. *Suppose that*

$$E(g_1, y)^2 < \infty, \quad y \in H, \qquad E(\|g_1\|^2 / \ln\ln \|g_1\|) < \infty$$

and

$$E(\|g_c\|^{\gamma_c} (\ln\ln \|g_c\|)^{-1/2}) < \infty, \quad c = 2, \ldots, m,$$

where $\gamma_c = 2c/(2c-1)$.

Then

$$\lim_{n\to\infty} \sup \|n(U_n - \theta)\| / a_n = m\sigma \quad a.s.$$

where

$$\sigma^2 = \sup\{E(g_1, y)^2 : \|y\| \leq 1\}.$$

Proof. Apply (9.1.1). It follows from a familiar result of Acosta and Kuelbs (1983) that with probability 1,

$$\lim_{n\to\infty} \sup \|g_1(X_1) + \cdots + g_1(X_n)\| / a_n = \sigma.$$

By analogy with (9.1.2)

$$R_n \to 0 \quad a.s.$$

as $n \to \infty$. This proves Theorem 9.2.1.

Theorem 9.2.2. *Assume that*

$$E\|g_1\|^2 < \infty,$$
$$E(\|g_c\|^{\gamma_c} (\ln\ln \|g_c\|)^{-1/2}) < \infty, \quad c = 2, \ldots, m,$$

where $\gamma_c = 2c/(2c-1)$.

Then the sequence $\{m^{-1} n(U_n - \theta)/a_n\}$ *satisfies the compact LIL with a compact set* $K = V^{1/2}(D)$ *where* $D = \{x \in H : \|x\| \leq 1\}$ *and* V *is a covariance operator corresponding to a random element* $g_1(X_1) \in H$.

Proof. According to Corollary 4 in Acosta and Kuelbs (1983) under condition $E\|g_1\|^2 < \infty$, the sequence $\{(g_1(X_1) + \cdots + g_1(X_n))/a_n\}$ satisfies the compact LIL. To estimate R_n in (9.1.1) we use relation (9.1.2). This gives the desired result.

9.3 Bounded LIL

At first we consider UB-statistics with non-degenerate kernel $\Phi : X^m \to B$.

Theorem 9.3.1. *Let B be a separable Banach space of type 2. Suppose that the following conditions are fulfilled*

$$E(x^*(g_1))^2 < \infty, \quad x^* \in B^*,$$
$$E(\|g_1\|^2/\ln\ln\|g_1\|) < \infty,$$
$$E\left(\|g_c\|^{\gamma_c}(\ln\ln\|g_c\|)^{-1/2}\right) < \infty, \quad c = 2, \ldots, m,$$

where $\gamma_c = 2c/(2c-1)$.

Then

$$\lim_{n\to\infty}\sup\|n(U_n - \theta)\|/a_n = m\ \sup_{x^*\in B_1^*}\{E(x^*(g_1))^2\}^{1/2} \quad a.s.$$

where B_1^* is the unit ball of B^*.

Proof. In (9.1.1) by the bounded LIL of Ledoux and Talagrand (1988, 1990)

$$\lim_{n\to\infty}\sup\|g_1(X_1) + \cdots + g_1(X_n)\|/a_n = \sup_{x^*\in B_1^*}\{E(x^*(g_1))^2\}^{1/2} \quad a.s.$$

Since B is a type 2 space then relation (9.1.2) takes place as well. From here we obtain the proof of Theorem 9.3.1.

In the following theorem we consider the bounded LIL for UB-statistics with degenerate kernel.

Theorem 9.3.2. *Let B be a separable Banach space of type 2. The symmetric kernel $\Phi : X^m \to B$ has rank $r = m$ and $\sigma^2 = E\|\Phi\|^2 < \infty$. Then we have*

$$\lim_{n\to\infty}\sup\Big\|\sum\nolimits_{1\le i_1\ne\cdots\ne i_m\le n}\Phi(X_{i_1}, \ldots, X_{i_m})\Big\|/a_n^m \le c\sigma \quad a.s. \tag{9.3.1}$$

where constant $c > 0$ depends only on B and m.

Proof. Following Arcones and Giné (1994) we will prove that

$$E\sup_{n\ge1}\Big(\Big\|\sum\nolimits_{1\le i_1\ne\cdots\ne i_m\le n}\Phi(X_{i_1}, \ldots, X_{i_m})\Big\|/a_n^m\Big) \le c\sigma. \tag{9.3.2}$$

Note that (9.3.2) implies (9.3.1). Indeed, the random variable on the left-hand side of (9.3.1) is invariant under finite permutations of $(X_1, X_2, \ldots, X_n, \ldots)$. Hence it is a constant with probability 1 by virtue of the Hewitt–Savage zero-one law. And by (9.3.2) this constant is no larger than $c\sigma$.

In the following we will take the notations: $\mathbf{i}$ for $(i_1, \dots, i_m)$, $\varepsilon_{\mathbf{i}}$ for $\varepsilon_{i_1} \cdots \varepsilon_{i_m}$, $\varepsilon_{\mathbf{i}}^{\mathrm{dec}}$ for $\varepsilon_{i_1}^1 \cdots \varepsilon_{i_m}^m$, $\Phi(X_{\mathbf{i}})$ for $\Phi(X_{i_1}, \dots, X_{i_m})$, and $\Phi(X_{\mathbf{i}}^{\mathrm{dec}})$ for $\Phi(X_{i_1}^1, \dots, X_{i_m}^m)$ ('dec' standing for 'decoupled'), and $I_n^m = \{(i_1, \dots, i_m) : 1 \le i_j \le n, \text{ and } i_j \ne i_r \text{ if } j \ne r\}$. So, the sequences of interest will be written as

$$\sum_{I_n^m} \Phi(X_{\mathbf{i}})/a_n^m, \quad n \ge 1.$$

Further, in the same way as in Arcones and Giné (1994) we will give the basic technical estimates in the following three lemmas. Let us introduce into consideration the Young functions of α-exponential type, $0 < \alpha \le 2$. Namely, for $\alpha \ge 1$, $\psi_\alpha(x) := \exp(x^\alpha) - 1$, $x \ge 0$. If $\alpha < 1$ then this function is only convex for $x > ((1-\alpha)/\alpha)^{1/\alpha}$. Therefore we modify it as follows: let φ_α coincide with $\exp(x^\alpha)$ for $x \ge ((1-\alpha)/\alpha)^{1/\alpha}$ and coincide with the tangent to this function at $(((1-\alpha)/\alpha)^{1/\alpha}, \exp((1-\alpha)/\alpha))$ for $0 \le x \le ((1-\alpha)/\alpha)^{1/\alpha}$; next, let $\psi_\alpha(x) = \varphi_\alpha(x) - \varphi_\alpha(0) = \varphi_\alpha(x) - \alpha \exp((1-\alpha)/\alpha)$. Thus, this function is convex, increasing and $\psi_\alpha(0) = 0$ i.e. it is a Young function. Further, the ψ_α-norm of a random variable ξ is defined as

$$\|\xi\|_{\psi_\alpha} := \inf\{c > 0 : E\psi_\alpha(|\xi|/c) \le 1\}.$$

This norm is homogeneous, i.e. $\|t\xi\|_{\psi_\alpha} = |t|\,\|\xi\|_{\psi_\alpha}$ for any $t \in R$.

There is finite constant $c_{\alpha p}$ for any $\alpha > 0$, $p > 0$, such that for all random variables ξ,

$$\|\xi\|_p \le c_{\alpha p}\|\xi\|_{\psi_\alpha}. \tag{9.3.3}$$

Lemma 9.3.1. *Let B be a type 2 space. Then for all finite sets $\{x_{i_1,\dots,i_m}\} \in B$ the following is valid*

$$\Big\|\Big\|\sum \varepsilon_{i_1}^1 \cdots \varepsilon_{i_m}^m x_{i_1,\dots,i_m}\Big\|\Big\|_{\psi_{2/m}} \le c\Big(\sum \|x_{i_1,\dots,i_m}\|^2\Big)^{1/2}$$

with some constant $c > 0$.

Proof. By Khintchine's inequality (2.6.4) or by Borell's hypercontractivity inequality (2.6.5), combined with the decoupling inequalities of Theorem 2.5.4, there are finite and positive constants c_1 and c_2 such that

$$E\exp\{c_1(\|z\|/\|z\|_2)^{2/m}\} \le c_2,$$

where

$$z = \sum_{\mathbf{i} \in I_n^m} \varepsilon_{\mathbf{i}}^{\mathrm{dec}} x_{\mathbf{i}}, \quad x_{\mathbf{i}} \in B.$$

The result follows by virtue of applying type 2 inequality (2.7.1) for one sequence $\{\varepsilon_i^j, i \geq 1\}$, $j = 1, \ldots, m$, at a time, because $E\|z\|^2 \leq A^m \sum \|x_i\|^2$.

Lemma 9.3.2. *For every sequence of random variables* $\{\xi_k, k \geq 1\}$, *there exists a finite constant* c_{ψ_α} *such that*

$$\left\| \sup_k \frac{|\xi_k|}{\psi_\alpha^{-1}(k)} \right\|_{\psi_\alpha} \leq c_{\psi_\alpha} \sup_k \|\xi_k\|_{\psi_\alpha} .$$

Proof. Note that the Young functions ψ_α, $\alpha > 0$, possess the following two properties

$$\psi_\alpha^{-1}(xy) \leq c\psi_\alpha^{-1}(x)\psi_\alpha^{-1}(y), \quad x, y \geq b_1 ,$$

and

$$\psi_\alpha^{-1}(x^2) \leq d\psi_\alpha^{-1}(x), \quad x \geq b_2$$

with some finite positive constants c, d, b_1, b_2. From homogeneity we will assume that $\sup_k \|\xi_k\|_{\psi_\alpha} \leq 1$. In what follows, we set $k_0 = \max\{8, \psi_\alpha(\psi_\alpha^{-1}(b_1)/d), b_2\}$, $\beta = \max\{\psi_\alpha^{-1}(b_2^2)/\psi_\alpha^{-1}(1/2), \psi_\alpha^{-1}(b_1)/\psi_\alpha^{-1}(1/2), d\}$ and $a = cd\beta$.

If $t \geq 1/2$, then we have

$$\begin{aligned}
P\left(\psi_\alpha\left(\sup_{k \geq k_0} \frac{|\xi_k|}{a\psi_\alpha^{-1}(k)}\right) \geq t\right) &\leq P\left(\sup_{k \geq k_0} \frac{|\xi_k|}{a\psi_\alpha^{-1}(k)\psi_\alpha^{-1}(t)} \geq 1\right) \\
&\leq \sum_{k=k_0}^{\infty} \frac{1}{\psi_\alpha(a\psi_\alpha^{-1}(k)\psi_\alpha^{-1}(k)(t))} \\
&\leq \sum_{k=k_0}^{\infty} \frac{1}{\psi_\alpha(d\psi_\alpha^{-1}(k))\psi_\alpha(\beta\psi_\alpha^{-1}(t))}
\end{aligned}$$

$$[\text{since } \psi_\alpha(\beta\psi_\alpha^{-1}(t)) \geq t^2]$$

$$\begin{aligned}
&\leq \sum_{k=k_0}^{\infty} \frac{1}{k^2 t^2} \\
&\leq \frac{1}{4t^2} .
\end{aligned}$$

With the help of this estimate we obtain

$$\begin{aligned}
E\left(\psi_\alpha\left(\sup_{k \geq k_0} \frac{|\xi_k|}{a\psi_\alpha^{-1}(k)}\right)\right) &= \int_0^\infty P\left(\psi\left(\sup_{k \geq k_0} \frac{|\xi_k|}{a\psi_\alpha^{-1}(k)}\right) \geq t\right) dt \\
&\leq \frac{1}{2} + \int_{\frac{1}{2}}^\infty P\left(\psi\left(\sup_{k \geq k_0} \frac{|\xi_k|}{a\psi_\alpha^{-1}(k)}\right) \geq t\right) dt \\
&\leq \frac{1}{2} + \frac{1}{4}\int_{\frac{1}{2}}^\infty \frac{dt}{t^2} = 1.
\end{aligned}$$

This proves Lemma 9.3.2.

From Lemma 9.3.2. it follows that

$$\left\| \sup_k \frac{|\xi_k|}{(\ln k)^{1/\alpha}} \right\|_{\psi_\alpha} \leq c_{\psi_\alpha} \sup_k \|\xi_k\|_{\psi_\alpha} . \tag{9.3.4}$$

By virtue of (9.3.3) the ψ_α-norm on the left-hand side of (9.3.4) can be replaced by the L^1-norm (up to multiplicative constants). In this way we will apply (9.3.4) to the random variables

$$\xi_k = \frac{1}{2^{km/2}} \sum_{\mathrm{i} \in I_{2^k}^m} \varepsilon_{\mathrm{i}}^{\mathrm{dec}} \Phi(X_{\mathrm{i}}^{\mathrm{dec}}) .$$

Therefore, it is necessary to majorize the L^1-norm of a sup over $n \in N$ by the L^1-norm of a sup over the subset $n = 2^k$, $k \in N$. This is a consequence of Lévy's inequality (see Lemma 2.4.1).

Lemma 9.3.3. *Let $x_{i_1,\dots,i_m} \in B$ and b_n be an increasing sequence, $b_n \to \infty$. Set $b_k^* = b_{2^k}$ for all k. Then,*

$$E \sup_n \| \frac{1}{b_n} \sum_{i_1,\dots,i_m \leq n} \varepsilon_{i_1}^1 \cdots \varepsilon_{i_m}^m x_{i_1,\dots,i_m} \| \leq$$

$$\leq 2^m E \sup_k \| \frac{1}{b_k^*} \sum_{i_1,\dots,i_m \leq 2^{k+1}} \varepsilon_{i_1}^1 \cdots \varepsilon_{i_m}^m x_{i_1,\dots,i_m} \| .$$

Proof. It suffices to show that for all $p < \infty$

$$E \max_{k \leq p} \max_{n \in N_k} \| \frac{1}{b_k^*} \sum_{|\mathrm{i}| \leq n} \varepsilon_{\mathrm{i}}^{\mathrm{dec}} x_{\mathrm{i}} \| \leq 2^m E \max_{k \leq p} \| \frac{1}{b_k^*} \sum_{|\mathrm{i}| \leq 2^{k+1}} \varepsilon_{\mathrm{i}}^{\mathrm{dec}} x_{\mathrm{i}} \| ,$$

where $N_0 = \{1, 2\}$, $N_k = (2^k, 2^{k+1}] \cap N$, $k \in N$, and $|\mathrm{i}| = \max\{i_1, \dots, i_m\}$. If $|\mathrm{i}| \in N_k$, $k = 0, 1, \dots, p$, then we define new variables x_{i}^* in $l^\infty(B)$ by $x_{\mathrm{i}}^* = (0, \dots, 0, x_{\mathrm{i}}/b_{k+1}^*, \dots, x_{\mathrm{i}}/b_p^*)$. Then we can write

$$\max_{k \leq p} \max_{n \in N_k} \| \frac{1}{b_k^*} \sum_{|\mathrm{i}| \leq n} \varepsilon_{\mathrm{i}}^{\mathrm{dec}} x_{\mathrm{i}} \| = \max_{n \leq 2^{p+1}} \| \sum_{|\mathrm{i}| \leq n} \varepsilon_{\mathrm{i}}^{\mathrm{dec}} x_{\mathrm{i}}^* \| ,$$

$$\max_{k \leq p} \| \frac{1}{b_k^*} \sum_{|\mathrm{i}| \leq 2^{k+1}} \varepsilon_{\mathrm{i}}^{\mathrm{dec}} x_{\mathrm{i}} \| = \| \sum_{|\mathrm{i}| \leq n} \varepsilon_{\mathrm{i}}^{\mathrm{dec}} x_{\mathrm{i}}^* \|,$$

where the norm in the second terms is defined as the norm of $l_{p+1}^\infty(B)$. Further, by Lévy's inequality applied conditionally on all the variables except ε_i^1, $i \geq 1$, we have

$$E \max_{n \leq 2^{p+1}} \| \sum_{|\mathrm{i}| \leq n} \varepsilon_{\mathrm{i}}^{\mathrm{dec}} x_{\mathrm{i}}^* \| \leq$$

$$\leq E \max_{n_1 \leq 2^{p+1}} \max_{n_j \leq 2^{p+1}, 2 \leq j \leq m} \| \sum_{i_1 \leq n_1} \varepsilon_{i_1}^1 \left(\sum_{i_2 \leq n_2, \dots, i_m \leq n_m} \varepsilon_{i_2}^2 \cdots \varepsilon_{i_m}^m x_{\mathrm{i}}^* \right) \| \leq$$

$$\leq 2E \max_{n_j \leq 2^{p+1}, 2 \leq j \leq m} \| \sum_{i_1 \leq 2^{p+1}} \varepsilon_{i_1}^1 \left(\sum_{i_2 \leq n_2, \dots, i_m \leq n_m} \varepsilon_{i_2}^2 \cdots \varepsilon_{i_m}^m x_{\mathrm{i}}^* \right) \| .$$

Performing these estimates for $j = 2, \dots, m$, we get the proof of Lemma 9.3.3.

Now we will prove inequality (9.3.2), by applying the mentioned lemmas. At first, let $\Phi_k^* = (0, \dots, 0, \Phi/a_k^m, \Phi/a_{k+1}^m, \dots) \in c_0(B)$. Then we have

$$\sup_n \frac{1}{a_n^m} \|\sum_{\mathrm{i} \in I_n^m} \Phi(X_\mathrm{i})\| = \|\sum_{\mathrm{i} \in I^m} \Phi_{|\mathrm{i}|}^*(X_\mathrm{i})\|,$$

where $I^m = \cup_n I_n^m$ and $|\mathrm{i}| = \max\{i_1, \dots, i_m\}$. Further, we apply the decoupling inequalities of Theorem 2.5.4 and symmetrize with respect to one group of variables $\{X_i^j\}_{i=1}^\infty$ at a time. We therefore obtain

$$E \sup_n \frac{1}{a_n^m} \|\sum_{\mathrm{i} \in I_n^m} \Phi(X_\mathrm{i})\| \le c\, E \sup_n \frac{1}{a_n^m} \|\sum_{\mathrm{i} \in I_n^m} \varepsilon_\mathrm{i}^{\mathrm{dec}} \Phi(X_\mathrm{i}^{\mathrm{dec}})\|$$

with some constant $c < \infty$. By Lemma 9.3.3 with $x_\mathrm{i} = \Phi(X_\mathrm{i}^{\mathrm{dec}})$

$$E \sup_n \frac{1}{a_n^m} \|\sum_{\mathrm{i} \in I_n^m} \varepsilon_\mathrm{i}^{\mathrm{dec}} \Phi(X_\mathrm{i}^{\mathrm{dec}})\| \le 2^m E \sup_k \frac{1}{a_{2^k}^m} \|\sum_{\mathrm{i} \in I_{2^{k+1}}^m} \varepsilon_\mathrm{i}^{\mathrm{dec}} \Phi(X_\mathrm{i}^{\mathrm{dec}})\| .$$

According to inequalities (9.3.3)–(9.3.4) with $\alpha = 2/m$ and by Lemma 9.3.1 we can write

$$E_\varepsilon \sup_k \frac{1}{a_{2^k}^m} \|\sum_{\mathrm{i} \in I_{2^{k+1}}^m} \varepsilon_\mathrm{i}^{\mathrm{dec}} \Phi(X_\mathrm{i}^{\mathrm{dec}})\| \le$$

$$\le 3 \cdot 2^{\frac{m}{2}} E_\varepsilon \sup_k \frac{1}{(\ln\, k)^{\frac{m}{2}}} \|\frac{1}{2^{\frac{km}{2}}} \sum_{\mathrm{i} \in I_{2^k}^m} \varepsilon_\mathrm{i}^{\mathrm{dec}} \Phi(X_\mathrm{i}^{\mathrm{dec}})\| \le$$

$$\le c \sup_k \left(\frac{1}{2^{km}} \sum_{\mathrm{i} \in I_{2^k}^m} \|\Phi(X_\mathrm{i}^{\mathrm{dec}})\|^2 \right)^{1/2},$$

where constant $c < \infty$ depends only on B and m. Hence, the following inequality is valid

$$E \sup_k \frac{1}{a_{2^k}^m} \|\sum_{\mathrm{i} \in I_{2^{k+1}}^m} \varepsilon_i^{\mathrm{dec}} \Phi(X_i^{\mathrm{dec}})\| \le c\; E \sup_k \left(\frac{1}{2^{km}} \sum_{\mathrm{i} \in I_{2^k}^m} \|\Phi(X_i^{\mathrm{dec}})\| \right)^{1/2} .$$

By Theorem 2.3.3

$$P(\sup_n \frac{1}{n^m} \sum_{\mathrm{i} \in I_n^m} \|\Phi(X_i^{\mathrm{dec}})\|^2 > t) \le \frac{E\, \|\Phi\|^2}{t}$$

and hence we can write

$$E \sup_k \left(\frac{1}{2^{km}} \sum_{\mathrm{i} \in I_{2^k}^m} \|\Phi(X_\mathrm{i}^{\mathrm{dec}})\|^2 \right)^{1/2} \le E \sup_n (\frac{1}{n^m} \sum_{\mathrm{i} \in I_n^m} \|\Phi(X_\mathrm{i}^{\mathrm{dec}})\|^2)^{1/2} =$$

$$= \int_0^\infty P \left(\sup_n \frac{1}{n^m} \sup_{\mathrm{i} \in I_n^m} \|\Phi_\mathrm{i}^{\mathrm{dec}}\|^2 > t^2 \right) dt \le (E\, \|\Phi\|^2)^{1/2} + \frac{1}{2} \int_{E\|\Phi\|^2}^\infty t^{\frac{1}{2}-1} \frac{E\, \|\Phi\|^2}{t} dt \le$$

$$\le 2(E\, \|\Phi\|^2)^{1/2} .$$

This completes the proof of inequality (9.3.2) and therefore that of Theorem 9.3.2.

9.4 Compact LIL

Non-degenerate kernels in type 2 Banach spaces.

Theorem 9.4.1. *Let B be a type 2 separable Banach space. Suppose that a non-degenerate kernel Φ satisfies the following conditions: the sequence $\{(g_1(X_1)+\cdots+g_1(X_n))/\sqrt{n}\}$ satisfies the central limit theorem and*

$$E\left(\|g_1\|^2/\ln\ln\|g_1\|\right)<\infty$$

and

$$E\left(\|g_c\|^{\gamma_c}(\ln\ln\|g_c\|)^{-1/2}\right)<\infty,\quad c=2,\ldots,m$$

where $\gamma_c=2c/(2c-1)$.

Then with probability 1, $\{m^{-1}n(U_n-\theta)/a_n;\ n=1,2,\ldots\}$ is relatively compact in B, and its limit set coincides with the closed unit ball K of H_T, where H_T is the reproducing kernel Hilbert space in B generated by the covariance function

$$T(x^*,y^*)=E(x^*(g_1(X_1))y^*(g_1(X_1))),\quad x^*,y^*\in B^*.$$

Proof. At first, for the sequence $\{(g_1(X_1)+\cdots+g_1(X_n))/a_n\}$ from (9.1.1) we apply the compact LIL of Heinkel (1979) and Goodman *et al.* (1981). Since B is of type 2, then relation (9.1.2) is valid. This proves Theorem 9.4.1.

Canonical kernels in $\boldsymbol{R^k}$. Consider UR^k-statistics with degenerate kernel $\Phi: X^m\to R^k$. Moreover $\Phi=(\Phi_1,\ldots,\Phi_k)$ and $\Phi_r: X^m\to R$ are P-canonical functions, $r=1,\ldots,k$.

Theorem 9.4.2. *Let $E\Phi_r^2<\infty$, $r=1,\ldots,k$. Then, with probability 1, the sequence*

$$\Big\{\frac{1}{a_n^m}\sum\nolimits_{(i_1,\ldots,i_m)\in I_n^m}(\Phi_1(X_{i_1},\ldots,X_{i_m}),\ldots,\Phi_k(X_{i_1},\ldots,X_{i_m}))\Big\}_{n=1}^{\infty}\tag{9.4.1}$$

is relatively compact and its limit set is

$$K:=\left\{E(g(X_1)\cdots g(X_m)(\Phi_1(X_1,\ldots,X_m),\ldots,\Phi_k(X_1,\ldots,X_m))):\|g\|_2\le 1\right\}$$

where $\|g\|_2^2=Eg^2(X_1)$.

Proof. We will follow Arcones and Giné (1994). For any $\delta>0$ there are functions $\varphi_1(x),\ldots,\varphi_N(x)$, $N<\infty$, such that $E\,\varphi_j(X_1)=0$ and for each $1\le r\le k$

$$E\left(\Phi_r(X_1,\ldots,X_m)-\bar{\Phi}_r(X_1,\ldots,X_m)\right)^2\le\frac{\delta^2}{9(1+c^2)}\tag{9.4.2}$$

where

$$\bar{\Phi}_r(X_1,\dots,X_m)=\sum\nolimits_{j=1}^{N}c_j(\Phi_r)\varphi_j(X_1)\cdots\varphi_j(X_m). \tag{9.4.3}$$

In addition $c_j(\Phi_r)$ are some real numbers, and c is a constant of inequality (9.3.1) for $B=R$ (see Arcones and Giné (1992)). Now we prove this theorem for canonical kernels $(\bar{\Phi}_1,\dots,\bar{\Phi}_k)$ with $\bar{\Phi}_r$ of the form (9.4.3). According to Strassen LIL in R^k (see e.g. Ledoux and Talagrand (1991)), the sequence

$$\{\frac{1}{a_n}\sum\nolimits_{i=1}^{n}(\varphi_1(X_i),\dots,\varphi_N(X_i))\}_{n=1}^{\infty}$$

is relatively compact and its limit set is

$$\{E\,(g(X_1)(\varphi_1(X_1),\dots,\varphi_N(X_1))) : E\,g^2(X_1)\le 1\}.$$

According to the second Waring formula, there is an m-th degree polynomial Q_m of m variables such that

$$\frac{1}{a_n^m}\sum\nolimits_{(i_1,\dots,i_m)\in I_n^m}\varphi_j(X_{i_1})\cdots\varphi_j(X_{i_m})=$$

$$=Q_m(\frac{1}{a_n}\sum\nolimits_{i=1}^{n}\varphi_j(X_i),\dots,\frac{1}{a_n^m}\sum\nolimits_{i=1}^{n}\varphi_j^m(X_i))$$

for all $n\in N$ (see e.g. Borovskikh and Korolyuk (1994b)).

Therefore, by Strassen LIL and by the Marcinkiewicz law of large numbers we conclude that, with probability 1, the sequence

$$\Big\{\frac{1}{a_n^m}\sum\nolimits_{(i_1,\dots,i_m)\in I_n^m}(\varphi_1(X_{i_1})\cdots\varphi_1(X_{i_m}),\ \dots,\ \varphi_N(X_{i_1})\cdots\varphi_N(X_{i_m}))\Big\}_{n=1}^{\infty}$$

is relatively compact and its limit set is

$$\Big\{(Q_m(E\,\varphi_1(X_1)g(X_1),0,\dots,0),\dots,Q_m(E\,\varphi_N(X_1)g(X_1),0,\dots,0)) :\ \|g\|_2\le 1\Big\}.$$

It is well known that, in a topological space, if $\{y_n\}$ is relatively compact with limit set A and if α is a continuous function, then $\{\alpha(y_n)\}$ is also relatively compact and its limit set is $\{\alpha(y) : y\in A\}$. From the second Waring formula we have $Q_m(x,0,\dots,0)=x^m$. This gives for all j

$$Q_m(E\,\varphi_j(X_1)g(X_1),0,\dots,0)=(E\,\varphi_j(X_1)g(X_1))^m=$$

$$=E\,(g(X_1)\cdots g(X_m)\varphi_j(X_1)\cdots\varphi_j(X_m)).$$

Hence, with probability 1, the sequence

$$\Big\{\frac{1}{a_n^m}\sum\nolimits_{(i_1,\dots,i_m)\in I_n^m}(\bar{\Phi}_1(X_{i_1},\dots,X_{i_m}),\dots,\bar{\Phi}_k(X_{i_1},\dots,X_{i_m}))\Big\}_{n=1}^{\infty}$$

is also relatively compact and its limit set is

$$\left\{E\left(g(X_1)\cdots g(X_m)(\bar{\Phi}_1(X_1,\dots,X_m),\dots,\bar{\Phi}_k(X_1,\dots,X_m))\right):\|g\|_2\le 1\right\}. \quad (9.4.4)$$

This proves the theorem for canonical kernels of the form (9.4.3).

Note that by Theorem 9.3.2 the sequence in (9.4.1) is almost surely relatively compact. Now we will show that if x is in K then x is almost surely a limit point of this sequence (9.4.1). Let $x \in K$. Then there is a function g with $E\,g^2(X_1) \le 1$ and

$$x = E\left(g(X_1)\cdots g(X_m)(\Phi_1(X_1,\dots,X_m),\dots,\Phi_k(X_1,\dots,X_m))\right).$$

With given function g, the previous arguments show that there is almost surely a subsequence $\{n_p\}$, dependent on ω, such that

$$|h_{n_p}(\bar{\Phi}_r) - E\left(g(X_1)\cdots g(X_m)\bar{\Phi}_r(X_1,\dots,X_m)\right)| \le \frac{\delta}{3} \quad (9.4.5)$$

for $1 \le r \le k$, where

$$h_n(\Phi) := \frac{1}{a_n^m}\sum\nolimits_{(i_1,\dots,i_m)\in I_n^m}\Phi(X_{i_1},\dots,X_{i_m}).$$

In addition, we also have by Theorem 9.3.2, for $1 \le r \le k$,

$$\lim_{n\to\infty}\sup|h_n(\Phi_r - \bar{\Phi}_r)| \le \frac{\delta}{3}$$

with probability 1, and by inequality (9.4.2)

$$|E\left(g(X_1)\cdots g(X_m)\Phi_r(X_1,\dots,X_m)\right) - \\ - E\left(g(X_1)\cdots g(X_m)\bar{\Phi}_r(X_1,\dots,X_m)\right)| \le \frac{\delta}{3}. \quad (9.4.6)$$

Combining (9.4.5)-(9.4.6) we obtain, for each $1 \le r \le k$,

$$|h_{n_p}(\Phi_r) - E\left(g(X_1)\cdots g(X_m)\Phi_r(X_1,\dots,X_m)\right)| \le \delta$$

with probability 1. This relation shows that x is almost surely a limit point of the sequence (9.4.1).

Now we will prove that the set of almost sure limit points of sequence (9.4.1) is contained in K. Let x be almost surely a limit point of sequence (9.4.1). Then by Theorem 9.3.2, there are functions $\bar{\Phi}_r$, $1 \le r \le k$, of the form (9.4.3), which satisfy (9.4.2) (also (9.4.6) for all g with $E\,g^2(X_1) \le 1$), such that

$$\lim_{p\to\infty}\|x - (h_{n_p}(\bar{\Phi}_1),\dots,h_{n_p}(\bar{\Phi}_k))\| \le \frac{\delta}{3}$$

with probability 1 for some subsequence $\{n_p\}$, dependent on ω. The limit set of the sequence $\{(h_n(\bar{\Phi}_1),\dots,h_n(\bar{\Phi}_k))\}$ almost surely coincides with (9.4.4). Then there is a function g with $E\,g^2(X_1) \le 1$ such that

$$\|x - E\left(g(X_1)\cdots g(X_m)(\Phi_1(X_1,\dots,X_m),\dots,\Phi_k(X_1,\dots,X_m))\right)\| \le \delta.$$

Hence, there is a sequence $\{g_p\}$ of functions in $L^2(P)$ such that

$$E\left(g_p(X_1)\cdots g_p(X_m)(\Phi_1(X_1,\dots,X_m),\dots,\Phi_k(X_1,\dots,X_m))\right)\to x$$

as $p\to\infty$. Further, the unit ball of $L^2(P)$ is weakly sequentially compact and if a sequence $\{g_p\}$ converges weakly in $L^2(P)$ to g then the sequence $\{g_p(y_1),\dots,g_p(y_k)\}$ converges weakly in $L^2(P^k)$ to $g(y_1)\cdots g(y_k)$. This denotes that there is a subsequence $\{g_{p_j}\}_{j=1}^{\infty}$ and a function g such that

$$E\left(g_{p_j}(X_1)\cdots g_{p_j}(X_m)\Phi_r(X_1,\dots,X_m)\right)\to E\left(g(X_1)\cdots g(X_m)\Phi_r(X_1,\dots,X_m)\right)$$

for each $1\le r\le k$. Hence,

$$x=E\left(g(X_1)\cdots g(X_m)(\Phi_1(X_1,\dots,X_m),\dots,\Phi_k(X_1,\dots,X_m))\right),$$

i.e. $x\in K$. This completes the proof of Theorem 9.4.2.

Canonical kernels in type 2 Banach spaces.

Theorem 9.4.3. *Let B be a type 2 Banach space and let $\Phi : X^m \to B$ be a P-canonical kernel such that $E\|\Phi\|^2<\infty$. Then the sequence*

$$\{\frac{1}{a_n^m}\sum\nolimits_{(i_1,\dots,i_m)\in I_n^m}\Phi(X_{i_1},\dots,X_{i_m})\}_{n=1}^{\infty} \qquad (9.4.7)$$

is almost surely relatively compact and its limit set is

$$K(\Phi)=\{E\left(g(X_1)\cdots g(X_m)\Phi(X_1,\dots,X_m)\right): E\,g^2(X_1)\le 1\}.$$

Proof. According to Lemma 4.2.1 for every $\varepsilon>0$ there is a simple canonical kernel $\bar\Phi$ such that $E\|\Phi-\bar\Phi\|^2\le\varepsilon$. This approximation and Theorem 9.3.2 show relative compactness of sequence (9.4.7) (see Arcones and Giné (1994)). Now we can apply Theorem 9.4.2 to the canonical simple function $\bar\Phi$ with values in B. By Theorem 9.4.2 the LIL limit set is $K(\bar\Phi)$. Further, by repeating the proof of Theorem 9.4.2 from (9.4.4) under these conditions with only formal changes, in the main B-valued $\bar\Phi$ for R^k-valued $(\bar\Phi_1,\dots,\bar\Phi_k)$, we get the proof of Theorem 9.4.3.

9.5 Functional LIL

Let $l^\infty([0,1],B)$ be a Banach space of B-valued functions on $[0,1]$, with sup-norm. The functional LIL for partial sum UB-processes can be proved by the same method as Theorem 9.4.2 and Theorem 9.4.3.

Theorem 9.5.1. *Let B be a type 2 Banach space and let $\Phi : X^m \to B$ be a canonical kernel such that $E\|\Phi\|^2 < \infty$.*

Then the sequence

$$\left\{\left(\frac{1}{a_n^m}\sum\nolimits_{(i_1,\dots,i_m)\in I^m_{[nt]}}\Phi(X_{i_1},\dots,X_{i_m})\right)_{0\le t\le 1}\right\}_{n=1}^{\infty}$$

is almost surely relatively compact in $l^\infty([0,1],B)$ and its limit set is

$$K(\Phi)=\left\{\int_0^t E\left(g(X_1,s)\cdots g(X_m,s)\Phi(X_1,\dots,X_m)\right)ds : \int_0^1 E\, g^2(X_1,s)ds \le 1\right\}$$

where $0\le t\le 1$.

Proof. Note that

$$\sup\nolimits_{n,t}\|\frac{1}{a_n^m}\sum\nolimits_{I^m_{[nt]}}\Phi(X_{\mathrm{i}})\| = \sup\nolimits_n\|\frac{1}{a_n^m}\sum\nolimits_{I^m_n}\Phi(X_{\mathrm{i}})\|\,.$$

The statement of this theorem is valid for finite-dimensions (see e.g. Dehling (1989a), Theorem 2.1). Therefore, it is sufficient to apply the method of the proof of Theorem 9.4.3 to obtain the result of Theorem 9.5.1.

9.6 Multi-sample *UB*-statistics

Now we consider multi-sample *UB*-statistics of the form (1.2.1). The general case is very complicated. Therefore, we will analyze two situations preserving some lines of general complexity. Denote $a_{\bar n} = (2c|\bar n|\ln\ln|\bar n|)^{1/2}$ with $|\bar n| = n_1\cdots n_c$.

Theorem 9.6.1. *Let B be a type 2 Banach space. Let $c=2$, $m_1=m_2=2$ and rank $r=2$. Suppose that kernel Φ satisfies the following conditions*

$$g_{20}(x,y)=0,\quad g_{11}(x,y)\ne 0,\quad g_{02}(x,y)=0$$

and

$$E\left(\|g_{11}\|^2\ln\|g_{11}\|(\ln\ln\|g_{11}\|)^{-1}\right)<\infty$$

and

$$E\left(\|g_{\bar d}\|^{d/(d-1)}(\ln\ln\|g_{\bar d}\|)^{-1/2}\right)<\infty,\quad d=3,4\,.$$

If $n_2n_1^{-1}\to\lambda$, $0<\lambda<\infty$, as $|\bar n|\to\infty$, then the sequence $\{|\bar n|(U_{\bar n}-\theta)/a_{\bar n}\}$ satisfies a bounded LIL

$$\limsup_{|\bar n|\to\infty}\||\bar n|(U_{\bar n}-\theta)/a_{\bar n}\|<\infty\quad a.s.$$

and also a compact LIL.

Proof. Apply formula (1.2.6). In the conditions of the theorem from (1.2.6) we write

$$|\bar{n}|(U_{\bar{n}} - \theta)/a_{\bar{n}} = \frac{4}{a_{\bar{n}}}\sum_{i=1}^{n_1}\sum_{j=1}^{n_2} g_{11}(X_{1i}, X_{2j}) + R_{\bar{n}},$$

where $R_{\bar{n}} \to 0$ with probability 1 by Corollary 3.2.1. Further, by Li and Wu (1989) the sequence

$$\{\frac{1}{a_{\bar{n}}}\sum_{i=1}^{n_1}\sum_{j=1}^{n_2} g_{11}(X_{1i}, X_{2j})\}$$

satisfies the bounded LIL and the compact LIL. This gives the required result.

Corollary 9.6.1. *If $B = R$ then we have*

$$\lim_{|\bar{n}|\to\infty} \sup |\bar{n}|\left|(U_{\bar{n}} - \theta)/a_{\bar{n}}\right| = 4\sqrt{2}(E\, g_{11}^2)^{1/2}.$$

The result in such form for the sequence is due to Wichura (1973) (see, e.g. Gut (1983)).

Theorem 9.6.2. *Let B be a type 2 Banach space. Let $c \geq 2$, and $m_1 = m_2 = \cdots = m_c = 1$ and rank $r = c$. Suppose that*

$$E\,(\|\Phi\|^2(\ln\|\Phi\|)^{c-1}(\ln\ln\|\Phi\|)^{-1}) < \infty.$$

Then the sequence

$$\{|\bar{n}|\|U_{\bar{n}}\|/a_{\bar{n}}\}$$

satisfies a bounded LIL and a compact LIL.

Proof. Since

$$|\bar{n}|U_{\bar{n}}/a_{\bar{n}} = \frac{1}{a_{\bar{n}}}\sum_{i_1=1}^{n_1}\cdots\sum_{i_c=1}^{n_c}\Phi(X_{1i_1},\ldots,X_{ci_c})$$

then the statements of this theorem follow from Li and Wu (1989).

Corollary 9.6.2. *By virtue of the compact LIL we have*

$$\lim_{|\bar{n}|\to\infty} \inf_{x\in K}\||\bar{n}|U_{\bar{n}}/a_{\bar{n}} - x\| = 0 \quad a.s.,$$

$$P(C\{|\bar{n}|U_{\bar{n}}/a_{\bar{n}};\bar{n}\in N^c\} = K) = 1$$

and

$$\lim_{|\bar{n}|\to\infty} \sup \||\bar{n}|U_{\bar{n}}/a_{\bar{n}}\| = \sqrt{c}\,\sup_{x\in K}\|x\| \quad a.s.$$

where K is the closed unit ball of the reproducing kernel Hilbert space H_T with $T(f,g) = E\,(f(\Phi)g(\Phi))$, $f,g \in B^*$ and $C\{x_{\bar{n}};\bar{n}\in N^c\}$ is the set of all the limit points of the sequence $\{x_{\bar{n}};\bar{n}\in N^c\}$.

Proof. Under the conditions of the compact LIL according to Wichura's LIL we can write

$$\lim_{|\bar{n}|\to\infty} \sup f(|\bar{n}|U_{\bar{n}}/a_{\bar{n}}) = (E\, f^2(\Phi))^{1/2} \quad a.s. \quad \forall\, f \in B^* .$$

The corollary then follows from the arguments provided in Kuelbs (1976) for $c = 1$.

Chapter 10

Dependent variables

It can be assumed that the random variables $X_1, \ldots, X_n$ in the definition of UB-statistics are dependent. For these kinds of UB-statistics we are able to study the whole spectrum of asymptotic problems which are considered for independent random variables. In order to obtain valuable results in this situation it is necessary to control the nature of dependence. In what follows, UB-statistics are studied for symmetrically dependent random variables, weakly dependent random variables, bootstrap variables and for order statistics.

10.1 Symmetrically dependent random variables

Let $X = (X_1, X_2, \ldots, X_n, \ldots)$ be a sequence of random variables. This sequence is said to be *symmetrically dependent* (or *permutable* or *exchangeable* or *interchangeable*) if arbitrary subsets $(X_{i_1}, \ldots, X_{i_k})$ with k different elements have the same joint distribution which depends only on k. Here the sequence X can be either finite or infinite.

Consider UB-statistics

$$U_n = n^{-[m]} \sum\nolimits_{I_n^m} \Phi(X_{i_1}, \ldots, X_{i_m}),$$

where $X_1, \ldots, X_n$ are symmetrically dependent random variables, and $\Phi : X^m \to B$ is a symmetric function such that $E\|\Phi(X_1, \ldots, X_m)\| < \infty$.

Let $\mathfrak{B}_n = \sigma\{\omega : U_n, U_{n+1}, \ldots\}$ be the σ-algebra generated by the sequence $U_n, U_{n+1}, \ldots$. Evidently,

$$\mathfrak{B}_n \supseteq \mathfrak{B}_{n+1} \quad \text{for all} \quad n = m, m+1, \ldots .$$

Denote

$$\mathfrak{B}_\infty = \bigcap\nolimits_{n=m}^{\infty} \mathfrak{B}_n .$$

Since $X_1, \ldots, X_n, \ldots$ are symmetrically dependent random variables, it is not necessary that this σ-algebra $\mathfrak{B}_\infty$ should be trivial.

Theorem 10.1.1. *If* $E\|\Phi(X_1,\ldots,X_m)\| < \infty$, *then*

$$U_n \to E(\Phi(X_1,\ldots,X_m)|\mathfrak{B}_\infty)$$

as $n \to \infty$ *with probability* 1 *and in* L^1.

Proof. Let $Y = h(X)$ be a bounded $\mathfrak{B}_n$-measurable random variable. Then the symmetrical dependence of X implies that for $1 \le i_1 \ne \ldots \ne i_m \le n$

$$E(\Phi(X_{i_1},\ldots,X_{i_m})h(X)) = E(\Phi(X_1,\ldots,X_m)h(X)).$$

Carrying out the summation over all indices $1 \le i_1 \ne \ldots \ne i_m \le n$ on both sides of this equality, we obtain

$$E(U_nY) = E(\Phi(X_1,\ldots,X_m)Y).$$

For any $A \in \mathfrak{B}_n$, Y may be taken as the indicator of A, so that

$$\int_A U_n dP = \int_A \Phi(X_1,\ldots,X_m)dP, \quad A \in \mathfrak{B}_n.$$

The integrand on the left is $\mathfrak{B}_n$-measurable, and therefore

$$U_n = E(\Phi(X_1,\ldots,X_m)|\mathfrak{B}_n).$$

The sequence $(U_n, \mathfrak{B}_n)$ is a reversed martingale.

Hence, the statement of Theorem 10.1.1 follows from the martingale limit theorem of Chatterji (1968).

10.2 Weakly dependent random variables

Assume that a sequence X_1, $X_2,\ldots,X_n,\ldots$, in the definition of *UB*-statistics, is a stationary (in the strict sense) sequence of random variables taking values on a measurable space $(X, \mathfrak{X})$.

Let $\mathfrak{F}_a^b$ denote the σ-algebra of events generated by $X_a, X_{a+1},\ldots,X_b$, where $1 \le a \le b < \infty$ and $\mathfrak{F}_m^\infty$ is the σ-algebra generated by $X_m, X_{m+1},\ldots$. If the σ-algebras $\mathfrak{F}_1^k$ and $\mathfrak{F}_{k+n}^\infty$ are independent, then $P(A \cap B) - P(A)P(B) = 0$ for all the events $A \in \mathfrak{F}_1^k$ and $B \in \mathfrak{F}_{k+n}^\infty$. In the general case, this formula does not hold, and the expression on the left-hand side serves as a measure of dependence between $\mathfrak{F}_1^k$ and $\mathfrak{F}_{k+n}^\infty$; moreover, the mixing coefficients may be regarded as specific numerical characteristics of this dependence.

We say that a stationary (in the strict sense) sequence $X_1, X_2,\ldots,X_n,\ldots$ possesses the property of *ψ-mixing* (or satisfies the condition of *strengthened uniformly strong mixing*) if

$$\psi(n) = \sup\left\{|P(A\cap B) - P(A)P(B)|/P(A)P(B) : A \in \mathfrak{F}_1^k,\ B \in \mathfrak{F}_{k+n}^\infty,\ k = 1,2,\ldots\right\}$$

tends to zero as $n \to \infty$. For $c = 0, 1, \ldots, m$, we define the functions

$$\Phi_c(x_1, \ldots, x_c) =$$

$$= \int_X \cdots \int_X \Phi(x_1, \ldots, x_c, x_{c+1}, \ldots, x_m) P(dx_{c+1}) \ldots P(dx_m)$$

and

$$g_c(x_1, \ldots, x_c) = \sum_{d=0}^{c} (-1)^{c-d} \sum_{1 \le j_1 < \ldots < j_d \le c} \Phi_d(x_{j_1}, \ldots, x_{j_d})$$

by setting $\Phi_0 = \theta$ and $\Phi_m = \Phi$, where P denotes the distribution of the random variable X_j.

Theorem 10.2.1. *Let $B = H$ and let the kernel Φ satisfy the following conditions:*

$$\int_X \cdots \int_X \|\Phi(x_1, \ldots, x_m)\|^2 P(dx_1) \ldots P(dx_m) < \infty,$$

$$\sum_{k=1}^{\infty} k^{m-1} \psi^{1/2}(k) < \infty,$$

$$\sigma^2 = E\|g_1\|^2 + 2 \sum_{j=2}^{\infty} E(g_1(X_1), g_1(X_j)) \neq 0.$$

Then the weak convergence

$$m^{-1}\sqrt{n}(U_n - \theta) \Rightarrow \tau,$$

as $n \to \infty$, takes place, where τ is a Gaussian random element in H with zero mean value, and the covariance operator V is defined by

$$(Vx, y) = E\{(x, g_1(X_1))(y, g_1(X_1))\} + \sum_{n=2}^{\infty} E\{(x, g_1(X_1))(y, g_1(X_n))\} +$$

$$+ \sum_{n=2}^{\infty} E\{(y, g_1(X_1))(x, g_1(X_n))\}$$

for all $x, y \in H$.

Proof. We now make the argument in Borovskikh (1994a) more precise. By using the Hoeffding decomposition we get

$$m^{-1}\sqrt{n}(U_n - \theta) = n^{-1/2}(g_1(X_1) + \ldots + g_1(X_n)) + R_n,$$

where

$$R_n = m^{-1}\sqrt{n} \sum_{c=2}^{m} \binom{m}{c} U_{nc}.$$

First we show that, under conditions of this theorem,

$$R_n \xrightarrow{P} 0$$

as $n \to \infty$. If $\{e_j\}$, $j \geq 1$ is an orthonormal system in H, then

$$E\|U_{nc}\|^2 = \sum_{j=1}^{\infty} E(e_j, U_{nc})^2 .$$

Here (e_j, U_{nc}) is a *UR*-statistic

$$(e_j, U_{nc}) = \binom{n}{c}^{-1} \sum_{1\leq i_1<\ldots<i_c\leq n} (e_j, g_c(X_{i_1}, \ldots, X_{i_c})),$$

for which the following inequality holds

$$E(e_j, U_{nc})^2 \leq c_\psi n^{-c} \int_X \cdots \int_X (e_j, \Phi_c(x_1, \ldots, x_c))^2 P(dx_1) \ldots P(dx_c)$$

according to Sen (1972). The constant c_ψ depends only on $\psi = (\psi(1), \psi(2), \ldots)$. By inserting this inequality into the preceding formula, we get

$$E\|U_{nc}\|^2 \leq c_\psi n^{-c} \int_X \cdots \int_X \|\Phi(x_1, \ldots, x_m)\|^2 P(dx_1) \ldots P(dx_m) .$$

Since the definition of R_n contains the elements U_{nc} for $c \geq 2$, the Chebyshev inequality gives $R_n \to 0$ in probability.

Let $S_n = n^{-1/2}(g_1(X_1) + \cdots + g_1(X_n))$ and $\sigma_n^2 = E\|S_n\|^2$. We have

$$\sigma_n^2 = E\|g_1\|^2 + 2\sum_{j=2}^{n}(1 - jn^{-1})E(g_1(X_1), g_1(X_j)),$$

and therefore, under conditions of this theorem, $\sigma_n^2 \to \sigma^2$ as $n \to \infty$. Hence, in order that the statement of this theorem is valid, it is sufficient to show that

$$S_n \Rightarrow \tau .$$

This relation takes place if the sequence of measures generated by the random elements $\{S_n\}$ is weakly compact and

$$\lim_{n\to\infty} E\exp\{i(a, S_n)\} = \exp\{-\frac{1}{2}(Va, a)\}$$

for all $a \in H$ and the covariance operator V. These two properties have been established in Borovskikh (1986).

A stationary (in the strict sense) sequence $X_1, \ldots, X_n, \ldots$ satisfies the condition of α-*mixing* (*strong mixing*) if

$$\alpha(n) = \sup\{|P(A \cap B) - P(A)P(B)| : A \in \mathfrak{F}_1^k, B \in \mathfrak{F}_{k+n}^{\infty}, k = 1, 2, \ldots\}$$

approaches zero as $n \to \infty$. This sequence satisfies the condition of φ-*mixing* (*uniform strong mixing*) if

$$\varphi(n) = \sup\{|P(B|A) - P(B)| : A \in \mathfrak{F}_1^k, B \in \mathfrak{F}_{k+n}^{\infty}, k = 1, 2, \ldots\}$$

tends to zero. If

$$\beta(n) = \sup_{1\le k<\infty} E \sup\{|P(B|\mathfrak{F}_1^k) - P(B)| : B \in \mathfrak{F}_{k+n}^\infty\}$$

tends to zero, then we say that this sequence satisfies the condition of *absolute regularity*. It follows from these definitions that

$$\begin{aligned} \alpha(n) &\le \beta(n) \\ &\le \varphi(n) \\ &\le \psi(n) \end{aligned}$$

for all $n \ge 1$.

Hence, Theorem 10.2.1 also holds for the sequences with α-, β-, and φ-mixing.

10.3 Bootstrap variables

Bootstrapping *UB*-statistics. Let us consider a random sample $X_1, \ldots, X_n$, where the random variables X_i are independent and identically distributed with common law P. Define the empirical measure

$$P_n = n^{-1} \sum_{i=1}^{n} \delta_{X_i}.$$

Fix P_n. Then we define a random *bootstrap* sample $X_1^*, \ldots, X_N^*$ of volume N from a population with distribution function P_n. Thus, if the vector $(X_1, \ldots, X_n)$ is fixed, then the random variables $X_1^*, X_2^*, \ldots$ are independent, identically distributed, and

$$P(X_1^* = X_j | X_1, \ldots, X_n) = \frac{1}{n}, \quad j = 1, \ldots, n$$

(i.e. every X_i^* is obtained by sampling with replacement, from the data).

On this bootstrap sample, we define *bootstrapping UB-statistics* (*U*B-statistics*) as

$$U_N^* = N^{-[m]} \sum_{1\le i_1 \ne \ldots \ne i_m \le N} \Phi(X_{i_1}^*, \ldots, X_{i_m}^*). \tag{10.3.1}$$

Below, P^* and E^* correspond to distribution P_n. Therefore,

$$\theta^* = E^* U_N^* =$$

$$= E^*\Phi(X_1^*, \ldots, X_m^*) = n^{-m} \sum_{i_1=1}^{n} \cdots \sum_{i_m=1}^{n} \Phi(X_{i_1}, \ldots, X_{i_m}).$$

Lemma 10.3.1. *Let $m = 2$. Then the following formula holds*

$$U_N^* = N^{-[2]} \sum_{1\le i\ne j\le n} \gamma_i\gamma_j \Phi(X_i, X_j) + N^{-[2]} \sum_{i=1}^{n} \gamma_i(\gamma_i - 1)\Phi(X_i, X_i),$$

where the random vector $(\gamma_1, \dots, \gamma_n)$ *has a multinomial distribution with the parameters* $(N; n^{-1}, \dots, n^{-1})$ *for fixed vector* $(X_1, \dots, X_n)$.

Proof. Let $X_1, \dots, X_n$ be fixed. The distribution of the bootstrap random sample $X_1^*, \dots, X_N^*$ is concentrated on $\{X_1, \dots, X_n\}$. Due to this fact, the conditional probability of the event, which consists in the fact that X_1 has occurred in the bootstrap sample γ_1 times, ..., and X_n has occurred γ_n times, is determined by the formula

$$\frac{N!}{\gamma_1! \dots \gamma_n!} \left(\frac{1}{n}\right)^N, \quad \gamma_1 + \dots + \gamma_n = N.$$

This coincides with the multinomial vector with parameters $(N; n^{-1}, \dots, n^{-1})$. Then, taking into account the definition of the vector $(\gamma_1, \dots, \gamma_n)$, one can write

$$\sum_{1 \le i \ne j \le N} \Phi(X_i^*, X_j^*) = \sum_{i=1}^{N} \sum_{j=1}^{N} \Phi(X_i^*, X_j^*) - \sum_{i=1}^{N} \Phi(X_i^*, X_i^*) =$$

$$= \sum_{i=1}^{n} \sum_{j=1}^{n} \gamma_i \gamma_j \Phi(X_i, X_j) - \sum_{i=1}^{n} \gamma_i \Phi(X_i, X_i).$$

This yields the desired formula.

Bootstrapped canonical functions. The *bootstrapped canonical functions*

$$g_{c*}(x_1, \dots, x_c), \quad c = 1, \dots, m$$

are defined by the formula

$$g_{c*}(x_1, \dots, x_c) = \int_X \cdots \int_X \Phi(y_1, \dots, y_m) \prod_{s=1}^{c} (\delta_{x_s}(dy_s) - P_n(dy_s)) \prod_{s=c+1}^{m} P_n(dy_s)$$

i.e. in the definition of $g_c(x_1, \dots, x_c)$ the distribution P is replaced by P_n. Therefore, the functions $g_{c*}(x_1, \dots, x_c)$ are P_n-*canonical functions.* In particular, if $m = 2$ then

$$g_{1*}(x) = \frac{1}{n} \sum_{j=1}^{n} \Phi(x, X_j) - \frac{1}{n^2} \sum_{i=1}^{n} \sum_{j=1}^{n} \Phi(X_i, X_j),$$

$$g_{2*}(x, y) = \Phi(x, y) - \frac{1}{n} \sum_{j=1}^{n} \Phi(x, X_j) - \frac{1}{n} \sum_{i=1}^{n} \Phi(X_i, y) +$$

$$+ \frac{1}{n^2} \sum_{i=1}^{n} \sum_{j=1}^{n} \Phi(X_i, X_j).$$

This function does not involve P.

The bootstrapped Hoeffding decomposition. In these conditions

$$U_N^* - \theta^* = \sum_{c=1}^{m} \binom{m}{c} U_{Nc}^*, \tag{10.3.2}$$

where

$$U_{Nc}^* = N^{-[c]} \sum_{1 \le i_1 \neq \ldots \neq i_c \le N} g_{c*}(X_{i_1}^*, \ldots, X_{i_c}^*).$$

In what follows, it assumed that $N = n$.

Bootstrapping law of large numbers for *UB*-statistics.

Theorem 10.3.1. *Let B be a separable Banach space of type 2. Assume that for each possible combination of integers $i_1, \ldots, i_m$,*

$$E\|\Phi(X_{i_1}, \ldots, X_{i_m})\|^{\#(i_1,\ldots,i_m)/m} < \infty,$$

where $\#(i_1, \ldots, i_m)$ denotes the cardinality of the set $(i_1, \ldots, i_m)$. Then

$$U_n^* \xrightarrow{P^*} \theta \quad a.s.$$

as $n \to \infty$.

Proof. In the bootstrapped Hoeffding decomposition θ^* is a von Mises' statistic V_n and, by the law of large numbers for V_n, as $n \to \infty$, $\theta^* \to \theta$ a.s. In (10.3.2) for $1 \le c \le m$, $U_{nc}^* = U_n(g_{c*}) = U_n(g_{c*}^1) + U_n(g_{c*}^2)$,

$$g_{c*}^1(x_1, \ldots, x_c) =$$

$$= \int_X \cdots \int_X \Phi(y_1, \ldots, y_m) I(\|\Phi\| \le \sqrt{n}) \prod_{s=1}^{c} (\delta_{x_s}(dy_s) - P_n(dy_s)) \prod_{s=c+1}^{m} P_n(dy_s)$$

and $g_{c*}^2(x_1, \ldots, x_c) = (g_{c*} - g_{c*}^1)(x_1, \ldots, x_c)$.

Now by Corollary 2.7.1

$$E^*\|U_n(g_{c*}^1)\|^2 \le A_{1c} \binom{n}{c}^{-1} E^*\|g_{c*}^1\|^2 \le$$

$$\le A_{2c} \binom{n}{c}^{-1} n^{-2m+c} \sum_{i,j,k=1}^{n} \|\Phi\| I(\|\Phi\| \le \sqrt{n})(X_{i_1}, \ldots, X_{i_c}, X_{j_1}, \ldots, X_{j_{m-c}}) \times$$

$$\times \|\Phi\| I(\|\Phi\| \le \sqrt{n})(X_{i_1}, \ldots, X_{i_c}, X_{k_1}, \ldots, X_{k_{m-c}})$$

$$\le A_{2c} \binom{n}{c}^{-1} n^{-m+1/2} \sum_{i,j=1}^{n} \|\Phi(X_{i_1}, \ldots, X_{i_c}, X_{j_1}, \ldots, X_{j_{m-c}})\| \to 0$$

with probability 1 as $n \to \infty$ by the law of large numbers for von Mises' statistics. Then we have

$$E^*\|U_n(g_{c*}^2)\| \leq$$
$$\leq 2^m E^*\|\Phi(X_1^*,\dots,X_m^*)\| I(\|\Phi(X_1^*,\dots,X_m^*)\| > n^{1/2}) =$$
$$= 2^m n^{-m} \sum_{i_1,\dots,i_m=1}^{n} \|\Phi(X_{i_1},\dots,X_{i_m})\| I(\|\Phi(X_{i_1},\dots,X_{i_m})\| > \sqrt{n}) \to 0$$

with probability 1 as $n \to \infty$ again by LLN. This proves Theorem 10.3.1.

Central limit theorem for U*B-statistics. In what follows, L^* denotes the conditional law with respect to the sample $X_1,\dots,X_n$, i.e. $L^*(\xi) = L(\xi|X_1,\dots,X_n)$.

Theorem 10.3.2. *Let Banach space B be of type 2. Suppose that kernel $\Phi : X^2 \to B$ satisfies the following conditions*

$$0 < E\|g_1(X_1)\|^2 < \infty, \quad E\|g_2(X_1,X_2)\|^{4/3} < \infty, \quad E\|g_2(X_1,X_1)\|^{1/2} < \infty.$$

Then, with probability 1

$$\left| L^*\left(\frac{\sqrt{n}}{2}(U_n^* - \theta^*)\right) - L\left(\frac{\sqrt{n}}{2}(U_n - \theta)\right) \right| \to 0$$

as $n \to \infty$.

Proof. By formula (10.3.2)

$$\frac{\sqrt{n}}{2}(U_n^* - \theta^*) = \frac{1}{\sqrt{n}}(g_{1*}(X_1^*) + \dots + g_{1*}(X_n^*)) + R_{n*},$$

where

$$R_{n*} = \frac{1}{(n-1)\sqrt{n}} \sum_{1 \leq i < j \leq n} g_{2*}(X_i^*, X_j^*).$$

According to Theorem 4.1.3

$$L\left(\frac{\sqrt{n}}{2}(U_n - \theta)\right) \to L(\tau) \quad \text{as } n \to \infty$$

where τ is a Gaussian random element with covariance function V, defined in this theorem. We also have by the conditional central limit theorem with probability one

$$L^*(\frac{1}{\sqrt{n}}(g_{1*}(X_1^*) + \dots + g_{1*}(X_n^*))) \to L(\tau)$$

as $n \to \infty$. Now we will show that the random variable R_{n*} possesses the property

$$R_{n*} \xrightarrow{P^*} 0 \quad a.s.$$

as $n \to \infty$. Let $R_{n*}(g_{2*}) := R_{n*}$. Then we present $g_{2*}(x,y)$ in the form of a sum $g_{2*} = g_{2*}^1 + g_{2*}^2$ with the functions g_{2*}^1 and g_{2*}^2, defined in Theorem 10.3.1. Hence,

$$R_{n*} = R_{n*}(g_{2*}^1) + R_{n*}(g_{2*}^2) .$$

For any $\varepsilon > 0$, by the Chebyshev inequality, this yields

$$P^*(\|R_{n*}\| > \varepsilon) \le 4\varepsilon^{-2}E^*\|R_{n*}(g_{2*}^1)\|^2 + (2\varepsilon^{-1})^{4/3}E^*\|R_{n*}(g_{2*}^2)\|^{4/3} .$$

The functions g_{2*}^1 and g_{2*}^2 possess the property of degeneracy. Therefore, the expectations on the right-hand side of this inequality can be estimated by use of Corollary 2.7.1. Thus, one can show that

$$E^*\|R_{n*}(g_{2*}^1)\|^2 \to 0 \quad a.s.$$

and

$$E^*\|R_{n*}(g_{2*}^2)\|^{4/3} \to 0 \quad a.s.$$

as $n \to \infty$. This completes the proof of Theorem 10.3.2.

Convergence in the degenerate case. Let kernel $\Phi : X^m \to B$ have rank $r = m$. Then we can write

$$\Phi(x_1, \dots, x_m) = \int_X \cdots \int_X \Phi(y_1, \dots, y_m) \prod_{s=1}^m (\delta_{x_s}(dy_s) - P(dy_s)).$$

Denote

$$\Phi^*(x_1, \dots, x_m) = \int_X \cdots \int_X \Phi(y_1, \dots, y_m) \prod_{s=1}^m (\delta_{x_s}(dy_s) - P_n(dy_s))$$

as the bootstrapped variant of this kernel Φ. Consider the bootstrapped U^*B-statistics

$$U_{n*}^* = n^{-[m]} \sum_{1 \le i_1 \ne \dots \ne i_m \le n} \Phi^*(X_{i_1}^*, \dots, X_{i_m}^*) .$$

Theorem 10.3.3. *Let Banach space B be of type 2. Suppose that kernel $\Phi : X^m \to B$ has rank $r = m$ and satisfies the conditions*

$$E\|\Phi(X_{i_1}, \dots, X_{i_m})\|^{2\#(i_1,\dots,i_m)/m} < \infty .$$

Then, with probability 1

$$|L^*(n^{m/2}U_{n*}^*) - L(n^{m/2}U_n)| \to 0$$

as $n \to \infty$.

Proof. It is sufficient to combine the methods of the proof of Theorem 4.2.5 and Theorem 4.2 in Arcones and Giné (1992).

Exchangeably weighted bootstrap of *UB*-statistics. Lemma 10.3.1 contains the idea of the *exchangeably weighted bootstrap.* Let us introduce into consideration the *weighted UB-statistics*

$$U_n^w = \sum_{1\le i\ne j\le n} \Big(w_{ni} - \frac{1}{n}\Big)\Big(w_{nj} - \frac{1}{n}\Big)\Phi(X_i, X_j),$$

where we assume that the weights $(w_{n1}, \dots, w_{nn})$ are exchangeable. For example, the multinomial weights $(nw_{n1}, \dots, nw_{nn})$ in Efron's resampling scheme have a multinomial distribution with parameters

$$(n; n^{-1}, \dots, n^{-1})$$

for fixed vector $(X_1, \dots, X_n)$.

Theorem 10.3.4. *Let Banach space B be of type 2. Suppose that the degenerate kernel $\Phi : X^2 \to B$ satisfies the following conditions*

$$E\|\Phi(X_1, X_2)\|^2 < \infty$$

and

$$E\|\Phi(X_1, X_2)\| < \infty.$$

In addition, assume the weights satisfy the conditions

$$w_{ni} \ge 0, \quad i = 1, \dots, n,$$

$$\sum_{i=1}^{n} w_{ni} = 1,$$

$$n\sum_{i=1}^{n}(w_{ni} - \frac{1}{n})^2 \xrightarrow{P} c^2, \quad n \to \infty \quad \textit{for some} \quad c > 0,$$

$$n \max\nolimits_{1\le i\le n}\Big(w_{ni} - \frac{1}{n}\Big)^2 \xrightarrow{P} 0, \quad n \to \infty.$$

Then, with probability 1

$$|L^*(nc^{-1}U_n^w) - L(nU_n)| \to 0$$

as $n \to \infty$.

Proof. By Theorem 4.2.5 $L(nU_n) \to L(K_m(\Phi))$. Further, combining, the arguments of Arcones and Giné (1992), Huskova and Janssen (1993) and the method of the proof of Theorem 4.2.5 we find that with probability 1 $L^*(nc^{-1}U_n^w) \to L(K_m(\Phi))$ as $n \to \infty$. This completes the proof of Theorem 10.3.4.

10.4 Order statistics

Let $X_{n1} \leq X_{n2} \leq \ldots \leq X_{nn}$ be the *order statistics* of the sample data $X_1, X_2, \ldots X_n$, having the function of distribution $F(x)$. Consider the generalized *UB*-statistics of such a form

$$U_n = \sum_{1\leq i_1 \neq \ldots \neq i_m \leq n} \Phi_{n,i_1,\ldots,i_m}(X_{ni_1}, \ldots, X_{ni_m}). \tag{10.4.1}$$

Here, if $m = 1$, then

$$U_n = \sum_{i=1}^{n} \Phi_{ni}(X_{ni})$$

is a well-known L-statistic.

The general problem of the asymptotics of random variables of type (10.4.1) is difficult enough. Therefore, here we consider the following simple case. Let α and β be fixed numbers such that $0 < \alpha < \beta < 1$.

Let

$$F(a) = \alpha, \quad F(b) = \beta,$$

$$\mu = \int_a^b x dF(x), \quad \sigma^2 = Y_j^2,$$

where

$$Y_j = h(X_j) - Eh(X_j), \quad j = 1, \ldots, n,$$

$$h(x) = \begin{cases} a, & x < a \\ x, & a \leq x \leq b \\ b, & x > b. \end{cases}$$

Theorem 10.4.1. *Suppose that $EX_1^2 < \infty$. Let m be fixed and $n \to \infty$. Then the following weak convergence takes place*

$$\sigma^{-m} n^{-m/2} \sum_{\alpha n \leq i_1 \neq \ldots \neq i_m \leq \beta n} \prod_{s=1}^{m} (X_{ni_s} - \mu) \Rightarrow H_m(\tau)$$

where τ is a standard normal random variable and H_m is a Hermite polynomial.

Proof. By Remark 4.5 in Borovskikh and Korolyuk (1994b) we have the following representation

$$\sigma^{-m} n^{-m/2} \sum_{\alpha n \leq i_1 \neq \ldots \neq i_m \leq \beta n} \prod_{s=1}^{m} (X_{ni_s} - \mu) =$$

$$= H_m(\tau_{n1}, \tau_{n2}) + \sum_{k=3}^{m} P_k(\tau_n^{(k)}) H_{m-k}(\tau_{n1}, \tau_{n2}) m^{[k]},$$

where $P_k(\tau_n^k)$ is the polynomial of degree k with variables $\tau_n^{(k)} := (\tau_{n3}, \ldots, \tau_{nk})$, which are defined in the asymptotic expansion

$$\exp\{-\sum_{k=3}^{\infty}(-1)^k z^k \tau_{nk}/k\} = 1 + \sum_{k=3}^{\infty} z^k P_k(\tau_n^{(k)}).$$

In addition,

$$\tau_{nk} = \sigma^{-k} n^{-k/2} \sum_{\alpha n \le j \le \beta n} (X_{nj} - \mu)^k , \quad k = 1, 2, \ldots .$$

By the central limit theorem for L-statistics

$$\tau_{n1} \Rightarrow \tau \quad \text{as} \quad n \to \infty .$$

By the strong law of large numbers

$$\tau_{n2} \to 1 \quad a.s.$$

and for $3 \le k \le m$

$$\tau_{nk} \xrightarrow{P} 0$$

as $n \to \infty$. Hence,

$$H_m(\tau_{n1}, \tau_{n2}) \Rightarrow H_m(\tau)$$

and for $3 \le k \le m$

$$P_k(\tau_n^{(k)}) \xrightarrow{P} 0$$

as $n \to \infty$. This proves Theorem 10.4.1.

Theorem 10.4.2. *Let $\mu \neq 0$ and $n \to \infty$,*

$$mn^{-1/2} \to \gamma, \quad \gamma > 0 .$$

Assume that $EX_1^2 < \infty$. Then the weak convergence

$$n^{-[m]} \sum_{\alpha n \le i_1 \neq \ldots \neq i_m \le \beta n} \prod_{s=1}^{m} (X_{ni_s}/\mu) \Rightarrow \exp\{\frac{\gamma\sigma}{\mu}\tau - \frac{1}{2}(\frac{\gamma\sigma}{\mu})^2\}$$

takes place.

Proof. It is sufficient to apply the method of the proof of Theorem 4.7 in Borovskikh and Korolyuk (1994b).

Bibliographical supplements and comments

This section contains additional information which shows the basic papers from which the author has taken the results that are presented in this book. After the paper of Hoeffding (1948a) many publications devoted to the theory of U-statistics appeared and therefore the author's intention is to cite here all of the most important of them. The absence of comments with respect to some results presented in the text of the monograph does not, of course mean that they were established by the author or belong to him.

Introduction

The beginning of the probabilistic theory of U-statistics dates back to the fundamental work of Hoeffding (1948a) in which the central limit theorem was proved and different examples of the estimation of unknown parametric functionals were given.

From that time interest in the theoretical investigation of probabilistic properties of this class of random variables constantly arose and their practical value was proven. Inside the theory of unbiased estimates, U-statistics play a fundamental role by virtue of the theorem that states that if there exists unbiased estimate $\varphi(X_1, \ldots, X_n)$ of the parameter $\theta = \theta(P)$ and U_n is a corresponding U-statistic, then $E\,(U_n - \theta)^2 \leq E\,(\varphi(X_1, \ldots, X_n) - \theta)^2$, where the sign of inequality takes place if and only if $\varphi(X_1, \ldots, X_n) = U_n$ a.s. It means that each unbiased estimate for θ in L^2 can be improved with the help of U-statistics.

Problems of the necessary and sufficient conditions of the existence of unbiased estimates for the regular functional $\theta(P)$ were studied by Halmos (1946). So, in this respect Hoeffding (1948a) defined U-statistics under the influence of the already mentioned paper of Halmos. The second work which is close to the spirit of Hoeffding's (1948a) paper and is one of the sources of the theory of U-statistics, is the paper of von Mises (1947) in which he studied functionals $V_n = \theta(P_n)$ from the empirical distribution function P_n. Random variable V_n is close in structure to U-statistics but it is not an unbiased estimate of the regular functional $\theta(P)$.

Until 1986 only *UR*-statistics were in the field of vision of mathematical explorers. First works concerning *UH*-statistics belong to Borovskikh (1986), Borovskikh and Korolyuk (1987a, 1987b) and Dehling *et al.* (1986). The asymptotic theory of *UB*-statistics was developed in Borovskikh (1988a,d,e), where basic probabilistic laws for that class of random variables were proved.

Further progress in exploring stochastic laws of *UB*-statistics can be found in the works of Arcones (1994), Arcones and Giné (1993a, 1993b, 1994), Borovskikh (1989c, 1991, 1994a), Borovskikh and Korolyuk (1994a), de la Peña (1992), de la Peña *et al.* (1994), Dehling (1989a).

Chapter 1

1.1. The definition of *UB*-statistics for $B = R$ and numerous examples can be found in Hoeffding (1948a, 1948b), Borovskikh and Korolyuk (1994a), Lee (1990), Serfling (1980), Ghosh (1984), Sen (1981), Puri and Sen (1979).

The canonical decomposition (1.1.9) was discovered by Hoeffding (1961) in connection with the proof of the strong law of large numbers for *UR*-statistics under the condition $E|\Phi(X_1, \ ..., \ X_m)| < \infty$. The dependence of the limit distribution of *UR*-statistics on rank r from (1.1.10) was found first by Hoeffding (1948a, 1948b). The martingale properties of *UR*-statistics were first established by Hoeffding (1961) and Berk (1966). Formula (1.1.16) was derived by Hoeffding (1963).

1.2. Multi-sample *UR*-statistics (1.2.1) were introduced by Lehmann (1951) and Dwass (1956). Various examples of these *UR*-statistics are collected in Puri and Sen (1971), Sen (1981), and Sugiura (1965). Formula (1.2.11) was also obtained by Hoeffding (1963).

1.3. Real-valued statistics of the form (1.3.1) were introduced by von Mises (1947). Representation via *U*-statistics (1.3.3) can be found in Bönner and Kirschner (1977). Multi-sample von Mises' statistics (1.3.6) were considered by Sen (1974c, 1977, 1981).

1.4. The contents of this section correspond to Borovskikh (1994a).

1.5. The permanent definitions connected with symmetric statistics are taken from Borovskikh and Korolyuk (1994b).

1.6. Here, we follow Borovskikh and Korolyuk (1994a). The basic notions of the theory of multiple stochastic integrals can be found in Engel (1982) and Major (1981). A multiple integral with respect to Gaussian random measures was first defined by Itô (1951). The definition of the integral $I_m(f)$ is given according to the paper Mandelbaum and Taqqu (1984). Relations (1.6.1)–(1.6.3) can be found in Major (1981). Multiple stable integrals were studied in Janicki and Weron (1994), Samorodnitsky and Szulga (1989), Samorodnitsky and Taqqu (1990), and Szulga (1991, 1992a, 1992b).

1.7. Probabilistic properties of polynomial chaos have been investigated in Acosta (1987), Jong (1987, 1990), Krakowiak and Szulga (1986, 1988), Kwapień (1987), Kwapień and Woyczynski (1986, 1992), McConnell and Taqqu (1987), Mathai and Provost (1992), Rajput and Rosinski (1990). The different types of convergence in (1.7.2) were studied in Kwapień and Woyczynski (1992). *UB*-statistics of the type (1.7.3) have been considered slightly in the literature.

Chapter 2

2.1. Example 2.1.1 can be found in Kwapień and Woyczynski (1992). Majorization Theorem 2.1.3 belongs to Kwapień and Woyczynski (1992). It improves a result of Berger (1991) by replacing the original constant 4 by the constant 2 (see also Pinelis (1994b)). Theorem 2.1.4 is also taken from Kwapień and Woyczynski (1992).

2.2. This section is presented according to Borovskikh (1994a), and Borovskikh and Korolyuk (1988e).

2.3. The inequalities of this section are also presented in accordance with Borovskikh (1994a), and Borovskikh and Korolyuk (1988e).

2.4. Lemmas 2.4.1–1.4.4 are well-known in probability theory (see, for example, Araujo and Giné (1980), Kwapień and Woyczynski (1992)). Symmetrization inequality (2.4.2) was established by Sherman (1994). Pollard (1982) discovered how to symmetrize using sign variables for the case $m = 1$. Nolan and Pollard (1987) developed the technique for $m = 2$.

2.5. The decoupling principle is descended from the theory of martingale transforms (see Burkholder (1986)). Beginning with Kwapień (1987), and McConnell and Taqqu (1986, 1987) many papers have appeared on decoupling inequalities for homogeneous random forms. The interest of these authors in decoupling inequalities was motivated by the investigation of stochastic integrals.

Now the theory of the decoupling principle has moved into several other directions and the sphere of its application has expanded. Inequality (2.5.4) improves a result of de la Peña (1992) by replacing the original constant 8 by the constant 7 and inequality (2.5.5) improves de la Peña's estimate by replacing his constant 1/4 by the constant 1/3. Basic Theorem 2.5.4 of this section also strengthens the corresponding decoupling inequalities of de la Peña (1992). Decoupling inequalities for multilinear forms and U-statistics were proved in de la Peña *et al.* (1994) with various types of comparisons, including tail distributions and rearrangement invariant norms (e.g. Lorentz and Orlicz norms). The conditions of measurability, under which the theorems of this section hold, contain mainly Pollard's ideas (1990) and were discussed in de la Peña (1992).

2.6. The inequalities of Khintchine are reviewed in the paper of Peskir and Shiryaev (1994). Applications of hypercontractive methods in probability theory started with Borell (1979, 1984). He proved that Bernoulli random variables are hypercontractive in Banach spaces. Definition 2.6.1 and inequality (2.6.1) are taken from Kwapień and Woyczynski (1992). The notion of a hypercontractive random variable was introduced in Krakowiak and Szulga (1988).

2.7. Definition 2.7.1 can be found in Hoffmann-Jørgensen (1977).

Chapter 3

3.1. Convergence with probability 1 for one-sample UR-statistics was first proved by Hoeffding (1961) under the condition $E|\Phi(X_1, ..., X_m)| < \infty$. After that this law was again proved by Berk (1966) and Kingman (1969) with the help of the theorem on the convergence of reverse martingales. Theorem 3.1.1 is presented according to Borovskikh (1988e, 1994a) (see also, Borovskikh and Korolyuk (1994a)). Marcinkiewicz type laws of large numbers for U-statistics were also studied by Giné and Zinn (1992a), and Teicher (1992). Sen (1974) considered L^p-convergence of UR-statistics. Theorem 3.1.4 belongs to McConnell (1987a, 1988).

3.2. Theorem 3.2.1 can be found in Borovskikh and Korolyuk (1988e). Theorem 3.2.2 belongs to McConnell (1987b). The laws of large numbers in the sector index set were investigated in Borovskikh and Korolyuk (1988e).

3.3. Analogous theorems for real-valued von Mises' statistics are in Borovskikh and Korolyuk (1994a).

3.4. Estimates for UB-statistics were proved by Borovskikh (1988d, 1994a). Estimates for UH-statistics were also presented in Borovskikh and Korolyuk (1994a). There exist many papers devoted to the laws of large numbers for UB-statistics under different conditions on Banach spaces B, kernel Φ and distribution P (see, Borovskikh (1988b, 1988d, 1989b), Christofides and Serfling (1990), Giné and Zinn (1992a), Henze and Voigt (1992), McConnell and Rieders (1989), Chen (1980), Denker (1985), Eagleson and Weber (1978), Kokic (1987), Sen (1974), Su (1984, 1987)).

Chapter 4

4.1. The formulation of the central limit theorem for non-degenerate *UR*-statistics under condition $E\Phi^2(X_1, \dots, X_m) < \infty$ belongs to Hoeffding (1948a). It is a classic result. The theorems of this section strengthen Hoeffding's theorem. Theorems 4.1.1–4.1.2 were first proved by Borovskikh (1988a, 1989a, 1994). Theorem 4.1.3 can also be found in Arcones (1994a). Previously this theorem was obtained by Borovskikh (1988a, 1991, 1994) under the uniform smoothness condition on a Banach space. Borisov (1991) remarked that this condition is not required.

Theorem 4.1.4 strengthens Borovskikh's Theorem (1994a).

4.2. Theorem 4.2.1 was obtained by Janson (1979), and Rubin and Vitale (1980). Permanent Lemma 4.2.1 can be found in Borovskikh and Korolyuk (1994b). The representation of the limit random variable $K_m(\Phi)$ in the form of stochastic integral (4.2.4) was established by Dynkin and Mandelbaum (1983). Theorem 4.2.2 belongs to Giné and Zinn (1994). Theorem 4.2.3 belongs to the author. Corollary 4.2.2 was proved by Gregory (1977) with the help of the method of orthogonal expansions, which can be found in Borovskikh and Korolyuk (1994a), Eagleson (1979), Janson (1979), Serfling (1980) and de Wet (1987). Theorems 4.2.4–4.2.7 were proved by the author. Lemma 4.2.2 belongs to Arcones (1994a).

4.3. Theorems in this section can be found in Borovskikh and Korolyuk (1988e). Proposition 4.3.4 agrees with the result of Eagleson (1979).

4.4. This section is written according to Borovskikh and Korolyuk (1994b).

4.5. Theorem 4.5.1 was proved in Borovskikh and Korolyuk (1991a, 1991b, 1994e). Theorem 4.5.2 belongs to Borovskikh and Korolyuk (1991c, 1994b).

4.6. Asymptotic properties of *U*-statistics with increasing degrees were investigated in Borovskikh and Korolyuk (1991a, 1994b, 1994e), Es (1986), Es and Helmers (1986), Frees (1989), Halasz and Szekely (1976), Szekely (1982). Theorems 4.6.1–4.6.2 can be found, for example, in Borovskikh and Korolyuk (1994e), Es and Helmers (1986). Theorems 4.6.3–4.6.7 are presented according to Borovskikh and Korolyuk (1994e).

4.7. Theorems 4.7.1–4.7.2 were proved by the author. For *UR*-statistics, Theorem 4.7.2 was obtained in Dynkin and Mandelbaum (1983). Theorems 4.7.4–4.7.5 and Examples 4.7.1–4.7.2 belong to Feldman and Rachev (1993).

4.8 Theorem 4.8.1 was first given in Borovskikh and Korolyuk (1984a, 1994a). Corollaries 4.8.1–4.8.3 can be found in Borovskikh and Korolyuk (1994a). Corollary 4.8.2 belongs to Hall (1984). Corollary 4.8.3 was proved by Rosen (1969). Theorem 4.8.2

belongs to Weber (1983). The example presented in Theorem 4.8.3 is taken from Rosen (1969). Theorem 4.8.4 is given according to Silverman and Brown (1978). Theorem 4.8.5 and Example 4.8.1 belong to Jammalamadaka and Janson (1986). Theorems 4.9.1–4.9.2 were obtained by Major (1994). Theorem 4.8.3 was established in Dehling *et al.* (1990).

Chapter 5

5.1. Theorems 5.1.1–5.1.2 sharpen the corresponding results of Miller and Sen (1972) (see also Sen (1981)).

5.2. Theorem 5.2.1 is presented according to Hall (1979).

5.3. Theorem 5.3.1 is a well-known result (see, for example, Borovskikh and Korolyuk (1994), Mori (1981), Mandelbaum and Taqqu (1984)). Theorem 5.3.2 was proved by the author. Theorems about multi-sample invariance principles are given according to Borovskikh and Korolyuk (1988e).

5.4. The results of this section are connected with Borovskikh and Korolyuk (1994a).

5.5. Theorem 5.5.1 belongs to Mandelbaum and Taqqu (1984). Theorem 5.5.2 was proved by Feldman and Rachev (1993).

5.6. Theorem 5.6.1 is also presented in Borovskikh and Korolyuk (1988i, 1994a). Theorem 5.6.2 is given according to Ronzhin (1986).

5.7. Basic knowledge about empirical processes can be found in Wellner (1992). Theorems in this section belong to Arcones and Giné (1993) and Arcones (1994a).

Chapter 6

6.1. The general inequalities connected with the method of characteristic functions can be found in Bhattacharya and Ranga Rao (1983), Paulauskas and Raĉkauskas (1989), Sazonov (1981). The paper of Arratia *et al.* (1990) was used for the description of the Chen–Stein method.

6.2. Theorem 6.2.1 with new constants c_1 and c_2 belongs to the author.

6.3. Theorem 6.3.1 was proved by the author. Theorem 6.3.2 was obtained by Borovskikh and Korolyuk (1994b, 1995).

6.4. The nonuniform estimate in Theorem 6.4.1 is given according to Borovskikh and Korolyuk (1994a).

6.5. Theorem 6.5.1 was proved by the author.

6.6. Theorem 6.6.1 is also presented in Borovskikh *et al.* (1995). Theorems 6.6.2–6.6.3 belong to the author.

6.7. Theorem 6.7.1 belongs to the author.

6.8. This section is written according to Borovskikh and Korolyuk (1994a). The nonuniform Theorem 6.8.2 in the case $p = 1$, that corresponds to the Hilbert space H, was obtained by Sazonov and Zalesskii (1983).

6.9. In this section, we follow Borovskikh and Korolyuk (1994b). The rate of Poisson approximation with the help of the Chen–Stein method was studied in Barbour and Eagleson (1983–1987) for some symmetric statistics, which can be expressed in terms of U-statistics.

Chapter 7

7.1. Theorem 7.1.1 was obtained by Callaert *et al.* (1980). A brief proof of this theorem can be found in Lin (1984). Theorems 7.1.2–7.1.3 are given in Maesono and Yamato (1994). Theorem 7.1.4 was established by Bickel *et al.* (1986).

7.2. The general method is given according to Götze (1985) (see also Borovskikh and Korolyuk (1994a)).

7.3. Theorem 7.3.1 was proved by Götze (1984).

7.4. Theorem 7.4.1 is a generalization of Theorem 7.3.1.

Chapter 8

8.1. Lemma 8.1.1 and Theorem 8.1.1 are given according to Christofides (1991). Theorem 8.1.2 and Lemma 8.1.2 were established in Arcones and Giné (1993b). Theorems for UH-statistics and for UB-statistics belong to the author.

8.2. Theorem 8.2.1 sharpens the corresponding theorem of Vandemaele and Veraverbeke (1982).

8.3. Theorem 8.3.1 can be found in Vandemaele (1982), and Vandemaele and Veraverbeke (1982). Theorem 8.3.2 belongs to Dasgupta (1984).

8.4. Theorem 8.4.1 was proved by Borovskikh (1994a,b).

Chapter 9

9.1. The Law of Iterated Logarithms for non-degenerate *UR*-statistics under the condition $E\Phi^2(X_1, ..., X_m) < \infty$ was proved by Serfling (1980). The same result is established by the author under the conditions of Theorem 9.1.1.

9.2. Theorem 9.2.1 belongs to the author.

9.3. Theorem 9.3.1 belongs to the author. Theorem 9.3.2 and its proof are given according to the paper of Arcones and Giné (1994).

9.4. Theorem 9.4.1 was obtained by the author. Theorems 9.4.2–9.4.3 belong to Arcones and Giné (1994).

9.5. Theorem 9.5.1 is formulated according to Arcones and Giné (1994).

9.6. The statements of this section belong to the author.

Chapter 10

10.1. Symmetrically dependent random variables were considered by Kingman (1978). The law of large numbers in Theorem 10.1.1 was studied in Eagleson and Weber (1978) and Kingman (1978) for *UR*-statistics.

10.2. Theorem 10.2.1 is given according to Borovskikh (1986, 1994a). The problems connected with the estimates of the rate of weak convergence and the invariance principle for *UR*-statistics of weakly dependent variables were studied by Denker (1985), Denker and Keller (1983), Denker *et al.* (1985), Eagleson (1979), Sen (1972), Yoshihara (1976, 1984, 1990, 1992), and Kanagawa and Yoshihara (1994).

10.3. The asymptotic normality of U_n^* in (10.3.1) with a non-degenerate kernel Φ under the conditions $E\Phi^2(X_1, X_2) < \infty$ and $E\Phi^2(X_1, X_1) < \infty$ was studied by Bickel and Freedman (1981). Later, this result was improved by Athreya *et al.* (1984) and Shi (1987). Bootstrapping general empirical measures were established by Giné and Zinn (1990), Præstgaard and Wellner (1993). The bootstrap of *UR*-statistics in the degenerate case was studied by Bretagnolle (1983), Arcones and Giné (1992), Huskova and Janssen (1993). Lemma 10.3.1 can be found in Borovskikh and Korolyuk (1994a). Theorem 10.3.1 for *UR*-statistics was proved by Arcones and Giné (1992). Theorem 10.3.4 for *UR*-statistics was obtained by Huskova and Janssen (1993). Janssen (1994) considered also the weighted bootstrapping of *U*-statistics.

10.4. This section is presented according to Borovskikh and Korolyuk (1994b).

Bibliography

Acosta, A. de (1981). Inequalities for B-valued random vectors with applications to the strong law of large numbers. *Ann. Probab.* **9**, 1, 157–161.

Acosta, A. de (1987). A decoupling inequality for multilinear functions of stable vectors. *Probability and Math. Statistics* **8**, 1, 71–76.

Acosta, A. de and Kuelbs, J. (1983). Some results on the cluster set $C(\{S_n/a_n\})$ and LIL. *Ann. Probab.* **11**, 102–122.

Aerts, M. (1985). Rate of convergence to normality for U-statistics with kernel of arbitrary degree. *Metrika* **32**, 6, 351–362.

Aerts, M. and Callaert, H. (1986). The exact approximation order in the central limit theorem for random U-statistics. *Sequent. Anal.* **5**, 1, 19–35.

Aerts, M. and Callaert, H. (1989). The accuracies of the normal approximation for U-statistics with a random summation index convergence to random variable. *Acta. Sci. Math.* **53**, 3, 385–394.

Aldous, D.J. (1985). Exchangeability and related topics. *Lecture Notes in Math.* **1117**, Springer, New York.

Aldous, D.J. (1989). *Probability Approximation via the Poisson Clumping Heuristic.* Springer, Berlin.

Aleshkyavichene, A.K. (1990). Probabilities of large deviations for U-statistics and von Mises' functionals. *Theory Probab. Appl.* **35**, 1, 3–14.

Aleshkyavichene, A.K. (1992). Large deviations for U-statistics. *Lithuanian Math. J.* **32**, 1, 7-19.

Andersen, N.T., Giné, E. and Zinn, J. (1988). The cent ral limit theorem for empirical processes under local conditions: The case of Random infinitely divisible limits without Gaussian component. *Trans. Amer. Math. Soc.* **308**, 603–635.

Andersen, N.T. and Dobrić, V. (1987). The central limit theorem for stochastic processes. *Ann. Probab.* **15**, 1, 164–177.

Araujo, A. and Giné, E. (1980). *The Central Limit Theorem for Real and Banach Valued Random Variables.* Wiley, New York.

Arcones, M.A. (1993). The law of the iterated logarithm for *U*-processes. *J. Multivariate Anal.* **47**, 1, 139–151.

Arcones, M.A. (1994a). Limits of canonical *U*-processes and *B*-valued *U*-statistics. *J. Theor. Probab.* **7**, 2, 339–349.

Arcones, M.A. (1994b). On the central limit theorem for U-statistics under absolute regularity. *Preprint.*

Arcones, M.A., Chen, Z. and Giné, E. (1994). Estimators related to *U*-processes with applications to multivariate medians: asymptotic normality. *Ann. Statist.* **22**, 3, 1460–1477.

Arcones, M.A. and Giné, E. (1992a). On the bootstrap of *U*- and *V*-statistics. *Ann. Statist.* **20**, 2, 655–674.

Arcones, M.A. and Giné, E. (1992b). On the bootstrap of *U*-estimators and other statistical functionals. In: *Exploring the Limits of Bootstrap,* R. Le Page and L. Billard (Eds). Wiley, New York, 14–17.

Arcones, M.A. and Giné, E. (1993a). On decoupling, series expansions and tail behavior of chaos processes. *J. Theor. Probab.* **6**, 101–122.

Arcones, M.A. and Giné, E. (1993b). Limit theorems for *U*-processes. *Ann. Probab.* **21**, 3, 1494–1542.

Arcones, M.A. and Giné, E. (1994). *On the law of the iterated logarithm for canonical U-statistic and processes.* Preprint.

Arcones, M.A. and Yu, B. (1994). Central limit theorems for empirical and *U*-processes of stationery mixing sequences. *J. Theor. Probab.* **7**, 1, 47–71.

Arratia, R., Goldstein, L. and Gordon, L. (1990). Poisson approximation and the Chen–Stein method. *Statist. Sci.* **5**, 403–434.

Athreya, K.B., Ghosh, M., Low, L.V. and Sen, P.K. (1984). Laws of large numbers for bootstrapped *U*-statistics. *J. Statist. Plan. Infer.* **9**, 2, 185–194.

Avram, F. and Taqqu, M.S. (1986). Symmetric polynomials of random variables attracted to an infinitely divisible law. *Probab. Theory Related Fields* **71**, 4, 491–500.

Avram, F. and Taqqu, M.S. (1987). Non-central limit theorems and Appell polynomials. *Ann. Probab.* **15**, 2, 767–775.

Baldi, P. and Rinott, Y. (1989a). Asymptotic normality of some graph-related statistics. *J. Appl. Probab.* **26**, 1, 171–175.

Baldi, P. and Rinott, Y. (1989b). On normal approximations of distributions in terms of dependency graphs. *Ann. Probab.* **17**, 5, 1646–1650.

Barbour, A.D. (1988). Stein's method and Poisson process convergence. *J. Appl. Probab.* **25**, 1, 175–184.

Barbour, A.D. and Eagleson, G.K. (1983). Poisson approximation for some statistics based on exchangeable trials. *Adv. Appl. Probab.* **15**, 3, 585–600.

Barbour, A.D. and Eagleson, G.K. (1984). Poisson convergence for dissociated statistics. *J. R. Statist. Soc.* **46**, 3, 397–401.

Barbour, A.D. and Eagleson, G.K. (1985). Multiple comparisons and sums of dissociated random variables. *Adv. Appl. Probab.* **17**, 8, 147–162.

Barbour, A.D. and Eagleson, G.K. (1986). Random association of symmetric arrays. *Stochastic Anal. Appl.* **4**, 2, 239–281.

Barbour, A.D. and Eagleson, G.K. (1987). An improved Poisson limit theorem for sums of dissociated random variables. *J. Appl. Probab.* **24**, 3, 586–599.

Barbour, A.D. and Hall, P. (1984). On the rate of Poisson convergence. *Math. Proc. Camb. Phil. Soc.* **95**, 3, 473–480.

Barbour, A.D., Holst, L. and Janson, S. (1992). *Poisson Approximation.* Oxford University Press.

Bartmanska, B. and Szynal, D. (1992). Berry–Esséen theorem for some functions of *U*-statistics. *J. Statist. Plann. Inference* **30**, 3, 385–393.

Bentkus, V. and Götze, F. (1993). On smoothness conditions and convergence rates in the CLT in Banach spaces. *Probab. Theory Related Fields* **96**, 2, 137–151.

Bentkus, V., Götze, F., Paulauskas, V. and Raĉkauskas, A. (1990). *The accuracy of Gaussian approximation in Banach spaces.*, Encyclopedia of Mathematical Sciences. Springer-Verlag.

Bentkus, V., Götze, F. and Zitikis, R. (1993). Asymptotic expansions in the integral and local limit theorems in Banach spaces with applications to ω^2-statistics. *J. Theor. Probab.* **6**, 4, 727–780.

Bentkus, V., Götze, F. and Zitikis, R. (1994). Lower estimates of the convergence rate for *U*-statistics. *Ann. Probab.* **22**, 4.

Bentkus, V.I. and Račkauskas, A.I. (1990). On probabilities of large deviations in Banach spaces. *Probab. Theory Related Fields* **86**, 131–154.

Berger, E. (1991). Majorization, exponential inequalities and almost sure behavior of vector valued random variables. *Ann. Probab.* **19**, 3, 1106–1226.

Berk, R.H. (1966). Limiting behavior of posterior distributions when the model is incorrect. *Ann. Math. Statist.* **37**, 1, 51–58.

Bhattacharya, R.N. and Ghosh, J.K. (1992). A class of *U*-statistics and asymptotic normality of the number of K−clusters. *J. Multivariate Anal.* **43**, 2, 300–330.

Bhattacharya, R.N. and Ranga Rao, R. (1983). *Normal Approximation and Asymptotic Expansion.* Wiley, New York.

Bhattacharya, R.N. and Waymire, E.C. (1990). *Stochastic Processes with Application.* Wiley, New York.

Bickel, P.J. (1974). Edgeworth expansions in nonparametric statistics. *Ann. Statist.* **2**, 1, 1–20.

Bickel, P.J., Götze, F. and van Zwet, W.R. (1986). The Edgeworth expansion for *U*-statistics to degree two. *Ann. Statist.* **14**, 4, 1463–1484.

Billingsley, P. (1968). *Convergence of Probability Measures.* Wiley, New York.

Bingham, N.H. (1986). Variants on the law of the iterated logarithm. *Bull. London Math. Soc.* **18**, pt.5, No. 74, 433–467.

Bolthausen, E. and Götze, F. (1993). The rate of convergence for multivariate sampling statistics. *Ann. Statist.* **21**, 4, 1692–1710.

Bonami, A. (1970). Étude des coefficients de Fourier des fonctions de $L^p(G)$. *Ann. Inst. Fourier* **20**, 335–402.

Borell, C. (1979). On the integrability of Banach space valued Walsh polynomials. *Lecture Notes in Math.* **721**, Springer, New York, 1–3.

Borell, C. (1984). On polynomial chaos and integrability. *Probability and Math. Statistics* **3**, 191–203.

Borisov, I.S. (1990). Exponential inequalities for the distributions of von Mises and *U*-statistics. *Probability theory and mathematical statistics*, **1**, (Vilnius, 1989), "Mokslas", Vilnius, 166–178.

Borisov, I.S. (1991). Approximation of distributions of von Mises statistics with multidimensional kernels. *Sib. Math. J.* **32**, 4, 20–35.

Borovskikh, Yu.V. (1976). Asymptotic analysis in Hilbert space and some problems of mathematical statistics. *Soviet Math. Dokl.* **17**, 3, 752–755.

Borovskikh, Yu.V. (1979a). Approximation of distributions of *U*-statistics. *Dokl. Akad. Nauk Ukrain. SSR, Ser. A*, 9, 695–698. (In Russian).

Borovskikh, Yu.V. (1979b). Approximation of distributions of *U*-statistics. Part 1. *Deposited in VINITI, 04.12.79* **41-79**, *Institute of Mathematics, Novosibirsk*, 1–24. (In Russian).

Borovskikh, Yu.V. (1979c). The nonuniform estimates in the central limit theorem in Hilbert space. *Lithuanian Math. J.* **19**, 4, 29–37.

Borovskikh, Yu.V. (1980a). The problem of approximation of distributions of *U*-statistics and von Mises' functionals. *Preprint. #80.6, Institute of Math. Ukr. Acad. of Sci., Kiev*, 1–52. (In Russian).

Borovskikh, Yu.V. (1980b). The problem of approximation of distributions of *U*-statistics and von Mises' functionals. *Preprint. #80.7, Institute of Math. Ukr. Acad. of Sci., Kiev*, 1–56. (In Russian).

Borovskikh, Yu.V. (1980c). Approximation of distributions of *U*-statistics. *Sib. Math. J.* **21**, 5, 180.

Borovskikh, Yu.V. (1980d). Approximation of distributions of von Mises' functionals. *Dokl. Akad. Nauk Ukrain. SSR, Ser. A*, 2, 7–9. (In Russian).

Borovskikh, Yu.V. (1980e). On the asymptotic theory of *U*-statistics. *Dokl. Akad. Nauk Ukrain. SSR, Ser. A*, 11, 6–9. (In Russian).

Borovskikh, Yu.V. (1980f). Analytical theory of rank criteria. *Preprint. #80.21, Institute of Math. Ukr. Akad. of Sci., Kiev*, 1–68. (In Russian).

Borovskikh, Yu.V. (1980g). An estimate of the rate of convergence of the Anderson–Darling criterion. *Theor. Probab. Math. Statist.* 19, 29–35.

Borovskikh, Yu.V. (1981a). Nonuniform estimation of the rate of convergence for *L*-statistics. *Ukr. Math. J.* **33**, 2, 127–132.

Borovskikh, Yu.V. (1981b). Asymptotics of a von Mises' functionals. *Preprint. #81.46, Institute of Math. Ukr. Acad. of Sci., Kiev*, 1–52. (In Russian).

Borovskikh, Yu.V. (1982a). Asymptotics of a multidimensional ω^2-statistics. *Dokl. Akad. Nauk Ukrain. SSR, Ser. A*, 5, 3–6. (In Russian).

Borovskikh, Yu.V. (1982b). Asymptotics of ω^2-statistics. *Soviet Math. Dokl.* **25**, 2, 554–558.

Borovskikh, Yu.V. (1983a). Exponential nonuniform estimations for ω^2-statistics. *Dokl. Akad. Nauk Ukrain. SSR, Ser. A*, 3, 3–7. (In Russian).

Borovskikh, Yu.V. (1983b). Asymptotical properties of the distributions of *U*-statistics and von Mises' functionals. *Theory Probab. Appl.* **28**, 1, 188–191.

Borovskikh, Yu.V. (1983c). Asymptotics in the central limit theorem in Banach spaces. *Dokl. Akad. Nauk Ukrain. SSR, Ser. A*, 1, 3–7. (In Russian).

Borovskikh, Yu.V. (1983d). Asymptotics of the Anderson–Darling ω^2-statistics. In: *Theory of random processes* 11, Naukova Dumka, Kiev, 9–12. (In Russian).

Borovskikh, Yu.V. (1983e). Approximation of probability distributions in infinite-dimensional spaces. *Preprint. #83.42, Institute of Math. Ukr. Acad. of Sci., Kiev*, 1–56. (In Russian).

Borovskikh, Yu.V. (1983g). Asymptotics of *U*-statistics and von Mises' functionals. *Sov. Math. Dokl.* **27**, 2, 303–308.

Borovskikh, Yu.V. (1984a). Estimates of characteristic functions of certain random variables with application to ω^2-statistics, I. *Theory Probab. Appl.* **29**, 3, 488–503.

Borovskikh, Yu.V. (1984b). Estimate of the rate of convergence in the central limit theorem in Banach spaces. *Theory Probab. Appl.* **29**, 2, 414–415.

Borovskikh, Yu.V. (1984c). On Gaussian approximation in Banach spaces. *Theory Probab. Appl.* **29**, 4, 838–839.

Borovskikh, Yu.V. (1984d). On normal approximation in infinite-dimensional spaces. *Sov. Math. Dokl.* **30**, 2, 548–553.

Borovskikh, Yu.V. (1985a). Two-dimensional approximation of *U*-statistics. *Ukr. Math. J.* 6, 648–651.

Borovskikh, Yu.V. (1985b). Estimates of characteristic functions of certain random variables with application to ω^2-statistics, II. *Theory Probab. Appl.* **30**, 1, 117-127.

Borovskikh, Yu.V. (1986). Theory of *U*-statistics in Hilbert space. *Preprint #86.78, Institute of Math. Ukr. Acad. of Sci., Kiev*, 1–56. (In Russian).

Borovskikh, Yu.V. (1987). Asymptotics of integral functionals of a multidimensional empirical process. In: *Theory of random processes, 15,* Naukova Dumka, Kiev, 10–16. (In Russian).

Borovskikh, Yu.V. (1988a). Central limit theorem for *UB*-statistics. *Dokl. Akad. Nauk Ukrain. SSR, Ser. A*, 2, 6–9. (In Russian).

Borovskikh, Yu.V. (1988b). The law of large numbers for *UH*-statistics. *Dokl. Akad. Nauk Ukrain. SSR, Ser. A*, 7, 6–9. (In Russian).

Borovskikh, Yu.V. (1988c). Central limit theorem for *UH*-statistics of weakly dependent variables. In: *Analytical Methods in the Investigation of stochastic Models*, Institute of Mathematics, Ukrain. Acad. of Sci., Kiev, 12–17. (In Russian).

Borovskikh, Yu.V. (1988d). Estimates in the law of large numbers for *UB*-statistics. *Dokl. Akad. Nauk Ukrain. SSR, Ser. A*, 12, 9–14. (In Russian).

Borovskikh, Yu.V. (1988e). Theory of *U*-statistics in Banach space. *Preprint #88.3, Institute of Math. Ukr. Acad. of Sci., Kiev*, 1–60. (In Russian).

Borovskikh, Yu.V. (1989a). Central limit theorem for *U*- statistics. *Ukrainian Math. J.* **41**, 2, 240–242.

Borovskikh, Yu.V. (1989b). Estimation of the rate of convergence in the laws of large numbers for *U*-statistics. In: *Modern Analysis and Its Application,* Naukova Dumka, Kiev, 9–16. (In Russian).

Borovskikh, Yu.V. (1989c). Asymptotics of *UB*-statistics. *Abstracts of 5th Internat. Vilnius Conf. in Probability Theory and Mathematical Statistics, Institute of Mathematics and Cybernetics, Lithuanian Acad.of Sci., Vilnius,* **3**, 82–83.

Borovskikh, Yu.V. (1990). Rates of convergence for *U*-statistics. *Abstracts of 11th Prague Conf. in Information Theory, Statistical Decision Functions, and Random Processes (August, 1990)*, Institute of Information Theory and Automation (Czech. Acad. of Sci.) and Dept. of Math. and Phus. of Charles Univ., Prague.

Borovskikh, Yu.V. (1991). *UB*-statistics. *Theory Probab. Appl.* **36**, 3, 429–446.

Borovskikh, Yu.V. (1993). Approximation of *UH*-statistics. *Abstracts of communication*, **1** (A–L). *Sixth International Vilnius conference on Probability Theory and Mathematical Statistics*, Vilnius. Institute of Mathematics and Informatics, 51–52.

Borovskikh, Yu.V. (1994a). *U*-statistics in Banach Spaces. *Acta Applicandae Mathematicae* **35**, 213–297.

Borovskikh, Yu.V. (1994b). Probability of large deviations of *UH*-statistics. *Ukr. Math. J.* **46**, 12, 1611–1620.

Borovskikh, Yu.V. (1995). Approximation of generalized *U*-statistics. In: *Exploring Stochastic Laws*, A.V. Skorokhod and Yu.V. Borovskikh (Eds). Festschrift in Honour of the 70th Birthday of Academician V.S. Korolyuk, VSP, Utrecht, The Netherlands, 41–54.

Borovskikh, Yu.V. and Korolyuk, V.S. (1979). Asymptotic expansions for the Wilcoxon criterion. *Dokl. Akad. Nauk Ukrain. SSR, Ser. A*, 10, 797–801. (In Russian).

Borovskikh, Yu.V. and Korolyuk, V.S. (1980). Additive problems of the asymptotics of distributions of rank statistics. *Soviet Math. Dokl.* **21**, 2, 451–455.

Borovskikh, Yu.V. and Korolyuk, V.S. (1981). *Analytic Problems of Asymptotics of Probability Distributions.* Naukova Dumka, Kiev. (In Russian).

Borovskikh, Yu.V. and Korolyuk, V.S. (1982a). Approximation of the distribution of Spearman's rank correlation coefficient. *Theor. Probab. Math. Statist.* 25, 51–59.

Borovskikh, Yu.V. and Korolyuk, V.S. (1982b). On mean square tests of symmetry. *Theor. Probab. Math. Statist.* 24, 83–85.

Borovskikh, Yu.V. and Korolyuk, V.S. (1983). Exponential nonuniform estimates for classical rank statistics. *Dokl. Akad. Nauk Ukrain. SSR, Ser. A*, 2, 12–17. (In Russian).

Borovskikh, Yu.V. and Korolyuk, V.S. (1984a). Martingale theory of *U*-statistics. *Preprint #84.63, Institute of Math. Ukr. Acad. of Sci., Kiev*, 1–56. (In Russian).

Borovskikh, Yu.V. and Korolyuk, V.S. (1984b). Characteristic functions and estimates of the moments of *U*-statistics and von Mises' functionals. *Dokl. Akad. Nauk Ukrain. SSR, Ser. A*, 11, 11–14. (In Russian).

Borovskikh, Yu.V. and Korolyuk, V.S. (1984c). *Asymptotic Analysis of Distributions of Statistics.* Naukova Dumka, Kiev. (In Russian).

Borovskikh, Yu.V. and Korolyuk, V.S. (1984d). Approximation of non-degenerate *U*-statistics. *Theory Probab. Appl.* **30**, 3, 439–450.

Borovskikh, Yu.V. and Korolyuk, V.S. (1985). Expansions for *U*-statistics and von Mises' functionals. *Ukr. Math. J.* **37**, 4, 358–364.

Borovskikh, Yu.V. and Korolyuk, V.S. (1986). The convergence rate for *U*-statistics. *Abstracts of 1th World Congr. of Bernoulli Society, Tashkent*, **2**, 789.

Borovskikh, Yu.V. and Korolyuk, V.S. (1987a). Central limit theorem for *U*-statistics in Hilbert space. *Dokl. Akad. Nauk Ukrain. SSR, Ser. A*, 3, 16–19. (In Russian).

Borovskikh, Yu.V. and Korolyuk, V.S. (1987b). Functional limit theorem for the mean square deviation of the density estimate. In: *Asymptotic Methods in the Investigation of Stochastic Models*, Institute of Mathematics, Ukrain. Acad. of Sci., Kiev, 81–84.

Borovskikh, Yu.V. and Korolyuk, V.S. (1987c). Limiting representation of *U*-statistics by the multiple Wiener–Itô integral. *Dokl. Akad. Nauk Ukrain.SSR, Ser. A*, 2, 16–18. (In Russian).

Borovskikh, Yu.V. and Korolyuk, V.S. (1987d). Weak convergence of *U*-statistics to a multiple stochastic integral in Hilbert space. *Dokl. Akad. Nauk Ukrain. SSR, Ser. A*, 4, 9–13. (In Russian).

Borovskikh, Yu.V. and Korolyuk, V.S. (1988a). Multidimensional *U*-statistical empirical process. *Dokl. Akad. Nauk Ukrain. SSR, Ser. A*, 2, 12–15.

Borovskikh, Yu.V. and Korolyuk, V.S. (1988b). Glivenko theorem for multisample *U*-statistical empirical distribution function. In: *Analytic Methods in the Investigation of Stochastic Models, Institute of Mathematics, Ukrain. Akad. of Sci., Kiev*, 60–62. (In Russian).

Borovskikh, Yu.V. and Korolyuk, V.S. (1988c). Approximation of non-degenerate *UH*-statistics. *Preprint #88.35, Institute of Mathematics, Ukrain. Akad. of Sci., Kiev*, 1–40. (In Russian).

Borovskikh, Yu.V. and Korolyuk, V.S. (1988d). A nonuniform estimate of the rate of convergence for *UH*-statistics. *Dokl. Akad. Nauk Ukrain. SSR, Ser. A*, 10, 20–22. (In Russian).

Borovskikh, Yu.V. and Korolyuk, V.S. (1988e). Approximation of multi-sample *UB*-statistics. *Preprint #88.72, Institute of Mathematics, Ukrain. Acad. of Sci., Kiev*, 1–48. (In Russian).

Borovskikh, Yu.V. and Korolyuk, V.S. (1988f). Von Mises' statistics with values in Hilbert space. *Dokl. Akad. Nauk Ukrain. SSR, Ser. A*, 11, 15–18. (In Russian).

Borovskikh, Yu.V. and Korolyuk, V.S. (1988g). Asymptotic theory of *U*-statistics. *Ukr. Math. J.* **40**, 2, 142–154.

Borovskikh, Yu.V. and Korolyuk, V.S. (1988h). Rate of convergence of degenerate von Mises' functionals. *Theory Probab. Appl.* **33**, 1, 125–135.

Borovskikh, Yu.V. and Korolyuk, V.S. (1988i). *Martingale Approximation*. Naukova Dumka, Kiev. (In Russian).

Borovskikh, Yu.V. and Korolyuk, V.S. (1989a). *Theory of U-statistics*. Naukova Dumka, Kiev. (In Russian).

Borovskikh, Yu.V. and Korolyuk, V.S. (1989b). The law of iterated logarithm for *U*-statistics. *Ukrain. Math. J.* **41, 1**, 97–100. (In Russian).

Borovskikh, Yu.V. and Korolyuk, V.S. (1989c). The compact law of iterated logarithm for *UH*-statistics. *Ukrain. Math. J.* **41**, 5, 581–584.

Borovskikh, Yu.V. and Korolyuk, V.S. (1989d). Approximation of distributions of *UH*-statistic. *Abstracts of 5th Internat. Vilnius Conf. in Probability Theory and Mathematical Statistics, Institute of Mathematics and Cybernetics, Lithuanian Academy of Sci., Vilnius*, **3**, 319–320.

Borovskikh, Yu.V. and Korolyuk, V.S. (1990a). *UH*-statistics. *Soviet Math. Dokl.* **40**, 432–435.

Borovskikh, Yu.V. and Korolyuk, V.S. (1990b). *U*-statistics with increasing degrees. *Docl. Acad. Nauk Ukrain. SSR, Ser. A*, 3, 3–7. (In Russian).

Borovskikh, Yu.V. and Korolyuk, V.S. (1990c). Random symmetric polynomial with increasing degree. *Docl. Acad. Nauk Ukrain. SSR, Ser. A*, 4, 10–13. (In Russian).

Borovskikh, Yu.V. and Korolyuk, V.S. (1990d). Elementary symmetric bootstrap polynomial. *Docl. Acad. Nauk Ukrain. SSR, Ser. A*, 5, 9–12. (In Russian).

Borovskikh, Yu.V. and Korolyuk, V.S. (1990e). Rate of convergence in the central limit theorem for *UH*-statistics. *Ukr. Math. J.* **42**, 8, 930–936.

Borovskikh, Yu.V. and Korolyuk, V.S. (1990f). *U*-statistics of a bootstrap sample. *Ukr. Math. J.* **42**, 10, 1245–1248.

Borovskikh, Yu.V. and Korolyuk, V.S. (1991a). *U*-statistics of increasing degrees. *Preprint #91.3, Institute of Mathematics, Ukrain. Acad. of Sci., Kiev*, 1–36. (In Russian).

Borovskikh, Yu.V. and Korolyuk, V.S. (1991b). Random permanents and symmetric statistics. *Preprint #91.12, Institute of Mathematics, Ukrain. Acad. of Sci., Kiev*, 1–61. (In Russian).

Borovskikh, Yu.V. and Korolyuk, V.S. (1991c). Weak convergence of *UH*-statistics to stable laws. *Theory Probab. Appl.* 36, 3, 609–612.

Borovskikh, Yu.V. and Korolyuk, V.S. (1991d). An estimate of the closeness of the moments of *UH*-statistics under the conditions of the central limit theorem. *Theor. Probab. Math. Statist.* Issue 42, 55–60.

Borovskikh, Yu.V. and Korolyuk, V.S. (1991e). Rate of convergence in the central limit theorem for *UH*-statistics. *Theor. Probab. Math. Statist.* Issue 43, 79–85.

Borovskikh, Yu.V. and Korolyuk, V.S. (1991f). *U*-statistics with increasing degrees. *Soviet Math. Dokl.* **42**, 2, 502–506.

Borovskikh, Yu.V. and Korolyuk, V.S. (1991g). *U*-statistics of increasing degrees with asymptotically Poisson distribution. *Docl. Acad. Nauk Ukrain. SSR, Ser. A* 4, 29–32. (In Russian).

Borovskikh, Yu.V. and Korolyuk, V.S. (1991h). *U*-statistics of increasing degrees with asymptotically Poisson distribution. *Report BS–R 9120, Dept. of Operat. Research, Statistics and System Theory*, Centre for Mathematics and Computer Science, Amsterdam, 1–4.

Borovskikh, Yu.V. and Korolyuk, V.S. (1991i). Weak convergence of *UH*-statistics to stable laws. *Theory Probab. Appl.* **39**, 3, 597–600.

Borovskikh, Yu.V. and Korolyuk, V.S. (1991j). Random permanents and symmetric statistics. *VI USSR-JAPAN Symposium on Probability Theory and Mathematical Statistics. Abstr. of Communic.*, Kiev, p. 83.

Borovskikh, Yu.V. and Korolyuk, V.S. (1992). Random permanents and symmetric statistics. *Proceedings of the Sixth USSR-JAPAN Symposium on Probability Theory and Mathematical Statistics (Kiev)*, World Scientific, 176–187.

Borovskikh, Yu.V. and Korolyuk, V.S. (1993a). *Random Permanents.* Institute of Math., Ukrain. Acad. of Sci., Kiev. (In Russian).

Borovskikh, Yu.V. and Korolyuk, V.S. (1993b). Random permanents. *Abstracts of communications, v.1 (A - L). Sixth International Vilnius conference on Probability Theory and Mathematical Statistics*, Vilnius Institute of Mathematics and Informatics, 200–201.

Borovskikh, Yu.V. and Korolyuk, V.S. (1994a). *Theory of U-statistics.* Mathematics and Its Applications **273**, Kluwer Academic Publishers, Dordrecht, The Netherlands, x + 552 p.

Borovskikh, Yu.V. and Korolyuk, V.S. (1994b). *Random Permanents.* VSP, Utrecht, The Netherlands.

Borovskikh, Yu.V. and Korolyuk, V.S. (1994c). On normal approximation in Banach space L^p. In: *Asymptotical Analysis of Random Evolutions,* Institute of Math., Ukrain. Acad. of Sci., Kiev, 168–180. (In Russian).

Borovskikh, Yu.V. and Korolyuk, V.S. (1994d). Permanent random measures and symmetric statistics. *Abstracts of ICM 94, Short Communications of International Congress of Mathematicians, Zürich 3-11 August 1994. 10. Probability and Statistics, 149.*

Borovskikh, Yu.V. and Korolyuk, V.S. (1994e). Random permanents and symmetric statistics. *Acta Applicandae Mathematicae* **36**, 227–288.

Borovskikh, Yu.V. and Korolyuk, V.S. (1995). Normal approximation of random permanents. *Ukrainian Math. J.* **47**, 7, 922–927.

Borovskikh, Yu.V., Puri, Madan L. and Sazonov, V.V. (1994). Normal approximation of *U*-statistics in Hilbert space. *Department of Information and Systems Management, Hong Kong University of Science and Technology, June 1994, 1–23.*

Bose, A. and Dasgupta, R. (1994). On some asymptotic properties of *U*-statistics and one sided estimates. *Ann. Probab.* **22**, 4.

Bönner, N. and Kirschner, H.P. (1977). Note on conditions or weak convergence of von Mises differentiable statistical functions. *Ann. Statist.* **5**, 405–407.

Bretagnolle, J. (1983). Lois limites du bootstrap de certaines fonctionnelles. *Ann. Inst. H. Poincaré Sect. B (N.S.)* **3**, 256–261.

Brown, B.M. and Eagleson, G.K. (1971). Martingale convergence to infinitely divisible laws with finite variance. *Trans. Amer. Math. Soc.* **162**, 449–453.

Brown, T.S. and Silverman, B.W. (1979). Rates of Poisson convergence for *U*-statistics. *J. Appl. Probab.* **16**, 2, 428–432.

Burkholder, D.L. (1986). *Martingales and Fourier Analysis in Banach Space.* Lecture Notes in Math. **1206**, 61–108, Springer, Berlin.

Cai, Zongwu. (1991). On Chernoff type large deviations for trimmed *U*-statistics. *J. Hangzhou Univ. Natur. Sci. Ed.* **18**, 1, 21–26.

Callaert, H. and Janssen, P. (1978). The Berry–Esséen theorem for *U*-statistics. *Ann. Statist.* **6**, 2, 417–421.

Callaert, H. and Veraverbeke, N. (1981). The order of the normal approximation for a studentized *U*-statistics. *Ann. Statist.* **9**, 1, 194–200.

Callaert, H., Janssen, P. and Veraverbeke, N. (1979). Some asymptotic results for the distribution function of *U*-statistics. *Proceedings of 2nd Prague Symp. on Asymptot. Statist. (Hradek Kralove, 1978) North-Holland, Amsterdam*, 135–146.

Callaert, H., Janssen, P. and Veraverbeke, N. (1980). An Edgeworth expansion for *U*-statistics. *Ann. Statist.* **8**, 2, 299–312.

Chan, Y.K. and Wierman, J. (1977). On the Berry–Esséen theorem for *U*-statistics. *Ann. Probab.* **5**, 1, 136–139.

Chatterji, S.D. (1968). Martingale convergence and the Radon–Nikodym theorem in Banach spaces. *Math. Scand.* **22**, 1, 21–41.

Chen, L.H.Y. (1975). Poisson approximation for dependent trials. *Ann. Probab.* **3**, 3, 534–545.

Chen, Xia (1993). On the law of the iterated logarithm for independent Banach space valued random variables. *Ann. Probab.* **21**, 4, 1991–2011.

Chen, Xia (1994). On strassen's law of the iterated logarithm in Banach space. *Ann. Probab.* **22**, 2, 1026–1043.

Chen, Xiru (1980). On limiting properties of *U*-statistics and von Mises statistics. *Scientia Sinica* **23**, 9, 1079–1091.

Chow, Y.S., de la Peña, V.H. and Teicher, H. (1993). Wald's equation for a class of denormalized *U*-statistics. *Ann. Probab.* **21**, 2, 1151–1158.

Christofides, T. (1991). Probability inequalities with exponential bounds for *U*-statistics. *Statist. Probab. Letters* **12**, 3, 257–261.

Christofides, T. and Serfling, R. (1990a). Maximal inequalities for multidimensionaly indexed submartingale arrays. *Ann. Probab.* **18**, 2, 630–641.

Christofides, T. and Serfling, R. (1990b). Maximal inequalities and convergence results for generalized *U*-statistics. *J. Statist. Plan. Infer.* **24**, 3, 271–286.

Daffer, P.Z. (1984). Central limit theorems for weighted sums of exchangeable random elements in Banach spaces. *Stochast. Anal. and Appl.* **2**, 3, 229–244.

Dasgupta, R. (1984). On large deviation probability of *U*-statistics in non i.i.d. case. *Sankhya, Ser. A*, **46**, 110–116.

Dehling, H. (1985). Limit theorems for sums of weakly dependent Banach space valued random variables. *Z. Wahr. verw. Gebiete* **63**, 3, 393–432.

Dehling, H. (1986). Almost sure approximation for *U*-statistics. In: *Dependence in Probability and Statistics,* E. Eberlein and M. Taqqu (Eds). Birkhäuser, Basel, 119–135.

Dehling, H. (1989a). The functional law of the iterated logarithm for von Mises functionals and multiple Wiener integrals. *J. Multivariate Anal.* **28**, 2, 177–189.

Dehling, H. (1989b). Complete convergence of triangular arrays and the law of the iterated logarithm for *U*-statistics. *Statist. Probab. Letters* **7**, 319–321.

Dehling, H. and Philipp, W. (1982). Almost sure invariance principles for weakly dependent vector-valued random variables. *Ann. Probab.* **10**, 3, 689–701.

Dehling, H. and Taqqu, M.S. (1989). The empirical processes of some long-range dependent sequences with an application to U-statistics. *Ann. Statist.* **17**, 4, 1767–1783.

Dehling, H., Denker, M. and Philipp, W. (1984). Invariance principles for von Mises and U-statistics. *Z. Wahr. verw. Gebiete* **67**, 2, 139–167.

Dehling, H., Denker, M. and Philipp, W. (1986). A bounded law of the iterated logarithm for Hilbert space valued martingales and its application to U-statistics. *Probab. Theory Related Fields* **72**, 1, 111–131.

Dehling, H., Denker, M. and Philipp, W. (1987). The almost sure invariance principle for the empirical process of U-statistic structure. *Ann. Inst. H. Poincare. Probab. Statist.* **23**, 2, 121–134.

Dehling, H., Denker, M. and Woyczynski, W.A. (1990). Resampling U-statistics using p-stable laws. *J. Multivariate Anal.* **34**, 1, 1–13.

Denker, M. (1985). *Asymptotic theory in nonparametric statistics.* Vieweg Braunschweig.

Denker, M. and Keller, G. (1983). On U-statistics and von Mises' statistics for weakly dependent processes. *Z. Wahr. verw. Gebiete* **64**, 4, 505–522.

Denker, M., Grillenberger, C. and Keller, G. (1985). A note on invariance principles for Mises' statistics. *Metrika* **32**, 3-4, 197–214.

Doukhan, P. (1994). *Mixing: Properties and Examples.* Lecture Notes in Statistics **85**, Springer-Verlag, New York.

Dudley, R.M. (1984). A course on empirical processes. *Lecture Notes in Math.* **1097**, Springer, New York, 1–142.

Dudley, R.M. (1987). Universal Donsker classes and metric entropy. *Ann. Probab.* **15**, 3, 1306–1326.

Dudley, R.M. (1989). *Analysis and Probability.* Wadsworth and Brooks Cole, Pacific Grove, CA.

Dudley, R. and Philipp, W. (1983). Invariance Principles for sums of Banach space valued random elements and empirical processes. *Z. Warsch. Verw. Gebiete* **62**, 509–552.

Dudley, R.M., Giné, E. and Zinn, J. (1991). Uniform and universal Glivenko–Cantelli classes of functions. *J. Theor. Probab.* **4**, 485–510.

Dynkin, E.B. and Mandelbaum, A. (1983). Symmetric statistics, Poisson point process and multiple Wiener integrals. *Ann. Statist.* **11**, 739–745.

Dwass, M. (1956). The large-sample power of rank tests in the two-samples problem. *Ann. Math. Statist.* **27**, 2, 352–374.

Eagleson, G.K. (1979). Orthogonal expansions and *U*-statistics. *Austral. J. Statist.* **21**, 3, 221–237.

Eagleson, G.K. and Weber, N.C. (1978). Limit theorems for weakly exchangeable arrays. *Math. Proc. Cambridge Phil. Soc.* **84**, 1, 123–130.

Efron, B. and Tibshirani, R.J. (1993). *An Introduction to the Bootstrap.* Chapman and Hall, New York.

Einmahl, U. (1988). Strong approximation for partial sums i.i.d. *B*-valued r.v.'s in the domain of attraction of a Gaussian law. *Probab. Theory Related Fields* **77**, 1, 65–85.

Einmahl, U. (1989). Stability results and strong invariance principles for partial sums of Banach space valued random variables. *Ann. Probab.* **17**, 333–352.

Einmahl, U. (1991). On the almost sure behavior of sums of i.i.d. random variables in Hilbert space. *Ann. Probab.* **19**, 1227–1263.

Einmahl, U. (1993). Toward a general law of the iterated logarithm in Banach space. *Ann. Probab.* **21**, 4, 2012–2045.

Engel, D.D. (1982). The multiple stochastic integral. *Mem. Amer. Math. Soc.* **38**, 265, 1–82.

Es, A.I. van (1986). On the weak limits of elementary symmetric polynomials. *Ann. Probab.* **14**, 2, 677–695.

Es, A.I. van, Helmers R. (1986). Elementary symmetric polynomials of increasing order. *Probab. Theory Related Fields* **80**, 1, 21–35.

Fazekas, I. (1983). Convergence of vector-valued martingales with multidimensional indices. *Publicationes Mathematicae (Debrecen)* **30**, 1, 157–164.

Fazekas, I. (1985). Convergence rates in the Marcinkiewicz strong law of large numbers for Banach space valued random variables with multidimensional indices. *Publicationes Mathematicae (Debrecen)* **32**, 3-4, 203–209.

Feldman, R.E. and Rachev, S. (1993). *U*-statistics of random-size samples and limit theorems for systems of Markovian particles with non-Poisson initial distributions. *Ann. Probab.* **21**, 4, 1927–1945.

Frees, E.W. (1989). Infinite order *U*-statistics. *Scand. J. Statist.* **16**, 1, 29–45.

Ghosh, M. (1984). Rank statistics and limit theorems. In: *Handbook of Statistics. 4: Non-parametric methods*, P.R.Krishnaian and P.K.Sen (Eds). North Holland, Elservier Sci. Publ., Amsterdam, 145–171.

Gikhman, I.I. and Skorokhod, A.V. (1971). *Theory of Random Processes*. Nauka, Moscow. (In Russian).

Gijbels, I., Janssen, P. and Veraverbeke, N. (1988). Weak and strong representations for trimmed *U*-statistics. *Probab. Theory Related Fields* **77**, 2, 179–194.

Giné, E. and Zinn, J. (1983). Central limit theorems and weak laws of large numbers in certain Banach spaces. *Z. Wahr. verw. Gebiete* **62**, 323–354.

Giné, E. and Zinn, J. (1984). Some limit theorems for empirical processes. *Ann. Probab.* **12**, 929–989.

Giné, E. and Zinn, J. (1986). Lectures on the central limit theorem for empirical processes. *Lecture Notes in Math.* **1221**, Springer, New York, 50–113.

Giné, E. and Zinn, J. (1990). Bootstrapping general empirical measures. *Ann. Probab.* **18**, 851–869.

Giné, E. and Zinn, J. (1991). Gaussian characterization of uniform Donsker classes of functions. *Ann. Probab.* **19**, 2, 758–782.

Giné, E. and Zinn, J. (1992a). Marcinkiewicz type laws of large numbers and convergence of moments for *U*-statistics. In: *Probability in Banach Spaces* **8**, Birkhäuser, Boston, 273–291.

Giné, E. and Zinn, J. (1992b). On Hoffmann-Jørgensen's inequality for *U*-processes. In: *Probability in Banach Spaces* **8**, Birkhäuser, Boston, 80–91.

Giné, E. and Zinn, J. (1994). A remark on convergence in distribution of *U*-statistics. *Ann. Probab.* **22**, 1, 117–125.

Goodman, V., Kuelbs, J. and Zinn, J. (1981). Some results on the LIL in Banach space with applications to weight empirical processes. *Ann. Probab.* **9**, 713–752.

Götze, F. (1979). Asymptotic expansions for bivariate von Mises' functionals. *Z. Wahr. verw. Gebiete* **50**, 3, 333–355.

Götze, F. (1984). Expansions for Mises functionals. *Z. Wahr. verw. Gebiete* **65**, 4, 599–625.

Götze, F. (1985). Asymptotic expansions in functional limit theorems. *J. Multivariate Anal.* **76**, 1, 1–20.

Götze, F. (1986). On the rate of convergence in the central limit theorem in Banach spaces. *Ann. Probab.* **14**, 3, 922–942.

Götze, F. (1987a). Approximations for multivariate *U*- statistics. *J. Multivariate Anal.* **22**, 2, 212–229.

Götze, F. (1987b). Approximations in functional limit theorems. *Proceedings of the 1 st World Congress of the Bernoulli Society* **1**, *(Tashkent, 1986)*, 669–677. *VNU, Sci. Press, Utrecht.*

Götze, F. (1989). Edgeworth expansions in functional limit theorems. *Ann. Probab.* **17**, 4, 1602–1634.

Götze, F. (1991). On the rate of convergence in the multivariate CLT. *Ann. Probab.* **19**, 724–739.

Gregory, G. (1977). Large sample theory for *U*-statistics and tests of fit. *Ann. Statist.* **5**, 1, 110–123.

Gut, A. (1983). Strong laws for independent identically distributed random variables indexed by a sector. *Ann. Probab.* **11**, 3, 569–577.

Halasz, G. and Szekely, G.J. (1976). On the elementary symmetric polynomials of independent random variables. *Acta Math. Acad. Sci. Hung.* **28**, 3-4, 397–400.

Hall, P. (1979). On the invariance principle for *U*-statistics. *Stoch. Proc. and Appl.* **9**, 2, 163–174.

Hall, P. (1984). Central limit theorem for integrated square error of multivariate nonparametric density estimators. *J. Multivariate Anal.* **14**, 1, 1–16.

Hall, P. (1992). *The Bootstrap and Edgeworth Expansion.* Springer-Verlag, New York.

Hall, P. and Heyde C.C. (1980). *Martingale limit theory and its application.* Acad. Press, New York.

Halmos, P. (1946). The theory of unbiased estimation. *Ann. Math. Statist.* **17**, 1, 34–43.

Harel, M. and Puri, M.L. (1989). Weak convergence of the *U*-statistics and weak invariance of the one-sample rang order statistic for Markov processes and ARMA models. *J. Multivariate Anal.* **31**, 2, 258–265.

Harel, M. and Puri, M.L. (1990). Weak invariance of generalized *U*-statistics for nonstationary absolutely regular processes. *Stoch. Proc. and Appl.* **34**, 2, 341–360.

Harel, M., O'Cinneide, Colm Art and Puri, M.L. (1994). Weak convergence of weighted empirical *U*-statistics processes for dependent random variables. *J. Multivariate Anal.* **48**, 2, 297–314.

Heinrich, L. and Wolf, W. (1993). On the convergence of *U*-statistics with stable limit distribution. *J. Multivariate Anal.* **44**, 2, 266–278.

Helmers, R. and Huskova, M. (1994). Bootstrapping multivariate *U*-quantiles and related statistics. *J. Multivariate Anal.* **49**, 1, 97–109.

Helmers, R. and van Zwet, W.R. (1982). The Berry–Esséen bound for *U*-statistics. In: *Statistical Theory and Related Topics III, 1* S.S.Gupta and J.O.Berger (Eds). Academic Press, New York, 497–512.

Helmers, R., Janssen, P. and Serfling, R. (1988). Glivenko–Cantelli properties of some generalized empirical DF's and strong convergence of generalized *L*-statistics. *Probab. Theory Related Fields* **79**, 1, 75–93.

Helmers, R., Janssen, P. and Serfling, R. (1990). Berry–Esséen and bootstrap results for generalized L-statistics. *Scand. J. Statist. Th. Appl.* **17**, 1, 65–77.

Helmers, R., Janssen, P. and Veraverbeke, N. (1990). Bootstrapping *U*-quantiles. In: *Exploring the Limits of Bootstrap,* E.Le Page and L.Billard (Eds). Wiley, New York, 145–155.

Henze, N. and Voigt, B. (1992). Almost sure convergence of certain slowly changing symmetric one-and multi-sample statistics. *Ann. Probab.* **20**, 1086–1098.

Hitczenko, P. (1990). Best constant in decoupling inequalities for nonnegative random variables. *Statist. Probab. Letters* **9**, 327–329.

Hoeffding, W. (1948a). A class of statistics with asymptotically normal distribution. *Ann. Math. Statist.* **19**, 3, 293–325.

Hoeffding, W. (1948b). A non-parametric test of independence. *Ann. Math. Statist.* **19**, 4, 546–557.

Hoeffding, W. (1961). The strong law of large numbers for *U*-statistics. *Inst. Statist. Mimeo Ser.* 302, 1–10.

Hoeffding, W. (1963). Probabilities inequalities for sums of bounded random variables. *J. Amer. Statist. Assoc.* **58**, 301, 13–30.

Hoffmann-Jørgensen, J. (1974). Sums of independent Banach space valued random variables. *Studia Math.* **52**, 159–186.

Hoffmann-Jørgensen, J. (1977). Probability in *B*-spaces. *Lecture Notes in Math.* **598**, 2–119.

Hoffmann-Jørgensen, J. (1984). Stochastic processes on Polish spaces. *Aarhus. Univ., Mat. Inst. Various Publications Series* 39, Aarhus, Denmark.

Hoffmann-Jørgensen, J. and Pisier, G. (1976). The law of large numbers and the central limit theorem in Banach spaces. *Ann. Probab.* **4**, 587–599.

Holst, L. and Janson, S. (1990). Poisson approximation using the Stein–Chen method and coupling: number of exceedances of Gaussian random variables. *Ann. Probab.* **18**, 2, 713–723.

Huskova, M. and Janssen, P. (1993). Consistency of the generalized bootstrap for degenerate *U*-statistics. *Ann. Statist.* **21**, 4, 1811–1823.

Inglot, T. and Ledwina, T. (1993). Cramer type large deviations for some *U*-statistics. *Theory Probab. Appl.* **38**, 4, 858–868.

Itô, K. (1951). Multiple Wiener integral. *J. Math. Soc. Jap.* **3**, 1, 157–169.

Itô, Y. (1988). Generalized Poisson functionals. *Probab. Theory Related Fields* **77**, 1, 1–29.

Itô, Y. and Kubo, I. (1988). Calculus on Gaussian and Poisson white noises. *Nagoya Math. J.* **111**, 1, 41–84.

Jacobs, K. (1983). *Einführung in die Kombinatorik.* Walter de Gruyter, Berlin.

Jammalamadaka, S. Rao and Janson, S. (1986). Limit theorems for a triangular scheme of *U*-statistics with applications to inter-point distances. *Ann. Probab.* **14**, 4, 1347–1358.

Janicki, A. and Weron, A. (1994). "Can one see α-stable variables an processes?", *Statist. Sci.* **9**, 1, 109–126.

Janson, S. (1979). The asymptotic distribution of U-statistics. *Preprint #5, Dept. of Math., University of Uppsala, Uppsala*, 1–17.

Janson, S. (1984). The asymptotic distributions of incomplete U-statistics. *Z. Wahr. verw. Gebiete* **66**, 4, 495–505.

Janson, S. (1990). A functional limit theorem for random graphs with applications to subgraph count statistics. *Random Structures and Algorithms* **1**, 1, 15–38.

Janson, S. and Nowicki, K. (1991). The asymptotic distributions of generalized U-statistics with applications to random graphs. *Probab. Theory Related Fields* 90, 341–375.

Janssen, P. (1980). Berry–Esséen rates for functionals of the empirical distribution function. *Bull. Soc. Math. Belg.* **32**, 1, 33–45.

Janssen, P. (1981). Rate of convergence in the central limit theorem and in the strong law of large numbers for von Mises statistics. *Metrika* **28**, 1, 35–46.

Janssen, P. (1994). Weighted bootstrapping of U-statistics. *J. Statist. Plann. Inference* **38**, 1, 31–41.

Janssen, P., Serfling, R. and Veraverbeke, N. (1987). Asymptotic normality of U-statistics based on trimmed samples. *J. Statist. Plann. and Inference* **16**, 1, 63–74.

Johnson, W.B. and Schechtman, G. (1989). Sums of independent random variables in rearrangement invariant function spaces. *Ann. Probab.* **17**, 789–808.

Johnson, W.B., Schechtman, G. and Zinn, J. (1985). Best constants in moment inequalities for linear combinations of independent and exchangeable random variables. *Ann. Probab.* **13**, 234–253.

Jong, Peter de (1987). A central limit theorem for generalized quadratic forms. *Probab. Theory Related Fields* **75**, 2, 261–277.

Jong, Peter de. (1990). A central limit theorem for generalized multilinear forms. *J. Multivariate Anal.* **34**, 2, 275–289.

Kallenberg, O. (1975). *Random Measures.* Akademie-Verlag, Berlin.

Kanagawa, S. and Yoshihara, K. (1994). The almost sure invariance principles of degenerate *U*-statistics of degree two for stationary random variables. *Stochastic Process. Appl.* **49**, 2, 347–356.

Kingman, J.F.C. (1969). An ergodic theorem. *Bull. London Math. Soc.* **1**, Part 3, 3, 339–340.

Kingman, J.F.C. (1978). Uses of exchangeability. *Ann. Probab.* **6**, 2, 183–197.

Kohatsu-Higa, Arturo (1991). Weak convergence of infinite order *U*-processes. *Statist. Probab. Letters* **12**, 2, 145–150.

Kokic, P.N. (1987). Rates of convergence in the strong law of large numbers for degenerate *U*-statistics. *Statist. Probab. Letters* **5**, 5, 371–374.

Kokic, P.N. and Weber, N.C. (1990). An Edgeworth expansion for *U*-statistics based on samples from finite populations. *Ann. Probab.* **18**, 1, 390–404.

Kokic, P.N. and Weber, N.C. (1991). Rates of strong convergence for *U*-statistics in finite populations. *J. Austral. Math. Soc. (A)* **50**, 468–480.

Kokic, P.N. and Weber, N.C. (1995). Large deviation probabilities for *U*-statistics based on samples from finite populations. In: *Exploring Stochastic Laws,* A.V. Skorokhod and Yu.V. Borovskikh (Eds). Festschrift in Honour of the 70th Birthday of Academician V.S. Korolyuk, VSP, Utrecht, The Netherlands, 175–182.

Krakowiak, W. and Szulga, J. (1986). Multilinear random forms. *Ann. Probab.* **14**, 955–973.

Krakowiak, W. and Szulga, J. (1988). Hypercontraction principle and random multilinear forms. *Probab. Theory Related Fields* **77**, 325–342.

Kubo, I. and Mustafid (1989). Law of small numbers and limit theorems of asymmetric statistics. *Abstracts of communications,* **1**, *Fifth International Vilnius conference on Probability Theory and Mathematical Statistics, Vilnius. Institute of Mathematics and Informatics,* 292–293.

Kuelbs, J. (1976). A strong convergence theorem for Banach space valued random variables. *Ann. Probab.* **4**, 5, 744–771.

Kuelbs, J. (1985). The LIL when X is in the domain of attraction of a Gaussian law. *Ann. Probab.* 13, 5, 825–859.

Kuelbs, J. and Mandrekar, V. (1974). Domains of attraction of stable measures on a Hilbert space. *Stud. Math.* **50**, 2, 149–162.

Kuelbs, J. and Philipp, W. (1980). Almost sure invariance principles for partial sums of mixing B-valued random variables. *Ann. Probab.* **8**, 6, 1003–1036.

Kuelbs, J. and Zinn, J. (1983). Some results on LIL behavior. *Ann. Probab.* **11**, 506–557.

Kwapień, S. (1987). Decoupling inequalities and polynomial chaos. *Ann. Probab.* **15**, 1062–1072.

Kwapień, S. and Szulga, J. (1991). Hypercontraction methods in moment inequalities for series of independent random variables in normed spaces. *Ann. Probab.* **19**, 369–379.

Kwapień, S. and Woyczynski, W.A. (1986). Decoupling of martingale transforms and stochastic integration for processes with independent increments. In: *Probability Theory and Harmonic Analysis,* M. Dekker, Inc., 139–148.

Kwapień, S. and Woyczynski, W.A. (1992). *Random Series and Stochastic Integrals: Single and Multiple.* Birkhäuser, Boston, pp. xvi + 360.

Lai, T.L. (1974). Convergence rates in the strong law of large numbers for random variables taking values in Banach spaces. *Bull. Inst. Math. Acad. Sinica* **2**, 67–85.

Le Cam, L. (1986). *Asymptotic methods in statistical decision theory.* Springer-Verlag, New York.

Ledoux, M. (1986). The law of the iterated logarithm in uniformly convex Banach spaces. *Trans. Amer.Math. Soc.* **294**, 351–365.

Ledoux, M. and Talagrand, M. (1988). Characterization of the law of the iterated logarithm in Banach space. *Ann. Probab.* **16**, 1242–1264.

Ledoux, M. and Talagrand, M. (1989). Comparison theorems, random geometry and some limit theorems for empirical processes. *Ann. Probab.* **17**, 596–631.

Ledoux, M. and Talagrand, M. (1990). Some applications of isoperimetric methods to strong limit theorems for sums of independent random variables. *Ann. Probab.* **18**, 754–789.

Ledoux, M. and Talagrand, M. (1991). *Probability in Banach Spaces.* Springer-Verlag, New York.

Lee, A.J. (1990). *U-statistics–Theory and Practice.* M. Dekker, Inc. Statistics textb. a monogr. ser. 110.

Lehmann, E.L. (1951). Consistency and unbiasedness of certain nonparametric test. *Ann. Math. Statist.* **22**, 1, 165–179.

Li, D. (1990). Convergence rates of the strong law of large numbers for fields of B-valued random variables. *Chinese Ann. Math.* **11(A)**, 6, 744–752.

Li, D. and Wu, Z. (1989). Law of the iterated logarithm for B-valued random variables with multidimensional indices. *Ann. Probab.* **17**, 760–774.

Li, D., Rao, M.B. and Wang, X. (1992). The law of the iterated logarithm for independent random variables with multidimensional indices. *Ann. Probab.* **20**, 2, 660–674.

Lin, Zhengyan (1984). A note on asymptotic expansion for U-statistics. *Acta. Math. Sinica* **27**, 5, 595–598.

Linde, W. (1987). *Probability in Banach spaces.–Stable and infinitely divisible distributions.* Wiley, New York.

Lindenstrauss, J. and Tzafriri, L. (1979). *Classical Banach Spaces.* Function Spaces **2**, Springer-Verlag.

Liptser, R.Sh. and Shiryayev, A.N. (1989). *Theory of Martingales.* Kluwer Academic Publishers, Dordrecht, The Netherlands.

Liu, R.V. (1990). On a notion of data depth based on random simplexes. *Ann. Statist.* **18**, 405–414.

Loynes, R.M. (1969). The central limit theorem for backwards martingales. *Z. Wahr. verw. Gebiete* **13**, 1, 1–8.

Loynes, R.M. (1978). On the weak convergence of U-statistic processes. *Math. Proc. Cambridge Phil. Soc.* **83**, 2, 269–272.

Maesono, Y. (1988). Edgeworth expansion for one-sample U-statistics. *Bull. Inform. Cybernet.* **22**, 3-4, 189–197.

Maesono, Y. and Yamato, H. (1994). U-statistics and related topics. *Sugaku Expositions* **7**, 1, 43–58.

Major, P. (1981). Multiple Wiener–Itô integrals. *Lecture Notes in Math.* **849**, Springer, Berlin.

Major, P. (1988). On the tail behaviour of the distribution function of multiple stochastic integrals. *Probab. Theory Related Fields* **78**, 3, 419–435.

Major, P. (1994). Asymptotic distributions for weighted *U*-statistics. *Ann. Probab.* **22**, 3, 1514–1535.

Mandelbaum, A. and Taqqu, M.S. (1984). Invariance principle for statistics. *Ann. Statist.* **12**, 2, 483–496.

Mandrekar, V. and Patterson, R.F. (1993). Limit theorems for symmetric statistics of exchangeable random variables. *Statist. Probab. Letters* **17**, 2, 157–161.

Mathai, A.M. and Provost, S.B. (1992). *Quadratic Forms in Random Variables. Theory and Applications.* Dekker, New York.

McConnell, T.R. (1987a). A two-parameter maximal ergodic theorem with dependence. *Ann. Probab.* **15**, 1569–1585.

McConnell, T.R. (1987b). Two-parameter strong laws and maximal inequalities for *U*-statistics. *Proc. Royal Soc. Edinburgh* **107A**, 133–151.

McConnell, T.R. (1988). On the strong maximal function and rearrangements. *Studia Math.* **88**, 1, 85–102.

McConnell, T.R. (1989). Decoupling and stochastic integration in UMD Banach space. *Probability and Math. Statistics* **10**, 283–285.

McConnell, T.R. and Rieders, E. (1989). The two-parameter strong law for partially exchangeable arrays. In: *Almost Everywhere Convergence,* G. Edgar and I. Sucheston (Eds). Academic Press, 281–296.

McConnell, T.R. and Taqqu, M.S. (1986a). Dyadic approximation of double integrals with respect to symmetric stable processes. *Stochastic Processes and Applications* **22**, 323–331.

McConnell, T.R. and Taqqu, M.S. (1986b). Decoupling inequalities for multilinear forms in independent symmetric random variables. *Ann. Probab.* **14**, 943–954.

McConnell, T.R. and Taqqu, M.S. (1987). Decoupling of Banach-valued multilinear forms in independent symmetric Banach-valued random variables. *Probab. Theory Related Fields* **75**, 499–507.

Menon, V.V. and Rai, V.N. (1991). *U*-statistics, conditional independence and graph theory. *Sankhya, Ser. A* **53**, 1, 51–59.

Mijnheer, Joop (1991). *U*-statistics and double stable integrals. *Selected Proceedings of the Sheffield Symposium on Applied Probability (Sheffield, 1989)*, IMS Lecture Notes Monograph Ser. 18, Inst. Math. Statist. Hayward, CA, 256–258.

Mikosch, T. (1993). A weak invariance principle for weighted *U*-statistics with varying kernels. *J. Multivariate Anal.* **47**, 1, 82–102.

Mikosch, T. (1994). Weak invariance principles for weighted *U*-statistics. *J. Theor. Probab.* **7**, 1, 147–173.

Miller, R.G. and Sen, P.K. (1972). Weak convergence of *U*- statistics and von Mises' differentiable statistical functions. *Ann. Math. Statist.* **43**, 1, 31–41.

Mises, R. von (1947). On the asymptotic distributions of differentiable statistical functions. *Ann. Math. Statist.* **18**, 2, 309–348.

Mori, T. (1981). On a Hoeffding-type problem. *Lecture Notes in Statistics* **8**, 174–181.

Mori, T.F. and Szekely, G.J. (1982). Asymptotic behaviour of symmetric polynomial statistics. *Ann. Probab.* **10**, 1, 124–131.

Mustafid (1990a). A limit theorem of symmetric statistics for nonidentically distributed independent random elements. *Hiroshima Math. J.* **20**, 1, 159–169.

Mustafid (1990b). The law of small numbers and the limit theorem for symmetric statistics with mixing conditions. *Hiroshima Math. J.* **20**, 1, 171–183.

Neuhaus, G. (1977). Functional limit theorem for *U*-statistics in the degenerate case. *J. Multivariate Anal.* **7**, 3, 424–439.

Nolan, D. and Pollard, D. (1987). *U*-processes: rates of convergence. *Ann. Statist.* **15**, 2, 780–799.

Nolan, D. and Pollard, D. (1988). Functional limit theorems for *U*-processes. *Ann. Probab.* **16**, 3, 1291–1298.

O'Neil, K.A. and Render, R.A. (1993). Asymptotic distributions of weighted *U*-statistics of degree 2. *Ann. Probab.* **21**, 2, 1159–1169.

Patterson, R.F. (1989). Strong convergence for *U*-statistics in arrays of row wise exchangeable random variables. *Stochast. Anal. and Appl.* **7**, 1, 89–102.

Paulauskas, V. and Račkauskas, A. (1989). *Approximation Theory in the Central Limit Theorem. Exact Results in Banach Spaces.* Kluwer Acad. Publ., Dordrecht.

Peña, V.H. de la (1992). Decoupling and Khintchine's inequalities for *U*-statistics. *Ann. Probab.* **20**, 4, 1877–1892.

Peña, V. de la (1993). Inequalities for tails of adapted processes with an application to wald's lemma. *J. Theor. Probab.* **6**, 2, 285–302.

Peña, V.H. de la (1994). A bound on the moment generating function of a sum of dependent variables with an application to simple random sampling without replacement. *Ann. Inst. Henri Poincare* **30**, 2, 197–211.

Peña, V.H. de la and Klass, M.J. (1994). Order-of-magnitude bounds for expectations involving quadratic forms. *Ann. Probab.* **22**, 2, 1044–1077.

Peña, V.H. de la, Montgomery-Smith, S.J. and Szulga, J. (1994). Contraction and decoupling inequalities for multilinear forms and *U*-statistics. *Ann. Probab.* **22**, 4, 1745–1765.

Peskir, Goran (1993). Best constants in Kahane–Khintchine inequalities in Orlicz spaces. *J. Multivariate Anal.* **45**, 2, 183–216.

Peskir, Goran (1994a). Maximal inequalities of Kahane–Khintchine's type in Orlicz spaces. *Math. Proc. Cambridge Philos. Soc.* **115**, 175–190.

Peskir, Goran (1994b). On the exponential Orlicz norms of stopped Brownian motion. *Aarhus University, Preprint series* 17, 1–20.

Peskir, Goran and Shiryaev A.N. (1994). The inequalities of Khintchine and expanding sphere of their action. *Aarhus University, Preprint series* 24, 1–57.

Pinelis, I. (1994a). Optimum bounds for the distributions of martingales in Banach spaces. *Ann. Probab.* **22**, 4.

Pinelis, I. (1994b). On a majorization inequality for sums of independent random vectors. *Statist. Probab. Letters* **19**, 2, 97–99.

Pisier, G. (1986). Probabilistic methods in the geometry of Banach spaces. *Lecture Notes in Math.* **1206**, Springer-Verlag, 167–241.

Pollard, D. (1982). A central limit theorem for empirical processes. *J. Austral. Math. Soc. Ser. A*, **33**, 235–248.

Pollard, D. (1984). *Convergence of Stochastic Processes.* Springer, New York.

Pollard, D. (1989). Asymptotics via empirical processes (with discussion). *Statist. Sci.* **4**, 341–366.

Pollard, D. (1990). *Empirical Processes: Theory and Application.* IMS, Hayward, CA.

Præstgaard, J. and Wellner, J.A. (1993). Exchangeably weighted bootstraps of the general empirical process. *Ann. Probab.* **21**, 4, 2053–2086.

Puri, M.L. and Sazonov, V.V. (1991). On Hilbert space valued *U*-statistics. *Theory Probab. Appl.* **36**, 3, 604–605.

Puri, M.L. and Sen, P. (1971). *Nonparametric Methods in Multivariate Analysis.* Wiley, New York.

Račkauskas, A.I. (1988). Probabilities of large deviations in the Linnik zones in Hilbert spaces. *Lithuanian Math. J.* **28**, 3, 520–533.

Rajput, B. and Rosinski, J. (1990). Complements on decoupling inequalities for multilinear functions in stable random vectors. *Probability and Math. Statistics* **11**, 1, 1–17.

Randles, R.H. and Wolfe, D.A. (1979). *Introduction to the Theory of Nonparametric Statistics.* Wiley, New York.

Rieders, E. (1991). Marcinkiewicz-type strong laws for partially exchangeable arrays. *J. Multivariate Anal.* **38**, 1, 114–140.

Ronzhin, A.F. (1986). Functional limit theorems for *U*-statistics. *Mat. Zametki* **40**, 5, 683–696.

Rosén, B. (1969). A note on asymptotic normality of sums of higher dimensionally indexed random variables. *Ark. Mat.* **8**, 1, 33–43.

Rubin, H. and Sethuraman, J. (1965). Probabilities of moderate deviations. *Sankhya. Ser. A* **27**, 2-4, 325–346.

Rubin, M. and Vitale, R.A. (1980). Asymptotic distribution of symmetric statistics. *Ann. Statist.* **8**, 1, 165–170.

Ruymgaart, F.H. and van Zuijlen, M.C.A. (1992). Empirical *U*-statistics processes. *J. Statist. Plann. Inference* **32**, 2, 259–269.

Samorodnitsky, G. and Szulga, J. (1989). An asymptotic evaluation of the tail of a multiple symmetric α-stable integral. *Ann. Probab.* **17**, 1503–1520.

Samorodnitsky, G. and Taqqu, M. (1990). Multiple stable integrals of Banach-valued functions. *J. Theor. Probab.* **3**, 267–287.

Sazonov, V.V. (1981). *Normal Approximation – Some Recent Advances.* Lecture Notes in Math. **879**, Springer-Verlag, Berlin.

Sazonov, V.V. and Zalesskii, B.A. (1983). On the central limit theorem in Hilbert space. *Techn. Report, Center Stochast. Process., Dept. Statist.* 35, Univ. North Carolina, Chapel Hill, 1–32.

Schneemeier, W. (1989). Weak convergence and Glivenko–Cantelli results for empirical processes of *U*-statistical structure. *Stochastic Process. Appl.* **33**, 325–334.

Schneemeier, W. (1993). Weak convergence and Glivenko–Cantelli results for weighted empirical *U*-processes. *Ann. Probab.* **21**, 2, 1170–1184.

Sen, P.K. (1972). Limiting behavior of regular functionals of empirical distributions for stationary $*$-mixing processes. *Z. Wahr. verw. Gebiete* **25**, 1, 71–82.

Sen, P.K. (1974a). On L^p-convergence of *U*-statistics. *Ann. Inst. Statist. Math.* **26**, 1, 55–60.

Sen, P.K. (1974b). Weak convergence of generalized *U*-statistics. *Ann. Probab.* **2**, 1, 90–102.

Sen, P.K. (1974c). Almost sure behavior of *U*-statistics and von Mises' differentiable statistical functions. *Ann. Statist.* **2**, 2, 387–395.

Sen, P.K. (1977). Almost sure convergence of generalized *U*-statistics. *Ann. Probab.* **5**, 2, 287–290.

Sen, P.K. (1981). *Sequential Nonparametrics: Invariance Principles and Statistical Inference.* Wiley & Sons, New York.

Serfling, R.J. (1980). *Approximation Theorems of Mathematical Statistics.* Wiley, New York.

Sheehy, A. and Wellner, J.A. (1992). Uniform Donsker classes of functions. *Ann. Probab.* **20**, 4, 1983–2030.

Sherman, R. (1994). Maximal inequalities for degenerate *U*-processes with applications to optimization estimators. *Ann. Statist.* **22**, 1, 439–459.

Shi, Zhan. (1990). Loi du logarithme itere pour le processus empirique associe aux U-statistique. *C. r. Acad. Sci., Ser.* 1, **310**, 10, 751–756.

Shi, Zhan. (1991). A generalization of the Chung–Mogulskii law of the iterated logarithm. *J. Multivariate Anal.* **37**, 2, 269–278.

Silverman, B.W. and Brown, T. (1978). Short distances flat triangles and Poisson limits. *J. Appl. Probab.* **15**, 4, 816–826.

Stein, C. (1972). A bound for the error in the normal approximation to the distribution of a sum of dependent random variables. *Proc. Sixth Berkeley Symp. Math. Statist. Probab.* **2**, 583–602. Univ. California Press, Berkeley.

Stein, C. (1986). *Approximate Computation of Expectations.* IMS, Hayward, California.

Stute, W. (1994). Integrated U-statistic processes: a martingale approach. *Ann. Statist.* **22**, 4.

Su, Chun (1984). On the rate of convergence in the r-th mean of U-statistics with m-dependent samples. *J. Math. (Wuhan)* **4**, 1, 1–6.

Su, Chun (1987). Some results on the L_p-convergence ($p \geq 1$) of U-statistics. *Comput. Statist. Data Anal.* **5**, 4, 321–326.

Sugiura, N. (1965). Multisample and multivariate nonparametric test based on U-statistics and their asymptotic efficiencies. *Osaka J. Math.* **2**, 2, 385–426.

Szekely, G.J. (1974). On the polynomials of independent random variables. In: *Coll. Math. Soc. J. Bolyai: Limit Theorems of Probability Theory,* P. Revesz (Ed.) Keszthely (Hungary), 365–371.

Szekely, G.J. (1982). A limit theorem for elementary symmetric polynomials of independent random variables. *Z. Wahr. verw. Gebiete* **59**, 3, 355–359.

Szekely, G.J. and Szeidl, L. (1995). On limit distributions of random permanents. In: *Exploring Stochastic Laws,* A.V. Skorokhod and Yu.V. Borovskikh (Eds). Festschrift in Honour of the 70th Birthday of Academician V.S. Korolyuk, VSP, Utrecht, The Netherlands, 443–456.

Szulga, J. (1991). Multiple stochastic integrals with respect to symmetric infinitely divisible random measures.*Ann. Probab.* **19**, 1145–1156.

Szulga, J. (1992a). Limit distributions of U-statistics resampled by symmetric stable laws. *Probab. Theory Related Fields* **94**, 1, 83–90.

Szulga, J. (1992b). Series expansions of multiple Lévy integrals. *J. Theor. Probab.* **5**, 4, 601–615.

Takahashi, H. (1988). A note Edgeworth expansion for the von Mises functionals. *J. Multivariate Anal.* **24**, 1, 56–65.

Talagrand, M. (1987). The Glivenko–Cantelli problem. *Ann. Probab.* **15**, 837–870.

Talagrand, M. (1989). Isoperimetry and integrability of the sum of independent Banach-space valued random variables. *Ann. Probab.* **17**, 4, 1546–1570.

Teicher, H. (1988). Distribution and moment convergence of martingales. *Probab. Theory Related Fields* **79**, 2, 303–316.

Teicher, H. (1992). Convergence of self-normalized generalized *U*-statistics. *J. Theor. Probab.* **5**, 391–405.

Tjur, T. (1980). *Probability Based on Random Measures.* Wiley, New York.

Vandemaele, M. (1982). On large deviation probabilities for *U*-statistics. *Theory Probab. Appl.* **27**, 3, 573–574.

Vandemaele, M. and Veraverbeke, N. (1982). Cramer type large deviations for linear combinations of order statistics. *Ann. Probab.* **10**, 2, 423–434.

Vandemaele, M. and Veraverbeke, N. (1985). Cramer type large deviations for studentized *U*-statistics. *Metrika* **32**, 3-4, 165–180.

Veraverbeke, N. (1995). The bootstrap estimator for the asymptotic variance of *U*-quantiles. In: *Exploring Stochastic Laws,* A.V. Skorokhod and Yu.V. Borovskikh (Eds). Festschrift in Honour of the 70th Birthday of Academician V.S. Korolyuk, VSP, Utrecht, The Netherlands, 481–490.

Vitale, R.A. (1992). Covariances of symmetric statistics. *J. Multivariate Anal.* **41**, 1, 14–26.

Weber, N.C. (1980). Rates of convergence of *U*-statistics with varying kernels. *Bull. Austral. Math. Soc.* **21**, 1, 1–5.

Weber, N.C. (1981). Incomplete degenerate *U*-statistics. *Scand. J. Statist. Theory and Appl.* **8**, 2, 120–123.

Weber, N.C. (1983). Central limit theorems for a class of symmetric statistics. *Math. Proc. Cambridge Phil. Soc.* **94**, 2, 307–313.

Wellner, J.A. (1992). Empirical processes in action: a review. *Internat. Statist. Rev.* **60**, 247–269.

Wet, T. (1987). Degenerate *U*- and *V*-statistics. *South African Statist. J.* **21**, 2, 99–129.

Woyczynski, W.A. (1975). Geometry and martingales in Banach spaces. *Part 1: Springer Lecture Notes in Math.* **472**, 229–275; (1978) *Part 2: Advances in Probability (Dekker)* **4**, 267–518.

Yoshihara, Kenichi (1976). Limiting behavior of *U*-statistics. *Z. Wahr. verw. Gebiete* **35**, 3, 237–252.

Yoshihara, Kenichi (1984). The Berry–Esséen theorem for *U*-statistics generated by absolutely regular processes. *Yokohama Math. J.* **32**, 1-2, 89–111.

Yoshihara, Kenichi (1990). Weak convergence to some processes defined by multiple Wiener integrals. *Yokohama Math. J.* **38**, 1, 57–75.

Yoshihara, Kenichi (1992). Limiting behavior of *U*-statistics for strongly mixing sequences. *Yokohama Math. J.* **39**, 2, 107–113.

Yurinskii, V.V. (1976). Exponential inequalities for sums of random vectors. *J. Multivariate Anal.* **6**, 4, 473–499.

Yurinskii, V.V. (1983). On the error of normal approximation. *Sib. Mat. Zh.* **24**, 6, 188–199. (In Russian).

Yurinskii, V.V. (1991). On asymptotics of large deviations in Hilbert space. I,II. *Theory Probab. Appl.* **36**, 1, 78-92; 3, 535–541.

Zalesskii, B.A. (1989a). Probabilities of large deviations in Hilbert space. *Theory Probab. Appl.* **34**, 4, 650–655.

Zalesskii, B.A. (1989b). Exactness of the Gaussian approximation in Banach spaces. *Theory Probab. Appl.* **34**, 4, 815–817.

Zalesskii, B.A., Sazonov, V.V. and Ulyanov, V.V. (1988). Normal approximation in Hilbert space I, II. *Theory Probab. Appl.* **33**, 2, 225–245; 3, 508–521.

Zalesskii, B.A., Sazonov, V.V. and Ulyanov, V.V. (1989). A precise estimate of the rate of convergence in the Central Limit Theorem in Hilbert space. *Mat. Sbornik* **180**, 12, 1587–1613.

Zhang, Bu Cheg. (1990). The asymptotic behavior of *U*-statistics based on exchangeable random elements. *Acta Sci. Natur. Univ. Amoien.* **29**, 2, 124–128.

Zhang, Yi. (1989). On Berry–Esséen bounds of *U*-statistics for dependent samples. *J. Hangzhou Univ. Nat. Sci. Ed.* **16**, 4, 392–398.

Zhao, Lincheng (1983a). Non-uniform bounds for *U*-statistics. *China Annal. Math., Ser. A.* **4**, 6, 699–716.

Zhao, Lincheng (1983b). The rate of the normal approximation for a Studentized *U*-statistics. *Sci. Exploration* **3**, 2, 45–52.

Zhao, Lincheng and Chen, Xiru (1982). Rate of nonuniform convergence of the distribution of *U*-statistics. *J. Math. Res. Exposition* **2**, 4, 109–112.

Zhao, Lincheng and Chen, Xiru (1983). Nonuniform convergence rates for distributions of *U*-statistics. *Sci. Sinica, Ser. A.* **26**, 8, 795–810.

Zhao, Lincheng and Chen, Xiru (1987). Berry–Esséen bounds for finite-population *U*-statistics. *Sci. Sinica, Ser. A.* **30**, 2, 113–127.

Zhao, Lincheng and Chen, Xiru (1990). Normal approximation for finite–population *U*-statistics. *Acta Math. Appl. Sinica (English Ser.)* **6**, 3, 263–272.

Zinn, J. (1985). Comparison of martingale difference sequences. *Lecture Notes in Math.* **1153**, Springer-Verlag, 453–457.

Index

Bei Fragen zur Produktsicherheit wenden Sie sich bitte an:
If you have any questions regarding product safety,
please contact:

Walter de Gruyter GmbH
Genthiner Straße 13
10785 Berlin
productsafety@degruyterbrill.com